A TAXONOMIC MONOGRAPH
OF THE GENUS AGLAIA Lour.
(Meliaceae)

ROYAL BOTANIC GARDENS, KEW

KEW BULLETIN ADDITIONAL SERIES XVI

A TAXONOMIC MONOGRAPH OF THE GENUS AGLAIA Lour. (Meliaceae)

by C.M. Pannell

LONDON: HMSO

© Copyright C.M. Pannell, 1992
Applications for reproduction should be made to HMSO

First published 1992

ISBN 0 11 250067 6

Printed in the United Kingdom for HMSO
Dd291790 11/92 C6 G3397 10170

General editor of Series J.M. Lock; Special editor of this number M.J.E. Coode. Cover design by Media Resources, Royal Botanic Gardens, Kew, after paintings by Rosemary Wise; illustrations by Rosemary Wise except for Figs 50, 77 (by Richard Wise) Fig. 74 (with Richard Wise) and Figs 1 & 113 (by Caroline Watterston).

Address of the Author: C.M. Pannell, Department of Plant Sciences, University of Oxford, South Parks Road, Oxford OX1 3RA, U.K.

The drawing on the cover represents leaves and a dehisced fruit of the *Aglaia erythrosperma* C.M. Pannell with *Anthracoceros malayanus* (Raffles), the Black Hornbill, which eats and disperses the bright orange-red arillate seeds of this species of *Aglaia*.

to

Christopher McCrudden

CONTENTS

	Page
Introduction	1
Taxonomic history	4
Aglaia and related genera	5
Taxonomic features of the tribe Aglaieae	5
The taxonomic status of *Amoora*	6
Section *Aglaia*	7
Section *Amoora*	7
The morphology of *Aglaia*	8
Habit	8
Latex	8
Leaves	8
Indumentum	9
Inflorescence	10
Floral structure	11
Fruits and seeds	11
Floral biology and pollination	14
Non-pollinating insects associated with *Aglaia*	16
Beetles	16
Ants	16
Insect nymphs	16
Insect larvae	17
Fruits and seed-dispersal	18
Germination	20
Cytology	21
Variation and distribution	23
Species concept	23
Variation in the flowers and fruits of *Aglaia*	23
Taxonomic groups	24
Endemism	25
Geographical patterns and evolution in *Aglaia*	28
Taxonomic revision	31
Typification	31
Terminology and conventions in the species descriptions	31
Leaflets	31
Indumentum	31
Inflorescences and infructescences	32
Other characters	32
Species distributions and citation of specimens	33
Abbreviations and meanings of words used in citation of localities	33

	Page
Notes	34
Acknowledgements	34
Generic description	34
Artificial key to the species of *Aglaia*	37
Section *Amoora*	58
Section *Aglaia*	111
Names not placed, for which no type specimen seen	350
Rejected names	353
Bibliography	359
Appendix 1. Regional species lists	363
Appendix 2. A new species from India, by N. Sasidharan	369
New Taxa	373
Index	374

INTRODUCTION

The family *Meliaceae* is mainly tropical, occurring in the New and Old Worlds in a variety of habitats including rain forest, mangrove swamp and semi-desert. It has been the subject of monographic study at Oxford since 1960, initially by F. White, B.T. Styles and T.D. Pennington. A generic monograph of the family (Pennington & Styles, 1975) includes 51 genera. *Chisocheton* has since been monographed (Mabberley, 1979) and includes two of the genera outlined in the family monograph (*Chisocheton* and *Megaphyllaea*) so that the total number of genera is reduced to 50. The African species are now well known (Hutchinson & Dalziel, revised by Keay, 1958; Staner & Gilbert, 1958; White & Styles, 1963; Keay et al., 1964; White, 1986; White & Styles, 1986; Styles & White, 1991). For the New World, the taxonomy and phenology of *Cedrela* (German-Ramirez & Styles, 1978) and a monograph of the New World species have been published (Pennington, 1981). In Asia and Australasia, *Vavaea* (Pennington, 1969) and *Chisocheton* (Mabberley, 1979) have been monographed and accounts of the family published for Flora Vitiensis (A.C. Smith, 1985), Flore de la Nouvelle-Calédonie et Dépendances (Mabberley, 1988) and the Tree Flora of Malaya (Mabberley and Pannell, 1989) and written by Mabberley & Pannell for Flora of Australia and A Revised Handbook to the Flora of Ceylon. *Aglaia*, with at least 105 species, is the largest genus in the family and has presented more taxonomic problems than any other genus in the Meliaceae, especially in species delimitation. Variation is reticulate, involving different combinations of relatively few herbarium characters, each one of which may be variable and is rarely diagnostic on its own. This revision is based on the examination of more than 10,000 herbarium collections, and on information gathered during four field trips to Western Malesia in which the author made comprehensive collections, including male and female flowering material and ripe fruits from the same population for several species. Twelve new species are described. The genus occurs in the tropics of S.E. Asia from Sri Lanka and India through Myanmar (Burma), S. China, and Taiwan, Indochina, Malaysia, Indonesia, the Philippine Islands, New Guinea, Solomon Islands, Vanuatu (New Hebrides), New Caledonia, Australia (Queensland, Northern Territory and Western Australia), Fiji, as far east as the island of Samoa in Polynesia and north to the Marianne Islands (Saipan, Roti and Guam) and the Caroline Islands (Palau and Ponape) in Micronesia.

Together with *Aphanamixis*, *Chisocheton* and *Dysoxylum*, species of *Aglaia* form an important component of the moist tropical forest in the Indo-Malesian region. Some species of *Aglaia* are tall trees and occasionally emergent, some are undergrowth treelets, which may be unbranched, and others are rheophytes. At least two species, *A. elaeagnoidea* and *A.*

brownii, are frequently and *A. lawii* is sometimes littoral; *A. cucullata* is found in tidal estuaries and mangrove swamps. All species appear to be dioecious. The name *Aglaia* is of Greek origin. It refers to one of the Graces who presided over the Olympic Games and means 'beauty' or 'lustre'. This probably alludes to the inflorescences which consist, in the male, of thousands of minute yellow, pink or white flowers, often with a strong perfume of citronella. With the glossy green, pinnate leaves and brick red or orange fruits, which dehisce in some species to reveal bright red or orange seeds, many species of *Aglaia* are attractive trees. When viewed from the canopy, flowering and fruiting trees of the taller species are quite conspicuous, but from the ground they are easily overlooked. The swamp forest emergent, *A. rubiginosa*, has a 'coppered and feathered crown' (Corner, 1988: xxi) which is unmistakable.

The wood of *Aglaia* is sometimes used locally for building or making furniture and the fleshy layer around the seeds of some species is edible, although inferior in quality to that of *Lansium domesticum* Correa *agg.*, the Langsat. A variety of one species, *Aglaia korthalsii*, is grown for its fruits in villages in the state of Kelantan, Peninsular Malaysia. The flowers of the ornamental species, *A. odorata* Lour., are used by the Chinese for scenting tea and in Java for scenting clothes. A detailed account of the uses of individual species is given by Burkill (1935). More recently investigation of the essential oil of the root bark of a species (incorrectly referred to as *A. odoratissima*) from Karnataka in India has revealed that the oil possesses antimicrobial and anthelmintic activities *in vitro* and that it has a depressant action on the central nervous system of mice (Nanda et al, 1987).

White (in Pennington & Styles, 1975) wrote of the need for field studies of the functional significance of the floral and fruit characters used in the classification of the *Meliaceae*. Such a study of *Aglaia* was carried out by the author during two years of field work, mainly in Peninsular Malaysia and Sumatra. This resulted in the identification of some pollinating insects, but the study was much more successful with respect to discovering dispersal agents and establishing a correlation between fruit morphology, aril biochemistry and seed dispersal by either birds or mammals (Pannell & Kozioδ, 1987). Additional studies of flowering and fruiting phenology and seedling recruitment and establishment extended over five years; data from the first two years appeared in Pannell (1980). Two features of the genus make it particularly well suited for ecological study. First, *Aglaia* trees are rarely grown by man, so it can be assumed that trees growing in the forest are of wild origin. Second, there are usually several, and often as many as 12, species of *Aglaia* in a single forest site, which makes it possible to compare the reproductive biology of co-existing species and contributes to an understanding of the plant and animal community of the forest as a whole. These studies of the reproductive biology of this large and common rain forest genus have therefore added to our understanding of the role of animals as pollinators and dispersers of rain forest trees and of the relevance of this both to conservation and to the regeneration of tropical forest after logging (Pannell, 1989a, 1989b). Some aspects of speciation in relation to geography and the availability of dispersers suggested by the ecological and taxonomic studies of *Aglaia* have also been discussed by Pannell & White (1988).

The study of the interdependence between plants and animals is one of the most important elements in understanding the forest ecosystem, crucial to both conservation and the sustainable management of forests, but it is an approach which has often been neglected. A thorough knowledge of the plants, their taxonomy and variation provides a foundation for such studies, so it is desirable that investigations of plant-animal interactions should be carried out in conjunction with a taxonomic monograph of a plant genus or family. It is to be hoped that this revision will provide a framework for further investigation of the ecology of individual species or groups of species of *Aglaia* and particularly of the variable and complex species for which improved field knowledge is required in order to resolve and interpret their taxonomic and ecological variation. It seems likely that the study of such complex species will prove to be particularly valuable in understanding the forest ecosystem, because they are often widespread and have apparently adapted readily to different physical and biotic conditions.

TAXONOMIC HISTORY

The genus *Aglaia* was first described by Loureiro in his Flora Cochinchinensis (1790), based on a single species, *Aglaia odorata*. No type specimen has survived, but Loureiro, in his description of *A. odorata*, refers to *Camunium sinense* of Rumphius; this is described and illustrated in Rumphius' Herbarium Amboinense 5: 28, t. 18 (1747) and the illustration is selected here as the type both of the genus *Aglaia* and of the type species, *Aglaia odorata*. Another species described by Rumphius, *Lansium silvestre* in Herbarium Amboinense 1: 153, t. 55 (1741) was taken up by Roemer in his Familiarum Naturalium Regni Vegetabilis Synopses Monographicae 1: 99 (1846) and transferred to *Aglaia* by Merrill in An Interpretation of Rumphius's Herbarium Amboinense (1917). Roxburgh included the name *Andersonia* in his Hortus Bengalensis (1814), but it was described and illustrated posthumously under the name, *Amoora* in Plants of the Coast of Coromandel (1820), based on a single species *Amoora cucullata*.

In 1868, Miquel published the first synthesis of the genera *Aglaia* and *Amoora* in his monograph of the Meliaceae of the Indian Archipelago, followed by Casimir de Candolle's monograph of the two genera throughout their ranges in A. & C. de Candolle, Monographiae Phanerogamarum (1878). Harms reviewed *Aglaia* and *Amoora* in Engler & Prantl, *Die Natürlichen Pflanzenfamilien* (1896 & ed 2, 1940). The flora accounts of Hiern (1875) for British India, King (1895) and Ridley (1922) for the Malay Peninsula and Backer & Bakhuizen van den Brink for Java (1965) provided keys and/or additional information for many of the species of those areas.

AGLAIA AND RELATED GENERA

Taxonomic features of the tribe Aglaieae

Family: Meliaceae
Subfamily: Melioideae
Tribe: Aglaieae
Genera: *Aglaia* Lour. (including *Amoora* Roxb.)
             Lansium Correa
             Aphanamixis Blume
             Reinwardtiodendron Koord.
             Sphaerosacme Wall.

The tribe Aglaieae was first described by Blume (1825: 169) and owes its present circumscription to the work of Pennington and Styles (1975). It encompasses the large genus *Aglaia* and several small, closely related genera (listed above).

Aglaia is separated from most other genera in the *Meliaceae* by its characteristic indumentum of peltate scales or stellate hairs. Simple hairs are rare and are never found on vegetative parts of the plant, but in some species there are a few on the staminal tube and anthers. The indumentum is usually present on the lower surfaces of the leaflets and is always dense on the apical buds. *Lepidotrichilia*, *Asterotrichilia*, *Pterorhachis* possess a stellate indumentum and *Trichilia* rarely has stellate or peltate scales; these genera might therefore be confused with *Aglaia* on indumentum alone, but they are easily separated on floral characters. When stellate hairs or scales occur in genera other than these, their structure is quite different. They are usually either forked hairs or clumps of simple hairs, e.g. *Melia*, *Aphanamixis cumingiana* (C. de Candolle) Harms (a synonym of *Aphanamixis polystachya* (Wallich) R.N. Parker, see Mabberley, 1985) and are nearly always interspersed with simple hairs. The shoot apices in these species are densely covered with simple hairs, not stellate hairs or peltate scales.

The inflorescence in *Aglaia* is nearly always axillary and a panicle, but it is occasionally spicate in the female. The flowers have 3 – 6 petals and, in common with the other genera in the tribe Aglaieae, they have no disk. (The term 'disk' is applied to various structures which occur between the base of the ovary and the staminal tube in most genera in the Meliaceae). The stamens in *Aglaia* form a staminal tube, with the anthers usually in a single whorl on the inside, but rarely in two overlapping whorls. The anthers are usually 5 or 6 in number, but occasionally up to 21, and lack appendages. The pollen is usually 3-colporate with a smooth exine, but it is said to be 4-colporate in one species (Pennington & Styles, 1975). There are 1, 2, 3 or rarely 4, 5 or 10 loculi. In all species examined, the cotyledons are superposed and no species is known to have united cotyledons.

Of the other genera in the Aglaieae, *Lansium* is characterised by its simple indumentum, 5-locular ovary and the structure of the style and style-head. In addition, the ramiflorous and cauliflorous spicate inflorescences characteristic of this genus are rarely found in *Aglaia*. *Aphanamixis* has spicate female inflorescences, 3 petals and 4-colporate pollen, usually with a scabrous exine. The fruit is a loculicidal capsule and the cotyledons are collateral and joined throughout their length. The indumentum is usually simple.

Reinwardtiodendron has a simple indumentum and a spicate female inflorescence. The anthers are arranged in two alternating whorls of 5, and each anther in the lower whorl has a short, acute appendage. The ovary is usually 5-locular and the fruit is 1- or 2-seeded. The male inflorescences of *Aphanamixis* and *Reinwardtiodendron* may be somewhat branched, but less so than in *Aglaia*.

Sphaerosacme is like *Aphanamixis* in that the pollen is 4-colporate, usually with a scabrous exine and the cotyledons are collateral and united throughout their length, but it differs in having 5 petals and 10 anthers, the latter in 2 whorls of 5.

The taxonomic status of Amoora

In the past, *Amoora* has been separated from *Aglaia* because of differences in the number of anthers (C. de Candolle, 1878), or confined to either those species with 3 petals (King 1895) or those with dehiscent fruits (Harms 1940). The inclusion of *Aphanamixis* and *Sphaerosacme decandra* (Wall.) Pennington in *Amoora* by some authors (e.g. C. de Candolle, 1878) added to the confusion. The reasons for accepting *Aphanamixis* and *Sphaerosacme* as separate genera are given above. The remaining species of *Amoora*, including the type, cannot be separated from *Aglaia* on any combination of characters, although some species can be removed on a single character. The present account follows Pellegrin (1911) and Pennington and Styles (1975) who reduced *Amoora* to *Aglaia*; it should be noted that most species included in section *Amoora* in this account have in the past either been assigned only to *Aglaia* or to both *Amoora* and *Aglaia*. Pellegrin (1911) also considered that *Lansium* should be included in *Aglaia* and, as recently as 1966, Kostermans combined the two genera. His *Aglaia* section *Lansium* contains 7 or 8 species of *Lansium*, 6 of *Reinwardtiodendron* and 1 or 2 of *Lepisanthes* (Sapindaceae; see Pennington & Styles, 1975; Mabberley, 1985). He correctly transferred *Lansium pedicellatum* Hiern to *Aglaia*.

Two sections, section *Aglaia* and section *Amoora*, are recognised in this account, based on the dehiscence of the fruit. Two species, *Aglaia lawii* and *Aglaia teysmanniana* have flower characters intermediate between the two sections, but they are placed in section *Amoora* because they have dehiscent fruits.

Section Aglaia

The flowers have 5, rarely 6, petals and 5, rarely 6 – 8, anthers. The fruits are 1-, 2-, rarely 3-locular and in one species (*A. costata*) 5- or 10-locular; the fruits are indehiscent; in some species there are one or more longitudinal lines of weakness encircling the fruit along which it may split open when pressure is applied. The twigs are either stout or slender and the leaflets are sometimes coriaceous. In some species the leaves are almost without indumentum (e.g. *A. odorata*, *A. oligophylla*, *A. leptantha*), but the indumentum is usually more conspicuous with the hairs or scales larger and more dense than in most of the species in section *Amoora*.

Section Amoora

The flowers usually have 3 petals and 6 or more anthers, up to a maximum of 21 in *Aglaia penningtoniana*; *A. lawii* usually has 4 petals and *A. teysmanniana* 3 – 5. The fruits are 3- or rarely 4-locular and dehiscent. The largest trees in the genus belong to this section and they often attain or emerge from the canopy layer of the forest. *A. lepidopetala*, *A. rugulosa*, *A. meridionalis* and *A. australiensis* are smaller, rarely exceeding 10 m, while *A. lawii* and *A. teysmanniana* grow to 30 m and 15 m respectively. The fruits of the larger species and of *A. rugulosa* are large and heavy and the twigs are stout. In *A. lepidopetala*, *A. meridionalis*, *A. australiensis*, *A. lawii* and *A. teysmanniana* the fruits are smaller and the twigs are relatively slender. In most species, the leaflets are coriaceous and the indumentum on the leaves is inconspicuous and easily overlooked without magnification. However, *A. rubiginosa*, *A. densitricha* and sometimes *A. penningtoniana* have a dense reddish-brown indumentum and *A. meridionalis* has numerous stellate hairs. *A. lawii* varies from having almost none to numerous peltate scales on the lower leaflet surface and there is similar variation in stellate hairs in *A. teysmanniana*. The flowers are relatively large in *A. rubiginosa* (up to 9 mm long) and *A. penningtoniana* (up to 10 mm long). They rarely exceed 6 mm in the rest of the genus.

THE MORPHOLOGY OF AGLAIA

Habit

All species of *Aglaia* are woody, ranging from treelets a few metres high to large trees up to 40 m high. Most species are sympodial with orthotropic branching. *A. odoratissima* is unusual in that the more common variant has plagiotropic branching. Some of the smaller species are monopodial; they are unbranched or have few lateral branches and belong to Corner's model (Hallé & Oldeman, 1970; Hallé & Mabberley, 1977). *A. angustifolia*, *A. tenuicaulis* and *A. coriacea* are examples of species which belong to this model. In all species for which information is available, an axillary bud grows to form a new apical meristem if the shoot apex is damaged.

Latex

There is nearly always some sticky white exudate, known as latex, which flows from the cut trunk of *Aglaia*; it is often also present in the twigs and the pericarp of unripe fruits. Latex occurs in several other genera of Meliaceae, including *Lansium*, *Chisocheton* and *Dysoxylum*.

Leaves

The leaves are usually imparipinnate, but occasionally simple or trifoliolate. Some species with imparipinnate leaves have a closely related simple-leaved species, for example *A. rivularis* has simple leaves, but otherwise resembles *A. yzermannii*, which has pinnate leaves, while *A. luzoniensis* has simple leaves, but otherwise resembles *A. odoratissima*, which has pinnate leaves.

The terminal leaflet is occasionally absent but not consistently so, even on one tree. It is likely in such cases that the terminal leaflet has been eaten by a herbivore at an early stage of development: young leaves are frequently found where some leaflets have been wholly or partly eaten by insect larvae. The terminal leaflet of *A. cucullata* is sometimes smaller than the lateral leaflets and has a longer petiolule and, in some specimens, there is a pocket at the base of the lamina (see fig. 3). The specific epithet refers to this. The indeterminate leaf, found in many species of *Chisocheton*, never occurs in *Aglaia*.

The lateral veins in the middle of the leaflet usually ascend at an angle of less than 65° to the midrib, and curve upwards near the margin. In most Fijian species and in some species from other areas, such as the Solomon Islands, the lateral veins in the centre of the leaflet are at an angle greater than 65° to the midrib. This gives the leaflets a characteristic appearance. Such leaflets usually also have a short, broad, rounded acumen instead of the narrower, more pronounced acumen which is found in the majority of those species with more acutely ascending veins. In some species, a shorter lateral vein may be present between the main lateral veins. A network of secondary veins, referred to here as the reticulation, is found between the lateral veins in all species. In some, it is almost invisible in the dry leaflet, in others it is visible, not or only slightly prominent and vein terminations are frequent; in a few species the reticulation is subprominent and most of the vein branches anastomose with other veins so that vein terminations are few.

Indumentum

An indumentum of peltate scales or stellate hairs is characteristic of the genus *Aglaia*.

According to Theobald et al. (1979: 40), hairs and scales, collectively known as trichomes, 'are easily observable and have often been found to have variation patterns which correlate with other features of the taxa under investigation,' while Metcalfe & Chalk (1950) point out that the presence of a particular type of trichome frequently delimits species, genera or even whole families. Studies of trichomes have been made by Bachmann (1886), Solereder (1908), Netolitzky (1932), Uphof (1962), Stace (1965) and Roe (1971). Bachmann surveyed the occurrence and structure of peltate scales in a number of plant families, including the Meliaceae. Solereder refers to the different types of trichome found in the Meliaceae. Netolitzky discusses at length the morphology, development and physiology of trichomes in general, while Uphof's account in the second edition of this work provides some extra information and is written in English. The cuticular studies of Stace includes suggestions for the use of trichomes in classification and Roe has devised a set of terms for hairs for the genus *Solanum*. When the structure of the trichomes of *Aglaia* has been studied in greater detail these or similar terms may be useful for describing them. For a critical discussion of the use of indumentum in the taxonomy of another genus, *Dombeya* (Sterculiaceae), see Seyani (1991: 23 – 28).

In this account, the trichomes in *Aglaia* are described as scales if they lie flat on the surface of the plant and as hairs if the arms project at various angles outwards from the plant surface. In peltate scales the arms are usually joined for all or most of their length, but not less than half; if they are free at the margin, the scale is referred to as fimbriate, if joined throughout their length, the margin is entire. In stellate scales the arms are free for a minimum of half their length, they are often joined only at the centre of the scale, where it is attached to the plant surface. In a few species, the trichomes are more complicated; one or more arms form a central rhachis and numerous additional arms project from this rhachis at various angle to it. In *A. grandis*, *A. ramotricha* and *A. pachyphylla*, the arms are in a series of about 4 whorls, whereas in *A. penningtoniana*, they are scattered all over the rhachis of the hair.

In most species, the hairs or scales are at least 0.15 mm in diameter;

in a few species in section *Amoora* they are less than this and may be very difficult to see at low magnification. In *Aglaia macrocarpa*, the scales sometimes appear minute; they are less than 0.15 mm in diameter and are coloured reddish-brown only at the centre, so that they resemble the reddish-brown pits (which seem to represent the former point of attachment of a trichome) which occur in this and some other species. In a few species, the scales are 0.25 mm or more in diameter and are visible to the naked eye. In species where this large size is consistent and the scales have a characteristic distribution, these features can be used as diagnostic characters without magnification. The same is occasionally true for species in which the scales are less than 0.25 mm in diameter. In a few species, the structure of the trichomes is visible without magnification. This is true of *Aglaia rufibarbis* which has stellate hairs with very long arms (to 4 mm). For the taxonomic descriptions of the species, sufficient detail was obtained at a magnification of ×20 using a Nikon binocular microscope, and in the field these characters can usually be seen with a ×20 hand lens.

The structure of the trichomes, their size, and their distribution on leaves, twigs, flowers and fruits, are often correlated with other characters. The indumentum on the leaves, together with leaflet number and venation usually makes identification to species possible from leaves alone. This is despite the fact that there is no clear dividing line between stellate hairs and peltate scales and that both sometimes occur, together with intermediates, on the same plant. *A. odoratissima*, for example, has a mixture of peltate and stellate scales, and *A. basiphylla* often has a mixture of stellate hairs and peltate scales. In some species, the younger parts have stellate hairs which have more numerous and longer arms than those on the rest of the plant, while in others the peltate scales have a fimbriate margin when young, which is later lost.

Inflorescence

The inflorescence is a panicle which varies from a large inflorescence, two-thirds of a metre long with profuse branching and an abundance of flowers to a small inflorescence 1 – 2 cm long. Inflorescences are usually borne in the axils of several leaves near the apex of the shoot. They sometimes appear to be ramiflorous when they develop in the axil of a leaf which falls before the inflorescence reaches maturity. Only in *A. macrostigma*, are there known to be truly ramiflorous inflorescences, but even in this species, some are borne in the leaf axils. Male inflorescences are usually larger with more branches and more flowers (up to several thousand) than in the female; in *A. leucophylla*, the male inflorescence may be as long as 60 cm and the female only 8 cm. Over a period of about three days, there is a sequential anthesis of the flowers of a male inflorescence from the base of the inflorescence outwards so that one inflorescence has mature flowers for each of two or three days. A similar sequence of maturation is found in female inflorescences.

Floral Structure

The flowers are small (1 – 10 mm long) and subglobose or ellipsoid. They are unisexual, but the structure of male and female flowers is similar, because each has the apparent structure of a hermaphrodite flower, in which either the androecium or gynoecium is sterile. The female flowers are often slightly larger than the male. The floral character which provides the most taxonomic information is the staminal tube. In some species, this is short and cup-shaped with a wide aperture in which the ovoid anthers are inserted subapically on the inside and are orientated towards the centre of the flower; the petals are closed over the apex of the flower until anthesis, when they open slightly, sometimes revealing the ring of anthers inside; examples of this type of staminal tube are found in *A. odoratissima*, *A. elliptica* and *A. sapindina* (see Figs 82 & 74). In other species, such as *A. silvestris* and *A. leptantha*, the staminal tube is obovoid and may have only a small opening at the apex (see Fig 53); the anthers are inserted longitudinally within the staminal tube and their apices may or may not protrude through the aperture; at anthesis the staminal tube protrudes beyond the petals and the apical pore is exposed, but the anthers are not visible. Intermediates betweeen these two extremes are common. Harms described sections in *Aglaia* based on the position of the anthers in the staminal tube, but this character is variable within species and there are no clear discontinuities between species, so that these sections cannot be maintained. The thickness of the staminal tube varies from being of uniform thickness throughout to being either thinner or thicker in the lower part and / or between the anthers. In two species, *A. euryanthera* and *A. puberulanthera*, the staminal tube is cup-shaped, but the anthers are inserted on the margin and the dorsal surface of each anther is continuous with the staminal tube; in these species, the margins of the anthers are densely hairy.

Fruits and seeds

The infructescence is axillary or supra-axillary, rarely ramiflorous, with as many as several hundred fruits. The calyx is accrescent. The fruit is ellipsoid, obovoid or pear-shaped, dehiscent or indehiscent, covered with stellate or peltate scales and sometimes glabrescent. Even when indehiscent, the fruits of some species, such as *A. elliptica*, *A. subsessilis* and *A. forbesii*, have a longitudinal ridge along which the pericarp splits when pressure is applied to the ripe fruit. The pericarp in some species is thick and pliable and it often contains latex, especially when the fruit is unripe. In other species, the pericarp is thin and brittle or leathery. *A. silvestris* is unusual in often having a flattened, obreniform fruit in the western part of its range. Pennington and Styles (1975) mention an undescribed species in which the pericarp is entirely dry and woody and the fruit is therefore a nut. This has not been confirmed in the present study. There are 1 – 4(– 10) loculi. The septa are thin and membranous and often disintegrate in the ripe fruit. There is one seed in each loculus or rarely 2 (*A. pachyphylla*); each seed is attached by a long, fibrous hilum.

If more than one seed is present in the fruit, the placentation is axile. In some fruits, especially the dehiscent ones, the inner layer of the pericarp becomes detached and forms a membrane around the seed.

The outer layer of the seed coat is usually fleshy; it is attached to the hilum and covers all or part of the seed. In an indehiscent fruit it is translucent and yellow, orange or white; and the inner and outer epidermis enclose a juicy or gelatinous flesh. In a dehiscent fruit there is a tough, bright orange, red, yellow or white outer skin and the flesh beneath is either solid or milky and contains a high percentage of lipid. This layer is called an arillode by Pennington and Styles (1975: 434) who consider it inadvisable to use the term 'aril' until the morphology is better understood. Kapil et al. (1980) recommend a return to general application of the term 'aril' to nearly all accessory seed outgrowths, whether they arise from the funicle, hilum, raphe, chalaza or exostome and their recommendation is followed here. The aril may be removed from the seed, leaving underlying layers of the seed coat intact. It seems to grow from the hilum over the raphe side of the seed to the antiraphe side, where the edges may or may not meet, and the epidermis continues from the outer surface to the inner surface. Developmental studies of the seeds of *Aglaia* are needed in order to establish the origin and morphological nature of the aril. Beneath the aril there appear to be two thinner, hard, papery or membranous layers, attached at the hilum. The outer layer is tougher, with a shiny surface; the vascularisation lies in the surface of the inner layer. If these two layers are morphologically distinct, they are probably testa and tegmen; it is possible, however, that they represent two tissue layers of a sarcotesta. Sections through the micropyle would demonstrate the extent of the two integuments and the presence or absence of a sarcotesta replacing them. The vascularisation runs from the fruit-stalk into the septum, through the hilum, over the apex of the seed, continuing as a stout bundle on the antiraphe side with branches over the surface of the seed which extend around the sides towards the hilum.

Occasionally not all the layers of the seed coat described above are present. In one population of *A. rufibarbis* (at Kuala Lompat, Peninsular Malaysia), the fruit is indehiscent with a thin brittle pericarp and one large seed in which the aril and outer membranous layer are reduced to a small area around the hilum. The vascularisation is therefore near the surface of the seed; it is continuous over the raphe and antiraphe sides. The aril in the Australian *A. elaeagnoidea* is reduced to a small piece of tissue attached to the hilum and it is possible that this is the case in other Australasian species of *Aglaia* for which I have not seen spirit material.

The cotyledons are free, large, transverse and usually equal in size, but sometimes unequal and with the junction between the two oblique; the radicle is included and points towards the hilum, and in one case at least (the cultivated form of *Aglaia korthalsii* found in Kelantan, Malaysia, Professor I. Enoch, pers. comm.), the micropyle is covered by the hilum. The plumule has a covering of stellate hairs or peltate scales. Endosperm is absent. The fruits of section *Aglaia* are indehiscent, with or without a line of weakness along which the pericarp splits when pressure is applied; in some of the larger West Malesian species, they are borne in large, hanging infructescences with hundreds of fruits. The seeds are usually arillate. If present, the aril in these species is complete, translucent, gelatinous and adheres firmly to the testa.

The fruits of section *Amoora* are borne on infructescences with stiffer peduncles and fewer fruits. The fruits in these species are dehiscent, revealing one seed in each locule. The colour of the aril contrasts with the colours of the outer and inner pericarp. The aril does not adhere firmly to the testa and is easily peeled off.

FLORAL BIOLOGY AND POLLINATION

White & Styles (1963) found that the majority of species of Meliaceae which they examined had functionally unisexual flowers. This had been overlooked by earlier workers because both male and female flowers have well-formed, though sterile, organs of the opposite sex. All the species of *Aglaia* in which sex distribution has been investigated are strictly dioecious (Pannell, 1980); the female flowers produce no pollen and the male trees produce no fruits. Male inflorescences are usually larger than the female, the flowers are smaller and more numerous (as many as 6000 per inflorescence) and smell strongly of citronella. The ovary resembles that of the female in structure, but is small and sterile. After anthesis, the male flowers usually soon drop, singly or in clumps, but they sometimes remain on the tree for several weeks until the complete inflorescence falls off. The female flowers, which are less fragrant, are in smaller, less conspicuous inflorescences with fewer flowers. They have a staminal tube and anthers like those in the male but the anthers produce no pollen and do not dehisce, although they do pass through colour changes similar to those of the functional anthers in the male flowers. The petals in flowers of both sexes almost always remain concave and separate only slightly at the apex of the flower at maturity.

If it has a wide aperture, the staminal tube remains entirely within the corolla, but if it has a narrow aperture of pin-prick size, the apex sometimes protrudes beyond the petals at maturity. Flowers usually exhibit three features associated with successful pollination: attractants, rewards and deterrents (Baker, 1978). The attractants bring pollinating insects to the flowers. In *Aglaia*, the bright yellow, pink or white colour of the inflorescences and their strong scent, particularly in the male, serve to attract insects. The reward is pollen and, in this, *Aglaia* exhibits partial 'deception' by offering a reward only in the male flowers so that female flowers are apparently visited 'by mistake'. Pollen is produced by the male flowers in excess of the requirements for pollination and some is probably taken for food by visiting insects, being collected and transferred to parts of the body which do not come into contact with the stigmas of female flowers. The deterrents in *Aglaia* are mechanical, the minuteness of the flower and the even smaller aperture formed either between the apices of the petals or by the staminal tube, prevent many insects from gaining access to the pollen. There are no published records of pollination of *Aglaia*. Many other genera and species in the predominantly tropical and woody families which belong to the orders Rutales and Sapindales of Takhtajan (1986) have small yellow, white or greenish flowers and their pollinators are also unknown. Knowledge of the pollen vectors and pollination mechanisms in these families is most likely to be improved through collaboration between botanists and entomologists, since the taxonomy of many of the plants and insects involved is poorly known. In genera of Meliaceae other than *Aglaia*, bees and moths are believed to be the pollinators of the flowers of some species (Styles & Khosla, 1976). Bawa & Opler (1975) have shown that small bees in the subfamily *Meliponinae* pollinate the flowers of *Trichilia* in Costa Rican forests. Moths have been seen to visit flowers of *Guarea* spp. (Styles & Khosla, 1976) and Ragonese & Garcia (1972) record a dipteran fly visiting flowers of *Melia azederach* L..

Nectar has not been found in the flowers of any species of *Aglaia*. Access to the pollen is restricted by the minuteness and the complicated structure of the flower. The possibility of wind pollination in *Aglaia* can almost certainly be ruled out, because the anthers and stigma are enclosed within the corolla and staminal tube; the amount of pollen produced by each male flower is small and it is unlikely that wind-borne pollen would be carried through the narrow floral apertures to the stigma. The limited evidence obtained from bagging *Aglaia* inflorescences to exclude insects suggests that pollination is required for fruits and seeds to develop (Pannell, 1980). When the stigmas from alcohol-preserved female flowers (which had not been bagged) of one species of *Aglaia*, *A. elliptica*, were stained with aniline blue, germinating pollen grains in which the pollen tubes had penetrated into the stigmatic tissue, were clearly visible when examined under a microscope with incident fluorescent light. Female inflorescences of *A. elliptica* and *A. odoratissima* at Kuala Lompat in Peninsular Malaysia set no fruits when they were enclosed within 'total exclusion' bags made from fine nylon net in which the maximum dimension of the apertures between the fibres was 0.3 mm. Both male and female inflorescences seemed to develop normally when bagged and anthesis occurred in the male flowers. Whole female flowers fell immediately after maturity, whereas, in those which had not been bagged, only the corolla and staminal tube were shed, leaving the calyx and the enlarged ovary intact.

The bright yellow, pink or white colour of the flowers in *Aglaia* and their strong perfume, particularly in the male, apparently attract insects. Visiting insects may obtain pollen from the male flowers, but there is no known reward in female flowers. In addition to this, in the smaller species, the female inflorescences are small and inconspicuous. Differences between species in flower size and structure, especially of the staminal tube, and the level of presentation of inflorescences in the forest are likely to be associated with differences in the details of pollination.

Seven species of stingless bees, belonging to the genus *Trigona*, have been captured and identified visiting the flowers of four species of *Aglaia*: *A. leptantha* (as *A.* aff. *glabrifolia*) *A. odoratissima*, *A. elliptica* (as *A. tembelingensis*) and *A. aspera* (as *A.* sp. G) (Pannell, 1980). The bees collect pollen by probing into the male flowers with their mouthparts and brushing it off among the bristles on their legs and body. The posterior pair of legs bear pollen baskets, on which a relatively large mass of pollen can be carried. These bees occasionally visit the female flowers too. Other *Hymenoptera*, including Halictine bees, Eumenine wasps and *Hymenoptera Parasitica*, visit *Aglaia* flowers. Some species of *Diptera*, belonging to the families *Syrphidae*, *Drosophilidae*, *Cecidomyiidae* and *Lauxaniidae*, were also captured visiting male flowers. Many small flies were caught on sticky traps (fly papers) placed close to both male and female flowers and some of these might prove to be the most effective pollinators. Thrips sometimes emerge in large numbers from male flowers after anthesis, but have not been detected visiting nor breeding in female flowers.

The suggestion that small insects, possibly dipterans, may be the most effective pollinators gains some support from the results of experiments in which female inflorescences of *Aglaia odoratissima* were enclosed in bags of nylon net which permitted access to the flowers by insects of different sizes. These 'partial exclusion' bags were made from nylon net of different mesh sizes. When a mesh size with gaps between the fibres of 0.2 mm × 0.8 mm (which would exclude trigonid bees but not small dipterans) was used, some fruit was set. When a mesh size with gaps between the fibres of 0.1 mm × 0.2 – 0.3 mm (which would exclude most dipterans, but not thrips) was used, no fruit was set. The positive result of fruit set inside some of the bags with the larger mesh size suggests that small insects carrying *A. odoratissima* pollen had entered through the holes in these bags and pollinated the female flowers enclosed within.

Non-pollinating insects associated with Aglaia

Beetles

Small *Coleoptera*, which are more active at night than during the day, are usually present on the inflorescences (Pannell, 1980). They eat small pieces of petals and staminal tube, but pollen has not been found on their bodies or mouthparts. The majority belong to the family *Chrysomelidae* (Leaf Beetles) and some to the *Mordellidae* (Tumbling Flower-Beetles) and *Curculionidae* (Weevils).

Ants

Large, biting, red ants and smaller, black ants are frequently found on all parts of the tree. The black ants build earthy tunnels on the trunk and twigs and black papery tunnels on the underside of leaves. Black ants have once been found inhabiting a cavity inside a twig of *A. elliptica* (Pannell, 1980). A sapling of *A. elliptica* about 175 cm tall in cultivation in Oxford bore glands along the petioles which secreted a sugary liquid. The function of these glands is unknown, but it may be that a young tree thereby attracts ants, which build their nests and remain on the tree throughout its life. In *Chisocheton* (Meliaceae), all the trees of *C. tomentosus* (Roxb.) Mabberley which I examined at Kuala Lompat, Krau Game Reserve, Pahang, Peninsular Malaysia, had hollowed twigs and petioles in which there were colonies of black ants with larvae (Pannell, 1980). Ants had not previously been recorded living inside the twigs of *Chisocheton* west of Borneo (see Mabberley, 1979: 308).

Insect nymphs

Insect nymphs are often found on an inflorescence or twig of *Aglaia* (Pannell, 1980). Those recorded belong to the family *Cicadellidae*. Some are covered with a white powdery substance and others with a mass of hairs from the species of *Aglaia* on which they are found. The empty exoskeleton of an unidentified insect found on the leaf of a herbarium specimen of *A. archboldiana* (*Degener* 13705 UC) bears long projections which resembled in morphology and colour the arms of the hairs on the plant specimen.

Insect larvae

The seeds of *Aglaia* are sometimes destroyed by larvae, which hatch from eggs laid in the young fruits (Pannell, 1980). They belong to several groups of insects, including *Heterocera* (Moths), *Diptera* (Flies) and *Coleoptera* (Beetles). At least two families of *Coleoptera* are involved, one of them being the *Curculionidae* (Weevils). A proportion of each fruit crop of *A. elliptica* examined at Kuala Lompat contained seeds infected in this way. Similar larvae were found in some seeds of *Aglaia erythrosperma* in Pasoh Forest Reserve, Negri Sembilan, Peninsular Malaysia. In addition to these and other insects, such as praying mantis, spiders were occasionally found on inflorescences and their eggs were sometimes laid between the surfaces of a folded-over leaf of a seedling.

FRUITS AND SEED-DISPERSAL

The fruit has one to three (rarely four to ten) locules, and each of these either contains a single seed or is empty, presumably because of failure of fertilization of the ovule or abortion of the developing seed. In species of *Aglaia* which occur on the Sunda Shelf there are two main types of fruits, associated with two main types of vertebrate disperser, birds and primates (Pannell, 1980; Pannell & Kozioł, 1987). The pericarp varies in thickness from 1 mm to more than 1 cm and is either thin and brittle or thick, tough and fibrous. The fruits are either dehiscent or indehiscent and the seeds are usually arillate. The morphology and chemical composition of the aril in those species investigated is associated with dehiscence of the fruit. The results of biochemical analysis of the major nutrients in the aril flesh using spirit-preserved and dried aril material showed a correlation with morphological characters of the fruits and the main dispersal agents (Pannell, 1980; Pannell and Kozioł, 1987). The seeds of dehiscent species are dispersed mainly by birds and the arils are red-skinned, easily removed from the testa, rich in lipid (28–61% dry weight) and odourless. The aril in these species can be easily separated from the rest of the seed, and this probably means that it is easily removed by the action of the gizzard or gut of the dispersing birds, which regurgitate or pass the seeds unharmed. The seeds of indehiscent species are dispersed by primates and the arils are white, yellow, orange or brown, gelatinous and translucent, firmly attached to the testa, high in sugars (16 – 26% dry weight) and in sweet-tasting amino acids (such as alanine, γ-aminobutyric acid, glycine, and asparagine or aspartic acid), relatively low in lipid (3 – 11%) and have a characteristic odour. The aril adheres firmly to the testa and this, together with its gelatinous nature, probably encourages primates to swallow the seeds or carry them away from the fruiting tree before eating the flesh and discarding the seed.

Several features in the position and architecture of the infructescence, and in the position, size and structure of the fruit are also related to the mode of dispersal. The red-skinned arils of dehiscent fruits contrast with the white inner pericarp which is exposed on dehiscence and the pink or brick-red outer surface of the pericarp; dehiscence of the fruit makes it possible for the seed to be extracted easily by the bill of a bird. Dehiscent fruits are borne in small numbers on infructescences which have a short, stiff peduncle, so that they remain close to the subtending twig or branch of the tree, which provides a perch for birds feeding on the arillate seeds. When the fruits are large, the infructescences are borne on stout shoots; the seeds are large, and only large birds such as hornbills are able to swallow them. In species with smaller fruits, the infructescences are borne on slender shoots; it is more difficult, although still possible, for large birds to reach the fruits, but seeds are more frequently taken by smaller birds. Field observations suggest that the large-seeded dehiscent species of *Aglaia* are dispersed by hornbills and possibly fruit pigeons, and that the smaller-seeded species are dispersed by hornbills and an array of smaller birds, including bulbuls, broadbills and barbets. The small birds probably effect mainly local dispersal, but hornbills and pigeons are known to fly distances of 100 km or more and could be important in longer distance dispersal. The seeds are accessible to primates as well as birds, but there are no confirmed observations of primates feeding on seeds from species of *Aglaia* which have dehiscent fruits.

In indehiscent fruits the pericarp is orange, pinkish-orange or brown and they are borne near the ends of slender branches or in large hanging infructescences, sometimes with a long peduncle, where they can be reached and manipulated by long-armed and dexterous primates. The primates break open and peel off the pericarp and remove the seed. At least two species of monkey, the Banded Leaf Monkey (*Presbytis melalophus* (Raffles)) and the Long-tailed or Crab Macaque (*Macaca fasicularis* (Raffles)) and three species of ape, the Orang Utan (*Pongo pygmaeus abelii* Lesson) the Siamang (*Hylobates syndactylus* (Raffles)) and the White-handed Gibbon, (*H. lar* (L.)), are known to ingest the seeds of *Aglaia*. In these indehiscent species, the seeds are inaccessible to most birds either because of the position of the infructescence or because the bird cannot manipulate the fruit to remove the pericarp and gain access to the seed. Other animals in addition to primates might occasionally be dispersers in Sundaland, although they appear to be relatively unimportant. Squirrels are mainly destructive but occasionally disperse seeds over short distances. Terrestrial rodents and ground feeding birds such as the Crested Wood Partridge (*Rollulus rouloul* (Scopoli)) may play a similar role. Seeds of *Aglaia elaeagnoidea* have been found in the faeces of *Paradoxurus hermaphroditicus* (Pallas) (Viverridae, the civet family). Elephants, rhinoceros, tapirs and some deer are known to feed on fallen fruits of other plant families and may also contribute to the dispersal of *Aglaia*. Bat-dispersal is a possibility for some species. The total geographical range of *Aglaia*, however, extends far beyond the eastern limits of the distribution of primates and of some of the bird families that are thought to be the main dispersal agents of *Aglaia* to the west of Wallace's Line. Species of *Aglaia* with dehiscent and indehiscent fruits occur east of Wallace's Line but the fruit structure, dehiscence and the aril type are known for only a few of these species. Fruit pigeons, which occur throughout the range of *Aglaia* and some Australasian bird families such as the Birds of Paradise (Paradisaeidae), Bower Birds (Ptilonorhynchidae) and the flightless Cassowaries (Casuariidae) are likely to be dispersers in Australasia (see Pannell & White, 1988). Pigeons may have transported the seeds of *Aglaia* species with capsular fruits across Wallace's Line. They might therefore have played an important part in bringing about the collective geographical range of *Aglaia*, which lies entirely within the area of distribution of *Ducula* and *Ptilinopus*, the principal genera of fruit-eating pigeons in the Far East. Primates, which are important dispersers of the indehiscent species of *Aglaia* in Sundaland, are absent from Australasia and their role in dispersal may be filled there by terrestrial or arboreal fruit-eating marsupials, rodents, bats or ground-dwelling birds.

GERMINATION

In the terminology of Ng (1978), seed germination in *Aglaia* is 'semi-hypogeal': 'the hypocotyl is undeveloped and the cotyledons are exposed. The exposed cotyledons lie on or in the ground, usually with the broken seed coat still adhering more or less loosely to their outer surfaces. This is evidently not a good position for photosynthesis because green cotyledons at ground level are rarely encountered. But with the weight of the cotyledons fully supported by the ground, conditions are specially favourable for the cotyledons to be developed into bulky storage organs.' This type of germination is apparently peculiar to tropical trees with large seeds. The seeds of *Aglaia* have no endosperm. The cotyledons are peltate with the shoot axis lying between the two. When the shoot axis begins to grow, the cotyledons are forced apart, the testa splits and the cotyledons are exposed. The cotyledons are not carried up by the plumule to form the first seedling leaves; they remain on the surface of the soil or become covered by litter. 10% of the tropical rain forest seeds investigated by Ng have this type of germination. The seeds of *Aglaia* are large (often more than 1 cm long) and, in the Malay Peninsula, usually germinate in fewer than six weeks. The early leaves are simple and the first two are opposite. Ng found that 75% of Malayan rain forest tree species have seeds 1 cm long or longer and that 65% of his sample germinated in fewer than twelve weeks (Ng, 1978).

CYTOLOGY

Chromosome counts for the *Meliaceae* have been recorded by Mehra & Khosla (1969), Styles & Vosa (1971), Mehra, Sareen & Khosla (1972) and Khosla & Styles (1975). They found a very wide range of base chromosomes numbers, from 2n = 16 in *Sandoricum indicum* Cav. (= *S. koetjape* (Burm. fil.) Merrill) to 2n = c. 360 in *Trichilia dregeana* Sond. Some other tropical and subtropical woody families (eg. *Rubiaceae, Lauraceae, Leguminosae* and *Ebenaceae*) have been found to have small and rather constant chromosome numbers (see Styles & Vosa, 1971). Polyploid series occur in a number of genera of *Meliaceae*, including *Aglaia* and *Aphanamixis* in the tribe *Aglaieae*. Khosla & Styles (1975) state that the *Aglaieae* is cytologically the most neglected tribe in the family. The following chromosome numbers for *Aglaia* have been published in the papers cited above. *Aglaia edulis* (Roxb.) Wallich, n = 40. India: E. Himalayas, Sukna, Darjeeling, *Mehra & Khosla* 3235. *Aglaia perviridis* Hiern, n = 20. India: E. Himalayas, Teesta, Darjeeling, *Mehra & Khosla* 551. *Amoora wallichii* King (= *Aglaia spectabilis*), n = 20. India: E. Himalayas, Sukna, Darjeeling, *Mehra & Khosla* 3233. *Aglaia* sp. 2n = 92. Papua and New Guinea, Lae, LAE s.n. As part of the present study of the genus, seeds or seedlings were collected in Peninsular Malaysia and grown in the Cambridge Botanic Garden. Root tips were prepared with the standard colchicine treatment, except that, instead of keeping them in 3 : 1 absolute alcohol : glacial acetic acid for a few hours, this was extended to a few days to increase breakdown of the cell walls which otherwise resisted squashing and impeded uptake of the stain. Several factors made it difficult to obtain accurate counts from the root tips examined. The meristem of all the root tips was small, with only one or two dividing cells. The chromosomes were minute and the numbers were high. The cytoplasm seemed to resist squashing, so that the chromosomes did not lie in a single focal plane in the preparations. With the help of C.G. Vosa, chromosome numbers of four species of *Aglaia* from Peninsular Malaysia have been obtained in addition to those already published (Table 1). Voucher specimens, taken either from

Table 1. Chromosome numbers in Aglaia

Species	Chromosome number	Voucher specimen number (all Peninsular Malaysia)
Aglaia leptantha Merrill	68	Pannell 1432 (FHO) Kuala Lompat
Aglaia odoratissima Blume	84	Pannell 1579 (FHO) Kuala Lompat
Aglaia elliptica Blume	68	Pannell 1334 (FHO) Kuala Lompat
Aglaia korthalsii Miquel	84	Pannell 1463 (FHO) Kg Dermit

seedlings germinated from seeds from the same mother tree or collected from the same group of young seedlings in the forest as the one from which chromosome counts were obtained have been deposited in the Forest Herbarium, Oxford (FHO).

VARIATION AND DISTRIBUTION

Species concept

In this treatment of the taxonomy of *Aglaia*, a wide species concept has been adopted, based mainly on vegetative characters, particularly the structure and distribution of the indumentum. Wherever possible, these have been correlated with flower and fruit characters but, for some species, it has been necessary to encompass a wide range of variation in these characters. Future work based on field studies and improved collections, especially of female flowers and ripe fruits, may provide information which could justify the division of some of these variable species into separate species or subspecies. Such division should, however, only follow extensive additional fieldwork and collecting which is specifically aimed at the resolution of taxonomic problems, preferably in association with ecological investigation. In some species, the structure of the trichomes is variable and the distribution and density may also vary (notably in *A. penningtoniana*, *A. lawii* and *A. tomentosa*). Texture and thickness of leaf, the colour when dry, and prominence of venation (especially reticulation) and presence or absence of pitting have sometimes provided specific characters. Most species may be recognised on vegetative characters alone and in many cases there is no alternative, since complete material exists for only a small number of species and then from only parts of their ranges.

Variation in the flowers and of fruits of Aglaia

Many of the species of *Aglaia* as circumscribed here show considerable variation in characters of the flower and fruit. In particular the degree of development of the staminal tube sometimes varies from cup-shaped to obovoid within the same species. This may sometimes be partly explained by the age of the flower since, in some species, the staminal tube does not complete its development until just before anthesis. The flower is often larger near the edge of the range of a species, especially in India and China (e.g. in *A. lawii*). The fruit may vary in shape from subglobose to pear-shaped in the same species (e.g. *A. exstipulata*, *A. lawii*, *A. tomentosa*), in size and thickness of the pericarp (e.g. *A. elliptica*, *A. korthalsii*, *A. lawii*). Several species which occur in the Philippine Islands have glabrescent fruits there, but rarely elsewhere in their ranges (e.g. *A. elaeagnoidea*, *A. exstipulata*, *A. tomentosa*). The New Guinea variants of nearly all species which occur on both sides of Wallace's Line differ slightly in vegetative characters and more so in the fruit, but the incompleteness of the herbarium material available means that it has not been possible to describe this fully. Further resolution of patterns of variation will require collections of male and female flowers, plus ripe fruits preserved in spirit, from representative populations throughout the ranges of individual species. In particular, for most species, it is unlikely that further progress can be made without ripe fruits in spirit, these to be accompanied by notes on colour and dehiscence, and descriptions of the layers of the pericarp and seed, especially the fleshy seed coat ('aril').

Taxonomic groups

There are descriptions of 105 species in this revision and at least four further species occur in New Guinea for which material is insufficient to delimit or describe them. Two sections are recognised, sect. *Aglaia* and sect. *Amoora*. Two species in sect. *Amoora*, *Aglaia lawii* and *Aglaia teysmanniana*, are intermediate in floral characters, but are placed in sect. *Amoora* because they have dehiscent fruits. Within each section the species can be informally grouped as in Table 2, based mainly on the structure, density and distribution of the indumentum. Members of a pair or group are (with a few exceptions) considered to be morphologically more similar to each other than to any species outside the group. In many cases, there is a progression from one group to the next, but the morphological relationships between the species and groups are not linear and occasionally one species is more similar to a species in a non-adjacent group than it is to the species in the intervening groups. For example, in section *Aglaia*, *A. odoratissima* (group 13) is sometimes difficult to separate from *A. korthalsii* (group 8) or *A. elliptica* (group 16), both of which exhibit a wide range of variation and include some variants which closely resemble *A. odoratissima*. The numerical order of species in the taxonomic revision follows the groupings in Table 2. Three different types of species are recognised among these groups, based on their degree of distinctness and their internal variability in relation to distribution and ecology. The terminology used here is modified from White (1962) and Pennington (1981 & 1990).

1. *Taxonomically isolated species*. These are morphologically distinct species which cannot be subdivided and occupy an isolated position within the section or genus without any close relatives. Their internal variability is usually small. Their geographical extent is also usually small (e.g. *A. coriacea*, confined to Peninsular Malaysia, except for one record, the type, from Kalimantan) or occasionally extensive (e.g. *A. cucullata* extending from Bangladesh to Papua New Guinea, but confined to riverine, estuarine and mangrove forests).

2. *Closely related pairs or larger groups of species*. The members of these pairs or groups are closely related and often separable only by using the combined variation of several overlapping characters. The members of these pairs or groups may be allopatric, e.g *Aglaia elliptica* (Burma to the Philippine Islands and Sulawesi) and *Aglaia conferta* (confined to New Guinea). Others are partially sympatric e.g. *A. korthalsii* (N.E. India to Sulawesi) and *A. speciosa* (Peninsular Malaysia to Sulawesi). Simple

geographical replacement sometimes occurs among the species of pairs or larger groups of closely related species. It is, however, more common for one member of the group to be a widespread, variable or complex species and for some of the other species in the group to occupy a smaller area within the range of that species, whereas others lie partly or wholly outside its range. A good example is provided by the complex species *A. tomentosa*, which extends from India to Australia. *A. palembanica* (Malay Peninsula, Sumatra, Borneo, Philippine Islands) and 9 other species lie entirely within the range of *A. tomentosa*, whereas the distribution of *A. brownii* (New Guinea and Australia) overlaps with that of *A. tomentosa*, and *A. fragilis* and *A. archboldiana* are both confined to Fiji, which is outside the range of *A. tomentosa*. This group of species is described on pp. 312 – 349.

3. *Variable or complex species*. The term variable is applied to those species in which the variation is relatively simple, usually involving two variants linked by intermediates; the differences appear in the herbarium to be small but may prove to be greater and more easily defined in the field, on either morphological or ecological grounds. The term complex is applied to species with a more extensive, complicated and reticulate pattern of variation, in which the extremes appear at first sight to belong to distinct species. It may prove possible to divide some of these species into two or more subspecies based on partial discontinuities in their variation, probably correlated with different geographical distributions or ecological conditions. Such infraspecific categories are difficult to define in *Aglaia* from existing herbarium material and an attempt to recognise them here would add little to our present understanding of these species. These are the species for which further investigation is most needed, especially in the field and especially in India, Borneo, the Philippine Islands and New Guinea. Such investigation should make it possible to describe the variation more accurately and to interpret it in the context of distribution and ecology. It seems likely that much of the variation will be found to be associated with different substrates, altitude, climate and animal pollinators and dispersers and that these will be only partly correlated with distribution. Variable species are marked with one asterisk and complex species with two asterisks in Table 2. One of the variable species, *A. penningtoniana*, is taxonomically isolated, but all the remaining variable or complex species are members of closely related pairs or groups.

Endemism

The number of endemic species in most parts of the range of *Aglaia* is low (Fig. 1). An exceptionally high proportion of endemic species, 80%, is found in Fiji, in New Guinea 39% of species are endemic and in India 30%. Of the 50 species found in Borneo, 10% are endemic and of the 35 in the Philippine Islands, nearly 9% are endemic. However, most of those areas delimited to the west of New Guinea in Fig. 2 have a relatively large number of complex species and resolution of some of these complexes may result in the recognition of more species of restricted range. This is especially true of the Philippine Islands, where there is great variation

Variation and distribution

Table 2. Species groups recognised in *Aglaia*

Variable species are marked with an asterisk (*) and complex species with two asterisks (**)

Section *Amoora*
1. A. cucullata
2. A. australiensis
3. A. flavida, *A. macrocarpa, A. malaccensis
4. A. rugulosa, A. erythrosperma, *A. spectabilis
5. A. multinervis
6. A. lepidopetala, A. meridionalis
7. A. densitricha
8. A. rubiginosa
9. *A. penningtoniana
10. **A. lawii, *A. teysmanniana

Section *Aglaia*
1. *A. grandis, A. ramotricha
2. A. pachyphylla, A. bourdillonii, *A. eximia, *A. argentea
3. A. squamulosa, A. densitricha, A. subcuprea
4. A. lancilimba, A. lepiorrhachis
5. A. chittagonga
6. **A. elaeagnoidea, A. smithii
7. A. variisquama, *A. rimosa, A. costata
8. *A. agglomerata, *A. speciosa, **A. korthalsii, *A. apiocarpa, A. scortechinii, A. glabrata
9. A. flavescens, A. rubrivenia, *A. samoensis, A. gracilis, *A. vitiensis, A. unifolia, A. leucoclada
10. *A. silvestris, A. perviridis, **A. leptantha, A. cremea, A. forbesii, *A. foveolata
11. A. crassinervia, *A. aspera
12. A. parviflora, A. heterotricha, *A. leucophylla, **A. edulis, A. macrostigma
13. *A. odoratissima, A. luzoniensis, A. yzermannii, A. rivularis, A. brassii
14. A. amplexicaulis
15. *A. puberulanthera, A. euryanthera, A. polyneura, A. sapindina, A. ceramica
16. A. parksii, *A. subminutiflora, *A. basiphylla, A. evansensis, A. subsessilis, **A. elliptica, A. conferta, A. aherniana, A. barbanthera
17. *A. saltatorum, *A. mariannensis
18. A. cumingiana
19. A. laxiflora
20. A. pyriformis
21. A. coriacea
22. A. odorata, A. pleuropteris
23. *A. oligophylla, *A. simplicifolia
24. A. monozyga
25. *A. tenuicaulis, A. membranifolia, A. rufinervis, *A. exstipulata, A. palembanica, A. fragilis, A. brownii, **A. tomentosa, A. integrifolia, A. angustifolia, *A. hiernii, A. cuspidata, A. rufibarbis, A. archboldiana

in *A. elliptica*, such that about 18 species had been recognised from the Philippine Islands, mainly by Merrill and Elmer (whose names were not validly published). All of these are here included in the synonymy of *A. elliptica*. Many of these putative species are represented either by the type collection only or by a small number of collections. Much more extensive field-based knowledge of their variation in the Philippine Islands is

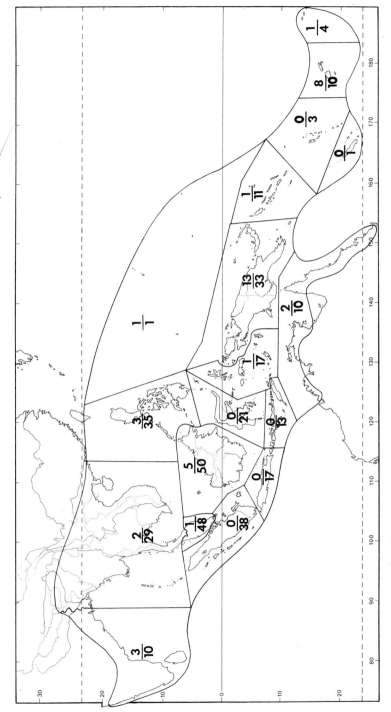

Fig. 1. Density map of *Aglaia*, showing the number of endemic species above the line and the total number of species below.

essential before attempts are made to split the complex and this, as with all complex species, should only be carried out if the variation throughout the range of the complex is also taken into consideration.

Geographical patterns and evolution in Aglaia

'Much of the tropical Far East is a gigantic archipelago, but one in which the number and size of the islands and the width of the water barriers have been constantly changing in response to changing sea levels throughout the Quaternary, especially on the continental shelves. This must have provided many opportunities both for speciation and the breakdown of incipient speciation, and has greatly influenced the total species richness of this region' (Pannell & White, 1988), Fig. 2. In addition, *Aglaia* is a genus with animal-dispersed fruits in which most species might either frequently or occasionally be dispersed from island to island (inter-island dispersal being probably mainly by birds). Dispersal from one island to another may therefore result in some divergence between the populations which occupy the different islands, but the frequency of inter-island dispersal in some species seems to be sufficient to bring different local variants of the same species into contact before reproductive or ecological isolation between them is complete. These factors could at least partly explain the reticulate variation observed in the variable and complex species of *Aglaia* and, consequently, the large number of species described which are here included in synonymy.

It also seems likely that the variation in *Aglaia* is due in part to the capacity of some species to adapt readily to new conditions and for this to be reflected in apparently minor morphological changes. Some of the diversification in the genus may be relatively recent and much of this may not yet have resulted in the evolution of recognisable species. These changes are often difficult to detect and may lie mainly in the fruits, where ripe collections are few. Improved collection of fruits is likely to lead to the discovery of correlations between fruit characters and the known vegetative variation in some species. Similarly, a better knowledge of the differences between male and female flowers from the same population and their variation between populations and of changes in flower morphology, especially in the staminal tube, during development would provide valuable information for the interpretation of the variation patterns within complex or variable species and between closely related species.

Some evolutionary insights are now emerging as the result of an approach in which the reproductive biology of several species has been studied in combination with a taxonomic revision of the entire genus throughout its range. The most marked geographical division in *Aglaia* is between species with a western distribution and those with an eastern distribution. 52 species are either confined to the Sunda Shelf or occur on the Sunda shelf and in Wallacea. 37 species are confined to Australasia and / or the Western Pacific or their distribution extends from the island of New Guinea westwards into Maluku only. 5 species are confined to Wallacea and *A. rimosa* occurs in New Guinea, Wallacea and the south of Taiwan (thereby just reaching the Sunda Shelf). Only 9 other species

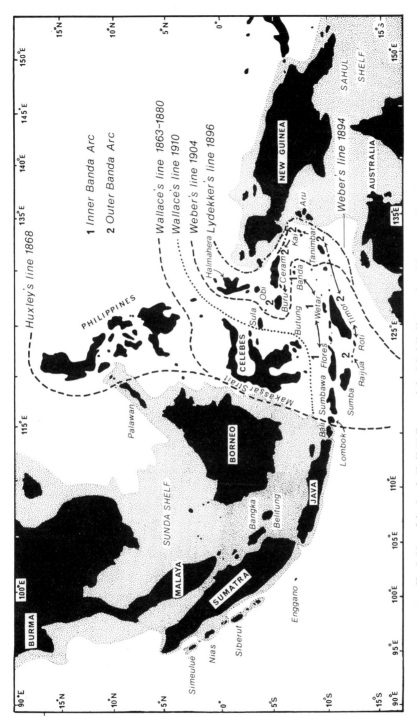

Fig. 2. Map of the Sunda Shelf, Wallacea and the Sahul Shelf (reproduced from Pannell & White (1988) with permission from the Missouri Botanical Garden).

occur in both Sundaland and Australasia and all but one of these (*A. cucullata*) are variable or complex species. The variation in these species lies at least partly in the fruits and these differences are likely to be associated with differences in the dispersal agents (see p. 19). After crossing Wallace's Line (in whichever direction), colonizing species of *Aglaia* must frequently have acquired new dispersal agents. Changes in fruit structure, chemistry and other dispersal-related features of these plants have probably arisen as a response to the structure, behaviour and nutritional requirements of their new dispersers. At present, however, we have very little information on this aspect of the diversity of the genus. Some aspects of speciation in relation to geography and occurrence of known or possible dispersers which have been suggested by the ecological and taxonomic studies of *Aglaia* are discussed by Pannell & White (1988).

The discussion in that paper includes the statement: 'cryptic variation in the fruits of vertebrate-dispersed plants may be much more widespread than has previously been supposed and until it has been detected and carefully described the taxonomy of the plants themselves will remain largely unresolved and our evolutionary understanding of them will be seriously impaired. In particular, a knowledge of dispersal of fruits by vertebrates may have much to contribute to such questions as the differences in floristic richness in faunistically different parts of the world, and to different patterns of speciation in relation to different patterns of vertebrate dispersal'.

TAXONOMIC REVISION

The descriptions are based on the collections of *Aglaia* at British Museum (Natural History) (BM), Bogor (BO), Cambridge (CGE), Edinburgh (E), Florence (FI) Forest Research Institute Malaysia (FRI), Geneva (G) Kew (K), Leiden (L), Paris (P), Oxford (FHO, OXF) and Singapore (SING), and on loans from Aarhus (AAU), Arnold Arboretum (A, GH), Berlin (B), Brisbane (BRI), California (UC), Canberra (CANB), Copenhagen (C), Darwin (DNA), Lae (LAE), Leningrad (LE), Michigan (MICH), New York (NY), Peradeniya (PDA), Perth (PERTH), Queensland (QRS), Smithsonian (US), Utrecht (U) and Vienna (W).

Typification

When there is no holotype, a lectotype has, if possible, been selected. Unless stated otherwise, all lectotypes cited are selected here. If the most likely candidate for the lectotype of a name is thought to be in an herbarium from which I have not been able to examine the material, I have refrained from formal lectotypification. This applies to names published by Chen (IBSC), Hayata (TI), King (CAL), Kanehira (FU), von Mueller (MEL), Schwartz (HBG), Turczaninow (KW). When the collection which either included the holotype of a name or from which a lectotype should have been selected has been destroyed and duplicates exist elsewhere, a lectotype has been selected from among these. This applies to names published by Harms, Perkins and O. Warburg (B) and to many of Merrill's names (PNH). The specimens thought to have been destroyed are referred to as isolectotypes and the acronym for the herbarium is followed by a dagger symbol (†).

Terminology and conventions used in the species descriptions

Leaflets
Since there is no confusion in the orientation of the leaflet surfaces, the terms 'upper' and 'lower' have been used to describe the 'adaxial' and 'abaxial' surfaces respectively.

Indumentum
The distribution of trichomes is described as *sparse* when the hairs or scales are scattered, *numerous* when they are sufficiently separated to be counted individually and the epidermis is still visible and as *densely covering* the surface when the epidermis is no longer or only barely visible; the

hairs or scales are then crowded together so that it is not possible to count them.

Inflorescences and infructescences

In many species, it has not been possible to distinguish with certainty between male and female inflorescences and they have both been included in a single description. In some cases, the female inflorescence is not represented among the available collections; in others the differences between male and female are slight and are difficult to detect unless flowering trees have been studied in the field and both sexes collected from the same locality.

Descriptions of the flower are taken mainly from dried herbarium material boiled in water for 10 to 60 seconds. In the first phase of this revision, descriptions of the flower were less detailed than latterly, especially in terms of fine measurements; the later descriptions give measurements of the flower parts in millimetres and fractions of a millimetre. Approximate measurements can be calculated for the remaining descriptions, from the dimensions given for the whole flower, the corolla being slightly shorter than the length of the whole flower. Male and female flowers are similar in structure, male flowers are usually more conspicuous in the field and are much more frequently collected than the female. Flowers in which the anthers have dehisced and contain pollen have usually been assumed to be male, but in three cases, the anthers of flowers taken from a specimen on which fruits were present on the same twig were found to contain pollen grains. The frequently deformed structure and poor staining qualities of these pollen grains suggested that they were infertile and no other indication has been found for departure from strict dioecy in any species. The sex, however, can often only be determined with certainty either by following the flowering and fruiting cycle of individual trees, as I have done for some species from Peninsular Malaysia, or, for the female, when flowers and fruits are present on the same twig.

Fruit measurements are taken from fresh or spirit material, when this is available and from dried carpological collections when it is not. Spirit-preserved material provides much more information on the morphology, texture and dimensions of the fleshy aril than does dried material and these are included where available. If only dried material has been seen, dimensions of the seed without aril are given, because when dry, the aril becomes either a paper-thin layer or disappears completely. If sufficient fresh or spirit material is available, the dimensions of the seed with aril and the approximate thickness of the aril are given. However, in most spirit collections, the amount of material is small and intact arillate seeds were not dissected to obtain dimensions of the seed without the aril. These collections provide valuable material for morphological and anatomical study and should be retained as whole seeds until such work is undertaken.

Other characters

Occasionally a character, not mentioned in most of the descriptions, is included, such as 'margin recurved' or colour of fresh or dried leaf; for these the general condition, if known, is given in the generic description and only included in an individual description if it is an exception to the rule or provides a distinctive character in recognising the species in either the fresh or dried state. Sometimes information is available for a few species only (e.g. crown shape and branching pattern), and is unknown for the majority of species; this is included in those descriptions where it is known, but no general statement for the remaining species is made in the generic description. The diameter or circumference of the bole is assumed to have been measured at 'breast height' (c. 1.3 m) or immediately above the buttresses, if they are higher than 1.3 m.

Species distributions and citation of specimens
　The species distributions are mapped by degree square and are based mainly on the herbarium collections at L, with some additions from the collections of other herbaria, especially for areas outside the Flora Malesiana area. Records are mapped by degree square. In some cases two or more species are included on the same map, represented by different symbols. If the degree square from which a collection was made cannot be located and there are no other collections from that part of the range, the symbol for the species is placed approximately and surrounded by a ring.
　For each species, illustrative specimens are cited after the description. These represent as far as possible male flowering, female flowering and fruiting specimens and are drawn from all the herbaria consulted. Localities are cited as they appear on the specimen labels except that altitudes given in feet are converted to an approximate equivalent in metres. The following abbreviations have been used to describe the fertility of the specimens: fl. = flower, ♂ = male, ♀ = female, fr. = fruit, yg = young, st. = sterile. Latitude and longitude are given when they appear on the herbarium label, but are not usually otherwise included, since the majority of species have been mapped. The author holds a record of all specimens seen in the course of preparing this revision and an identification list for the specimens held at L is available from her.

Abbreviations and meanings of words used in citation of localities
Bt = Bukit = hill
F.R. = Forest Reserve
Gg = G. = Gunong = Gunung = Goenoeng = mountain
K. = Kuala = junction of two rivers, or an estuary
Kg = Kampong = village
nr = near
P. = Pulau = Poeloe = island
Sg. = Sungei = Sungai = R. = river
Ulu = Source of a river
V.J.R. = Virgin Jungle Reserve

Notes

Each species description is followed by notes on its distribution and any available information on ecology, uses and vernacular names. Vernacular names are given by country, state or island followed by the names for each language or dialect (the language or dialect being given in brackets). Names for which the language or dialect is not known appear at the end of the list for each country, state or island.

The citation of specimens is followed by a note on recognition of the species, its distinction from other, similar, species and its variability. This section is occasionally concluded by field observations made by the author.

Acknowledgements

I wish to express my appreciation to all those who have helped in the preparation of this monograph, especially G. Coppen, E.J.H. Corner, L.L. Forman, B.H. & R. Kiew, Y. & E. Laumonier, D.J. Mabberley, S.K. Marner, T.D. Pennington, G.T. Prance, P.D. Sell, E. Soepadmo, B.T. Styles, Kalang bin Tot, P.C. van Welzen, F.R. Whatley, the directors, curators and staff of the many herbaria that I have visited and that have sent specimens on loan to Oxford and C. Styles who typed the first draft of the manuscript. The success of my field work in Sumatra was largely due to the detailed information and local contacts provided by the late M. Jacobs. I am greatly indebted to R. Palmer for his translation into Latin of the diagnoses of the twelve new species.

This revision would not have been started without the interest and support of R.R. Pannell, E. Pannell, R.E. Pannell and J.U. de la Tour and it would not have been completed without the help and encouragement of J.C. McCrudden, A. Strugnell (who prepared the final copies of all the distribution maps and who has given me a great deal of assistance during the last 5 years of the project) and F. White.

The Malaysian Government, the Chief Game Warden of Malaysia and the Director of the Forest Research Institute of Malaysia granted permission for me to carry out research in Malaysia. The Indonesian Government, the Indonesian Institute of Sciences and the Indonesian Wildlife Department granted permission for me to carry out research in Indonesia. For financial support, I would like to thank the Leverhulme Foundation, the Royal Society, the Natural Environment Research Council, the Studley College Trust, the Druce Bequest, Oxford, Lincoln College, Oxford, the Phyllis and Eileen Gibbs Travelling Fellowship of Newnham College, Cambridge and the Department of Plant Sciences, Oxford.

Generic description

Aglaia Loureiro, Fl. Cochinch. 173 (1790) nomen conservandum. Holotype species: *A. odorata* Loureiro (*Aglaia* Loureiro is conserved against *Aglaia* Allamand, Nova Act. Acad. Leop.-Carol 4: 93 (1770), Cyperaceae and against *Nialel* Adanson, Fam. Pl. 2: 582 (1763)); Harms in Engl. & Prantl, Nat. Pflanzenfam. III 4: 298 – 300, t. 163, figs A – Q (1896);

Harms in Engl. & Prantl, Nat. Pflanzenfam., ed. 2, 19b1: 140 – 147, t. 31, figs A – Q (1940).

Camunium Rumph. [Herb. Amb. 5: 26: t. 18, fig. 1 (1747)] ex Roxb., Fl. Ind. 2: 425 (1824), non Adanson, Fam. Pl. 2: 166 (1763) (= *Murraya* Koen. ex L.). Holotype species: *Camunium sinense* Rumph. ex Roxb. (= *Aglaia odorata* Lour.).

Amerina Noronha, Verh. Bat. Gen. 5(2): 64 (1790) nom. nud., non Rafin., Alsog. Amer.: 15 (1838) (= *Salix* Tourn.), nec A.P. de Candolle in Meissn. Gen.: 278 (1840) (= *Aegiphila* Jacq., Verbenaceae).

Andersonia Roxb., [Hort. Beng.: 87 (1814), nom. nud.], pro parte quoad *A. cucullata* Roxb. tantum; Fl. Ind., ed. 2, 2: 212 (1832), pro parte, non R. Brown, Prodr.: 553 (1810) (= Epacridaceae), nec Willd. ex J.J. Roemer & J.A. Schultes, Syst. Veg. 5: 21 (1819) (= *Gaertnera* Lam., Loganaceae). Lectotype species (designated here): *Andersonia cucullata* Roxb. (= *Aglaia cucullata* (Roxb.) Pellegrin).

Milnea Roxb., [Hort. Beng.: 18 (1814), nom. nud.] Fl. Ind. 2: 430 (1824), non Rafin. (1838). Holotype species: *Milnea edulis* Roxb. (= *Aglaia edulis* (Roxb.) Wall.).

Amoora Roxb., Pl. Coromandel 3: 54, t. 258 (1820). Holotype species: *Amoora cucullata* Roxb. (= *Aglaia cucullata* (Roxb.) Pellegrin); Harms in Engl. & Prantl, Nat. Pflanzenfam., ed. 2, 19b1: 128–129 (1940).

Nemedra A. Juss. in Bull. Sci. Nat. Géol. 23: 239 (1830). Holotype species: *Nemedra elaeagnoidea* A. Juss. (= *Aglaia elaeagnoidea* (A. Juss.) Benth.).

? *Argophilum [Argopyilum]* Blanco, Fl. Filip. 186 (1837), non *Argophyllum* Forst. (1776) (= Saxifragaceae). Holotype species: *Argophilum pinnatum* Blanco (? = *Aglaia tomentosa* Teijsm. et Binn.).

? *Selbya* M.J. Roemer, Synops. Hesp. Monogr. 1: 89, 126 (1846). Holotype species: *Selbya montana* (Jack ex Sprengel) M.J. Roemer (= ?).

Nimmoia Wight in Calc. Jour. Nat. Hist. 7: 13 (1847), non Wight in Madras Journ. Sci. (1837). Holotype species: *Nimmonia lawii* Wight (= *Aglaia lawii* (Wight) Saldanha ex Ramamoorthy).

Oraoma canarana Turcz. in Bull. Soc. Nat. Mosc. 31: 411 (1858). Holotype species: *Oraoma canarana* Turcz. (= *Aglaia lawii* (Wight) Saldanha ex Ramamoorthy).

Beddomea Hooker fil. in Benth. & Hooker fil., Gen. Pl. 1: 336 (1862). Holotype species: *Beddomea indica* Hooker fil. (= *Aglaia edulis* (Roxb.) Wall.).

Hearnia von Muell., Fragm. Phyt. Austr. 5: 55 (1865). Holotype species: *Hearnia sapindina* F. Muell. (= *Aglaia sapindina* (von Muell.) Harms).

Aglaiopsis Miq., Ann. Mus. Bot. Lugd. Bat. 4: 58 (1868). Lectotype species (designated here): *Aglaiopsis glaucescens* Miq. (= *Aglaia sapindina* (von Mueller) Harms).

Merostela Pierre, Fl. Forest. Cochinch. Fasc. 21, ante t. 334 (1895). Holotype species: *Merostela grandis* (Korth. ex Miq.) Pierre (= *Aglaia grandis* Korth. ex Miq.).

Lepiaglaia Pierre, Fl. Forest. Cochinch. Fasc. 21, ante t. 334 (1895). Holotype species: *Lepiaglaia pyramidata* (Hance) Pierre (= *Aglaia silvestris* (M.J. Roemer) Merrill).

Usually small to large trees, rarely bushes; dioecious; trunk often with buttresses, bosses often present where branches have fallen off and these frequently bearing leafy shoots; crown sympodial; branched throughout or in the upper part only, or unbranched with a crown of leaves, branches usually ascending or patent or sometimes arching. Bark smooth or somewhat rough, sometimes (especially in the larger species) deciduous in squarish scales, usually with longitudinal rows of lenticels. Latex often present, sometimes flowing rapidly when the trunk is cut. Twigs stout or slender, apical bud without bud scales, made up of 2 – 4 unexpanded leaves which are spike-like, always with dense stellate hairs or peltate scales. Leaves borne in spirals, widely separated on the apical shoots or close together with the petiole bases overlapping, usually imparipinnate with lateral and terminal leaflets usually similar, the basal pair of leaflets rarely markedly smaller in size, rarely simple. Leaflets (1 –)3 – 25, lanceolate, oblanceolate, ovate, obovate, elliptical or oblong, the lamina of moderate thickness, the surface usually smooth, sometimes rugose or rugulose, the margin entire in plants of all ages, usually planar, sometimes recurved or slightly wavy, apex acuminate to caudate with the acumen obtuse or acute, base rounded, subcordate, cuneate or attenuate, usually asymmetrical, one or both surfaces may be rugulose or pitted, in some species almost without indumentum but usually the lower surface has few, numerous or dense hairs or scales like those on the twigs; midrib and lateral veins usually prominent, subprominent or depressed and the secondary veins occasionally subprominent on both surfaces, sessile or with a petiolule up to or rarely exceeding 25 mm.

Inflorescences usually axillary or supra-axillary, occasionally ramiflorous or cauliflorous, often several on an apical shoot. Flowers unisexual with well developed rudiments of the opposite sex. Male inflorescence large, much divaricately branched, more or less triangular in outline, with small triangular or linear bracts which are often deciduous before maturity. Flowers up to 10,000, terminal on branchlets, solitary or in sessile clusters, sometimes with bracteoles similar to the bracts, usually smelling of citronella, minute, 1 – 6(– 10) mm long, subglobose, ellipsoid or obovoid. Female inflorescence similar to the male but usually smaller and less-branched, sometimes a narrow spike-like raceme with few flowers; flowers often larger than in the male. Calyx cup-shaped, often thickened at the base, shallowly or deeply 3 – 5 (or 6)-lobed, aestivation open or imbricate, the lobes unequal and sometimes patent at anthesis. Corolla aestivation imbricate or quincuncial, petals 3 – 5 (or 6), free or united at the base, free from the staminal tube or partially united to it, usually yellow, sometimes pink or white, subrotund, elliptical or obovate, unequal, concave and usually thickest in the centre, often hooded at the apex when in bud, separating at anthesis, occasionally with stellate hairs or peltate scales on the outside. Stamens united to form a tube 0.5 – 8 mm long which is more or less truncate at the base, usually subglobose, obovoid, cup-shaped, the apex incurved or rarely shortly cylindrical, without appendages, sometimes with stellate or simple hairs on the inner surface, aperture small to large with an entire, crenate or shallowly lobed margin; anthers (3 –)5 – 10(– 21), usually in a single whorl, rarely two or more overlapping whorls, occasionally with stellate or simple hairs, broadly or narrowly ovoid, dehiscing by two longitudinal slits, inserted on the inner surface of the tube either just below and protruding through the aperture and pointing towards the centre of the flower or more or less vertical against the inner surface of the tube, curved to follow the shape of the tube and partially or completely included, rarely inserted on the margin of the tube; anthers in the female flowers similar but sterile, usually not dehiscing and without pollen, rarely with a few misshapen pollen grains. Pollen 3 (? or 4)-colporate, 3 – 30m long, subprolate or prolate, exine- smooth or rarely minutely scabrous, thickened at the apertures. Disk absent. Ovary 1 – 3(– 10) locular, superior, depressed globose or ovoid with dense stellate hairs or peltate scales; loculi with 1 or 2 collateral or superposed ovules, where carpels more than 1, placentation axial; style a very short constriction between the ovary and style or absent; stigma ovoid, more or less cylindrical or depressed-globose, often dark and shiny when mature, entire at the apex or with 2, 3 or rarely 4 small lobes; ovary and stigma in the male either poorly developed or similar to the female but sterile.

Infructescences often several on a shoot with one to several hundred fruits. Fruit subglobose, obovoid or ellipsoid, indehiscent or a loculicidal capsule with 1 – 3 or rarely 4 loculi each with one seed or rarely 2. Seeds large, usually with an aril or sarcotesta nearly or completely covering the seed. Embryo with thick plano-convex superposed or rarely oblique cotyledons, radicle included, the shoot axis with dense stellate hairs or peltate scales; endosperm absent. Germination semi-hypogeal, with hypocotyl undeveloped. First two leaves simple and opposite, subsequent leaves spirally arranged, simple at first, later 2- or 3-foliolate and increasing to or exceeding the number of leaflets present on the leaves of the mature plant.

Artificial key to the species of Aglaia

The only characters for which sufficient comprehensively comparative information is available for use in a general key lie in the indumentum and the flowers; both of these usually require magnification. When using this key, a binocular dissecting microscope should be used and dried flowers should be boiled and dissected. In the absence of a microscope, a ×20 hand lens should be sufficient to reveal most of the characters in the key. Leaflet number and size have only been used near the final breaks in the key, where they reliably separate two closely related species. Characters of the flowers or fruits have only been used near the final breaks; if these are absent from the specimen being identified, the species descriptions should be referred to for deciding between the remaining small number of species.

1 Leaf always a single blade   **2**

1a Leaves trifoliolate or imparipinnate   **9**

2 Lower surface of leaf with numerous reddish-brown stellate hairs which are not deciduous   **99. A. integrifolia**

Artificial key to the species of Aglaia

2a Lower surface of leaf with few or no hairs or scales (if numerous on young leaves, soon deciduous) — **3**

3 Leaves subsessile and amplexicaul at the base — **65. A. amplexicaulis**

3a Leaves not amplexicaul — **4**

4 Leaves linear-lanceolate, more than 4 times longer than wide — **63. A. rivularis**

4a Leaves not linear-lanceolate, less than 4 times longer than wide (rarely one leaf on a specimen is slightly more than 4 times longer than wide) — **5**

5 Leaf apex rounded; midrib on lower surface of leaf densely covered with purplish-brown or reddish-brown peltate scales. Lateral veins spreading at an angle of 65 – 70° to the midrib in the middle of the leaflet — **45. A. unifolia**

5a Leaf apex acuminate; midrib on lower surface of leaf not densely covered with purplish-brown or reddish-brown peltate scales. Lateral veins ascending at an angle of less than 65° to the midrib or curved throughout their length — **6**

6 Indumentum of stellate hairs, sometimes with peltate scales interspersed — **7**

6a Indumentum of stellate and / or peltate scales only, stellate hairs absent — **8**

7 Leaves with stellate hairs and peltate scales few to numerous on the midrib below; staminal tube deeply lobed, the anthers inserted on the inside of the lobes, densely covered with simple white hairs on the margins of the lobes and on the anthers — **66. A. puberulanthera**

7a Leaves with reddish-brown stellate hairs and occasionally peltate scales; the indumentum dense on the twig apices and sparse elsewhere; staminal tube not deeply lobed, anthers and staminal tube not densely hairy — **89. A. simplicifolia**

8 Leaves brown or greenish-brown when dry, indumentum of stellate and peltate scales, mainly on the midrib and veins on the lower surface of the leaves; inflorescence densely covered with peltate scales — **61. A. luzoniensis**

8a Leaves yellowish-green, pale green or pale
brown when dry, with few small reddish-brown
peltate scales on the midrib below; inflorescence
with few to numerous peltate or stellate scales **46. A. leucoclada**

9 Leaflets with few or no hairs or scales on the
lower surface, the reticulation continuous and
subprominent on one or both surfaces **10**

9a Leaflets with at least some scales or hairs on
the lower surface, although these may be few and
difficult to see, reticulation not continuous and
subprominent on either surface, or if
subprominent, then with indumentum on lower
surface of leaflet **14**

10 Leaflets with reticulation subprominent on lower
surface **82. A. cumingiana**

10a Leaflets with reticulation subprominent on both
surfaces **11**

11 Petiole, rhachis and petiolules flattened on the
adaxial side, part or all of the rhachis and some or
all of the petiolules with narrow wings **12**

11a Petiole, rhachis and petiolules terete or with
a groove on the adaxial side but not flattened,
without wings **13**

12 Leaflets 3 – 5 **86. A. odorata**

12a Leaflets 11 – 17 **87. A. pleuropteris**

13 Leaflets 11 – 14, dull when dry **83. A. laxiflora**

13a Leaflets 3 – 11, slightly shiny when dry **88. A. oligophylla**

14 Leaflets linear-lanceolate or narrowly elliptical,
most being at least 5 times longer than wide **15**

14a Leaflets ovate, elliptical, oblong, obovate,
lanceolate or oblanceolate, most less than 5 times
longer than wide **18**

15 Leaflets 3 – 5 **62. A. yzermannii**

15a Leaflets 9 – 27 **16**

Artificial key to the species of Aglaia

16 Leaflets at least 10 times longer than wide, with reddish-brown stellate hairs numerous on the rest of the lower surface	**100. A. angustifolia**
16a Leaflets about 5 times longer than wide, with stellate hairs or scales or peltate scales few on or densely covering the midrib and few or absent on the lower surface	**17**
17 Leaflets with reddish-brown stellate hairs or scales densely covering the midrib below, with 9 – 19 pairs of lateral veins	**76. A. elliptica**
17a Leaflets with few to densely covered with peltate scales which have a fimbriate margin on the midrib below; with 20 – 50 pairs of rather indistinct and widely spreading lateral veins	**9. A. multinervis**
18 Indumentum dense, of white or pale brown hairs or scales which totally conceal the lower surface of leaflet	**19**
18a Indumentum reddish-brown or, if pale, not totally concealing the lower surface of leaflet	**24**
19 Indumentum of hairs which have a central rhachis and several whorls of arms radiating from it	**20**
19a Indumentum of stellate hairs or peltate scales	**21**
20 Indumentum on lower surface of leaflet so dense that the surface is not visible between the hairs even when using a hand lens; twigs and rhachis channelled	**19. A. pachyphylla**
20a Indumentum dense but the lower surface visible between the hairs when using a hand lens; twigs and rhachis terete	**17. A. grandis**
21 Upper surface of leaflet rugose	**23. A. squamulosa**
21a Leaflet surfaces not rugose	**22**
22 Indumentum dense on lower surface of leaflet, consisting of white peltate scales with few to many brown peltate scales interspersed	**22. A. argentea**
22a Indumentum dense on lower surface of leaflet, consisting of white stellate scales with few to many brown stellate hairs interspersed	**23**

Artificial key to the species of Aglaia

23 Upper surface of leaflet not shiny and reticulation not subprominent when dry   **21. A. eximia**

23a Upper surface of leaflet shiny and reticulation subprominent when dry   **19. A. pachyphylla**

24 Lower surface of leaflet densely covered with reddish-brown or orange- brown hairs or scales, the surface not or barely visible between them   **25**

24a Hairs or scales absent from the lower surface or, when present, the lower surface of leaflet readily visible between them   **29**

25 Lower surface of leaflet with numerous stellate hairs, with paler hairs which have one or few ascending arms interspersed, the surface barely visible in between   **101. A. hiernii**

25a Lower surface of leaflet densely covered with stellate or peltate scales, sometimes with stellate hairs interspersed, the surface not visible between them   **26**

26 Flowers pentamerous; fruits indehiscent   **27**

26a Flowers trimerous; fruits dehiscent   **28**

27 Leaflets with dark reddish-brown entire peltate scales, upper surface with midrib and lateral veins markedly depressed and densely covered with reddish-brown scales which contrast with the pale yellowish-green of the rest of the surface when dry   **24. A. densisquama**

27a Lower surface of leaflet densely covered with orange-brown fimbriate peltate scales, the upper surface with the midrib and lateral veins not markedly depressed and scales, when present, not contrasting in colour with that of the surface   **20. A. bourdillonii**

28 Twigs often densely covered with hairs which have a central rhachis to 1 cm long and numerous whorls of arms radiating from it; lower surface of leaflet with dense stellate hairs and scales, sometimes completely or partly deciduous   **14. A. penningtoniana**

28a Twigs without hairs which have a central rhachis to 1 cm long and numerous whorls of arms radiating from it; lower surface of leaflet densely covered with persistent stellate scales or peltate scales which have a fimbriate margin   **13. A. rubiginosa**

Artificial key to the species of Aglaia

29 Indumentum of peltate scales, sometimes with stellate scales interspersed	**30**
29a Indumentum of stellate hairs or scales; peltate scales absent	**88**
30 Indumentum of peltate scales only	**31**
30a Indumentum of peltate and stellate scales (or with at least some of the scales with a long fimbriate margin)	**74**
31 Scales densely covering lower surface of leaflet	**32**
31a Scales ± absent to numerous on lower surface of leaflet	**36**
32 Leaflets densely covered with dark reddish-brown peltate scales which have a very dark, depressed centre (so that the scales resemble a minute volcano)	**30. A. smithii**
32a Peltate scales orange-brown, pale brown or almost white	**33**
33 Twigs, inflorescence branches and lower surface of leaflet with scales all less than 0.25 mm in diameter, not shiny, scales pale brown	**53. A. crassinervia**
33a Twigs, inflorescence branches and lower surface of leaflet thickly covered with large, shiny, peltate scales at least some of which are 0.25 mm or more in diameter	**34**
34 Upper surface of leaflet rugose	**23. A. squamulosa**
34a Upper surface of leaflet not rugose	**35**
35 Leaflets (1 –)3 – 7, scales orange-brown, pale brown or nearly white throughout	**29. A. elaeagnoidea**
35a Leaflets 7 – 11, with numerous to densely covered with pale orange-brown or reddish-brown peltate scales which have a dark centre and pale margin	**25. A. subcuprea**
36 Scales few to numerous on lower surface of leaflet	**37**

36a Scales ± absent from lower surface of leaflet but may densely cover the midrib below and immediately adjacent to it and occasionally on the lateral veins	**56**
37 Scales orange-brown, pale brown or almost white throughout	**38**
37a Scales at least partly reddish-brown	**48**
38 Peltate scales evenly distributed on the lower surface of leaflet and visible to the naked eye as tiny dots	**39**
38a Peltate scales not evenly distributed on the lower surface of leaflet or if they are then not visible to the naked eye	**45**
39 Leaflets markedly asymmetrical and often curved	**1. A. cucullata**
39a Leaflets only slightly asymmetrical and not curved	**40**
40 Leaflets grey or brownish-green when dry, often coriaceous; twigs often stout	**41**
40a Leaflets orange, pale green or reddish-brown when dry, not coriaceous; twigs slender	**43**
41 Flowers trimerous; fruit dehiscent	**3. A. flavida**
41a Flowers pentamerous; fruit indehiscent	**42**
42 Leaflets with a recurved margin, with numerous large dark orange-brown peltate scales on the lower surface and visible to the naked eye as evenly distributed spots	**31. A. variisquama**
42a Leaflet margin not recurved, with numerous small yellowish-brown or orange-brown peltate scales which have an entire or fimbriate margin or almost stellate scales on the lower surface	**53. A. crassinervia**
43 Leaflets 5 – 7, with very pale orange-brown peltate scales of uniform size and colour numerous on the lower surface and visible to the naked eye as evenly distributed, pale, shiny spots	**28. A. chittagonga**

Artificial key to the species of Aglaia

43a Leaflets usually 9 – 11, with peltate scales which have a dark orange-brown or reddish-brown centre and pale entire or fimbriate margin numerous on the lower surface interspersed with a few larger scales which are dark throughout	**44**
44 Fruit 2-locular, without prominent ribs	**32. A. rimosa**
44a Fruit 5- or 10-locular, with 10 prominent longitudinal ribs	**33. A. costata**
45 Leaflets green when dry, with tiny pale orange-brown peltate scales; flower pentamerous	**56. A. heterotricha**
45a Leaves yellowish-green, yellowish-brown or pale brown when dry; or if green then flower not pentamerous	**46**
46 Fruit dehiscent	**15. A. lawii**
46a Fruit indehiscent	**47**
47 Leaflets with few to numerous small orange-brown peltate scales, reddish-brown, pale brown or orange-brown stellate hairs or scales or peltate scales on the midrib below; fruit large, up to 3.2 cm long and 3.8 cm in diameter, usually subglobose with an apical depression and 3-locular but sometimes the seed fails to develop in 1 or 2 of the locules; pericarp thick and woody	**58. A. edulis**
47a Leaflets with large pale brown or orange-brown peltate scales which are entire or have a fimbriate margin, few to numerous on the lower leaflet surface; fruits small, up to 2.5 cm long and 1.4 cm in diameter, subglobose, ellipsoid or obovoid without an apical depression, with 1 or two locules, the pericarp thin	**29. A. elaeagnoidea**
48 Staminal tube with a narrow pin-prick aperture c. 0.3 mm across, with an entire margin, anthers included	**47. A. silvestris**
48a Staminal tube with the aperture wider than 0.3 mm, the margin shallowly lobed, anthers protruding through the aperture	**49**
49 Leaflets usually markedly ovate, scales almost absent from the lower surface, veins often black or red when dry; fruit asymmetrically ellipsoid with one loculus	**48. A. perviridis**

49a Leaflets elliptical, ovate or obovate, scales few to numerous on the lower leaflet surface, veins neither black nor red when dry; fruits symmetrical with 2 or 3 loculi   **50**

50 Leaflets narrowly elliptical or narrowly obovate, scales numerous on or densely covering lower surface of leaflet   **51**

50a Leaflets neither narrowly elliptical nor narrowly obovate, scales few to numerous on lower leaflet surface   **52**

51 Leaflets coriaceous   **37. A. apiocarpa**

51a Leaflets not coriaceous   **35. A. speciosa**

52 Peltate scales with a dark purplish-brown centre and a pale margin; flowers trimerous (unknown whether fruit dehiscent or indehiscent)   **2. A. australiensis**

52a Peltate scales dark reddish-brown, with or without a paler margin; flowers pentamerous (fruit indehiscent)   **53**

53 Lower surface of leaflet with few to numerous peltate scales dark reddish-brown peltate scales which have a fimbriate margin and with paler scales interspersed; leaflet apex with a parallel-sided obtuse acumen   **50. A. cremea**

53a Lower surface of leaflet with peltate scales all of one type, or if paler scales interspersed then acumen not parallel-sided   **54**

54 Leaflets usually 9 – 13   **34. A. agglomerata**

54a Leaflets fewer than 9   **55**

55 Scales few and distributed all over lower surface of leaflet   **36. A. korthalsii**

55a Scales mainly on the midrib of lower surface of leaflet or if on rest of lower surface of leaflet, then densely covering it   **37. A. apiocarpa**

56 Scales densely covering the midrib on lower surface of leaflet and immediately adjacent to the midrib, occasionally also on the lateral veins   **57**

56a Scales ± absent from lower surface of leaflet   **67**

Artificial key to the species of Aglaia

57 Scales large (many 0.2 mm across), reddish-brown or almost white, with a tendency to flake off	**58**
57a Scales less than 0.2 mm across or if larger, then dark reddish-brown or purplish-brown and adhering closely to the leaflet	**59**
58 The lateral veins with bright reddish-brown peltate scales densely covering the midrib above and below and numerous to dense on the lateral veins above and below	**41. A. rubrivenia**
58a Scales rare on the midrib and veins of upper surface of leaflet and mainly on the midrib below	**61**
59 Leaflets with pale orange-brown or grey peltate scales few to numerous on the midrib on the lower surface	**27. A. lepiorrhachis**
59a Leaflets with numerous to densely covered with reddish-brown peltate scales on the midrib on the lower surface	**60**
60 Peltate scales dark reddish-brown, often with a long fimbriate margin	**40. A. flavescens**
60a Peltate scales large, shiny and reddish-brown, either entire or with a short fimbriate margin	**26. A. lancilimba**
61 Leaflet apex rounded, lateral veins spreading, (at an angle of more than 70° to the midrib in the middle of the leaflet)	**62**
61a Leaflet apex acuminate, veins ascending	**63**
62 Leaflets with dark reddish-brown peltate scales which are usually entire or have a short fimbriate margin densely covering the midrib on both surfaces; inflorescences always axillary	**44. A. vitiensis**
62a Leaflets with very dark reddish-brown or purplish-brown peltate scales which have a fimbriate margin few on or densely covering the midrib below; some inflorescences axillary and some ramiflorous	**43. A. gracilis**
63 Anthers and / or staminal tube with simple white hairs	**64**
63a Anthers and staminal tube without hairs	**65**

64 Flowers with the staminal tube deeply lobed, the anthers inserted on the inside of the lobes, densely covered with simple white hairs on the margins of the lobes and the anthers	**67. A. euryanthera**
64a Staminal tube neither deeply lobed nor densely hairy, but with some simple white hairs on the anthers	**42. A. samoensis**
65 Leaflets with purplish-brown fimbriate peltate scales densely covering the midrib below and ± absent from the rest of lower surface of leaflet	**39. A. glabrata**
65a Leaflets with dark reddish-brown peltate scales numerous on the midrib below	66
66 Leaflets (7 –)9 – 13(– 15), stellate scales absent	**38. A. scortechinii**
66a Leaflets 5 – 7, some stellate scales interspersed among the peltate scales	**60. A. odoratissima**
67 Veins usually black when dry; petals 5; fruit indehiscent	**49. A. leptantha**
67a Veins rarely black when dry; petals usually fewer than 5; fruit dehiscent	68
68 Leaflets markedly curved and asymmetrical	**1. A. cucullata**
8a Leaflets only slightly asymmetrical	69
69 Ripe fruits up to 3.5 cm in diameter maximum	70
69a Ripe fruits more than 3.5 cm in diameter	72
70 Leaflets brown when dry, with few pale brown peltate scales which have a fimbriate margin and compact reddish-brown stellate hairs interspersed on the lower surface; colour of fruits when fresh not known, dark reddish-brown when dry	**10. A. lepidopetala**
70a Leaflets pale green or orange-brown when dry, with few to numerous almost white or pale orange peltate scales (rarely with a few hairs interspersed) on the lower surface; fruits pink, yellow, orange or brown when fresh, pale brown or grey when dry	71
71 Fruits pink or yellow when fresh, dehiscent; seeds with an opaque aril	**15. A. lawii**

Artificial key to the species of Aglaia

71a Fruits orange or brown when fresh, indehiscent; seeds with a translucent aril	**80. A. saltatorum**
72 Leaflets 15 – 25, lanceolate, veins 20 – 50 on each side of the midrib, indistinct	**9. A. multinervis**
72a Leaflets 7 – 15, ovate, obovate, elliptical or oblanceolate, lateral veins 6 – 20 on each side of the midrib	73
73 Leaflet surfaces reddish-brown or orange-brown and often rugulose when dry, with few reddish-brown or grey peltate scales which have a fimbriate margin on the midrib below and often interspersed with minute reddish-brown scales or pits	**4. A. macrocarpa**
73a Leaflets pale brown and not rugulose when dry, with small orange, orange-brown or pale brown peltate scales which have a dark central spot numerous on the midrib below, sometimes numerous on lower surface of leaflet but usually absent	**3. A. flavida**
74 Leaves subsessile or with a petiole less than 1 cm long	75
74a Leaves not subsessile, the petiole more than 1 cm long	76
75 Leaves simple or with 3(– 7) leaflets, the lateral leaflets much smaller (about one quarter the length and breadth) than the terminal leaflet	**74. A. evansensis**
75a Leaves with 5 – 9 leaflets, the lateral leaflets only slightly smaller than the terminal leaflet	**73. A. basiphylla**
76 Leaflets rounded at apex, lateral veins spreading (at an angle of more than 70° to the midrib in the middle of the leaflet)	**73. A. basiphylla**
76a Leaflet apex acuminate, lateral veins ascending	77
77 Leaflets with numerous or densely covered with dark reddish-brown peltate scales and sometimes paler stellate scales on the midrib below, few to numerous on the rest of lower surface of leaflet; the staminal tube deeply lobed, the anthers inserted on the inside of the lobes, densely covered with simple white hairs on the margins of the lobes and the anthers	**67. A. euryanthera**

Artificial key to the species of Aglaia

77a Leaflets without or with few dark reddish-brown peltate scales, but with numerous pale brown or orange-brown peltate and stellate scales on the midrib below, few to numerous on the rest of lower surface of leaflet; the staminal tube neither deeply lobed nor densely hairy — **78**

78 Leaflets with pale brown or reddish-brown peltate scales which have a fimbriate margin numerous on or densely covering the midrib below, always with a few much darker peltate scales and dark reddish-brown pits interspersed, the inflorescence with reddish-brown stellate scales — **79**

78a Leaflets without dark reddish-brown peltate scales and dark reddish-brown pits, the indumentum usually brown or orange-brown, pits if present orange-brown — **80**

79 Leaflets pale bluish-green or pale green on both surfaces when dry — **69. A. sapindina**

79a Leaflets dark brown when dry — **68. A. polyneura**

80 Leaflets pale yellowish-green when dry or with whitish-green or orange veins — **81**

80a Leaflets not pale yellowish-green when dry — **85**

81 Fruit dehiscent (indicated in unripe fruits by the presence of three longitudinal ridges on the pericarp) — **15. A. lawii**

81a Fruit indehiscent (any longitudinal markings on the fruits are slight depressions rather than ridges) — **82**

82 Fruit subglobose with an apical depression or pyriform, pericarp thick (c. 3 mm) and woody — **83**

82a Fruits subglobose without an apical depression or obovoid, pericarp thin (c. 1 mm) — **84**

83 Leaflets pale brown or yellowish-brown when dry; fruit usually subglobose and 3-locular but sometimes the seed fails to develop in 1 or 2 of the locules — **58. A. edulis**

83a Leaflets yellow or yellowish-green when dry; fruits pear-shaped and 2-locular — **57. A. leucophylla**

Artificial key to the species of Aglaia

84 Leaflets with midrib rugulose, stigma depressed globose with an central depression and lobed margin	**55. A. parviflora**
84a Leaflets with midrib not rugulose; anthers inserted on margin; stigma ovoid	**80. A. saltatorum**
85 Staminal tube obovoid with a narrow aperture, anthers included or just protruding	**64. A. brassii**
85a Staminal tube cup-shaped, anthers protruding	86
86 Leaflets brown when dry, with numerous orange-brown peltate scales on the midrib below and some hairs interspersed when young	**81. A. mariannensis**
86a Leaflets bluish-green above and brown below when dry; stellate hairs absent	87
87 Inflorescences always axillary; fruit ellipsoid, up to 2 cm long	**60. A. odoratissima**
87a Inflorescences ramiflorous and axillary; fruit pyriform, up to 4.5 cm long	**59. A. macrostigma**
88 Leaflets with few to densely covered with stellate hairs or scales on the lower surface; when sparse, some hairs or scales occur evenly distributed between the veins and their presence visible with the naked eye	89
88a Leaflets without or with few hairs on the lower surface, with scales visible only with a lens or densely covered with hairs on the midrib only, few and unevenly scattered on the rest of the lower surface	109
89 Reticulation subprominent on both surfaces of leaflet when dry; with some hairs (on the twigs and sometimes elsewhere) which have a central rhachis and several whorls of arms radiating from it	**18. A. ramotricha**
89a Reticulation not subprominent; all hairs lacking a central rhachis	90
90 Hairs pale yellowish-brown or if reddish-brown then flower trimerous	91
90a Hairs reddish-brown or brown, or if pale brown then flower pentamerous	93

Artificial key to the species of Aglaia

91 Leaflets 11 – 13	**12. A. densitricha**
91a Leaflets 1 – 9, rarely 11	92
92 Peltate scales absent	**16. A. teysmanniana**
92a Peltate scales present, at least on the shoot apex	**15. A. lawii**
93 Hairs on lower leaflet surface numerous and with the arms of adjacent hairs overlapping, but leaving the surface of the leaflet visible	94
93a Hairs on lower surface of leaflet few or at least with the arms not usually overlapping	100
94 Hairs compact with arms all ± equal in length c. 0.5 mm, brown, densely covering the midrib, densely covering or scattered on the rest of the lower surface of leaflet	95
94a Hairs large and spreading, arms unequal in length up to 1 mm long and in some species (e.g. *A. rufibarbis*) up to 4 mm, usually reddish-brown and numerous on lower surface of leaflet	96
95 Lower surface of leaflet densely covered with persistent brown stellate hairs and scales	**54. A. aspera**
95a Lower surface of leaflet densely covered with brown stellate hairs which are deciduous and leave numerous pits and an uneven indumentum most dense near the midrib and veins	**51. A. forbesii**
96 Stellate hairs with long arms, up to 4 mm	**103. A. rufibarbis**
96a Stellate hairs with arms not usually exceeding 1 mm in length	97
97 Leaflets (1 –)3 – 5; apex rounded, lateral veins spreading (at an angle of more than 70° to the midrib in the middle of the leaflet)	**96. A. fragilis**
97a Leaflets 5 or more, apex acuminate, lateral veins ascending	98
98 Inflorescence sessile or subsessile	**104. A. archboldiana**
98a Inflorescence with a peduncle 1 – 5 cm long	99

Artificial key to the species of Aglaia

99 Lower surface of leaflet with numerous stellate hairs, the arms of adjacent hairs overlapping, interspersed with some pale brown hairs which have one or few ascending arms; fruit up to 5 cm long and 3.5 cm wide, with a hard woody pericarp 2 – 4 mm thick **101. A. hiernii**

99a Lower surface of leaflet with numerous stellate hairs, the arms of adjacent hairs overlapping, interspersed with pale brown stellate hairs which have several ascending arms; fruit up to 2.5 cm long and 1.7 cm wide, with a thin brittle pericarp less than 2 mm thick **98. A. tomentosa**

100 Leaflets with indumentum of few fimbriate peltate scales and stellate hairs on the lower surface; fruits small, c. 5 mm in diameter, hairs few or absent **95. A. palembanica**

100a Peltate scales absent from lower surface of leaflet; fruits 1 cm or more in diameter, with dense indumentum **101**

101 Leaflets usually more than 11, elliptical or oblong, the arms of adjacent hairs on lower surface not overlapping **102**

101a Leaflets rarely more than 11, most obovate, if more than 11 then the indumentum dense and arms of adjacent hairs on lower surface overlapping **104**

102 Leaflet surfaces not rugulose and without pits, leaflets up to 4 cm wide **94. A. exstipulata**

102a Upper surface of leaflet rugulose and pitted, lower surface with numerous pits, leaflets up to 7.5 cm wide **103**

103 Hairs and scales reddish-brown **93. A. rufinervis**

103a Hairs and scales orange-brown **57. A. leucophylla**

104 Usually unbranched tree **105**

104a Tree branching several or many times when mature **106**

105 Leaflets usually pale green when dry, with numerous reddish-brown stellate hairs on lower surface, not glabrescent; inflorescences axillary **91. A. tenuicaulis**

105a Leaflets usually bluish-green when dry, with numerous to densely covered with orange-brown stellate hairs on the midrib below, interspersed on the lower surface with paler stellate hairs which have fewer arms, glabrescent; inflorescences ramiflorous — **70. A. ceramica**

106 Leaflets with 23 – 32 pairs of veins, with few stellate hairs on the lower surface — **92. A. membranifolia**

106a Most leaflets with fewer than 23 lateral veins, with numerous hairs or scales on the lower surface — **107**

107 Leaflets with numerous pale stellate scales on the lower surface, interspersed with reddish-brown stellate hairs which have arms of different lengths; veins not white or pale brown when dry — **98. A. tomentosa**

107a Leaflets with numerous pale brown or reddish-brown stellate hairs which have arms of similar lengths on the lower surface — **108**

108 Leaflets with reddish-brown stellate hairs interspersed with paler brown stellate hairs on the lower surface, reticulation same colour as rest of lower surface of leaflet — **91. A. tenuicaulis**

108a Leaflets with pale brown stellate hairs interspersed with pale brown stellate scales on the lower surface, reticulation white or pale brown on the lower surface when dry — **97. A. brownii**

109 Lower surface of leaflet with numerous stellate or peltate scales — **110**

109a Leaflets with hairs or scales few on the lower surface between the veins when mature, but sometimes densely covering the midrib — **114**

110 Upper and lower surfaces of mature leaflets with numerous stellate scales which have a pale margin — **84. A. pyriformis**

110a Scales stellate of uniform colour, scales usually absent from upper surface of mature leaflets — **111**

111 Stellate scales interspersed with compact stellate hairs — **54. A. aspera**

111a Scales peltate, often with a fimbriate margin, if scales stellate then stellate hairs absent — **112**

Artificial key to the species of Aglaia

112 Fruit dehiscent	**15. A. lawii**
112a Fruit indehiscent	**113**
113 Scales all of one type	**53. A. crassinervia**
113a Stellate and peltate scales present together	**57. A. leucophylla**
114 Stellate hairs or scales more than 0.15 mm in diameter, numerous on or densely covering the midrib, sometimes also on the lateral veins, almost absent elsewhere	**115**
114a Stellate hairs or scales either very small, less than 0.15 mm in diameter, or almost totally absent from the midrib below and from the rest of the lower surface of leaflet	**135**
115 Leaves ± sessile or with a short peduncle of not more than 1 cm	**116**
115a Leaves not sessile, the basal leaflets only slightly smaller than the rest and of similar shape	**117**
116 Basal pair of leaflets much smaller than the rest and subrotund	**75. A. subsessilis**
116a Basal pair of leaflets only slightly smaller than and of similar shape to the rest	**73. A. basiphylla**
117 Leaflets with a short broad rounded acumen, lateral veins spreading (at an angle of more than 70° to the midrib in the middle of the leaflet)	**118**
117a Apex acuminate, lateral veins ascending	**119**
118 Leaflets densely covered with reddish-brown peltate scales which have a fimbriate margin, sometimes concealed by reddish-brown stellate hairs	**73. A. basiphylla**
118a Leaflets densely covered with orange-brown or reddish-brown stellate hairs and scales densely covering the midrib below, peltate scales absent	**71. A. parksii**
119 Reticulation subprominent on lower surface and often on upper surface of leaflet	**120**
119a Reticulation may be visible, but not subprominent	**121**

120 Leaflets 11 – 14, with pale brown stellate scales few to numerous on the midrib below, reticulation subprominent above and below	**83. A. laxiflora**
120a Leaflets 3 – 11, with pale yellowish-brown stellate hairs absent to numerous on the midrib below, reticulation visible on the upper surface, subprominent below	**88. A. oligophylla**
121 Petals 3 densely covered with stellate scales on the outside; fruits dehiscent	**122**
121a Petals 5 without scales on the outside, fruits indehiscent	**123**
122 Leaflets with reddish-brown stellate hairs densely covering the midrib below and few to numerous on the rest of the lower surface, without peltate scales	**11. A. meridionalis**
122a Leaflets with compact reddish-brown stellate hairs and pale peltate scales, which have a fimbriate margin, few on the midrib below and scattered pale and brown scales on the rest of the lower surface	**10. A. lepidopetala**
123 Tree unbranched; leaflets shiny	**85. A. coriacea**
123a Tree branched; leaflets not shiny	**124**
124 Fruit c. 0.5 cm in diameter, with few stellate scales	**78. A. aherniana**
124a Ripe fruit 1 cm or more in diameter, with dense indumentum	**125**
125 Leaflet apex with a parallel-sided acumen	**126**
125a Leaflet apex with a tapering acumen	**127**
126 Leaflets coriaceous	**51. A. forbesii**
126a Leaflets not coriaceous	**49. A. leptantha**
127 Both surfaces of leaflet with prominent pits	**52. A. foveolata**
127a Leaflets not prominently pitted although may be faintly so	**128**
128 Stellate hairs with long arms up to 6 mm, conspicuous on stems, few or absent elsewhere	**102. A. cuspidata**

Artificial key to the species of Aglaia

128a All arms of hairs less than 4 mm long	**129**
129 Fruit with one or more longitudinal ridges	**130**
129a Fruit without longitudinal ridge(s)	**132**
130 Fruit with three longitudinal ridges running base to apex, dehiscent	**15. A. lawii**
130a Fruit with two or ten longitudinal ridges from base to apex, indehiscent	**131**
131 Fruit obovoid or ellipsoid with two longitudinal ridges (i.e. one ridge completely encircles the fruit longitudinally)	**76. A. elliptica**
131a Fruit narrowly ellipsoid with several longitudinal ridges	**75. A. subsessilis**
132 Fruit up to 5.5 cm long and 4 cm wide	**77. A. conferta**
132a Fruits less than 2.5 cm long and 2.5 cm in diameter	**133**
133 Leaflets with dark orange or reddish-brown stellate hairs or scales few on or densely covering the midrib below, few to numerous on the lateral veins	**72. A. subminutiflora**
133a Leaflets rarely with hairs or scales on the lateral veins	**134**
134 Leaflets 3 – 7, with numerous orange-brown or pale brown stellate scales on the midrib below, interspersed with some hairs when young	**81. A. mariannensis**
134a Leaflets (7 –)9 – 11, densely covered with dark reddish-brown stellate scales on the midrib below and few elsewhere, stellate hairs absent from the leaflets	**79. A. barbanthera**
135 Petals 5; leaflets yellow or yellowish-green or the veins black or dark brown when dry	**136**
135a Petals 3; leaflets brown, greenish-brown, purplish-brown or orange-brown, veins usually the same colour as the leaflet surface	**140**
136 Leaflets greyish-brown or black when dry, particularly the veins; scales pale grey or greyish-brown	**137**

136a Leaflets yellow or yellowish-green when dry; hairs or scales pale brown, reddish-brown or golden brown   **138**

137 Flowers with shallow, cup-shaped staminal tube which has a wide aperture; leaflets usually slightly pitted, pale brownish-green when dry, subcoriaceous with stellate scales on lower surface, few on or densely covering the midrib and few elsewhere, veins 13 – 18 pairs, often black but reticulation not conspicuous   **51. A. forbesii**

137a Flowers with obovoid staminal tube which has a minute apical pore; leaflets smooth, brownish-green or blackish-green when dry, thin; veins 8 – 14 pairs, the midrib, lateral veins and reticulation black when dry   **49. A. leptantha**

138 Inflorescence delicate, peduncle, rhachis and branches flattened, these and the calyces with very few pale brown or nearly white stellate scales   **82. A. cumingiana**

138a Inflorescence robust, the peduncle, rhachis and branches terete, with numerous to densely covered with orange-brown or reddish-brown hairs or scales   **139**

139 Leaflets 9 – 15, rounded or cuneate at the base, margins not recurved when dry, with sparse to frequent golden-brown scales on the lower surface   **57. A. leucophylla**

139a Leaflets (1 –)3 – 5, cuneate or attenuate at the base, margins recurved, with reddish-brown stellate hairs on the twig apices and ± absent from the leaflets   **90. A. monozyga**

140 Leaflet surfaces dull and pitted, with tiny stellate hairs sometimes frequent on lower surfaces, reticulation subprominent   **8. A. spectabilis**

140a Leaflet surfaces smooth or rugulose but not pitted, stellate hairs almost absent from lower surface, reticulation not subprominent   **141**

141 Leaflets 2 – 4 cm wide, lower surface usually purplish-brown and smooth when dry   **5. A. malaccensis**

141a Leaflets 6 – 12 cm wide, lower surface not purplish-brown when dry   **142**

Section Amoora

142 Both leaflet surfaces smooth or only slightly
rugulose, base usually rounded, sometimes cuneate
on some of the leaflets **7. A. erythrosperma**

142a Upper and lower surfaces of leaflet rugulose,
base cuneate or attenuate **143**

143 Small tree, up to 12 m; twigs with a few white
or reddish-brown stellate hairs; leaflets 7 – 9, base
attenuate **6. A. rugulosa**

143a Tall tree, up to 45 m; twigs densely covered
with hairs which have a central rhachis and
numerous whorls or arms radiating from it, these
often deciduous, leaving a dense covering of pale
brown or dark reddish-brown stellate hairs and
scales; leaflets 11 – 17, rounded or slightly cuneate
at the base **14. A. penningtoniana**

Section **Amoora** (Roxb.) stat. et comb. nov.

Basionym: *Amoora* Roxb., Pl. Coromandel 3: 54, t. 258 (1820). Holotype species: *Amoora cucullata* Roxb. = *Aglaia cucullata* (Roxb.) Pellegrin.

Amoora section *Amoora*, Miq., Ann. Mus. Bot. Lugd. Bat. 4: 37 (1868), as section '*Otamoora*' [N.B. section *Euamoora* Miq. (1868) = *Aphanamixis*].

Amoora section *Amoora*, C. de Candolle in A. & C. de Candolle, Monographiae Phanerogamarum 1: 583 (1878), as section *Pseudo-Aglaia*.

Amoora section *Amoora*, Pierre, Fl. Forest. Cochinch. Fasc. 22, ant e t. 343 (1896), as section '*Euamoora*'.

Amoora section *Neoamoora* Pierre, Fl. Forest. Cochinch. Fasc. 22, ante t. 343 (1896). Lectotype species (designated here): *Amoora spectabilis* Miq. (= *Aglaia spectabilis* (Miq.) Jain & Bennet).

Petals 3(– 5); anthers 6 – 21; fruit dehiscent.

1. Aglaia cucullata (Roxb.) Pellegrin in Lecomte, Fl. Gén. Indo-Chine, 1: 771 (1911); Pannell in Ng, Tree Flora of Malaya 4: 214 (1989).

[*Andersonia cucullata* Roxb., Hort. Beng.: 87 (1814) nom. nud.]
Amoora cucullata Roxb., Pl. Coromandel 3: 54, t. 258 (1820). Lectotype (designated here): India, H.B.C., November 1809, *Roxburgh* '1238' (BM!); Miq., Ann. Mus. Bot. Lugd. Bat. 4: 37 (1868); Hiern in Hooker fil., Fl. Brit. India 1: 560 (1875); C. de Candolle in A. and C. de Candolle, Monog. Phan. 1: 583 (1878); Pierre, Fl. Forest. Cochinch. Fasc. 22, ante t. 344A (1 July 1896); King in Jour. As. Soc. Bengal 64: 55 (1895); Ridley, Fl. Malay Penins. 1: 399 (1922); I.H. Burkill, Dictionary of Economic Products of the Malay Peninsula 1: 138 (1935); Backer & Bakhuizen, Fl. Java 2: 126 (1965).
[*Sphaerosacme rohituka* Wallich, Cat. 1278 (1829) quoad specim. *Anon.* in *Andersonia rohituka* (= *Aphanamixis polystachya* (Wallich) R.N. Parker).
[*Sphaerosacme laxa* Wallich, Cat. 4894 (1831-2) nom. nud.].

[*Sphaerosacme paniculata* sensu Miq., Ann. Mus. Bot. Lugd. Bat. 4: 37 (1868) non Wallich Cat. quoad specim. i.e. *Aphanamixis*].
[*Amoora auriculata* Miq., Ann. Mus. Lugd. Bat. 4: 37 (1868) nom. in syn.]
Amoora aherniana Merrill in Philipp. Gov. Lab. Bur. Bull. 17: 24 (1904). Lectotype (designated here): Philippine Islands, Luzon, Prov. of Bataan, Mt Mariveles, Lamao River, fr., June 1904, *Borden* For. Bur. 823 (NY!; isolectotypes: BM!, K!, PNH†, SING!); I.H. Burkill, Dictionary of Economic Products of the Malay Peninsula 1: 137 (1935).
Aglaia tripetala Merrill in Jour. As. Soc. Straits 76: 88 (1917). Lectotype (designated here): British North Borneo [Sabah, Sandakan], fl., Jan – Mar. 1916, *Villamil* 184 (PNH†, photo A!).
Aglaia conduplifolia Elmer in Leafl. Philipp. Bot. 9: 3324 (1937), sine diagn. lat. Type no.: *Elmer* 12195, Philippine Islands, Sibuyan, Magallanes, Mt Giting-giting, April 1910 (A!, BM!).

Tree up to 15(– 30) m, with a broad rounded crown. Bole up to 10 m, up to 100 cm in diameter, sometimes with plank buttresses upwards up to 3 m, sometimes with pneumatophores up to 7 m away from the bole and up to 60 cm high. Branches arching. Bark smooth, brown, pinkish-grey or pale orange-brown, sometimes flaking in small brittle or papery scales; inner bark pink, fibrous; sapwood pale yellowish-brown, pink or orange-brown; latex white. Twigs slender, longitudinally wrinkled, greyish-brown, densely covered with pale brown or almost white peltate scales which are darker in the centre and have a paler, sometimes (always in New Guinea) fimbriate, margin.

Leaves imparipinnate, up to 45 cm long and 40 cm wide, obovate in outline; petiole up to 15 cm, the petiole, rhachis and petiolules with the surface and a few peltate scales like those on the twigs. Leaflets 5 – 9, the laterals subopposite, all 4 – 20 cm long, 1.5 – 6 cm wide, subcoriaceous, ovate, asymmetrical and curved, acuminate at apex with the obtuse acumen up to 5 mm, rounded at the markedly asymmetrical base, the terminal leaflet sometimes reduced in size to c. 4 cm long and 1.5 cm wide, with a petiolule up to 4 cm long and the lamina folded at the base of the leaflet to form a pocket on the upper surface, lower surface rugulose and faintly pitted, with a few scales on the midrib and veins like those on the twigs and sometimes scattered on the surface in between; veins 8 – 13 on each side of the midrib, curved upwards, midrib depressed on upper surface, midrib prominent, lateral veins subprominent and longitudinally wrinkled on lower surface, secondary veins visible on both surfaces; petiolules up to 10 mm on lateral leaflets, up to 15 mm on terminal leaflet.

Inflorescence up to 30 cm long and 35 cm wide; peduncle up to 8 cm, the peduncle rhachis and branches flattened with few to numerous scales like those on the twigs. Flowers up to 3.5 mm long and 3.5 mm wide, subglobose; pedicel up to 3 mm with numerous scales like those on the twigs. Calyx c. $^1/_3$ length of the corolla, cup-shaped, with few to numerous white stellate scales on the outside, divided up to c. $^1/_2$ way into 3 – 4 obtuse lobes which have a fimbriate margin. Petals 3, aestivation imbricate, 2.5 – 3 mm long, 1 – 1.5 mm wide, yellow, obovate, glabrous. Staminal tube slightly shorter than the corolla, obovoid, the aperture up to 1.5 mm across and shallowly 6-lobed; anthers 6, $^1/_2 - ^1/_3$ the length of the tube, ellipsoid, protruding slightly through the aperture. Ovary

Section Amoora

depressed-globose, loculi 3; stigma ellipsoid with 3 apical lobes and 6 longitudinal ridges; ovary and stigma together c. $\frac{1}{2}$ the length of the staminal tube.

Infructescence with few fruits; peduncle c. 5 cm. Fruits c. 7 cm long and 6 cm wide, yellow, obovoid; pericarp leathery, thin, brittle and moulded around the seeds when dry, densely covered with reddish-brown peltate scales which have a fimbriate margin. Loculi (2 or) 3, each containing 0 or 1 seed. Seeds c. 5 cm long and 3 cm wide with a shiny, reddish-brown, yellow or white aril covering about half the seed. Fig. 3.

DISTRIBUTION. Bangladesh (Sundarbans: delta of the river Ganges), Thailand, Vietnam (delta of the Mékong), Peninsular Malaysia, Singapore, Sumatra, Borneo, Philippine Islands, Java, New Guinea. Fig. 4.

ECOLOGY. Riverine forest, tidal estuaries, mangrove, nipah swamp. Scarce to rather common. Alt.: sea level up to 20 m. The pouch at the base of the terminal leaflet is sometimes occupied by ants.

VERNACULAR NAMES. Borneo: Bengang (Iban lassa); Nyireh Batu (Brunei); Batu-batu (Bessaya); Jalongan (Malay); Merak, Polie. New Guinea: Didagou (Bian); Edjewoera (Argoeni).

USES. Wood is used for building in Brunei and building boats in Irian Jaya (Merau river).

Representative specimens. BANGLADESH. Sundarbans: fl., 20 Sep. 1809, *Buchanan-Hamilton* 2351 (E!); [22°10′N 89°20′E], fl., Sundarbans Inventory Project 1 (FHO!).

BURMA. S. Tenasserim, Maynuge, Tenasserim River, sea level, fl., 8 Mar. 1926, *Parkinson* 2005 (K!).

VIETNAM. *Thorel* 988 (BM!). S. Cochinchina, R. Cay Lai, fr., Feb. 1889?, *Pierre* 425 (L!).

THAILAND. Bangkok, *Marcan* 606 (BM!); Bangkok, Palanam, *Kerr* 6754 (BM!).

PENINSULAR MALAYSIA. Perak: Dindings, Bruas, fl., March 1896, *Ridley* 7960 (K!, SING!); Larut, sea coast, *Wray* 2505 (SING!). Johore: Sg. Tekau, *Ridley* 11507 (BM!, SING!); Pontian, Kukup, *Ahmad* KEP 96210 (FRI!).

SINGAPORE. Mandai Road, fl., Nov 1917, *Nur* s.n. (CGE!, BM!). Selitar, *Ridley* 3778 (BM!). Pulau Sakeng, fr., 22 Sep. 1950, *Sinclair* SFN 39011 (E!). Tanjong Penjuru, Pandan Forest Reserve, fr., 8 Aug. 1941, *Corner* SFN 38161 (K! large sheet).

SARAWAK: Lawas, ± sea level, *Drahman* 1 (FHO!), Limbang Sg. Melais, *Muhammad* 2 (FHO!).

BRUNEI. Pulau Siarau, K. Temburong, fl., 21 Jan. 1959, *Ashton* BRUN 5125 (FHO!). Ulu Brunei, *Ashton* BRUN 5077 (FHO!).

SABAH. Pulau Kembong, *Winkler* 3441 (BM!). Klias River, 1.5 m, fl., 26 March 1933, *Melegrito* BNB 3042 (FHO!).

IRIAN JAYA. Merauke District, E. bank of Merau river, S of Senajo [8°15′S 140°44′E], 20 m, fl., 12 Aug. 1954, *van Royen* 4673 (K!, L!).

PAPUA NEW GUINEA. Gulf Division, Uramu Island, 3 miles below Kinomeri village, st., 29 June 1955, *Floyd & Gray* NGF 8006 (K!). Gulf District, Purari delta, near Ravikivau, 3 m, fl., 11 Feb. 1966, *Craven & Schodde* 798 (K!, L!).

The leaflets of *A. cucullata* are curved, markedly asymmetrical and

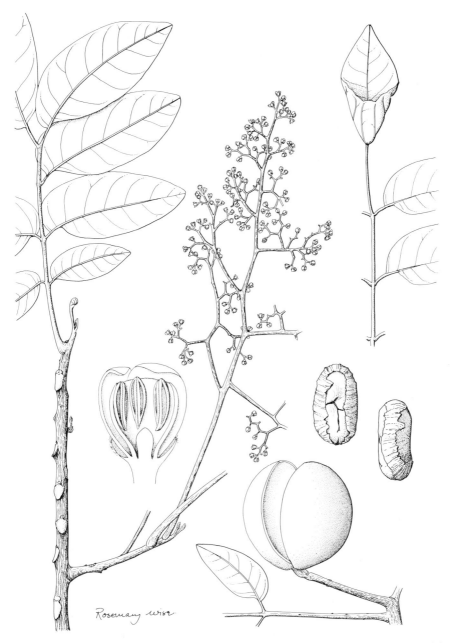

Fig. 3. *A. cucullata*. Habit with inflorescence x½. Half flower x10. Part of leaf, on which the lamina of the terminal leaflet forms a pocket at the base (top right) x½. Fruit and seeds x½.

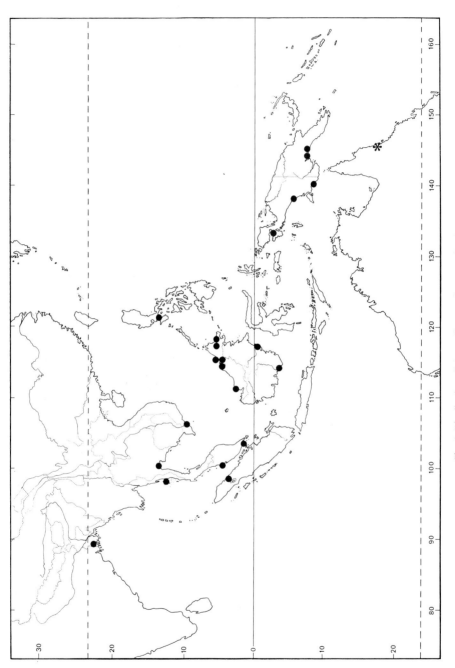

Fig. 4. Distribution of A. cucullata ● and A. australiensis *.

almost without indumentum. The terminal leaflet sometimes has a pouch at the base of the lamina which gives the species its name and resembles that found on the leaf of *Ficus benghalensis* L. var. *krishnae* (C. de Candolle) Corner. *A. cucullata* is the only species of *Aglaia* which grows in estuaries and mangrove swamps and the only species for which pneumatophores are recorded (*van Royen* 4673). The presence of a fleshy aril around the seed suggests that dispersal is by animals (probably birds) but, given the habitat and widespread distribution of the species, the possibility of dispersal by water should also be investigated.

2. Aglaia australiensis C.M. Pannell spec. nova *Aglaiae meridionali* C .M. Pannell similis, sed squamis peltatis (nec stellatis) distinguitur. Accedit etiam ad *Aglaiam cucullatum* (Roxb.) Pellegrin, sed foliolis nitidis in sicco, non valde asymmetricis nec curvis differt; squamae obscure badiae (nec pallidissime brunneae). Holotype: Australia, Queensland, Boonjie, near Malanda [17°16′S 145°28′E], 930 m, fl., 12 Aug. 1947, *L.S. Smith* 3274 (BRI!; isotypes: CANB!, K!, L!).

Small tree up to 10 m high. Bole up to 12.5 cm in diameter. Outer bark brown; middle bark dark red; inner bark pale reddish-brown to pale yellow; latex white. Sapwood pale yellowish-brown, heartwood pinkish-brown. Twigs reddish-brown, longitudinally wrinkled, with large round lenticels, with numerous or densely packed dark reddish-brown peltate scales which have a fimbriate margin.

Leaves imparipinnate, 42 – 72 cm long, 25 – 50 cm wide, obovate in outline; petiole 9 – 23 cm; peduncle, rhachis and petiolules greyish-brown with a few scales like those on the twigs. Leaflets 9 – 11, the laterals subopposite, all 7.5 – 29 cm long, 3.5 – 8 cm wide, elliptical or obovate, acuminate at apex, the broad obtuse acumen up to 15 mm long, cuneate at the slightly asymmetrical base, with scales which have a dark purplish-brown centre and paler fimbriate margin scattered on the lower surface; veins 6 – 14 on each side of the midrib, curved upwards; petiolules of lateral leaflets 7 – 15 mm, up to 20 mm on the terminal leaflet.

Inflorescence 8 – 19 cm long, 3 – 11 cm wide; peduncle 0.3 – 2.5 cm, peduncle rhachis, branches and pedicels densely covered with very dark reddish-brown peltate scales which have a fimbriate margin, with a few stellate scales interspersed. Flowers (not known whether male or female) 3 – 4 mm long, 3 – 4 mm wide subglobose; pedicels up to 2 mm. Calyx 1 – 2 mm long, green, divided into 3 shallow lobes. Petals 3, pale yellow, aestivation imbricate. Staminal tube 2 – 2.5 mm long and 2 – 3 mm wide, subglobose, pale yellow, the aperture 1.5 – 1.8 mm in diameter, anthers 6, 1.5 – 1.7 mm long and 0.5 – 0.8 mm wide, inserted $\frac{1}{4}$ to $\frac{1}{3}$ of the way up the tube and just protruding through the aperture. Ovary c. 0.7 mm long and 1.4 mm wide, depressed-globose, densely covered with reddish-brown stellate hairs; loculi 3, each containing 1 ovule; stigma c. $\frac{1}{2}$ the length of the staminal tube, 3 apical lobes and 6 longitudinal ridges.

Ripe fruits not seen. Immature fruits 1.5 cm long and 1.6 cm wide, subglobose, longitudinally ribbed.

DISTRIBUTION. Australia only: Queensland. Fig. 4.

ECOLOGY. Occurs on soil derived from basalt in lowland rain forest, coastal rain forest and complex mesophyll forest.

Representative specimens.
AUSTRALIA. Queensland, Cook District: State Forest Reserve 99 [17°S 145°E], 4 Sep. 1957, *L.S. Smith* 10169 (L!); nr Lammond's Hill, 12 km a little S. of Malanda [17°22'S 146°36'E], 900 m, 9 Sep. 1959, *L.S. Smith* 10837 (K!, L!).

Aglaia australiensis resembles *Aglaia meridionalis*, but it has peltate scales rather than stellate. It resembles *A. cucullata*, but the leaflets are shiny when dry and they are not markedly asymmetrical and not curved; the scales are dark reddish-brown in colour, not very pale brown. *Aglaia australiensis* resembles *A. macrocarpa*, but it is a smaller tree and the scales are larger and more numerous on the lower surface of the leaflets than in *A. macrocarpa*.

This species is endemic to Australia and of restricted range in the Cook District of Queensland. The distribution of *A. meridionalis* overlaps with *A. australiensis* but it extends further north on the east coast of Cape York.

3. Aglaia flavida Merrill & Perry in Jour. Arn. Arb. 21: 320 (1940). Holotype: Northeastern New Guinea, Morobe District, Sattelberg [6°30'S 147°46'E], 3000 ft [c. 900 m], fl., 19 Oct. 1935, *Clemens* 490 (A!; isotypes: G!, L!).

Aglaia cucullata sensu Henty in Bot. Bull. Lae 12: 100, t. 59 (1980), auct. non (Roxb.) Pellegrin.

Tree up to 36 m, girth up to 3 m; buttresses broad, steep, outwards to 60 cm and upwards to 3 m or narrow, plank-like and equal. Bark white, grey, pale, dark or reddish-brown, firmly fibrous, smooth or fissured and pitted, with scales either adherent or sloughing in medium-sized pieces which leave scroll marks on the bole; sapwood hard or soft, pale reddish-brown, reddish-brown, deep red or yellowish-brown; sometimes with white latex. Twigs greenish-brown, stout, longitudinally wrinkled, densely covered with small orange, orange-brown or pale brown peltate scales which have a fimbriate margin and often have a dark central spot, sometimes with a few darker scales interspersed.

Leaves up to 82 cm long and 55 cm wide, obovate in outline; petiole 10 – 22 cm, the petiole, rhachis and petiolules densely covered with scales like those on the twigs. Leaflets 7 – 15, the laterals subopposite, all 9.5 – 30 cm long, 4 – 10.5 cm wide, ovate, obovate or elliptical, acuminate or rounded at the apex, the acute or obtuse acumen up to 10 mm, cuneate or rounded at the sometimes markedly asymmetrical base, with numerous scales like those on the twigs on the midrib below and scattered on the rest of the lower leaflet surface, sometimes with numerous pale scales which are visible with the naked eye on dried specimens and a few brown scales like those on the twigs; 6 – 20 on each side of the midrib, ascending and curved upwards near the margin, the midrib prominent, the lateral veins subprominent and the reticulation just prominent below; petiolules 5 – 20 mm on lateral leaflets, up to 25 mm on terminal leaflet. Inflorescence 26 – 42 cm long, c. 24 cm wide; peduncle 1.5 – 19 cm, the peduncle, rhachis and branches greenish-brown and with numerous scales like those on the twigs. Flower 2.5 mm long, 2 mm wide, subglobose, fragrant; sessile or with a pedicel up to 0.5 mm. Calyx c. 1 mm, divided into 3 broad rounded lobes densely covered on the outside with brown scales like those on the twigs. Petals 3, yellow, with some stellate scales on the outside, aestivation imbricate. Staminal tube c. 2 mm long, subglobose, white, anthers 6, pale brown, c. $^2/_3$ the length of the tube and just protruding. Ovary depressed globose densely covered with stellate scales; stigma with three apical lobes, pale yellow.

Infructescence c. 20 cm long and 7 cm wide; peduncle c. 7 cm, the peduncle, rhachis and branches densely covered with pale and reddish-brown scales which have a fimbriate margin. Fruits up to 8 cm long, 2.5 – 5.5 cm in diameter, subglobose or obovoid, orange or brown, dehiscent, densely covered on the outside with scales like those on the twigs; fruit-stalk 1 – 4 cm. Loculi 3, each containing 0 or 1 seed; aril white.

DISTRIBUTION. New Guinea, New Britain, Solomon Islands. Fig. 5.

ECOLOGY. Found in primary lowland and hill forest, secondary forest, on occasionally inundated soil to well-drained soil, on clay, peat, coral. Alt.: sea level up to 1300 m. Common. Fruits eaten by birds.

USES. Wood is used for houses, paddles (Irian Jaya: Asmat); axe handles (Solomon Is.: New Georgia); canoes (Solomon Is.: Guadalcanal).

NOTE. The sap is irritant and may cause severe dermatitis ('*A. cucullata*' in Henty, 1980).

VERNACULAR NAMES. New Guinea: Amsam (Asmat); Oewaa (Kebar); Sjopob (Nemo). Solomon Islands: Maoa, Mawa, Norisinwani, Noriswani (Kwara'ae).

Representative specimens. PAPUA NEW GUINEA. Morobe District: Wau Subdistrict, 5 miles N.W. of Bulolo [7°07'S 146°37'E], c. 1150 m, fr., 26 Jan. 1971, *Kairo* NGF 44570 (K!, L!); Morobe District, Lae – Bulolo road, near Garagos [7°10'S 146°52'E], 450 m, fl., 3 May 1962. *Havel & Kairo* NGF 11195 (K!). West New Britain District, Kombe Subdistrict, near Linga Linga [5°40'S 149°50'E], 500 m, fl., 29 May 1973, *Henty & Lelean* NGF 49497 (K!, L!).

SOLOMON ISLANDS. E. Guadalcanal, Makina River Area, 85 m, fl., 11 May 1968, *Boraule & collectors* BSIP 9401 (K!, L!). N.E. Cristobal, Banks of Pegato River, near confluence with Waharito, 60 – 120 m, fr., 23 July 1965, *Whitmore* BSIP 6148 (K!).

A. flavida is similar to *A. macrocarpa*. They differ in details of the trichome structure, pigmentation and distribution and in the texture of the leaflet surfaces. *A. macrocarpa* does not occur east of Maluku, whereas *A. flavida* is confined to New Guinea and the Solomon Islands.

4. Aglaia macrocarpa (Miq.) C.M. Pannell comb. nova

Epicharis macrocarpa Miq., Fl. Ind. Bat. Suppl. 1: 196 (1860), 505 (1861).
  Lectotype (designated here): Sumatra, [Priaman], *Diepenhorst* 3090 H.B. no. 15 (U!: isolectotypes: BO!, L!).
Aglaia pycnocarpa Miq., Ann. Mus. Bot. Lugd. Bat. 4: 45 (1868), superfl. nom. illegit. pro *Epicharis macrocarpa*; C. de Candolle in A. & C. de Candolle, Monog. Phan. 1: 625 (1878).
Amoora rubescens Hiern in Hooker fil., Fl. Brit. India 1: 561 (1875). Lectotype (Pannell, 1982): Singapore, *Maingay* 3351 (Kew Dist. 355) (K!); C. de Candolle in A. & C. de Candolle, Monog. Phan. 1: 589

(1878); King in Jour. As. Soc. Bengal 64: 57 (1895); Ridley Fl. Malay Penins. 1: 399 (1922); I.H. Burkill, Dictionary of Economic Products of the Malay Peninsula 1: 138 (1935).

Amoora trichanthera Koord. & Val. in Meded. 'S Lands Plantent. 16: 123 (1896). Lectotype (designated here): Java, Preanger, Takoka, c. 1100 m, 28 July 1891, *Koorders* 4717β (L!; isolectotype: K!); Koorders & Valeton, Atlas der Baumarten von Java, 1: t. 163 (1913); Backer and Bakhuizen, Fl. Java 2: 126 (1965).

Aphanamixis trichanthera (Koord. & Val.) Koord., Exkursionsfl. Java 2: 4 44 (1912).

Aglaia trimera Ridley in Kew Bull. 1930: 368 (1930), non Merrill (1929) (= *Aglaia lawii*). Lectotype (designated here): Borneo, Sarawak, Kuching, fl ., 24 March 1893, *Haviland* 2847 (K!; isolectotype: SING!).

Aglaia triplex Ridley in Kew Bull. 1938: 215 (1938), nom. nova pro *A. trimera* Ridley.

Aglaia rubescens (Hiern) Pannell in Malaysian Forester 45: 455 (1982); Pannell in Ng, Tree Flora of Malaya 4: 223 (1989).

Tree up to 35 m. Bole up to 22 m, 150 cm in circumference, with buttresses outwards up to 100 cm. Bark reddish-brown and grey or pale, flaking in large irregular scales and numerous reddish-brown lenticels, some in longitudinal rows; inner bark dark red or pinkish-brown; sapwood pale yellow, pale or dark yellowish-brown or pale reddish-brown; latex white. Twigs fairly stout, brown, with longitudinal wavy ridges and densely covered with minute reddish-brown or grey peltate scales which have a fimbriate margin and are deciduous, leaving dark reddish-brown pits. Leaves imparipinnate, up to 70 cm long and 50 cm wide, oblong or slightly obovate in outline; petiole up to 20 cm, flattened on the adaxial side, the petiole, rhachis and petiolules with surface and scales like the twigs. Leaflets 11 – 15, the laterals subopposite, all 5 – 25 cm long, 2 – 7 cm wide, pale reddish-brown when dry, coriaceous, elliptical, oblong or oblanceolate, acuminate or caudate at apex with the obtuse or acute acumen up to 15 mm, rounded or cuneate at the asymmetrical base, upper surface rugulose, shiny, lower surface rugulose with few minute reddish- brown scales or numerous pale grey peltate scales on the midrib and lateral veins; veins 6 – 11 on each side of the midrib, curved upwards, midrib and lateral veins depressed on upper surface, lower surface with midrib and lateral veins with longitudinal wavy ridges, midrib prominent and lateral veins slightly prominent; petiolules up to 15 mm on lateral leaflets, up to 20 mm on terminal leaflet.

Inflorescence up to 30 cm long and 20 cm wide; peduncle up to 11 cm, the peduncle, rhachis, branches and pedicels with surface and indumentum like the twigs. Flowers dark yellow, up to 3.5(– 5) mm long, obovoid; sessile or with pedicel up to 1(– 3) mm. Calyx $1/3 - 1/2$ length of the corolla, densely covered on the outside with dark brown stellate scales, divided up to c. $1/2$ way into 3 lobes. Petals 3, pinkish-yellow, subrotund or obovate, aestivation imbricate, densely covered on the outside when young with dark brown stellate scales which are deciduous at maturity. Staminal tube shorter than the corolla, subglobose, c. 1.5 mm in diameter with the aperture c. 1 mm across, and shallowly and irregularly lobed; anthers 6 – 10, and $1/2$ to as long as the tube, ellipsoid or narrowly ovoid,

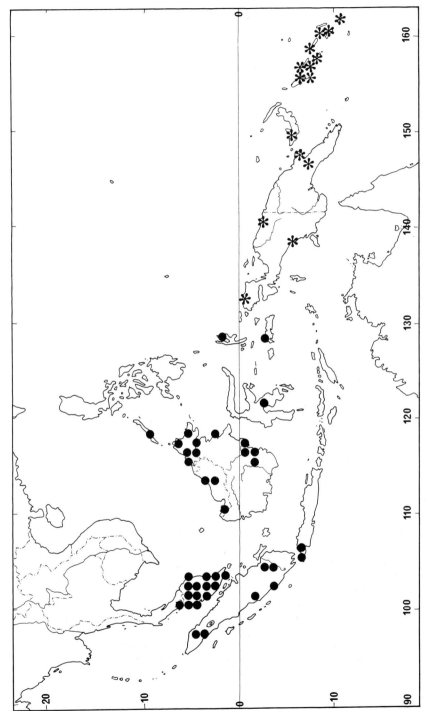

Fig. 5. Distribution of *A. macrocarpa* ● and *A. flavida* *.

curved with the tube and just protruding beyond the aperture, usually with scattered simple hairs. Ovary depressed-globose densely covered with stellate scales; stigma ovoid; ovary and stigma together $^3/_4$ to as long as the staminal tube. Infructescence with about 4 fruits; peduncle 3.5 – 8 cm, the peduncle, rhachis and branches stout and with surface and indumentum like the twigs. Fruits up to 6 cm long and 5.5 cm wide, obovoid, the pericarp longitudinally wrinkled and moulded around the seeds when dry, densely covered with dark reddish-brown stellate scales; fruit-stalk up to 1 cm. Loculi 3, each containing 0 or 1 seed; seed with a complete red, orange or white aril, c. 5.5 cm long, 2.5 cm wide and 1.5 cm thick; testa brown. Fig. 6.

DISTRIBUTION. Vietnam, Peninsular Malaysia, Singapore, Sumatra, Borneo, ? Philippine Islands (Palawan), Java, Sulawesi, ? Maluku (Ceram & Halmaheira). Fig. 5.

ECOLOGY. Found in lowland, hill and ridge forest, primary forest, secondary forest, pole forest and along rivers and paths; on sandy clay, sandstone, loam, limestone and basalt. Alt.: sea level up to 1750 m.

VERNACULAR NAMES. Peninsular Malaysia: Bekak. Sumatra: Balam Pelapah, Bekih, Kajoe Tenoe, Kembalau, Manehwoeh Boengo, Manehwoeh Falah, Balam Pelapah (Malay), Setoer Setoer, Soerin Halah. Borneo: Lantupak (Dusun Kinabatangan); Segera (Iban); Birajang.

Representative specimens. VIETNAM. S. Annam, Province de Chatrang, Mount Hon-ba, 1000 – 1500 m, 28 – 31 Aug. 1918, *Chevalier* 38876 (P!).

PENINSULAR MALAYSIA. Perak, Gopeng, c. 150 – 240 m, fr., April 1884, *King's Coll.* 5944 (BM!, CGE!, K!). Kelantan, Ulu Sat F.R., Compt 31, valley, fr., 3 Feb. 1970, *Kochummen* FRI 2957 (FRI!, L!). Pahang, Cameron Highlands, Boh Plantations valley, c. 1200 m, fl., 9 April 1937, *Nur* SFN 32589 (FRI!, E!). Johore, Mersing F.R., 8 miles W. of Jemaluang, 5 miles from road, c. 60 m, fr., 19 Dec. 1963, *Pennington* 8032 (FHO!).

SINGAPORE. *Maingay* 3351 (Kew Dist. 355) (K!). *Cantley's coll.* s.n. (SING!).

SUMATRA. W., Priaman, st., *Diepenhorst* 58, H.B. Bogor, 3090 (L!). Benkoelen, 1100 m, *Endert* Proefst. E 1042 (L!). Palembang, 75 m, Boschproefst. nr T 869 (L!).

SABAH. Ulu Biah Keningau District, fr., 7 March 1980, SAN 92178 (K!, L!). Kunak District, Virgin Jungle Reserve, *Madai Mostyn* SAN 77020 (L!).

KALIMANTAN. E. Kutei Reserve [0°24'N 117°16'E], tree no. TH17. 1552, 60 m, fr., 5 Aug. 1979, *Leighton* 914 (FHO!). C. Kutei, R. Pedohon, near Tabang, *Kostermans* 10623 (L!). Peak of Balikpapan, *Kostermans* 7388 (L!). Kalimantan Selatan, Djaro Dam, c. 10 km N.E. of Muara Uja [10°50'S 115°40'E], 260 m, 24 Nov. 1971, *de Vogel* 1028 (K!, L!).

JAVA. Preanger: Takoka, Region II, c. 1100 m, damaged fl., 28 July 1891, *Koorders* 4718β (K!, L!); Palabuan ratu, Region I, yg fr., 7 June 1890, *Koorders* 4996β (BO!, G!, K!, L!). S.W., *Kostermans* UNESCO 172 (K!, L!). S.W., *Kostermans* UNESCO 72 (K!, L!).

SULAWESI. Malili, fl., 21 Jan. 1931, *Boschproefst*. Cel./II-490 (L!).

It is sometimes difficult to separate *A. macrocarpa* from *A. malaccensis* and they may represent a single variable species. The leaves of *A. macrocarpa* are usually orange-red when dry and the leaflet base is more cuneate than

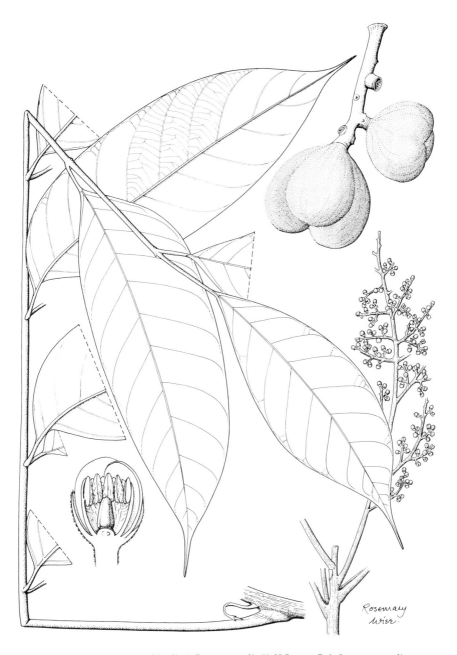

Fig. 6. *A. macrocarpa*. Habit x½. Inflorescence x½. Half flower x7. Infructescence x½.

in *A. malaccensis*. The upper and lower surfaces are rugulose and there are fewer pairs of veins. The fruit is obovoid. High-altitude specimens have shiny, coriaceous leaves. Low-altitude specimens exhibit two different indumentum types: minute reddish-brown peltate scales or grey peltate scales which have a fimbriate margin; either of these may be deciduous and leave dark reddish-brown pits. Trees with the different indumentum types are distinguishable in the field when they occur at the same site (e.g. East Kutei, Kalimantan, M. Leighton, pers. comm.). It has not, however, been possible to reliably distinguish them in the herbarium, except from the minute scale character. Further field work and improved collections may result in the discovery of more easily observed distinguishing characters.

5. Aglaia malaccensis (Ridley) Pannell in Malaysian Forester 45: 455 (1982); Pannell in Ng, Tree Flora of Malaya 4: 219 (1989).

Amoora malaccensis Ridley in Jour. As. Soc. Straits Branch 75: 16 (1917). Lectotype (Pannell, 1982): Malaysia, Malacca, Ayer Panas, 1894, *Ridley* 1797 (K!); Ridley, Fl. Malay Penins. 1: 399 (1922).

Tree up to 27 m. Bole up to 165 cm in circumference, with a few thick, shallow buttresses upwards up to 50 cm. Bark smooth, pale brown with longitudinal rows of lenticels, flaking in large scales of irregular size up to 65 cm long and 35 cm wide; sapwood soft, fibrous, pink, yellow or white; latex white. Twigs fairly stout, pale brown, surface longitudinally wrinkled, densely covered with small pale brown or almost white stellate hairs or scales.

Leaves imparipinnate, up to 50 cm long and 20 cm wide, oblanceolate in outline; petiole up to 20 cm, flattened on the adaxial side, the petiole, rhachis and petiolules with surface and indumentum like the twigs. Leaflets 11 – 15, the laterals subopposite, all 7 – 15 cm long, 2 – 4 cm wide, green above and dark purplish-brown below when dry, usually lanceolate, sometimes oblong or elliptical, acuminate at apex with the obtuse or acute acumen up to 15 mm long, rounded at the asymmetrical base with very small scales like those on the twigs on the lower surface, which are numerous on or densely covering the midrib and few to numerous on the lower leaflet surface; veins 10 – 16 on each side of the midrib, midrib and lateral veins depressed on upper surface, lower surface with midrib and lateral veins both longitudinally wrinkled, the former prominent, the latter slightly or not at all prominent; petiolules up to 15 mm on lateral leaflets, up to 20 mm on terminal leaflet.

Inflorescence up to 25 cm long and 15 cm wide; peduncle up to 10 cm, peduncle, rhachis and branches somewhat flattened longitudinally, wrinkled and with indumentum like the twigs. Flowers up to 3 mm long and 2.5 mm wide, obovoid; pedicel c. 1 mm. Calyx c. $1/3$ the length of the corolla, densely covered with stellate scales on the outside, divided up to c. $1/2$ way into 3 (or 4) acute lobes. Petals 3, white, obovate, densely covered with stellate scales on the outside, aestivation imbricate. Staminal tube c. $2/3$ the length of the corolla, obovoid, cup-shaped, the apical margin shallowly divided into c. 6 acute lobes; anthers 6(– 7), as long as the tube, narrowly ellipsoid, just protruding beyond the aperture of the tube. Ovary depressed-globose densely covered with stellate hairs; stigma ovoid with 3 apical lobes and 6 longitudinal ridges.

Fig. 7. *A. malaccensis*. Habit with male inflorescence x½. Half flower, male x10. Larger leaf x½. Very young infructescence x½.

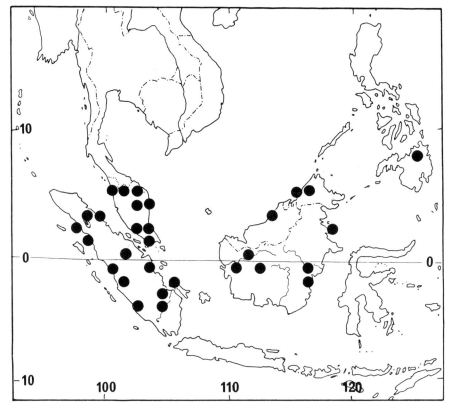

Fig. 8. Distribution of *A. malaccensis*.

Infructescence with few fruits. Fruits up to 6 cm long and 7 cm wide, reddish-brown, depressed globose, densely covered with hairs and scales like those on the twigs on the outside; latex white. Loculi 3 – 4, each containing 0 or 1 seed. Seed 3.5 – 3.8 cm long, c. 2.1 cm wide and 1.5 cm through, completely covered with a red or yellow aril 1 – 2 mm thick. Fig. 7

DISTRIBUTION. Peninsular Malaysia, Sumatra, Borneo, Philippine Islands. Fig. 8.

ECOLOGY. Found in primary and secondary forest on clay, shale, sand and loam. Alt.: 12 to 700 m. Rare and scattered to common. ? Fruits eaten by monkeys.

VERNACULAR NAMES. Peninsular Malaysia: Bekak, Kalbang, Kasai, Memberas. Sumatra: Goels, Palankoetan, Pyoeta. Borneo: Madensat (Dyak-Sembakung); Bilayang, Boenjoe, Buno, Bunyan, Lapak, Parak Keloewang, Ramo, Tebaul.

Representative specimens. PENINSULAR MALAYSIA. Trengganu, Ulu Trenggan, nr Kg Pertang, 170 m, fr., 3 June 1968, *Cockburn* FRI 8423 (FRI!). Pahang, Taman Negara, trail from Terenggan-Kumbang Salt Lick, c. 90 m, fr., 30 April 1975, *van Balgooy* 2599 (L! [spirit coll. no. 6774 L!], NY!). Selangor, Bt. Lagong F.R., fl., 12 Dec. 1962, *Matan* KEP 97747 (FRI!).

SUMATRA. Billiton, *van Rossum* 55 (BO!).

KALIMANTAN. E. Kutai Reserve [0°24'N 117°16'E], tree no. OB13. 00E30, fr., 28 July 1979, *Leighton* 859 (L!).

When dry, the leaflets of *A. malaccensis* are usually green above and purplish-brown below, they are rarely rugulose and the lateral veins are more numerous and more prominent than in *A. macrocarpa*. *A. malaccensis* has pale stellate scales or hairs whereas *A. macrocarpa* has minute peltate scales which have a fimbriate margin and these leave dark reddish-brown pits after they are shed. The fruit of *A. malaccensis* is depressed globose, whereas that of *A. macrocarpa* is obovoid.

6. Aglaia rugulosa C.M. Pannell spec. nova *Aglaiae erythrospermae* C.M. Pannell similis sed arbor minor, foliolis supra minus nitidis in sicco, utrimque conspicue rugulosis; basi plerumque valde attenuatis. Holotype: [Peninsular Malaysia], Pahang, Taman Negara, Ulu Sat, West Bank near Kuala Kelapa, 370 ft [c. 110 m], fl., 9 July 1970, *Whitmore* FRI 15226 (K!; isotype: L!); Pannell in Ng, Tree Flora of Malaya 4: 227 (1989) as *Aglaia* sp. 1.

Small tree up to 12 m. Bole up to 50 cm in circumference. Bark smooth, brownish-grey, lenticellate. Sapwood pale pink or white; latex white. Twigs stout, dark brown, with longitudinal wavy ridges and a few white or reddish-brown stellate hairs.

Leaves imparipinnate, up to 130 cm long and 100 cm wide, obovate in outline; petiole up to 35 cm, the petiole, rhachis and petiolules with surface and hairs like the twigs. Leaflets 7 – 9(– 15), the laterals opposite or subopposite, all 16 – 48 cm long, 6 – 12 cm wide, pale brown when dry, coriaceous, obovate or oblanceolate, recurved at the margin, acuminate or shortly caudate at apex with the acute acumen up to 15 mm long, attenuate or occasionally rounded at base, upper surface rugulose, lower surface more markedly so and with a few stellate hairs on the midrib and veins; veins 7 – 20 on each side of the midrib, ascending and markedly curved upwards near the margin, midrib and lateral veins subprominent with longitudinal wavy ridges on upper surface, prominent and more markedly ridged on lower surface; petiolules of lateral leaflets up to 25 mm, of terminal leaflet up to 40 mm.

Inflorescence up to 30 cm long and 15 cm wide; peduncle up to 14 cm, the peduncle and branches with surface and indumentum like the twigs. Flowers (not known whether male or female) c. 3 cm long and 3 cm wide, subglobose; pedicels c. 1 mm, each with one bracteole c. 2.7 mm long and 1 mm wide, ovate, the pedicels and bracteoles densely covered with stellate scales. Calyx with few to numerous stellate scales, shallowly divided into 3 acute lobes. Corolla c. 2 mm long and 2.5 mm wide, depressed-globose, aestivation imbricate; petals 3, yellow or pale yellow, subrotund. Staminal tube shorter than the corolla, cup-shaped, the apical margin shallowly

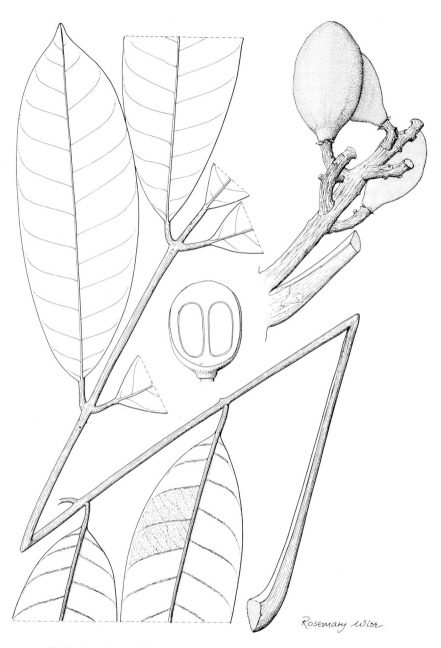

Fig. 9. *A. rugulosa*. Leaf x1/2. Infructescence x1/2. Longitudinal section of fruit x1/2.

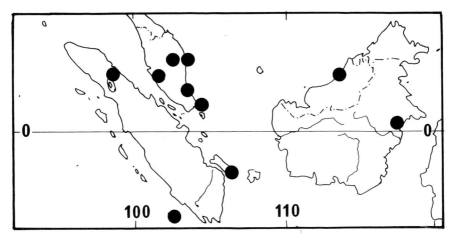

Fig. 10. Distribution of *A. rugulosa*.

lobed; anthers 7 – 9, as long as the tube and just protruding beyond the aperture. Ovary c. 0.3 mm long, c. 0.8 mm wide, depressed globose, densely covered with pale brown stellate hairs; loculi 3, each containing 1 ovule; stigma c. 0.4 mm long, c. 0.2 mm wide, ovate.

Infructescence with up to 5 fruits, very stout, the peduncle and branches with surface and indumentum like the twigs. Fruits 6 – 7 cm long and 5 – 6 cm wide, ellipsoid or obovoid, red or orange-red, dehiscent, densely covered with compact reddish-brown stellate hairs on the outer surface. Pericarp 2 – 5 mm thick, thickest at the apex, with white latex. Locules 3, each containing 1 seed; the seed 4.5 cm long, 1.5 cm wide and 1.8 cm thick, completely covered with a red aril. Cotyledons c. 2.5 cm long and 1.7 cm wide, unequal, transverse, peltate. Fig. 9.

DISTRIBUTION. Peninsular Malaysia, Sumatra, Borneo. Fig. 10.

ECOLOGY. Found in primary lowland and hill forest, secondary forest, riverine forest and swamps; on clay. Alt.: 10 to 830 m.

VERNACULAR NAMES: Peninsular Malaysia: Bekak, Memberas. Sumatra: Beko, Ehoehauwe.

Representative specimens. PENINSULAR MALAYSIA. Johore, Gg Panti F.R., compt. 64, c. 300 m, fr., 2 Mar. 1968, *Cockburn* 7749 (FRI!, K!). SUMATRA. Aceh: Simalur, fl., 15 Oct. 1918, *Achmad* 669 (K! L!); Gunong Leuser Nature Reserves, c. 75 km N.W.N. of Medan, Sikundur Forest Reserve, Besitang River, [c. 3°55'N 98°05'E], 50 – 100 m, *de Wilde & de Wilde- Duyfjes* 19463 (K!). Enggano Island, c. 100 m, st., 5 June 1936, *Lütjeharms* 4264 (NY!). S.E. Bangka Island, Lobok-besar, 2 m, ♀ fl., 26 Aug. 1949, *Kostermans & Anta* 141 (L!).

KALIMANTAN. C. Kutei, Belajan F.R., Gg Kelopok, nr Tabang, 50 m, *Kostermans* 10441 (K!).

Section Amoora

Aglaia rugulosa is a small tree with stout twigs and very large leaves.

Aglaia rugulosa resembles *Aglaia erythrosperma*, but it is a smaller tree, the upper surface of the leaflets is less shiny when dry and both surfaces are markedly rugulose; the base of the leaflet is usually markedly attenuate. The fruits are ellipsoid or obovoid, whereas they are subglobose in *A. erythrosperma*.

7. Aglaia erythrosperma C.M. Pannell spec. nova *Aglaiae spectabili* Miq. similis, sed foliolis magis coreaceis, margine recurvis, costa nervisque lateralibus elevatis et longitudinaliter porcatis, subtus undulatis differt. Holotype: Peninsular Malaysia, Negri Sembilan, Pasoh F.R., fr., April 1978, *Pannell* 1175 (FHO!); Pannell in Ng, Tree Flora of Malaya 4: 228 (1989), as *Aglaia* sp. 2.

Tree up to 35 m, with a rounded crown. Bole up to 23 m, up to 150 cm in circumference, with small L-shaped buttresses outwards up to 70 cm. Outer bark pale pinkish-brown with reddish-brown and grey patches, lenticellate and with longitudinal cracks, flaking off in irregular scales up to 15 cm in diameter; inner bark pinkish-brown, red or green. Sapwood pink or pinkish-brown; with watery greenish-brown exudate. Branches ascending patent. Twigs very stout, greyish-brown or dark brown, with pustules, cracks and longitudinal wavy ridges and densely covered with brown stellate scales.

Leaves imparipinnate, up to 60 cm long and 40 cm wide, obovate in outline; petiole up to 25 cm, the petiole, rhachis and petiolules with longitudinal wavy ridges and scales like those on the twigs. Leaflets 7 – 19, the laterals usually opposite or subopposite, sometimes alternate, all 11 – 24 cm long, 7.5 – 12 cm wide, shiny dark green on upper surface, coriaceous, elliptical, slightly recurved at margin, shortly acuminate at apex with the acute acumen up to 5 mm long, rounded at the asymmetrical base, lower surface dull, with few to numerous pale brown scales like those on the twigs on the midrib and veins and sparse on the surface in between; veins 7 – 13 on each side of the midrib, ascending and markedly curved upwards near the margin, midrib depressed and the lateral veins raised at the sides and slightly depressed in the centre on upper surface, both midrib and lateral veins raised and longitudinally ridged and wavy on lower surface; petiolules up to 15 mm on lateral leaflets, up to 30 mm on terminal leaflet.

Inflorescence up to 20 cm long; the peduncle, branches and pedicels densely covered with stellate scales. Flowers (not known whether male or female) c. 4.5 mm long and 3.5 mm wide, obovoid; pedicels c. 1 mm. Calyx half to nearly as long as the flower, divided up to c. $^1/_2$ way into 3 broad, obtuse lobes, densely covered with stellate scales on the outer surface. Petals 3, yellow, sub-rotund, with few to densely covered with stellate scales on the outside, aestivation imbricate. Staminal tube shorter than the corolla, cup-shaped; anthers 6(– 8), about two thirds the length of the tube, narrowly ovoid, curved with the tube and just protruding beyond the aperture, with simple hairs on the inner surface of the tube and on the anthers. Ovary depressed globose or ovate, densely covered with stellate scales; loculi 3, each containing 1 ovule; stigma as long as the staminal tube, obovoid, with 6 longitudinal lobes.

Fig. 11. *A. erythrosperma*. Habit x¹/₂. Male inflorescence x¹/₂. Male flower x6. Female inflorescence x¹/₂. Female flower x6. Fruit and seeds x¹/₂.

Section Amoora

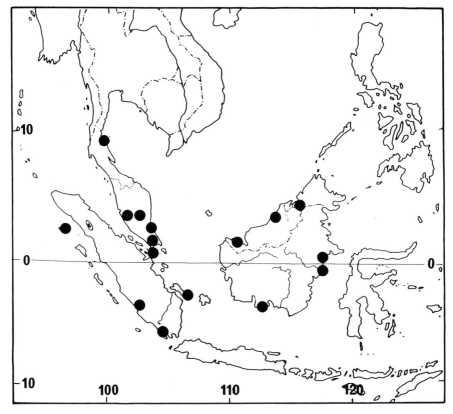

Fig. 12. Distribution of *A. erythrosperma*.

Infructescence up to 25 cm long with few fruits. Fruits up to 10 cm in diameter, bright orange or red, densely covered with reddish-brown stellate hairs on the outside, subglobose, dehiscing into 3; pericarp up to 1.5 cm thick, inner pericarp white, innermost layer in each loculus a detachable membrane surrounding the seed. Loculi 3, each containing 1 seed; seeds up to 5 cm long, 3.5 cm wide and 2 cm thick, completely surrounded with a bright shiny orange-red (? or sometimes yellow) aril, which is easily detached from the rest of the seed; testa shiny chestnut brown. Cotyledons large, oblique. Fig. 11 & 18.

DISTRIBUTION. Thailand, Peninsular Malaysia, Sumatra, Borneo. Fig. 12.

ECOLOGY. Found in evergreen forest, primary forest and kerangas; on granitic sand, sand, clay. Alt.: 5 to 1300 m. USES: Wood may provide good timber.

VERNACULAR NAMES: Peninsular Malaysia: Bekak. Sumatra: Boka-boka Abang, Parak, Parak Daun Besar, Parek, Oemawa, Oemawa Batoe,

Oemawa Boeloeh. Borneo: Bunya, Segera (Iban); Lantupak (Dusun Kinabatangan); Boenjau.

Representative specimens. THAILAND. Nakkon, Si Thammarat, Khao Luang, 300 – 350 m, *van Beusekom & Phengkhlai* 1014 (K!, L!).
PENINSULAR MALAYSIA. Negri Sembilan, Pasoh F.R., fr., April 1978, *Asri* F RI 25506 (FRI! (collected from same tree as holotype)).
SUMATRA. S.E. Bangka, Lobok-besar, 20 m, ♀ fl., 26 Aug. 1949, *Kostermans & Anta* 141 (L!). Simaloer [Simalur], fl., *Achmad* 669 (L!).
SARAWAK. 3rd Division, Bukit Raya, 2½ hours upstream from Kapit, c. 210 m, fl., 25 Nov. 1963, *Pennington* 7997 (FHO!).
KALIMANTAN. mouth of Makakam River, *Kostermans* 9557 (K!); E. Kutei Sg. Bambangan, S.E. of Samarinda, 20 m, *Kostermans* 6116 (K!).

A. erythrosperma is a tall, sometimes emergent, tree with very stout twigs. *A. erythrosperma* resembles *A. spectabilis* Miq., but the leaflets are more coriaceous and are recurved at the margin and the midrib and lateral veins are raised and longitudinally ridged and wavy on the lower surface of the leaflets. The upper surface of the leaflets is dark shiny green, and the lower surface dull has tiny stellate scales. The fruit is the size of a cricket ball and dehisces into three, revealing one bright orange-red arillate seed in each loculus. The seeds contrast with the white of the inner pericarp and its brick red outer surface.

8. Aglaia spectabilis (Miq.) Jain & Bennet, Indian Journal of Forestry, 1986, 9(3): 271 (1987).

[*Sphaerosacme rohituka* Wallich, Cat. 1278 (1829) quoad specim. *Anon.* in *Herb. E.I.C.* 1278 (K!), 1278.1 (K-W!)] nec syn. ie. *Andersonia rohituka* (=*Aphanamixis polystachya* (Wallich) R.N. Parker).
Amoora spectabilis Miq., in Ann. Mus. Bot. Lugd Bat. 4: 37 (1868). Types: Cult. in Hortus Calcuttensi, (? *Anon.* in *Herb. E.I.C.* 1278 (K!), 1278.1 (K-W!)); Hiern in Hooker fil., Fl. Brit Ind. 1: 561 (1875) pro parte; C. de Candolle in A. & C. de Candolle, Monog. Phan. 1: 585 (1878).
Amoora gigantea Pierre in Laness., Pl. Util. Col. Franc.: 311 (1886). Lectotype (designated here): [Cochinchina, Laos], in montibus Bassae, Feb. 1877, *Harmand* 108 in *herb. Pierre* 1444 (P!); Pierre, Fl. Forest. Cochinch. Fasc. 22, ante t. 343A (1 July 1896); I.H. Burkill, Dictionary of Economic Products of the Malay Peninsula, 1: 137 (1935).
Amoora ridleyi King in Jour. As. Soc. Bengal 64: 56 (1895). Syntypes: Malay Peninsula, Pahang, Sg. Rompin, 1883, *Ridley* 5027 (SING! [incorrectly designated the lectotype by Pannell, 1982]); Perak, Larut, Gopeng, 300 – 500 ft [c. 90 – 150 m], fr., April 1884, *King's Coll.* 5918 (CGE!, K!, SING!); Perak, Larut, Gopeng, 300 – 500 ft [c. 90 – 150 m], fr., May 1884, *King's Coll.* 6060 (G!, K!, L!, P!, SING!); Perak, *King's Coll.* 7917 (BM!, G!, L!, P!); Perak, *Wray* 2107 (K!, SING!); Ridley, Fl. Malay. Penins. 1: 398 (1922); I.H. Burkill, Dictionary of Economic Products of the Malay Peninsula 1: 137, 138 (1935).
Amoora wallichii King in Jour. As. Soc. Beng. 64: 56 (1895). Syntypes: Cult. H.B. Calcutta (origin: Assam, Goalpara), *Anon.* in *Herb. E.I.C.* 1278 (K!), 1278.1 (K-W!, excluding inflorescence at bottom right of the sheet (= *Aglaia cucullata*)); Andaman Islands, fl., 1884, *King's Coll.* 475 (BM!,

G!, K!); I.H. Burkill, Dictionary of Economic Products of the Malay Peninsula 1: 137 (1935).

Aglaia gigantea (Pierre) Pellegrin in Lecomte, Fl. Gén. L'Indo-Chine, 1: 769 (1911).

Aglaia ridleyi (King) Pannell in Malaysian Forester 45: 455 (1982); Pannell in Ng, Tree Flora of Malaya 4: 223 (1989).

Aglaia hiernii Viswanathan & Ramachandran in Bull. Soc. Surv. Ind. 24: 212 (1983), nom. superfl. nom. illegit. pro *Amoora wallichii*, non King (1895), nec Koord. (1898).

Aphanamixis wallichii (King) Haridasan & R.R. Rao, For. Fl. Meghalaya 1: 206 (1985).

Amoora stellatosquamosa [*stellato-squamosa*] C.Y. Wu, Flora Yunnanica **1**: 233 (1977). Type: [China], Yunnan, Meng-la, Jenn-yeh Hsien, 950 m, fl., Nov. 1936, *C.W. Wang* 80662 (KUN!).

Tree up to 40 m, rarely flowering at 8 m, with a broad rounded crown. Bole up to 18 m, up to 150 cm in diameter, with plank buttresses upwards up to 200 cm and outwards up to 370 cm. Bark greyish-white, pale yellowish-brown or brown, flaking in squarish scales up to 30 cm across, sometimes with large orange lenticels up to 3 mm in diameter; inner bark pink, reddish-orange or brown; sapwood pale brown, pink, white or magenta; latex white. Branches ascending. Twigs stout, sometimes greater than 1 cm in diameter, with longitudinal wavy ridges, greyish-brown with tiny lenticels and densely covered with reddish-brown or pale brown stellate hairs or scales, or peltate scales which have a fimbriate margin, and white latex. Leaves imparipinnate in dense spirals, the leaf bases almost overlapping, 50 – 135 cm long, 28 – 70 cm wide, obovate in outline; petiole 14 – 25 cm, flattened on the adaxial side, the petiole, rhachis and petiolules with surface and indumentum like the twigs. Leaflets (3 –)11 – 21, the laterals subopposite, all 8 – 40 cm long, 2.5 – 12.5(– 17) cm wide, coriaceous, lanceolate, oblong or elliptical, acuminate at apex with the acute acumen up to 15(– 40) mm, rounded at the asymmetrical base, upper surface rugulose and sometimes pitted, lower surface pitted, with few to densely covered with pale brown or reddish-brown stellate hairs and scales on the midrib and a few on or occasionally densely covering the lateral veins and surface of the lamina, sometimes with a few darker peltate scales which have a fimbriate margin scattered on the rest of that surface; veins 9 – 19 on each side of the midrib, ascending and markedly curved upwards near the margin, midrib and lateral veins depressed on upper surface, midrib and lateral veins prominent, secondary veins subprominent on lower surface; petiolules 8 – 20 mm.

Inflorescence up to 50 cm long and 30 cm wide; peduncle up to 18 cm, the peduncle, rhachis and branches stout, with longitudinal wavy ridges and indumentum like the twigs. Flowers fragrant. Male flowers 2 – 5 mm long and 2 – 3 mm wide; female flowers up to 7 mm long and up to 6 mm wide, ellipsoid; pedicels up to 3 mm, with stellate hairs like those on the twigs. Calyx $1/2 - 2/3$ the length of the corolla, cup-shaped, usually densely covered with stellate hairs on the outside, divided up to c. $1/2$ way into 3 obtuse lobes. Petals 3, up to 4 mm long and 2.5 mm wide, pinkish-yellow or white, elliptical, hooded and membranous at apex (forming a short tube at the base in the Solomon Islands), the lower $2/3$ with numerous stellate scales on the outside, aestivation imbricate. Staminal tube slightly shorter than the corolla, cup-shaped, membranous at base; anthers (5 or) 6(– 10), usually c. $\frac{1}{2}$ the length but sometimes as long as the tube, ovoid, c. $\frac{1}{4}$ of their length protruding beyond the aperture, sometimes with a few simple or forked hairs. Ovary subglobose densely covered with stellate hairs; stigma ellipsoid with 3 apical lobes and 6 longitudinal ridges, black and shiny; ovary and stigma together usually c. $\frac{1}{2}$ the length but sometimes as long as the staminal tube. Infructescence up to 13 cm long. Fruits 6 – 9 cm long and 5.5 – 9 cm wide, subglobose or obovoid, brown, red or yellow, densely covered with reddish-brown or pale yellowish-brown stellate hairs; pericarp up to 1 cm thick with white latex, shiny reddish-brown inside. Loculi 3 (or 4), each containing 0 or 1 seed. Seeds with aril 3.5 – 5 cm long and 2 – 2.7 cm wide, 1.5 – 2.2 cm thick; the aril entire, 2 – 4 mm thick, with a red, orange-red, yellow or white skin; raphe white. Fig. 13.

DISTRIBUTION. India, Sikkim, Burma, Laos, Cambodia, Vietnam, China, Thailand, Peninsular Malaysia, Sumatra, Borneo, Philippine Islands, Nusa Tenggara (Sumba), Sulawesi, New Guinea, New Britain, Solomon Islands, Santa Cruz Islands, Australia (Cape York Peninsula). Fig. 14.

ECOLOGY. Found in secondary forest, riverine forest, primary forest, alluvial flats, coastal swamp and along the seashore; on sandy clay, sand, loam, sandstone, alluvial, coral. Alt.: sea level up to 650 m. Scattered to common. In Australia grows on red soils derived from a mixture of basic rocks and ferruginous sandstone. Found in rain forest, gallery rain forest, coastal riverine forest, deciduous mesophyll vine forest.

NOTE. Causes dermatitis when milled (*Womersley* NGF 9064), see Henty (1980) for notes on this property in other species in Papua New Guinea.

VERNACULAR NAMES: Peninsular Malaysia: Bekak. Borneo: Balim (Kedayan); Langsat-langsat (Malay); Lantupak (Dusun Kinabatangan); Merasam (Banjar- Malay); Enggoha, Hgaling, Lans-abouti Mepoeloe, Mea Mepoeloe, Woema. New Guinea: Aban (Bilia); Aiba, Bagaibi (Oriomi & Kiwai); Boewa (Karoon); Bowwie, Lowdokwa, Lowtoekwa, Tessaai (Manikiong); Damasewou (Amele); Dowa (Kebar); Dzumpiem (Dumpu); Kamoengoto, Kwo (Mooi); Makin (Noemfoor); Mogwe (Faita); Mokken (Biak); Ruhogowo, Trugum, Tumpamoi. Solomon Islands: Chichetche (Guadalcanal); Maoa, Mawa (Kwara'ae).

Representative specimens. INDIA. E. Himalaya, Kalighara, 300 m, fr., 8 Aug. 1923, *Cave* s.n. (E!). Assam, Nambar Forest, fl., May 1884, *Mann* s.n. (K!).

ANDAMAN ISLANDS. fl., 1884, *King's Coll.* 475 (K!). fl., 1884, *King's Coll.* s.n. (BM!, G!).

BURMA. Tavoy District, Zimba Valley, fr., 20 Nov. 1924, *Parker* 2248 (K!). Rangoon, *McClelland* s.n. (K!).

VIETNAM. Annam, Langklisai (Luang tri), fl., *Poilane* 10739 (L!).

CHINA. Hainan: Yaichow, fl., 21 July 1933, *H.Y. Liang* 62229 (E!); fr., 1939, *Fenzel* 238 (NY!).

PENINSULAR MALAYSIA. Perak, Taiping, plains, fl., June 1888, *Wray* 2107 (K!, SING!). Pahang, Temerloh, Kemasul F.R., fr., 23 Feb. 1966, *Kochummen* Kep 98588 (L!). Selangor: Kepong, Forest Research

Fig. 13. *A. spectabilis*. Leaf x$^1/_2$. Inflorescence x$^1/_2$. Fruit x$^1/_2$. Half flower x6. Stellate hair x100.

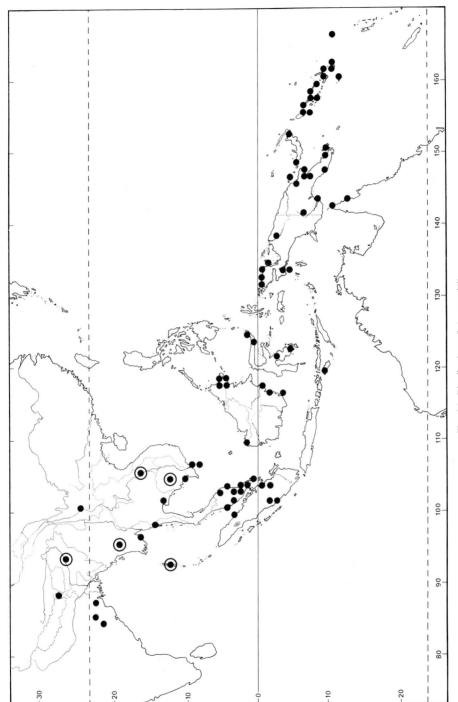

Fig. 14. Distribution of *A. spectabilis*.

Section Amoora

Institute, c. 245 m, fr., 7 Oct. 1963, *Pennington* 7863 (FHO!); Bt. Lagong F.R., c. 85 m, fr., 12 June 1975, *Putz* FRI 023647 (FRI!).

SUMATRA. Sumatera Barat, 30 km along road from Tapan to Bengkulu, S. of Lunang [2°15'S 101°10'E], c. 50 m, fl., 3 Oct. 1983, *Pannell* 2083 (FHO!).

SABAH. nr Sandakan, Sepilok F.R., c. 15 m: ♂ fl., 18 Oct. 1963, *Pennington* 7903 (FHO!) & ♀ fl. & fr., 18 Oct. 1963, *Pennington* 7902 (FHO!).

PHILIPPINE ISLANDS. Luzon, Tayabas Province, Kagascas, fr., Apr. 1929, *Oro* For. Bur. 30886 (UC!). Mindanao, Surigao Province, Manyayang, 150 m, fr., 30 July 1927, *Wenzel* 3012 (UC!).

PAPUA NEW GUINEA. Morobe District, near Lae, Busu [7°20'S 147°20'E] logging area, 60 m, fr., 14 Jan. 1964, *Pennington* 8055 (FHO!). Gulf District, Hill crest about 2 miles N.E. of junction of Vailala and Lohiki river, c. 250 m, fl., 31 Jan. 1966, *Schodde (& Craven)* 4398 (K!). Milne Bay District, Esa'ala Subdistrict, Normanby Island, near Miadeba airstrip [9°50'S 150°55'E], 5 m, fl., 22 Nov. 1976, *Croft et al* LAE 68849 (K!) . New Britain District: Jacquinot Bay, fl., April 1945, *Mair* NGF 1863 (K!); Keravat [4°20'S 152° 00'E] logging area, c. 30 m, fr., 31 Jan. 1964, *Pennington* 8103 (L!); Keravat, Vudal Forest, c. 30 m, fr., 2 Feb. 1964, *Pennington* 8115 (L!).

AUSTRALIA. Queensland: Claudie River, fr., 3 Jan. 1973, *Hyland* 6641 (BRI !, QRS!); near Lockerbie, fl., 1 Feb. 1980, *Hyland* 10231 (FHO! ex QRS).

SOLOMON ISLANDS. N.E. Kolombangara, Kokove Area, 60 m, fr., 10 Jan. 1968, *Mauriasi & collectors* BSIP 7604 (K!, L!). Three Sisters, Malaupaina Island, 15 m, fl., 8 Dec. 1969, *Mauriasi & collectors* BSIP 17991 (K!).

In Peninsular Malaysia, the upper surface of the leaflets of *A. spectabilis* is rugulose and pitted; in other parts of the range, the leaflet surfaces are often smooth and the upper surface slightly shiny. The raised secondary venation and minute pitting give the leaflet undersurface a characteristic appearance. Numerous or sparse, tiny pale yellow stellate scales can always be found under magnification on the lower leaflet surface. The presence of the scales is a useful character for identifying sterile specimens and, in particular, distinguishes this species from *A. lawii* in Australasia, where the leaflet number in *A. spectabilis* is lower than in the west and overlaps with the number of leaflets found in *A. lawii*.

Aglaia spectabilis is a widespread species with large dehiscent fruits. It may be dispersed by fruit pigeons (see Pannel & Kozioł, 1987).

9. Aglaia multinervis C.M. Pannell nom. novum.

Amoora lanceolata Hiern in Hooker fil., Fl. Brit. India 1: 560 (1875), non *Aglaia lanceolata* Merrill (1910) (= *Aglaia rimosa*). Lectotype (designated here): Malacca [Malaya], fl., 29 March 1865 – 66, *Maingay* 1610 (Kew Dist. 343) (K!); C. de Candolle in A. & C. de Candolle, Monog. Phan. 1: 584 (1878); King in Jour. As. Soc. Bengal 64: 55 (1895); Ridley, Fl. Malay Penins. 1: 399 (1922); Pannell in Ng, Tree Flora of Malaya 4: 228 (1989), as *Aglaia* sp. 3.

Tree up to 35 m. Bole up to 225 cm in circumference. Bark smooth, pale brown, reddish-brown or pinkish-grey with numerous reddish-brown lenticels and with deciduous scales which expose reddish-brown patches underneath; inner bark green or red; sapwood pale brown, pale pink, pale yellow or reddish-brown; latex white. Twigs stout, pale or dark brown, longitudinally wrinkled, densely covered with peltate scales which have a dark reddish-brown centre and pale fimbriate margin, sometimes with reddish-brown stellate hairs interspersed near the apex.

Leaves imparipinnate, up to 40 cm long and 20 cm wide, oblong in outline; petiole up to 10 cm, the petiole rhachis and petiolules with surface like the twigs and a few stellate hairs. Leaflets 15 – 25, the laterals subopposite, all 6 – 15 cm long, 1 – 3.5 cm wide, coriaceous, lanceolate, acuminate at apex with the obtuse acumen up to 5 mm (– 20 mm) long, rounded or sub-cuneate at the asymmetrical base, lower surface rugulose, with few to densely covered with peltate scales on the midrib and scattered peltate or stellate scales elsewhere; veins 20 – 50 on each side of the midrib, ascending, some not reaching the margin, the rest curved upwards near the margin, midrib slightly depressed on upper surface, midrib longitudinally wrinkled and lateral veins hardly prominent on lower surface; secondary veins sometimes visible; petiolules up to 10 mm. Inflorescence c. 20 cm long and 10 cm wide; peduncle up to 6 cm, the peduncle, rhachis, branches and pedicels with indumentum like the twigs. Flowers up to 3 mm long and 2.5 mm wide, fragrant, obovoid or subglobose; pedicel up to 0.5 mm. Calyx c. $1/2$ length of the corolla, cup-shaped densely covered with stellate scales on the outside, divided up to c. $1/2$ way into 3 (or 4) acute lobes. Petals 3, white or yellow, obovate or elliptical, densely covered with stellate scales on the outside, aestivation imbricate. Staminal tube shorter than the petals, subglobose, the aperture shallowly divided into c. 6 acute lobes; anthers 6, as long as or longer than the tube, narrowly ellipsoid, with a few simple hairs, inserted near the base of the tube and just protruding beyond the aperture. Ovary depressed globose densely covered with stellate scales or hairs; stigma c. $1/4$ the length of the staminal tube, ovoid with 3 apical lobes and 6 longitudinal ridges.

Infructescence with few fruits which ripen at different times; peduncle up to 6 cm, with surface and indumentum like the twigs. Fruits up to 6 cm long and 5 cm wide, brown, bright red or yellow, subglobose or obovoid with a small beak, densely covered with minute reddish-brown stellate hairs, dehiscent. Loculi 3, each containing one seed.

DISTRIBUTION. Peninsular Malaysia, Singapore, Sumatra, Borneo. Fig. 15.

ECOLOGY. In forest, often on hillsides. Alt.: up to 400 m.

VERNACULAR NAMES. Peninsular Malaysia: Beka. Sumatra: Parak, Parek Api. Borneo: Langsat-langsat (Malay); Lantupak, Manggi (Dusun Kinabatangan); San Kuang (Baj E.B.); Embunjau, Mendjelenoe, Mulak, Tebaul.

Representative specimens. PENINSULAR MALAYSIA. Johore, mile 6.5 Kota Tinggi – Mawai road, fr., 4 Feb. 1935, *Corner* SFN 21347 (K!, L!).
SINGAPORE. Bt. Timah, *Liew* SFN 37278 (K!, SING!).
SABAH. Sandakan, Lahad Datu, *Cuadia* A 218 (FRI!).

The epithet *lanceolata* is already occupied in the genus *Aglaia*. A new

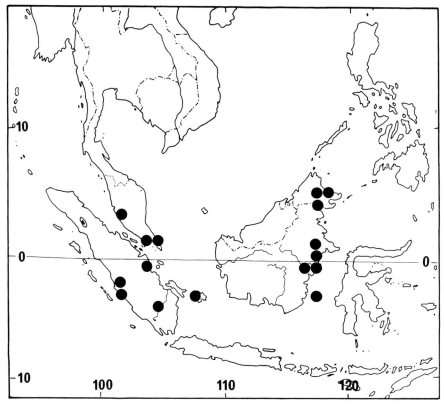

Fig. 15. Distribution of *A. multinervis*.

name is thus required; *multinervis* has been chosen because of the unusually high number of lateral veins on the leaflets. The leaves of *Aglaia multinervis* have up to 25 leaflets; the lateral veins are numerous and indistinct.

10. Aglaia lepidopetala Harms in Engl. Bot. Jahrb. 72: 177 (1942). Lectotype (designated here): New Guinea, near Prauwenbivak, [Gebiet des Flusses Mamberamo], 100 m, fl., 18 Aug. 1920, *H.J. Lam* 809 (L!; isolectotypes: B†, BO!).

Small tree up to 5 m, bole up to 3 m, 2 cm in diameter. Twigs sometimes with few, numerous or densely covered with compact reddish-brown stellate hairs (some of which have clumps of arms around a central rhachis) and/or pale brown peltate scales which have a fimbriate margin.

Leaves imparipinnate, 60 – 105 cm long, petiole 18 – 40 cm, the petiole with numerous scales like those on the twigs. Leaflets (7 –)9 – 17, the laterals subopposite, all 7.5 – 26 cm long, 3 – 7 cm wide, elliptical or narrowly obovate, cuneate at the base, acuminate at the apex, the acumen obtuse and 3 – 12 mm long, pale brown when dry, the midrib below with few pale scales and reddish-brown hairs like those on the twigs and with scattered to numerous inconspicuous, pale and brown stellate and peltate scales on the rest of the lower surface, the leaflet surfaces matt, the reticulation visible on the lower leaflet surface; veins 9 – 16 on each side of the midrib, with shorter lateral veins in between, midrib prominent, lateral veins barely subprominent; petiolule 0.5 – 1 cm.

Inflorescence 13 – 19 cm long, 4 – 8 cm wide, with one or two orders of branches only; peduncle 1.5 – 2.5 cm, the peduncle, rhachis and branches with numerous to densely covered with hairs and scales like those on the twigs. Flower 3.5 – 5.5 mm long, 3.5 – 6 mm wide; pedicel c. 2 mm densely covered with stellate hairs or scales. Calyx 1.5 mm, cup-shaped and shallowly divided into 3 (or 4) shallow lobes, densely covered with stellate hairs or scales on the outside. Corolla tube 4.5 mm long, 4 – 4.5 mm wide, divided almost to the base into 3 subrotund to broadly ovate lobes which are densely covered with brown stellate scales on the outside. Staminal tube 2.5 – 4 mm long, 2 – 3.5 m wide, the aperture 1 – 1.5 mm across, entire or shallowly lobed; anthers 9 – 10, 2.5 mm long, 1 mm wide, inserted near the base of the tube, included or just protruding; ovary c. 0.5 mm long, depressed globose, densely covered with stellate scales on the outside, loculi 3, each containing 1 ovule; the stigma c. 0.5 mm long, with 3 distinct lobes; the ovary and stigma together about 1 mm long.

Infructescence up to 9 cm long and 6 cm wide; peduncle 0.5 – 1 cm long, the peduncle, rhachis and branches with numerous hairs and scales like those on the twigs. Fruits 1 – 9, 3 – 3.2(– 4) cm long, 2.9 – 3 cm in diameter, obovoid, dehiscent, orange or reddish-brown, densely covered with reddish-brown peltate scales which have a fimbriate margin and with numerous compound stellate hairs interspersed; the pericarp thin and moulded around the seeds when dry, dehiscing into three when ripe, white inside. Loculi 3, each containing 0 or 1 seed which is almost or completely covered with a red aril c. 1 mm thick; seed 2.1 – 3.1 cm long, 1 – 1.8 cm wide, c. 1.6 cm through.

DISTRIBUTION. New Guinea. Fig. 16.

ECOLOGY. Found in oak forest, primary forest, fresh water swamp forest, secondary forest and along roads. Mainly in understorey. On volcanic clay, sandy clay, loamy clay. Alt.: up to 1050 m. Rare.

Representative specimens. IRIAN JAYA. W., Sukarnapura [= Hollandia = Djajapura] [2°37′S 140°39′E], c. 100 m, fl., 27 July 1966, *Kostermans & Soegang* 63 (L!).

PAPUA NEW GUINEA. E. Sepik District, Wewak subdistrict, nr Dagna [3°25′S 143°20′E], c. sea level, fr., 16 June 1971, *Stone & Streiman* LAE 53579 (K!, L!). Madang District, Usino Subdistrict, Amiaba River, c. 170 m, [5°25′S 145°25′E], fr., 8 Jan. 1970, *Foreman, Farley & Noble* NGF 45878 (BO!). Morobe District, Lae Subdistrict, near waterfall at Bisama, 15 m, fr., 1 Jan. 1972, *Stevens* s.n. (LAE!).

Aglaia lepidopetala is separated from *A. macrocarpa* by its compound stellate hairs (*A. macrocarpa* has no stellate hairs), the small size of the tree, smaller fruit and smooth lower leaflet surface. The fruits are smaller than in most species in section *Amoora*, except for *A. meridionalis*, ? *A. australiensis*, *A. lawii* and *A. teysmanniana*.

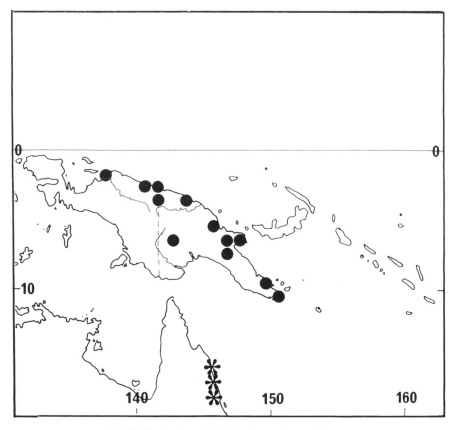

Fig. 16. Distribution of *A. lepidopetala*● and *A. meridionalis**.

11. Aglaia meridionalis C.M. Pannell nom. novum.

Amoora ferruginea C.T. White in Proc. Roy. Soc. Queensl. 53: 210 (1942). Holotype: Australia, Queensland, Cook District, Thornton Peak (Mt Alexander), 250 m, [? c. 15°29′S 145°15′E], fl., 20 Sep. 1937, *Brass & White* 262 (BRI: photo FHO!; isotype: K!).

Slender understorey tree up to 10 m high, unbranched or with few branches. Bole up to 13 cm in diameter. Bark mid greyish-brown with vertically elongated reddish-brown lenticels; inner bark pink or pale yellow. Twigs reddish-brown with round orange lenticels up to 2 mm densely covered with very dark reddish-brown stellate scales near the apex.

Leaves 48 – 84(– 93) cm long and 26 – 50(– 54) cm wide, petiole 13.5 – 28.5 cm, the petiole, rhachis and petiolules densely covered with reddish-brown or orange brown hairs stellate and some lenticels like those on the twigs. Leaflets 11 – 13, the laterals subopposite, all 10 – 25(– 28) cm long 2.5 – 7.5 cm wide, oblanceolate or narrowly elliptical or rarely lanceolate, slightly acuminate at apex, with the triangular, rounded acumen up to 7 mm, cuneate at the slightly asymmetric base; with reddish- brown stellate hairs densely covering the midrib below and few to numerous on the rest of the lower surface; veins 11 – 23(– 25) on each side of the midrib, usually shallowly ascending at c. 70° angle to the midrib, sometimes 50° to the midrib, and curved upwards near the margin; petiolules 5 – 15(– 17) mm.

Inflorescences crowded near the apex of the shoot in axils of about 5 leaves, sessile or with a peduncle up to 1.5 cm, 10 – 22 cm long and 4 – 14 cm wide, with 1 – 4 branches from the base, rhachis and branches densely covered with stellate hairs like those on the twigs. Flower 2 – 5 mm long, 2 – 5 mm wide; pedicel 2 mm, the pedicel, calyx and petals densely covered with stellate scales, calyx cup-shaped, shallowly divided into (2 or) 3 lobes, petals (2 or) 3, white, pink or pale yellow, densely covered with stellate scales on outside of petals, aestivation imbricate; the petals often curving away from the staminal tube to expose the anthers which protrude from the staminal tube. Staminal tube c. 2 mm long and 2 mm wide, white, obovoid, aperture 0.8 mm in diameter, with shallow obtuse lobes, anthers 6, c. 1.5 mm long and 1.5 mm wide, or as long as the staminal tube, dark with a pale border, protruding through the aperture. Ovary ovoid or depressed globose, densely covered with stellate hairs, loculi 2 or 3, each containing 1 (? or 2) ovule; stigma c. 1.5 mm or more long, nearly or as long as the anthers, cylindrical or obovoid, longitudinally ridged, with two small lobes at the flattened apex and a constriction between the base of stigma and the ovary.

Infructescences about 5 in the axils of leaves near the apex of the stem, 5.5 – 12 cm long and 5.5 – 9 cm wide; peduncle 1 – 2 cm. Fruits 2.5 – 3 cm long, 2.5 – 3 cm wide, obovoid; the pericarp dark reddish-brown outside, pale pink inside, thin and moulded around the seeds when dry. Loculi 3, each containing 0 or 1 seed; seeds 1.5 – 2.3 cm long, 1 – 1.3 cm wide, 0.7 – 1 cm through, completely enclosed in a thin orange opaque aril; testa brown.

DISTRIBUTION. Australia only, endemic to the eastern side of the Cape York Peninsula (Cook District) from Atherton Tableland south to Mt Bartle Frere. Fig. 16.

ECOLOGY. Found in montane, hillside and ridge-top rain forest, simple, complex or mixed meso-notophyll vine forest. Grows on sand, sandy clay, sand with pumice layers, soil derived from granite or granidiorite or basaltic krasnozem, red loamy clay. Alt.: 40 – 1100 m. Flowers July – Oct.; fruits Nov. – Feb.

Representative specimens. AUSTRALIA. Queensland: Daintree Mission Area [16°S 145°E], 7 Sept. 1948, *L.S. Smith* 04030 (L!); 24 km N.W.W. of Daintree [16°03′S 145°13′E], fr., 19 Oct. 1967, *Boyland (& Gilleatt)* 4 72 (BRI!, CANB!, K!); Mossman, [16°27′S 145°22′E] 31 May 1948, *L.S. Smith* 03943, (L!); N., Mt Lewis range, 1.6 km S. of Mt Lewis, main ridge montane forest, c. 1100 m, fl., 8 Oct. 1964, *Schodde* 4147 (CANB!, L!); Mt Lewis range, c. 12 miles N. of Mt Molloy, 65 m, fl., 19 Aug. 1963, *Schodde* 3334 (L!); S.F.R. 251, Tableland L.A., fl., 760 m, fl., 9 Oct. 1978, *Gray* 103 5 (QRS!); S.F.R. 143, North Mary, L.A., 1100 m, fr., 3 Feb. 1981, *Hyland* 10977 (QRS!); Tully Falls, 720 m, fl., 1 Oct. 1939, *Flecker* 6341 (QRS!).

The epithet *ferruginea* is already occupied in the genus *Aglaia*. A new name is thus required.

The indumentum on the stems differs from *Aglaia basiphylla* and *Aglaia parksii* in consisting of stellate hairs only, not a mixture of stellate scales and long-armed stellate hairs. *Aglaia meridionalis* resembles *A. lepidopetala* from New Guinea in the size of the tree and the size and form of the fruit, but it is separated from *A. lepidopetala* on the structure and distribution of its hairs.

12. Aglaia densitricha C.M. Pannell spec. nova ab *Aglaia rubiginosa* (Hiern) C.M. Pannell pagina inferiore foliolorum multis pilis stellatis pallide badiis vestita (in *Aglaia rubiginosa* squamis obscure badiis dense obtecta) dignoscitur. Holotype: Peninsular Malaysia, Trengganu, Trengganu – Besut road, 23rd mile, Belara F.R., fl., 13 July 1953, *Sinclair & Kiah bin Salleh* SFN 39933 (L!; isotype: E!).

Tree c. 5 m, with a narrow bole. Twigs c. 1.5 cm in diameter, densely covered with pale reddish-brown stellate hairs which have arms of different lengths.

Leaves imparipinnate, c. 115 cm long and 60 cm wide, obovate in outline; petiole c. 48 cm, the petiole, rhachis and petiolules densely covered with hairs like those on the twigs. Leaflets 11 – 13, the laterals subopposite, all 23 – 30 cm long, 8 – 10 cm wide, oblong or ovate, the apex acuminate with the acute acumen up to 10 mm long, rounded at the slightly asymmetrical base, with hairs like those on the twigs densely packed on the midrib and numerous on the rest of the lower surface; veins c. 18 on each side of the midrib, ascending, curved near the margin and not anastomosing; petiolules c. 1.5 cm.

Male inflorescence 23 – 36 cm long, 18 – 24 cm wide; peduncle 3 – 10 cm; the peduncle, rhachis and branches densely covered with hairs like those on the twigs. Male flowers c. 4.5 mm long and 3 mm wide; pedicels 2 – 3 mm, the pedicels and calyx densely covered with hairs like those on the twigs. Calyx 1.5 mm long, cup-shaped, shallowly divided into 3 broad lobes. Corolla c. 4.5 mm long and 3 – 4 mm wide, divided up to nearly halfway into 3 subrotund lobes; aestivation imbricate. Staminal tube c. 4 mm long and 3 mm wide, obovoid, the aperture c. 1 mm in diameter with the margin shallowly lobed; anthers 6, c. 2.3 mm long and 0.8 mm wide, inserted c. $1/3$ of the way up the tube and included. Ovary c. 0.5 mm high and 1 mm in diameter, depressed-globose densely covered with stellate hairs; loculi ? 3; stigma c. 1.2 mm long and 0.8 mm wide, longitudinally ridged and with 3 apical lobes. Female flowers and fruits not seen.

DISTRIBUTION. One locality in Peninsular Malaysia. Fig. 17.

Known only from the type collection, *Aglaia densitricha* differs from *A. rubiginosa* in having numerous pale reddish-brown stellate hairs on the lower surface of the leaflets, whereas the lower surface of the leaflets of *A. rubiginosa* are densely covered with dark reddish-brown scales. *A. densitricha* bears some resemblance to *A. rugulosa*, but lacks the rugulose leaflet surfaces and attenuate leaflet base.

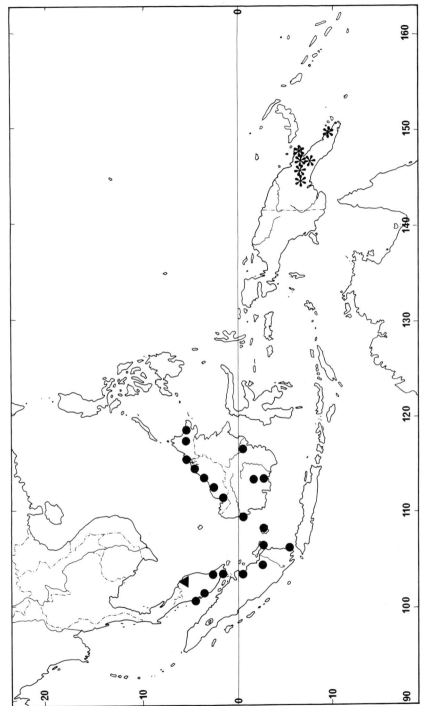

Fig. 17. Distribution of *A. rubiginosa* ●, *A. densitricha* ▲ and *A. penningtoniana* *.

13. Aglaia rubiginosa (Hiern) C.M. Pannell in Malaysian Forester 45: 455 (1982); Pannell in Ng, Tree Flora of Malaya 4: 225 (1989).

Amoora rubiginosa Hiern in Hooker fil., Fl. Brit. India 1: 561 (1875). Lectotype (Pannell, 1982): Malaysia, Malacca, *Griffith* 1050 (K!); C. de Candolle in A. & C. de Candolle, Monog. Phan. 1: 585 (1878); King in Jour. As. Soc. Bengal 64: 54 (1895); Ridley, Fl. Malay Penins. 1: 398 (1922); Corner in Gard. Bull. Suppl. 1: 131, 198, pl. 36 (1978).

Aglaia ignea Valeton ex K. Heyne in Nutt. Fl. Ned. Ind. 3: 59 (1917). Lectotype (designated here): [Sumatra], Billiton, fl., 1912, *(Heyne) van Rossum* 49 (BO!; isolectotypes: K!, L!, P!); I.H. Burkill, Dictionary of Economic Products of the Malay Peninsula, 1: 137 (1935).

Large tree up to 35 m, sometimes with buttresses upwards up to 1 m, with an open crown formed by a few ascending branches terminating in up to 40 subcrowns. Bark pale pinkish-brown or greyish-brown, flaking in squarish or long narrow scales 2 – 3 cm wide; inner bark pale pinkish-brown; sapwood yellowish-brown, pale yellow or red; latex white. Twigs stout dark brown, longitudinally wrinkled, with large leaf scars, densely covered with reddish-brown or dark brown stellate hairs.

Leaves imparipinnate, up to 80 cm long and 50 cm wide, obovate in outline; petiole up to 20 cm, slightly flattened on the adaxial side, the petiole, rhachis and petiolules with bark and indumentum like the twigs. Leaflets 15 – 21, the laterals subopposite, all 5 – 25 cm long, 2 – 7 cm wide, dark shiny green above, coriaceous, lanceolate or ovate, acuminate at apex with the acute acumen up to 10 mm long, rounded or cordate at the asymmetrical base, upper surface pitted, lower surface densely covered with reddish-brown stellate scales which have a darker, depressed centre, midrib and veins with similar but fewer scales; veins 11 – 24 on each side of the midrib, ascending and curved upwards near the margin, midrib and veins depressed on upper surface, prominent and with longitudinal wavy ridges on lower surface; petiolules up to 10 mm on lateral leaflets, up to 20 mm on terminal leaflet.

Inflorescence up to 70 cm long and 70 cm wide; peduncle up to 20 cm, the peduncle, rhachis and branches stout, flattened, longitudinally wrinkled, indumentum like the twigs. Flowers up to 9 mm long and 5 mm wide, slightly fragrant; pedicels up to 4 mm, with indumentum like the twigs. Calyx c. $1/2$ length of the corolla, cup-shaped, shallowly 3-lobed, with indumentum like the twigs. Petals 3, pinkish-yellow, white, yellow, red or purple, ovate, glabrous, aestivation imbricate. Staminal tube ellipsoid, the aperture up to 1.5 mm in diameter and shallowly 3-lobed; anthers 6, c. $3/4$ length of the tube, narrowly ovoid, usually included but sometimes just protruding through the aperture. Ovary depressed-globose, densely covered with stellate hairs; stigma ellipsoid with 3 apical lobes and 6 longitudinal ridges; the ovary and stigma together c. $1/2$ the length of the tube.

Infructescence c. 20 cm long. Fruits c. 6 cm long and 5 cm wide, ellipsoid or obovoid, red; pericarp thick, with indumentum like the twigs. Loculi 3, each containing 1 seed. Seeds with a complete red aril; testa brown; cotyledons pale yellow. Fig. 18.

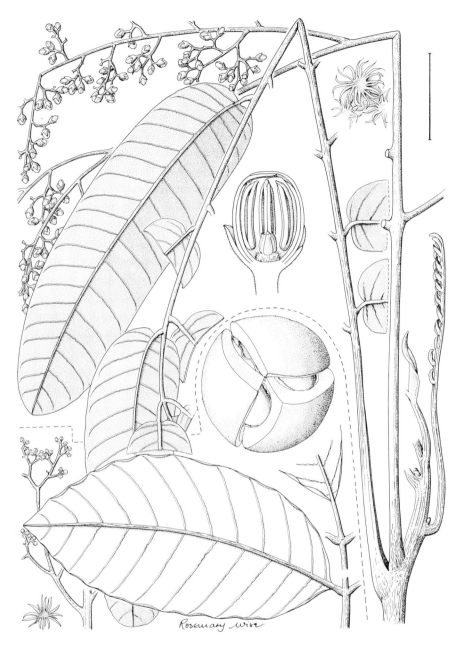

Fig. 18. *A. rubiginosa* (top). Habit with female inflorescence x^1/$_2$. Half flower x5. Stellate scales from lower leaflet surface x60. *A. erythrosperma* (bottom). Part of leaf x^1/$_2$. Part of inflorescence x^1/$_2$. Fruit viewed from apex x^1/$_2$. Stellate scale x70. (Reproduced from Pannell (1989) with permission from the Forest Research Institute Malaysia).

DISTRIBUTION. Peninsular Malaysia, Singapore, Sumatra, Borneo. Fig. 17.

ECOLOGY. Found in freshwater peat swamp forest, dry heath forest, kerangas, and less frequently in primary lowland and hill forest, secondary forest and along road-sides; also on granitic sand. Alt.: 2 to 300 m. Common.

VERNACULAR NAMES. Sumatra: Parah Ajer, Parak, Parak Api, Parak Talang, Pasak. Borneo: Lantupak (Dusun Kinabatangan); Upie (Malay); Be(r)sangai, Bunjau, Jelungan Sasak, Parak, Parak Api, Semparak.

Representative specimens. PENINSULAR MALAYSIA. Johore, mile 15 Muar-Yong Peng road, Ayer Hitam, 15 m, fr., 29 Jan. 1955, *Jaafar* KEP 74115 (FRI!).

SINGAPORE. Bt Timah, fl., 4 July 1941, *Corner* SFN 37277 (FRI!).

SUMATRA. Riow [Riouw] Island, 10 m, st., 14 May 1923, bb 5363 (BO!). Bengkalis Island: Panglang 222, Pulau Rangsand, c. 3 m, fr., 23 Nov. 1919, *Beguin* 494 (U!); Bengkalis Island, west coast, c. 2 m, 7 Oct. 1924, *Boschproefst* bb 5786 (U!).

SARAWAK. 1st Division, path to Gunong Periggi from Serian, Lundu, c. 15 m, fr., 16 Nov. 1963, *Pennington* 7980 (FHO!).

BRUNEI. Belait District: Sg Ingei, 30 m, fl., June 1958, *Brunig* S 4401 (T 510) (K!); Badas Sawmill, sea level, fr., 11 March 1958, *Ashton* BRUN 949 (K!).

SABAH. Nunukan, N. of Tarakan, Sg. Simengkadu, km 19, low, fl., Nov. – Dec. 1953, *Meijer* 2365 (L!).

A. rubiginosa is a conspicuous emergent tree in swamp forest. It has few ascending branches terminating in clumps of large pinnate leaves which form up to 40 subcrowns. The flower is one of the largest found in *Aglaia*, the calyx is cup-shaped and only shallowly 3-lobed.

14. Aglaia penningtoniana C.M. Pannell spec. nova *Aglaiae rugulosae* C .M. Pannell similis, sed ramunculi, petiolus, rhachis et pedunculus ramique inflorescentiae infructescentiaeque talibus pilis vestiuntur quales in *Aglaia grandi* Miq. occurrunt. Rhachis centralis pili longior, usque 1 cm, verticilli brachiorum inde radiantium crebriores; qui pili vel densum indumentum efficiunt vel interdum decidui sunt. Holotype: Papua New Guinea, Morobe District, near Lae, Bumbu logging area, 200 ft [c. 60 m], fl., 13 Jan. 1964, *Pennington* 8049 (FHO!).

Tree up to 45 m. Bole up to 20 m, diameter up to 150 cm, with buttresses upwards to 3 m and outwards to 2 m. Bark scaly, brown or pale greyish- brown; inner bark pinkish-brown. Sapwood hard, yellowish-brown or reddish- brown; latex white, copious, sticky. Twigs massive, up to 2 cm in diameter at apex, greyish-brown, longitudinally wrinkled, densely covered with reddish-brown hairs which have a central rhachis up to 1 cm long and 1 mm wide and numerous whorls of arms radiating from it, hairs sometimes deciduous and exposing numerous pale brown or dark reddish-brown stellate hairs and scales; latex white and copious.

Leaves in dense terminal spirals, imparipinnate, 35 – 100 cm long, 36 – 70 cm wide, obovate in outline; petiole 15 – 35 cm; the petiole, rhachis and petiolules with indumentum like the twigs. Leaflets 11 – 13, the laterals subopposite, all 8 – 40 cm long, 4.5 – 12 cm wide; shining dark green on upper surface, glossy pale green on lower surface, coriaceous, oblong, ovate or obovate, asymmetrical and curved, rounded or slightly cuneate at the asymmetrical base, rounded at the apex, the upper and lower surfaces of the leaflets rugulose, with few to numerous white stellate scales on the midrib and occasionally on the rest of the lower surface; veins 13 – 21 on each side of the midrib, shallowly ascending, curved upwards near the margin, not anastomosing; midrib prominent on the lower surface and depressed on the upper, the lateral veins subprominent on the lower surface; petiolules 5 – 10 mm.

Inflorescence c. 22 cm long and 7 cm wide; peduncle up to 8 cm, the peduncle, rhachis, branches and pedicels with hairs and scales like those on the twigs, deciduous, leaving densely packed pale brown stellate scales. Flowers (not known whether male or female) c. 8 mm long and 6.5 mm wide; pedicels up to 3 mm. Calyx up to 4 mm, cup-shaped and shallowly divided into 3(or 4) broad rounded lobes, densely covered with pale brown stellate scales on the outer surface. Petals 3, aestivation imbricate. Staminal tube obovoid, the margin of the aperture slightly lobed, anthers 17 – 21, $1/3 - 2/3$ the length of the staminal tube packed together and alternating in two rows, all included. Ovary depressed globose, densely covered with stellate scales; loculi 3, each containing one ovule; stigma c. 1 mm wide, with 3 narrow longitudinal lobes at the apex.

Fruits about 8 cm long and 7 cm wide, dark orange-brown, subglobose or obovoid, dehiscent, the outer surface rough but with few or no stellate hairs or scales; pericarp up to 1.5 cm thick, with copious white latex. Locule 1, containing 1 large seed c. 7 cm long, 4.5 cm wide and 4.5 cm thick; aril up to 4 mm thick, hard and with a dark red outer skin. Cotyledons transverse, slightly unequal. Shoot axis up to 7 mm long and 5 mm wide, plumule densely covered with orange-brown stellate hairs. Fig. 19.

DISTRIBUTION. Papua New Guinea only. Fig. 17.
ECOLOGY. Low and montane rain forest. Alt.: 30 – 1550 m.
VERNACULAR NAME. Bumbumdara.
Representative specimens. PAPUA NEW GUINEA. Morobe District, Lae Subdistrict: Sankwep logging area [6°35'S 147°05'E], 300 m, fl., 6 January 1975, *Barker* LAE 66771 (E!, L!) & Bumbu logging area, c. 60 m, fl., 13 Jan. 1964, *Pennington* 8048 (FHO!); Milne Bay District, Kaibo, near Kaporika [10°18'S 150°14'E], c. 75 m, 5 June 1964, *Henty* NGF 16969 (FHO!). Eastern Highlands, Okapa Patrol Post [6°35'S 145°40'E], c. 2000 m, ♀ fl. & fr., 18 Nov. 1957, *White* NGF 9585 (BO!).

Aglaia penningtoniana resembles *Aglaia rugulosa*, but the twigs, petiole, rhachis, and the peduncle and branches of the inflorescence and infructescence of *Aglaia penningtoniana* have hairs which resemble those found in *Aglaia grandis*. The central rhachis of the hair is longer than in *A. grandis*, up to 1 cm long, and the whorls or arms radiating from it are more numerous; these hairs may be densely packed, but they are sometimes deciduous.

The lower surface of the leaflets in this species is either densely covered with reddish-brown stellate hairs (as in the holotype) or almost without

Fig. 19. *A. penningtoniana*. Habit x½. Inflorescence x½. Half flower x5. Fruit x½. Stellate hair from leaflet midrib x120. Trichome from apex of twig x3.

hairs; in the former, the indumentum is sometimes deciduous on part of the leaflet, but it is not clear whether the lower surface of the leaflets of the latter would also have been densely covered with hairs when young and whether these were deciduous before maturity.

15. Aglaia lawii (Wight) Saldanha ex Ramamoorthy in Saldanha & Nicolson, Flora of Hassan District: 392 & pl. 76 (1976).

Epicharis exarillata Nimmo in J. Graham, Cat. Pl. Bombay: 31, no.227 (1839), non *Epicharis exarillata* Arn. in Wight & Arn., Prod.: 120 (1834) superfl. nom. illegit. pro *Guarea binectarifera* Roxb. Neotype (designated here): [India], Bombay, *J.S. Law* s.n. (K!; isoneotype: CGE!).

Nimmonia lawii Wight in Calc. Jour. Nat. Hist. 7: 13 (1847) nom. nov. pro *Epicharis exarillata* J. Graham, non Arn..

Oraoma canarana Turcz. in Bull. Soc. Nat. Mosc. 31: 411 (1858). [n.b. type species, at end of description of new genus, *Oraoma* Turcz.] Lectotype (designated here): prope Mercara, [in terra Canara Indiae orientalis], fl., Feb., *Anon*. in herb. *Hohenacker* 509 (K!; isolectotypes: G!, U!).

Nemedra nimmonii Dalzell in Dalzell & Gibson, Bombay Flora: 37 (1861) superfl. nom. illegit. pro *Nimmonia lawii* Wight.

Amoora korthalsii Miq., Ann. Mus. Bot. Lugd. Bat. 4: 36 (1868). Lectotype (designated here): Borneo, G. Sakoembang, *Korthals* (L!); C. de Candolle in A. & C. de Candolle, Monog. Phan. 1: 587 (1878).

Aglaia submonophylla Miq., Ann. Mus. Bot. Lugd. Bat. 4: 40 (1868). Lectotype (designated here): Borneo [S.], Prarawawin [Mountain], fl., *Korthals* (L!; isolectotypes: L!, U!); C. de Candolle in A. & C. de Candolle, Monog. Phan. 1: 620 (1878).

Aglaia oligocarpa Miq., Ann. Mus. Bot. Lugd. Bat. 4: 45 (1868). Lectotype (designated here): Sumatra, Hochankola, [Angkola province] 1 – 3000 ped. [c. 300 – 900 m], *Junghuhn* 41 (L!; isolectotype: L!); C. de Candolle in A. & C. de Candolle, Monog. Phan. 1: 626 (1878); Pannell in Ng, Tree Flora of Malaya 4: 221 (1989).

Aglaia littoralis Zippelius ex Miq., Ann. Mus. Bot. Lugd. Bat. 4: 45 (1868). Lectotype (designated here): Nov. Guinea, fl., *Zippelius* 212-C (L!); C. de Candolle in A. & C. de Candolle, Monog. Phan. 1: 621 (1878).

Amoora lawii (Wight) Beddome, Fl. Sylv.: t. 133 (1871); Hiern in Hooker fil., Fl. Brit. Ind. 1: 561 (1875); C. de Candolle in A. & C. de Candolle, Monog. Phan. 1: 585 (1878).

[*Amoora dysoxyloides* Kurz in Jour. As. Soc. Beng. 44: 147 (1875), nom. nud.]

Aglaia andamanica Hiern in Hooker fil., Fl. Brit Ind. 1: 555 (1875). Lectotype (designated here): South Andaman Island, [in the jungle between Port Monat & Homfray's Ghat], fr., *Kurz* s.n. (K!); as *Milnea* sp. in Kurz, Rep. Veg. Andaman Isl.: 33 (1870). C. de Candolle in A. & C. de Candolle, Monog. Phan. 1: 622 (1878); King in Jour. As. Soc. Bengal 64: 79 (1895).

Lansium pedicellatum Hiern in Hooker fil., Fl. Brit. India 1: 558 (1875) , non Kostermans in Reinwardtia 7: 31, t. 11 (1965) (= *Lansium domesticum* Correa). Lectotype (designated here): Malacca [Malaya], *Maingay* 3406 (Kew Dist. 356) (K!).

Amoora canarana (Turcz.) Hiern in Hooker fil., Fl. Brit. Ind. 1 : 560

(1875); C. de Candolle in A. & C. de Candolle, Monog. Phan. 1: 586 (1878).

Amoora maingayi Hiern in Hooker fil., Fl. Brit. India 1: 562 (Feb. 1875). Lectotype (designated here): Malacca [Malaya], 25 May 1865 – 1866, *Maingay* 1910 (Kew Dist. 342) (K!); C. de Candolle in A. & C. de Candolle, Monog. Phan. 1: 588 (1878); Ridley, Fl. Malay Penins. 1: 400 (1922); I.H. Burkill, Dictionary of Economic Products of the Malay Peninsula, 1: 138 (1935).

Amoora dysoxyloides Kurz in Jour. As. Soc. Beng. 44: 200 (1876). Lectotype (designated here): Burma, Martaban, Yoonzeleen, 900 ft [c. 270 m], *Brandis* s.n. (K!); C. de Candolle in A. & C. de Candolle, Monog. Phan. 1: 589 (1878).

Aglaia turczaninowii C. de Candolle in A. & C. de Candolle, Monog. Phan. 1: 623 (1878). Lectotype (designated here): Philippine Islands, Luzon, Tayabas, *Cuming* 772 (K!; isolectotypes: CGE!, L!, W!).

Aglaia beccarii C. de Candolle in Bull. Herb. Boiss. 2: 579 (1894). Holotype: Borneo [Sarawak], fl., 1865 – 68, *Beccari* 3297 (G!; isotype: G!).

Aglaia maingayi (Hiern) King in Jour. As. Soc. Bengal 64: 79 (1895).

Aglaia eusideroxylon Koord. & Valet. in Meded. 'S Lands Plantent. 16: 12 8 (1896). Lectotype (designated here): Java, Besoeki, Tjoermanis, 8 Dec. 1889, *Koorders* 4692β (L!; isolectotype: K!); Koorders & Valeton, Atlas der Baumarten von Java 1: t. 97 (1913). Backer and Bakhuizen, Fl. Java 2: 127 (1965).

Lepiaglaia tetrapetala Pierre, Fl. Forest. Cochinch. Fasc. 22, ante t. 337 (1 July 1897). Lectotype (designated here): in montibus Bay ad Chaudoc austro Cochinchina, fl. & fr., Dec. 1867, *Pierre* 281 (P!; isolectotypes : E!, K!, NY!).

Aglaia tetrapetala Pierre, Fl. Forest Cochinch. Fasc. 22 sub t. 337A (1 July 1897). Rehder in Jour. Arn. Arb. 18: 210 (1937); How & Chen in Acta Phytotax. Sin. 4: 23 (1955); Lauener in Notes from the Royal Botanic Garden Edinburgh 27: 270 (1967).

Amoora lepidota Merrill in Philipp. Gov. Lab. Bur. Bull. 17: 23 (1904). Lectotype (designated here): Philippine Islands, Luzon, Province of Bataan, Mount Mariveles, Lamao river, [100 m], fl., Oct. 1903, *Merrill* 3173 (NY!; isolectotype: PNH†).

Ficus ouangliensis ["*ouangliense*"] Léveillé in Fedde, Rep. Sp. Nov . 4: 66 (1907). Holotype: China, [Kouy-Tchéou: ouest de Lo Fou], Ouang-li, [Nov. 1905], *Cavalerie* 2568 (E!). Lauener in Notes from the Royal Botanic Garden Edinburgh 27: 270 (1967); C.Y. Wu, Flora Yunnanica 1: 237 (1977).

Ficus vanioti Léveillé in Fedde, Rep. Sp. Nov. 7: 258 (1909). Holotype: China, Kouy-Tcheou, Lo-Fou, yg fr., April 1908, *Cavalerie* 2984 (E!). Lauener in Notes from the Royal Botanic Garden Edinburgh 27: 270 (1967).

Aglaia brachybotrys Merrill in Philipp. Jour. Sci., Bot. 7: 274 (1912). Lectotype (designated here): Philippine Islands, Luzon, Province of Cagayan, fl., April 1910, *Bernado* For. Bur. 15497 (A!; isolectotypes: G !, PNH†).

Aglaia cagayanensis Merrill in Philipp. Jour. Sci., Bot. 7: 275 (1912). Lectotype (designated here): Philippine Islands, Luzon, Province of

Cagayan, fl., Jan. 1912, *Ramos* Bur. Sci. 13801 (US!; isolectotypes: BM!, G!, K!, PNH†).

[*Aglaia euryphylla* Koord. & Val. ex Koorders-Schumacher, Syst. Verz. 1, Fam. 140, Meliaceae: 37 (1912), nom.nud.]

Aglaia korthalsii (Miq.) Pellegrin in Lecomte, Fl. Gén. Indo-Chine 1: 771 (1911), non *Aglaia korthalsii* Miq. (1868).

Aglaia sclerocarpa C. de Candolle in Meded. Herb. Leid. 22: 9 (1914). Holotype: Celebes [Sulawesi], Insel Kabaëna, Eempuhu, Landschaft Balo (östlicher Teil der Insel), 26 Oct. 1909, *Elbert* 3343 (L!; isotypes: BO!, G!).

Amoora curtispica L.S. Gibbs in Jour. Linn. Soc., Bot. 42: 63 (1914). Lectotype (designated here): British North Borneo, Darut Province, Tenom, 700 ft [c. 210 m], ♀ fl. & fr., Jan. 1910, *Gibbs* 2801 (BM!; isolectotype : K!).

Aglaia alternifoliola Merrill in Philipp. Jour. Sci., Bot., 1914 9: 532 (1915). Lectotype (designated here): Philippine Islands, Basilan, fl., Oct. 1912, *Miranda* For. Bur. 18996 (PNH†, US!).

Aglaia grandifoliola Merrill in Philipp. Jour. Sci., Bot., 13: 293 (1918). Lectotype (designated here): Philippine Islands, Luzon, Province of Tayabas, *Ramos & Edaño* Bur. Sci. 28981 (US!; isolectotypes: K!, P!, PNH†).

Aglaia haslettiana Haines in Jour. As. Soc. Bengal. N.S. 15: 312 (1920). Lectotype (designated here): India, [Orissa], Puri, Selang pena to Barbara, yg fl., 21 April 1917, *Haines* 5546(b) (K!).

Aglaia sibuyanensis Elmer ex Merrill, Enum. Philipp. Fl. Pl. 2: 376 (1923), in obs. pro syn.; Elmer in Leafl. Philipp. Bot. 9: 3308 (1937), sine diagn. lat.. Type no.: *Elmer* 12243, Philippine Islands, Sibuyan, Magallanes, Mt Giting-giting, fr., April 1910, (W!).

Aglaia trimera Merrill in Univ. Calif. Publ., Bot. 15: 128 (1929). Lectotype (designated here): British North Borneo, Elphinstone Province, Tawao, fl., Oct. 1922 – March 1923, *Elmer* 20560 (UC!; isolectotypes: A!, BO!, L!, NY!, SING!).

Aglaia tsangii Merrill in Lingnan Sci. Journ. 6: 280 (1928). Lectotype (designated here): China, Kwangtung Province, Hainan, [Taam-chau District], Nga Ping Mt, fl., 23 Sep. 1927, *Tsang Wai Tak* 928, L.U.16427 (UC!; isolectotypes: A!, G!, K!, NY!); How & T.C. Chen in Acta Phytotax. Sinica 4: 24 (1955).

Aglaia racemosa Ridley in Kew Bull. 1930: 367 (1930). Lectotype (designated here): Borneo, [Sarawak], nr Kuching, ♀ fl., 6 Oct. 1892, *Haviland* 1781 (K!; isolectotype: K!).

Aglaia attenuata H.L. Li in Jour. Arn. Arb. 25: 303 (1944). Lectotype (designated here): China, Yunnan, Szemao, in forests, 4500 ft [c. 1370 m], fl., *Henry* 12228 (A!; isolectotype: K!).

Aglaia tenuifolia H.L. Li in Jour. Arn. Arb. 25: 304 (1944). Holotype: China, Yunnan Province, Che-li-Hsien [Che-li District], Dah-meng-lung, 1100 m, fl., Aug. 1936, *Wang* 77903 (A!).

Aglaia yunnanensis H.L. Li in Jour. Arn. Arb. 25: 305 (1944). Holotype: China, Yunnan Province, Fo-hai, 1300 m, fl., June 1936, *C.W. Wang* 74830 (A!).

Aglaia wangii H.L. Li in Jour. Arn. Arb. 25: 304 (1944). Holotype: China,

Yunnan Province, Che-li Hsien [District], Sheau-meng-yeang, 960 m, fr., Aug. 1936, *Wang* 75593 (A!).
Aglaia wangii var. *macrophylla* H.L. Li in Jour. Arn. Arb. 25: 304 (1944). Holotype: China, Yunnan Province, Nan-chiao, 1400 m, June 1936, *Wang* 751 31 (A!).
Amoora tetrapetala (Pierre) Pellegrin in Suppl. Fl. Gén. Indo-Chine, ed. Humbert 1: 717 (1948); C.Y. Wu, Fl. Yunnanica 1: 235 (1977).
Aglaia pedicellata (Hiern) Kostermans in Reinwardtia 7: 226, 264 (1966).
Amoora yunnanensis (H.L. Li) C.Y. Wu, Flora Yunnanica 1: 231 (1977).
Amoora yunnanensis var. *macrophylla* (H.L. Li) C.Y. Wu, Flora Yunnanica **1**: 231 (1977).
Amoora tetrapetala var. *macrophylla* (H.L. Li) C.Y. Wu, Fl. Yunnanica 1: 235 (1977).
Amoora ouangliensis ["*ouangliense*"] (Léveillé) C.Y Wu, Fl. Yunnanica 1: 237 (1977).
Amoora calcicola C.Y. Wu & H. Li ex C.Y. Wu, Fl. Yunnanica 1: 234 (1977). Type: [China, Yunnan, Mong-la, Xishuang Banna, 24°N 100°E], 1200 m, fl., 16 Sep. 1959, ? coll. 59-10292 (KUN!).
Aglaia jainii Viswanathan & Ramachandran in Bull. Bot. Surv. India (1982) 24: 212 (30 Nov. 1983), nom. nov. pro *Oraoma canarana* Turcz. and *Amoora canarana* (Turcz.) Hiern, non *Aglaia canarensis* Gamble (1915) (= *A. perviridis*); Jain & Gaur in Jour. Economic and Taxonomic Botany (1985) 7: 466 (1986).
Aglaia tamilnadensis Nair & Rajan in Nair & Henry, Fl. Tamilnadu India ser. vol. 1: 66 (Dec. 1983), nom. nov. pro *Amoora canarana* (Turcz.) Hiern.
Aglaia stipitata P.T. Li & X.M. Chen in Acta Phytotaxonomica Sinica 22: 495 (1984), nom. nov. pro *Lansium pedicellatum* Hiern.
Aphanamixis chittagonga (Miq.) Haridasan & Rao, Forest Flora of Meghalaya 1: 205 (1985), non *Aglaia chittagonga* Miq. (1868).
Amoora tsangii (Merrill) X.M. Chen in Jour. Wuhan Botanical Research 4: 180 (1986).

Tree up to 30 m, sometimes flowering as an unbranched treelet c. 1.6 m high. Bole up to 75 cm in diameter, fluted and with concave or tall narrow buttresses, upwards up to 1.8 m, outwards up to 1 m. Bark reddish-brown, orange brown, yellowish-brown or pale pinkish-brown, rough and flaking in large thin irregular scales, sometimes with large round orange lenticels, or bark grey or greenish-brown and smooth; inner bark green, cambium white, sapwood pale orange, orange-brown or yellowish-brown, sometimes turning magenta pink on exposure to air; latex white. Twigs usually slender, sometimes up to 9 mm across, slender, greyish-brown or pinkish-brown, with longitudinal wavy ridges and sometimes with numerous elliptical brown lenticels, densely covered with pale brown or pale orange brown usually peltate, scales which have an irregular or fimbriate margin and may have a dark brownish-black central spot, sometimes densely covered with stellate scales and sometimes with stellate hairs interspersed. Leaves imparipinnate, 7 – 66 cm long and 5 – 60 cm wide, ovate or obovate in outline; petiole 1.5 – 16 cm, the petiole, rhachis and petiolules longitudinally wrinkled and without, with a few or densely covered with scales like those on the twigs, in Borneo the rhachis sometimes ridged or with narrow foliolate wings up to 3 mm wide. Leaflets (1 or) 2 – 7(– 11), the laterals alternate or subopposite, all 4 – 30 cm long, 1.5 – 11.5 cm wide, often orange-brown or whitish-green when dry, especially the veins, sometimes subcoriaceous, asymmetrically elliptical, ovate or obovate, acuminate or acuminate-caudate at apex with the obtuse acumen often parallel-sided and 5 – 15(– 25) mm long, usually broadly cuneate but occasionally rounded, attenuate or (sometimes in Borneo) cordate at the asymmetrical base, sometimes rounded on the distal side and cuneate on the proximal side of the petiolule, sometimes shiny on the upper surface and often with numerous pits on the upper and lower surfaces, without hairs or scales or with occasional or numerous scales like those on the twigs on the lower surface; veins 5 – 21 on each side of the midrib, curved upwards or ascending and curved upwards near the margin, not or sometimes in Borneo quite anastomosing, sometimes with shorter lateral veins in between, midrib and lateral veins slightly depressed on upper surface, midrib prominent, lateral veins subprominent and secondary veins barely visible or subprominent on the lower surface; sometimes (but not in Peninsular Malaysia) with a depression in the axil between the lateral vein and the midrib which is surrounded by a dense tuft of stellate which have long arms or simple hairs; sessile or with petiolules up to 20 mm.

Inflorescences in the axils of c. 5 leaves near the apex of the shoot, in Borneo sometimes ramiflorous, 2.5 – 22 cm long and 1.5 – 20 cm wide, triangular or ovate in outline; sessile or with a peduncle up to 10.5 cm long, the peduncle, rhachis and branches longitudinally wrinkled and with numerous or densely covered with scales or hairs like those on the twigs. Flowers 1.5 – 4.5 mm long, 1.5 – 5 mm wide, obovoid or subglobose; pedicel 0.5 – 5 mm, the pedicels and calyx with few to densely covered with pale brown or orange peltate scales which have a fimbriate margin or occasionally with stellate hairs. Calyx cup-shaped 1 – 2.5 mm long, c. $1/3$ the length of the corolla, shallowly divided into 3 or 4(– 6) obtuse lobes. Corolla a short tube connate with the base of the staminal tube, divided into 3 or 4(– 6) subrotund lobes, yellow or white, sometimes with a few scales like those on the twigs on the outside, aestivation imbricate. Staminal tube shorter than the corolla, yellow, 0.5 – 3.5 mm long and 1 – 3.5 mm wide, either obovoid with the aperture entire and 0.3 – 0.5 mm in diameter or cup-shaped with the apical margin incurved and shallowly lobed and 0.6 – 1.5 mm in diameter; anthers (5 or) 6 – 10 (or 11), ovoid, 0.5 – 2.5 mm long, 0.25 – 0.7 mm wide, $1/3 - 3/4$ the length of the tube, inserted in the uppermost $1/3 - 1/2$ of the tube, included or just protruding through the aperture, sometimes with a few simple hairs on the anthers and the inside of the staminal tube. Ovary subglobose or ovoid, 0.25 – 0.8(– 1.2) mm high, 0.3 – 1 mm wide densely covered with peltate or stellate scales or stellate hairs, loculi (2 or) 3 (or 4), each containing 1 or 2 superposed ovules, which sometimes have a small dark brown appendage on one side; stigma 0.25 – 1 mm long, 0.3 – 0.5 mm wide, either ovoid, with (2 or) 3 small apical lobes, dark and shiny and the stigma and ovary together c. $1/3$ the length of the staminal tube, or the stigma up to 1 mm long and 0.3 mm wide, columnar, filling the space between the anthers, paler in colour, expanded and truncate at apex, the

Section Amoora

apex level with the apices of the anthers and visible through the aperture of the staminal tube.

Infructescence 3.5 – 15 cm long with 1 – 20 fruits which ripen at different times; sessile or with a peduncle up to 5.5 cm with surface and indumentum like the inflorescence. Fruits 1.7 – 2.8(– 6) cm long, 1.2 – 2.3(– 3.5) cm in diameter, subglobose, obovoid, ellipsoid or pear-shaped, sometimes with a small beak, asymmetrical if a seed does not develop in each locule, dehiscent, fruit-stalk 1.5 – 15 mm; pericarp usually c. 2.5 mm thick but sometimes in Borneo thinner so that it is moulded around the seeds in the dried fruit, outer pericarp pink or sometimes carmine red or yellow, densely covered with scales like those on the infructescence branches, inner pericarp white; loculi (2 or) 3 (or 4), each containing 0 – 1 arillate seed; aril 1 – 3.5 mm thick, the edges nearly meeting or meeting and overlapping on the antiraphe side, easily peeled off the testa, the outer skin red or white, the flesh soft, white and oily; seed with aril removed c. 14 mm long, 7 mm wide and 6 mm thick; testa shiny dark brown. Fig. 20 & 82.

DISTRIBUTION. India, Bhutan, Burma, Andaman Islands, Great Cocos Island, Laos, Vietnam, China, Taiwan, Thailand, Peninsular Malaysia, Sumatra, Borneo, Philippine Islands, Java, Nusa Tenggara, Sulawesi, Maluku, New Guinea, New Britain, Solomon Islands. Fig. 21.

ECOLOGY. Seashore, primary lowland and hill forest up to 1650 m in wet evergreen forest, semi-evergreen forest, deciduous forest, peat swamp forest, riverine forest; up to 500 m in secondary forest. On limestone, sandstone, granite, clay, sandy loam and alluvial river soil. Rare to common. USES. Leaves are applied for headache (Philippine Islands: Mindanao); wood is used for construction (Philippine Islands: Palawan).

VERNACULAR NAMES. Peninsular Malaysia: Bekak. Sumatra: Lasih, Lasoen Balah, Mottou, Sampai Kurakura. Borneo: Kanomogon (Kudas.); Langsat-langsat (Malay); Lantupak, Lasat-Lasat (Dyak Kinabatangan); Ngitonok (Dayak); Segera (Iban); Garamar. Philippine Islands: Bebaga (Sub.); Lambanaws-bagit (Tagbanua); Lambunao Bagit (Palawan). Sulawesi: Kajoe Djangan, Molomehoelo. Nusa Tenggara: Kadju Worok, Sentoel, Weloe Ketaka. New Guinea: Aisnepapir (Noemfoers); Herrib, Herrib Poetih (Manikiong); Kabrori (Ambai); Kamoengolon, Kwamesom (Mooi); Manienif, Naskajene, Soesbok, Wofdee (Biak); Moim (Kasimin); Sant(e)raauw, Tatarah, Tatraw, Tatro (Kebar); Teng (Moejoe); Wass (Karong); Beraai, Biaja, Maheo. Solomon Islands: Ulukwalo, Ulukwalobala (Kwara'ae).

Representative specimens. INDIA. Concan, fl. & fr., *Stocks* s.n. (CGE!, E!, K!). Carnatic, Tiruchi, Thuraiyur Taluk district, Pacchaimalais, Kannimar Skola, fl., RHT 27377 (K!). Mysore, Hassan District, Shiradi Ghat, near border, fr., 7 Aug. 1969, *Saldanha* 14447 (E!). Kerala State, Lady Smith R.F., yg fr., 30 May 1978, *Pascal* HIFP 1331 (FHO!).

BHUTAN. Gaylegphug District: slopes below Sham Khara [27°01'N 90°34'E], 1250 m, fr., 29 March 1982, *Grierson & Long* 4090 (E!, K!); north of Gaylegphug, 4 km below Sham Khara [27°01'N 90°34'E], 1450 m, fr. 3 June 1979, *Grierson & Long* 1583 (E!, FHO!, K!). Tongsa District, near Pertimi [27° 13'N 90° 41'E], 1400 m, yg fr., 3 April 1982, *Grierson & Long* 427 5 (E!).

GREAT COCOS ISLAND. fl., *Prain* s.n. (K!).

ANDAMAN ISLANDS. South Andaman, Port Monat, fl., 4 Oct. 1890, *King's Coll.* s.n. (K!). Middle Andaman, Bakultala, 6 km W. of Rayat, c. 40 m, fl., 6 Nov. 1977 *Bhargawa & Nooteboom, Kramer & Nair* 6408 (K!).

BURMA. Kachin State, Sumprabum Sub-division, eastern approaches from Suprabum to Kumon range [26°40′N 97°20′E], between Ning W'Krok and Kanang, on the eastern aspect of Gwe-Kya-Kat-Bum, c. 120 – 150 m, yg fr., 21 Jan. 1962, *Keenan, Tun Aung & Tha Hla* 3363 (E!). Maymyo Plateau, c. 1000 m, fr., 21 June 1913, *Lace* 6230 (E!, K!).

LAOS. *Poilane* 20614 (K!)

VIETNAM. Tonkin: Sai Wong Mo Shan (Sai Vong Mo Leng), Lomg Ngong Village, Dam-ha, fl., 18 July – 9 Sep., *Tsang* 30248 (E!, K!); Sai Wong Mo Shan (Sai Vong Mo Leng), Lomg Ngong Village, Dam-ha, fl., 18 July – 9 Sep., *Tsang* 30340 (E!, K!). Annam, Province Nhatrang, Massif de la Mère et L'Enfant, yg fr., 24 May 1923, *Poilane* 6727 (K!).

CHINA. Hainan: Yaichow, fr., 5 July 1933, *F.C. How* 70948 (NY!); Hung Mo Shan & vicinity, Lai (Loi) area, fl., 21 Jun. 1929, *Tsang & Fung* 347, L. U. 17881 (UC!). Yunnan: Szemao, fl., *Henry* 12228B (NY!); Szemao, 1220 m, fr., *Henry* 12170 (A!, E!, K!: paralectotypes of *A. attenuata*); Szemao, 1525 m, fl., *Henry* 12228A (K!: paralectotype of *A. attenuata*); Che-li, Dah -meng- lung, 1100 m, fl., Aug. 1936, *Wang* 77893A (A!: paralectotype of *A. tenuifolia*); Che-li district, 800 m, Aug. 1936, *Wang* 78043A (A!: paralectotype of *A. tenuifolia*): Fo-hai, 1000 m, June 1936, *Wang* 74823A (A!: paralectotype of *A. yunnanensis*).

TAIWAN. Crose road, Orchid Island, fl., 31 Aug. 1969, *M.T. Kao* 5242 (L!).

THAILAND. N., between Pong Pho and Khun Klong, N.W. of Doi Chieng Dao, 1200 m, fl., 3 July 1960, *K. Larsen, Santisuk & Warncke* 2949 (AAU!, L!). S.W., Utai Thani Province, Huai Ka Kaeng Game Reserve, Ban Rai [15°00′N 99°14′E], 600 – 700 m, fr., 23 Feb. 1970, *van Beusekom & Santisuk* 2939 (AAU!, C!). S.E., Chanthaburi Province, Doi Soi Dao [12°45′N 102°10′E], 300 m, fr., 12 May 1974, *Geesink, Hattink & Phengkhlai* 6683 (AAU!, C!). Saraburi Province, Muang District, Sahm Lahn forest, 125 m, fl. 28 July 1974, *Maxwell* 74-743 (AAU!).

PENINSULAR MALAYSIA. Pahang, Krau Game Reserve, Kuala Lompat, fr., 22 Aug. 1990, *Pannell* 1622 (FHO!). Kuala Lumpur, Weld's Hill Reserve, fl., 5 Dec. 1918, *Hamid* FMS 1834 (K!). Selangor, Sungei Buloh F.R., 60 m, fr., 11 Sep. 1963, *Pennington* 7806 (FHO!, L!).

SUMATRA. Aceh, Takigeum, c. 1100 m, fr., 12 Jan. 1932, *W.N. & C.M. Bangham* 855 (K!, NY!). Atjeh [Aceh], Gunung Leuser Nature Reserve, Gunung Bandahara, c. 6 km N.E. of Kampung Seldok (Alas Valley), c. 25 km N. of Kutatjane, 800 – 1000 m, ♂ fl., 20 March 1975, *de Wilde & de Wilde-Duyfjes* 15622 (L!, UC!). Sumatra Barat, between Sijunjung and Sungaidareh, central part of Barisan mountain range, nr Sigirik mountain, Bt Sebelah [00°45′S 101°10′E], 550 m, ♀ fl., 20 May 1983, *Pannell* 1864 (same tree as *Laumomier* TFB 4394) (FHO!).

SARAWAK. 1st Division, Stapok F.R., 5 miles S.W. Kuching, c. 150 m, fr., 27 April 1974, *Mabberley* 1625 (FHO!, L!); 4th Division, Lambir National Park, Miri, fl., 18 Sep. 1978. *George* S 40267 (FHO!). Ulu Kuyong, Muput Kanan, Anap, [2°45′N 112°50′E], c. 130 m, fr., 10 Oct. 1963, *Chai* S 19320 (FHO!).

SABAH. Tenom, Darut, 210 m, ♂ fl., *Gibbs* 2808 (BM!, K!: paralectotypes of *A. curtispica*). Sepilok F.R., nr Sandakan, Compt. 5, c. 15 m, fr., 23 Oct. 1963, *Pennington* 7925 (FHO!, SAN!). Mile 32 – 33 Ranau Road, Tenempok F.R., fl., 29 Oct. 1963, *Pennington* 7943 (FHO! SAN!). Ranau District, about 8 miles east of Kampung Merungin [6°05'N 117°10'E], c. 300 m, fr., 20 Nov. 1975, *Leopold & Saikeh* SAN 82668 (SAN!).

KALIMANTAN. Sungei Blaene?, fl., July 1897, *Jaheri* (Expedition Nieuwenhuis) 534 (BO!, L!). Sungei Pary, fr., July 1897, *Jaheri* (Expedition Nieuwenhuis) 1129 (BO!).

PHILIPPINE ISLANDS. Sibuyan, Province of Capiz, Magallanes (Mt Giting-giting) [11°48'N 122°03'E], fr., April 1910, *Elmer* 12243 (A!, G!, GH!, L!, NY!, U!, UC!, US!, W!). Luzon: Ilocos Norte Province, Sagiao Bangui, 200 m, fr., 27 June 1929, *Felix* 31101 (NY!); Province of Bataan, Mount Mariveles [14°32'N 120°30'E], Lamao River, fl., Oct. 1903, *Merrill* 317 3 (K!, L!); Rizal Province, fl., Nov. 1909, *Ramos* 45 (U!). Palawan, Bindoyan, vicinity of Puerto Princesa, fl., 4 – 8 March 1940, *Ebalo* 617 (NY!, UC!).

JAVA. Besoeki: Pantjoer-Idjen, fl., 5 June 1889, *Koorders* 4686β (L!); Rogodjampi, fr., 7 Sep. 1897, *Koorders* 28986β (L!).

NUSA TENGGARA. Soemba, Meniara, fl., 8 May 1925, *Iboet* 436 (L!, U!).

SULAWESI. Minahasa (Menado) Province, 50 m, st., 23 Feb. 1895, *Koorders* 19710β (BO!).

MALUKU. Key, fr., *Jaheri* 345 (L!). Amboina, fl., July – Nov. 1913, *Robinson* 1991 (L!, NY!).

IRIAN JAYA. Radjah Ampat District, West of Sorong, Batanta Island, Marchesa bay, beach east of Amdoei Village, fl., 2 April 1954, *van Royen* 3534 (K!, L!).

PAPUA NEW GUINEA. Central District, Abau Subdistrict, Cape Rodney, near P.I.T. Sawmill [10°07'S 148°18'E], c. 60 m, fr., 18 June 1968, *Henty* N GF 38523 (K!, L!). Morobe District: near Yalu, c. 8 m, fl., July 1944, *White, Dadswell & Smith* NGF 1609 (K!, L!); above Bulolo, Dengaloo Valley, c. 1000 m, fr., 17 Jan. 1964, *Pennington* 8064 (FHO!, L!). New Britain, Keravat, Vudal forest, 30 m, fr., 2 Feb. 1964, *Pennington* 8115 (FHO!, L!).

SOLOMON ISLANDS. Santa Ysabel, Barora Ite Island, Madagha Bay Area, fl., 27 Aug. 1969, *Mauriasi & collectors* BSIP 16172 (K!, L!).

Aglaia lawii is one of the most widespread, variable and ecologically versatile species of *Aglaia*. To the west, *A. lawii* occurs up to the limits of the range of *Aglaia* in India (Bombay District) and S.W. China. It is common throughout the equatorial range of the genus. To the east, it occurs in New Guinea, New Britain and Solomon Islands, but is not recorded from Australia, Fiji, New Caledonia, Samoan Is. or Micronesia. In the Philippine Islands and Irian Jaya, the habitats in which *A. lawii* occurs include the seashore and in Borneo it is known from coastal forests, peat swamp, stream-side, alluvial river soil, secondary forest and forest on limestone and sandstone. *Aglaia lawii* and its close relative, *A. teysmanniana*, can usually be recognised by their small (c. 2 cm diameter in most parts of the range), dehiscent, usually pink fruits and the red-skinned arillate seeds. The fruits of *A. lawii* are, however, sometimes much larger in China, Java and New Guinea (see below).

The variation patterns in *A. lawii* are partly geographical: for example the flower may be larger and the staminal tube may have a smaller aperture in India and China; some specimens from India, Borneo and the Philippine Islands, have stellate hairs as well as peltate scales; in Borneo the petiole and rhachis are frequently winged and in both drier and littoral localities the scales may be more numerous on the lower surface of the leaflets, this being especially marked in some specimens from China, Taiwan and Java. In India, China and the Philippine Islands, depressions surrounded by simple or stellate hairs (domatia), are often found in the axils of the lateral veins and midrib on the lower surface of the leaflets.

Variation also occurs in leaflet number and in the size of flowering and fruiting individuals. In Borneo, this species may flower as a small, unbranched treelet with simple leaves. Simple leaves are also found, together with compound leaves, on taller, branched individuals of this species. The small treelets may occur in the same site as the tall tree and they appear in the field to belong to different, clearly distinguishable species (e.g. Gunong Palung, Kalimantan; M. van Balgooy, pers. comm.); further study of the variation within *A. lawii* may result in its division into two or more species which can be reliably distinguished. I attempted to distinguish the following separate species, which have in the past been recognised under the following names: *Aglaia lawii* sensu stricto (India), *Amoora canarana* (Turcz.) Hiern (India), *Aglaia oligocarpa* Miq. (India, Bhutan, China, Indochina, Thailand, Peninsular Malaysia, Sumatra, Borneo, Philippine Islands), (*Amoora ouangliensis* Léveillé) (China, Thailand), *Aglaia tsangii* Merrill (China, Taiwan), *Aglaia eusideroxylon* Koord. et Val. (Java), *Aglaia turczaninowii* C. de Candolle (Philippine Islands), *Aglaia littoralis* Miq. (New Guinea) and *Aglaia beccarii* C. de Candolle (Borneo, Thailand, Philippine Islands). These species were based on variation in the structure and density of the indumentum, the flower size and structure and the fruit size and shape, but it has not been possible to consistently recognise these as separate species based on herbarium specimens nor from my field knowledge of *A. lawii* in Peninsular Malaysia and Sumatra.

The most variable character is the structure and density of the indumentum and the frequency of pits on the leaflet surfaces. Near the equator, the leaves are almost without pits and scales, but in more seasonal climates, the upper and lower leaflet surfaces often have numerous pits and there are numerous scales on the lower leaflet surface. With the exception of the largest of the fruit variants in *A. lawii*, the fruits of this species, along with *Aglaia teysmanniana*, are the smallest among the dehiscent species of *Aglaia* and the pericarp of the ripe fruit is nearly always pink. The fruit of *A. lawii* varies in shape from subglobose to pear-shaped and in China the largest fruits found in this species have a long stipe and, in the one seed examined, there was no aril. With few exceptions (such as the last-mentioned Chinese form) the seed has an entire red-skinned aril. Many bird species of different sizes and mobility feed on and presumably disperse the seeds (Pannell & Kozioł, 1987) and this may in part explain the frequency and wide distribution of this species. *Aglaia teysmanniana* is maintained as distinct from *Aglaia lawii* because its indumentum consists exclusively of stellate

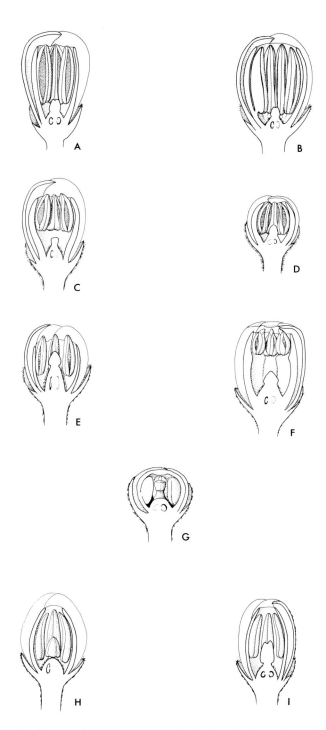

Fig. 20. *A. lawii*. Half flowers. A, B, H, I x7. C, D x10. E, F, G x15.

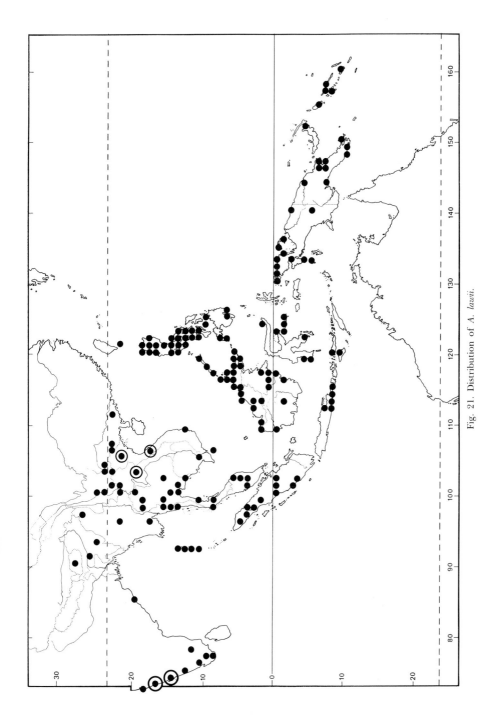

Fig. 21. Distribution of *A. lawii*.

hairs which usually have long arms and may densely cover the lower leaflet surface, but the distinction between this species and *Aglaia lawii* when the latter has stellate hairs in Borneo and the Philippine Islands is sometimes not very clear. In this treatment, any specimen which has some peltate scales is assigned to *A. lawii*. The fruit of *A. teysmanniana* is similar to *A. lawii* except for its indumentum of stellate hairs.

Biochemical analyses of seeds collected in Peninsular Malaysia have shown that the aril of *A. lawii* is lipid-rich (61% dry weight) while observations of vertebrates feeding on the seeds in Peninsular Malaysia have shown that seeds are swallowed whole and are probably dispersed by birds ranging in size from bulbuls to magpies and hornbills (Pannell & Kozioł, 1987). Some of these, especially hornbills, may occasionally disperse seeds over long distances.

16. Aglaia teysmanniana (Miq.) Miq., Ann. Mus. Bot. Lugd. Bat. 4: 48 (1868). C. de Candolle in A. & C. de Candolle, Monog. Phan. 1: 623 (1878); Pannell in Ng, Tree Flora of Malaya 4: 226 (1989).

[*Milnea montana* Teijsm. & Binn., Cat. Hort. Bog.: 211 (1866), ? non Jack, nom. nud.]

Amoora teysmanniana Miq., Fl. Ind. Bat. Suppl. 1: 196, 503 (1861). Lectotype (designated here): Sumatra, [S.], Lamp [Lampong], [River] Tarabangi, *Anon.* 4423 HB (U!; isolectotypes: BO!, L!).

Aglaia subgrisea Miq., Ann. Mus. Bot. Lugd. Bat. 4: 54 (1868). Lectotype (designated here): ? Java, '*Milnea montana* Jack', 1861, *Teijsmann* (U!; isolectotype: BO!).

Aglaia heptandra Koord & Valet. in Meded. 'S Lands Plantent 16: 132 (1896). Lectotype (designated here): Java, Djember, Poeger-Watangan, Res. Besoeki, fl., *Koorders* 4688β (BO!; isolectotypes: G!, K!, L!); Koorders & Valeton, Atlas der Baumarten von Java 1: t. 157 (1913); Backer and Bakhuizen, Fl. Java 2: 126 (1965).

Amoora stellata C.Y. Wu, Flora Yunnanica 1: 234 (1977). Type: [China], Yunnan, [Chin'ping], 360 m, fr., 29 May 1957, ? coll. 192 (KUN!).

Tree up to 15 m, with rounded crown. Bole up to 50 cm in circumference. Bark greyish-brown or pale brown with longitudinal cracks and lenticels; inner bark dark brown; sapwood pale yellowish-brown; latex white. Twigs slender, pale greyish-green with longitudinal wavy ridges and densely covered with pale yellowish-brown stellate hairs which have arms up to 1.5 mm long.

Leaves imparipinnate up to 60 cm long and 50 cm wide, obovate in outline; petiole up to 9.5 cm long, the petiole, rhachis and petiolules densely covered with stellate hairs like those on the twigs. Leaflets (1 –)5 – 9, usually subopposite, sometimes alternate, 4.8 – 25 cm long, 2 – 8.2 cm wide, usually elliptical, sometimes obovate, acuminate or shortly caudate at apex with the obtuse or acute acumen up to 15 mm long, cuneate or rounded at the asymmetrical base, densely covered with pale yellow stellate hairs on the midrib and numerous on the rest of the lower surface where they are often deciduous; veins 9 – 21 on each side of the midrib, ascending and markedly curved upwards near the margin, midrib prominent and lateral veins subprominent on lower surface; petiolules up to 10 mm on lateral leaflets, up to 20 mm on terminal leaflet.

Inflorescence up to 15 cm long and 15 cm wide; peduncle c. 10 mm, the peduncle, rhachis, branches and pedicels rather angular and longitudinally ridged with numerous pale yellowish-brown stellate scales or hairs with short arms. Flowers 1.5 – 2 mm in diameter, subglobose, fragrant; pedicels up to 2 mm. Calyx cup-shaped densely covered with scales like those on the branches on the outer surface, shallowly divided into 4 or 5 rounded lobes. Petals 3 – 5, yellow, obovate, glabrous, aestivation quincuncial. Staminal tube shorter than the corolla, cup-shaped with the apical margin incurved; anthers (6 or) 7(– 9), c. $^{1}/_{2}$ the length of the tube, ellipsoid, just protruding beyond the aperture. Ovary subglobose densely covered with stellate scales; stigma black when dry, glabrous, with three small apical lobes.

Infructescence with 1 – 4 fruits, up to 10 cm long and 6 cm wide; peduncle up to 6 cm, the peduncle, rhachis, branches and fruit-stalks with longitudinal ridges and hairs like the twigs. Fruits 1 – 2.2 cm long, 1.3 – 2 cm wide, pink, subglobose, dehiscing loculicidally into 3, pericarp with shallow longitudinal wrinkles, and densely covered with pale brown stellate scales, c. 2 mm thick, white, turning pink on exposure to air, containing some latex; innermost layer of the pericarp in each loculus a detachable membrane surrounding the seed. Loculi 3; seeds 1 – 3, (0.7 –)1.1 – 1.4 cm long, (0.4 –)0.6 – 0.9 cm wide, obovoid; aril orange or red, the edges almost or quite meeting on the antiraphe side.

DISTRIBUTION. China, Thailand, Peninsular Malaysia, Sumatra, Borneo, Philippine Islands, Java, Sulawesi, ? Maluku, ? New Guinea. Fig. 22.

ECOLOGY. Found in evergreen forest, primary forest and secondary forest; on limestone, sandy soil, clay, loam. Alt.: 3 to 1670 m. Rare to common.

USES. firewood (Borneo, Keningau).

VERNACULAR NAMES. Thailand: Rawngkhai, Mut Sang. Java: Langsat-loetoeng, Pantjal Kidang. Borneo: Bibilad (Malay); Malangsat (Bajau E.C.); Mumutah (Dusun). New Guinea: Boewei (Hattam); Herrib (Manikiong); Kamoentoeloe (Karoon); Manienif (Biak); Mewit (Mooi).

Representative specimens. THAILAND. S.W., Surat, Yanyao, c. 50 m, fl., 23 Feb. 1930, *Kerr* 18219 (C!, L!).

PENINSULAR MALAYSIA. Langkawi, P. Langgun, Telok Dalam, fr., 23 Feb. 1975, *van Balgooy* 2367 (AAU!, L!, NY!). Pahang: Taman Negara, Ulu Tembeling, N.W. of Tg Bungkal, *Whitmore* FRI 15373 (FRI!); Gg Tahan, *Symington* FMS 39581 (FRI!); Krau Game Reserve, K. Lompat, c. 50 m, fr., 21 Aug. 1979, *Pannell* 1404 (FHO!).

SARAWAK. Ulu Mayeng, Kakus [c. 2°N 113°E], c. 200 m, fl., 1 Aug. 1964, *Sibat ak Luang* S 21844 (FHO!)

SABAH. Mt Kinabalu, c. 1200 – 1500 m, fr., 4 Jan 1933, *J. & M.S. Clemens* 30717 (NY!). Ranau, Kg Pinawantai, SAN 76757 (L!).

PHILIPPINE ISLANDS. Luzon, Province of Bataan, Lamao River, fl., July 1904, *Ahern's collector* 150 (UC!).

The following specimens may also belong here, but they are more robust and have larger flowers than in the rest of this species and they may represent a distinct species:

PAPUA NEW GUINEA. Northern Division, Kokoda, fl., 25 Sep. 1951,

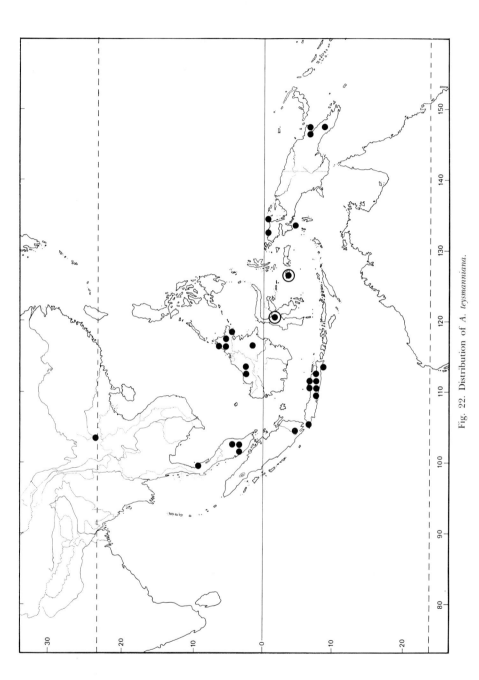

Fig. 22. Distribution of *A. teysmanniana*.

Womersley NGF 4494 (FHO!). New Britain, Gasmata, S.W. of Subdidi village, c. 180 m, fr., 7 April 1966, *Frodin* NGF 26564 (L!).

A. teysmanniana bears a superficial resemblance to *Dysoxylum grande* Hiern, but it is distinguished by its stellate hairs, whereas *Dysoxylum* has simple hairs. The hairs are pale brown and may be sparse or numerous; in the latter case, the arms of adjacent hairs overlap to form a continuous indumentum on the lower surface of the leaflets. The fruit is small, pink and dehiscent with 1 – 3 seeds, the seeds have a complete aril which has an orange skin. The stellate indumentum distinguishes *A. teysmanniana* from *A. lawii*.

Section **Aglaia**

Holotype species: *Aglaia odorata* Loureiro, Fl. Cochinch. 173 (1790).
Aglaia section *Hearnia* (von Muell.) Harms in Engl. & Prantl, Nat. Pflanzenfam. III 4: 298 (1896); Harms in Engl. & Prantl, Nat. Pflanzenfam., ed. 2, 19b1: 146 (1940).
Aglaia section *Euaglaia* Harms in Engl. & Prantl, Nat. Pflanzenfam. III 4: 298 (1896); Harms in Engl. & Prantl, Nat. Pflanzenfam., ed. 2, 19b1: 144 (1940).
Aglaia section *Neoaglaia* Harms in Engl. & Prantl, Nat. Pflanzenfam. III 4: 300 (1896); Harms in Engl. & Prantl, Nat. Pflanzenfam., ed. 2, 19b1: 146 (1940).
Aglaia section *Beddomea* (Hooker fil.) Harms in Engl. & Prantl, Nat. Pflanzenfam. III 4: 300 (1896).
Aglaia section *Macroaglaia* Harms in Engl. & Prantl, Nat. Pflanzenfam. III 4: 300 (1896).

Petals 5 (or 6); anthers 5(– 10); fruit indehiscent.

17. Aglaia grandis Korth. ex Miq., Ann. Mus. Bot. Lugd. Bat. 4: 56 (1868). Lectotype: Borneo [S., G. Sakoembang], fl., *Korthals* s.n. (U!; isolectotypes: BO!, K!, L!); C. de Candolle in A. & C. de Candolle, Monog. Phan. 1: 617 (1878); Pannell in Ng, Tree Flora of Malaya 4: 217 (1989).

Aglaia lanuginosa King in Jour. As. Soc. Bengal 64: 71 (1895). Syntypes: Malay Peninsula, Perak: Larut, Gg Booboo [Bubu], 300 – 600 ft [c. 90 – 180 m], March 1885, *King's Coll.* 7381 (K!, SING!); Larut, 500 – 800 ft [c. 1 50 – 250 m], June 1885, *King's Coll.* 7714 (CGE, K!); *Scortechini* 1682 (non vidi); Ridley, Fl. Malay Penins. 1: 407 (1922); Ridley, Fl. Malay Penins. 1: 407 (1922); Corner in Gard. Bull. Suppl. 1: 131 (1978).
Aglaia hemsleyi [*helmsleyi*] Koord. in Meded. 'S Lands Plantent. 19 : 383, 635 (1898). Lectotype (designated here): Insula Celebes [Sulawesi], Provincia Minahassa, Menado, 500 m, fl., 5 Mar. 1895, *Koorders* 17900β (BO!; isolectotypes: BO!, K!, L!, P!).
Merostela grandis (Korth. ex Miq.) Pierre, Fl. Forest. Cochinch. Fasc. 2 1, ante t. 334 (1 July 1895).
Merostela grandifolia Pierre, Fl. Forest. Cochinch. Fasc. 22, t. 342 (1 July 1896). Lectotype (designated here): ad montem Chiua Chiang, in prov. Bien Hoa, austro Cochinchina [Cambodia], fl., Sep. 1865, *Pierre* 1812 (P! fig. based on this collection; isolectotypes K!, L! [fl. only]).

Aglaia merostela Pellegrin in Lecomte, Fl. Gén. L'Indo-Chine, 1: 761 (1911), superfl. nom. illegit. pro *Merostela grandifolia*.
? *Aglaia bernardoi* Merrill in Philipp. Jour. Sci., Bot. 9: 302 (1914). Type: Philippine Islands, Luzon, *Bernado* For. Bur. 15205 (PNH†).
Aglaia stellatotomentosa [*stellato-tomentosa*] Merrill in Jour. Sci., Bot. 1914, 9: 535 (1915). Lectotype (designated here): Philippine Islands, Basilan, *Miranda* For. Bur. 20085 (PNH†, US!).
Aglaia perfulva Elmer in Leafl. Philipp. Bot. 9: 3302 (1937) sine diagn. lat. Type no. *Elmer* 13121 Philippine Islands, Island of Palawan, Province of Palawan, Brooks Point, Addison Peak, fr., March 1911 (A!, BM!, G!, GH!, K!, L!, NY!, US!).

Tree up to 27 m, sometimes small and unbranched. Bole up to 17 m, up to 75 cm in circumference. Bark smooth, grey, with shallow longitudinal fissures; inner bark brown or dark brown; sapwood pinkish-brown, brown or yellow; latex white. Twigs stout, up to 2 cm in diameter, with many leaf scars, densely covered with brown hairs which have a central rhachis and 2 – 4 whorls of arms radiating from it; apical bud up to 2.5 cm in diameter. Leaves in spirals towards the ends of the twigs where they are very close together and the expanded bases of the petioles overlap, imparipinnate, up to 135 (– 200) cm long and 80 cm wide, obovate in outline; petiole up to 20 cm long, up to 2 cm across at the base, the petiole, rhachis and petiolules clothed like the twigs; latex white. Leaflets 11 – 21 (– 25), the laterals subopposite, all 8.5 – 58 cm long, 4 – 18 cm wide, upper surface shiny, coriaceous, usually elliptical, sometimes obovate or oblong, acuminate at apex with the acute acumen up to 15 mm long, narrowed to a usually subcordate, asymmetrical base, sometimes cuneate, especially on the terminal leaflet, lower surface densely covered with pale brown hairs up to 1 mm long, like those on the twigs, with the surface visible between the hairs; veins 14 – 46 on each side of the midrib, ascending, markedly curved upwards near the margin and anastomosing, the midrib and lateral veins depressed on upper surface and prominent on lower surface, secondary veins visible on both surfaces; petiolules up to 20 mm.

Inflorescence up to 30 cm long and 15 cm wide, the final branches up to 20 mm long and densely packed with sessile flowers; peduncle up to 7 cm, the peduncle, rhachis and branches clothed like the twigs. Flowers up to 2 mm in diameter, depressed globose, fragrant. Calyx deeply divided into 5 narrow lobes, with hairs on the outer surface like those on the twigs. Petals 5, c. 1.5 mm long and 0.6 mm wide, glabrous, yellow, elliptical, obtuse at apex, aestivation quincuncial. Staminal tube c. 1 mm long, pale yellow, shorter than the corolla, subglobose with the aperture wide and deeply 5-lobed, glabrous; anthers 5, ovoid, $\frac{1}{2}$ to nearly as long as the tube and just protruding through the aperture. Ovary small, depressed-globose, with numerous stellate hairs; stigma c. $\frac{2}{3}$ length of the staminal tube, black and shiny, cylindrical, narrowed slightly to the obtuse apex and protruding through the aperture of the tube.

Infructescence up to 40 cm long and 24 cm wide, with c. 5 fruits. Fruits brown, up to 5 cm long and 4.5 cm in diameter, obovoid, sometimes with a small beak, with a thick indumentum of hairs up to 4 mm long like those on the twigs, fruit-stalks up to 2 cm. Fig. 23.

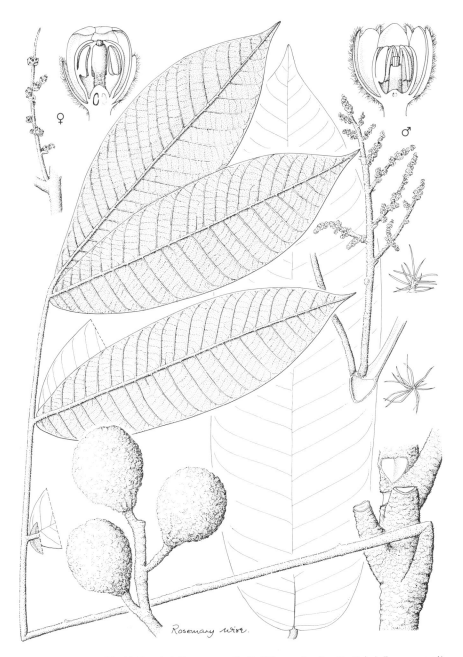

Fig. 23. *A. grandis*. Habit x½. Female inflorescence x½. Half flower, female x15. Male inflorescence x½. Half flower, male x15. Part of infructescence x½. Hairs x50.

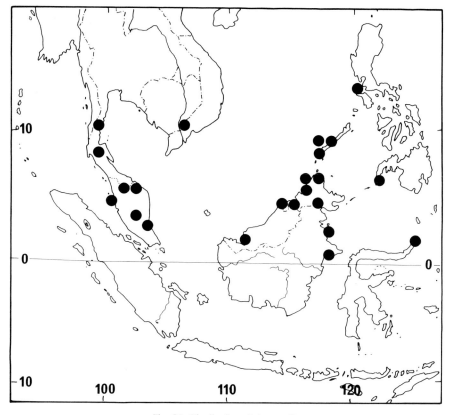

Fig. 24. Distribution of *A. grandis*.

DISTRIBUTION. Vietnam, Thailand, Peninsular Malaysia, Borneo, Philippine Islands, Sulawesi. Fig. 24.

ECOLOGY. Primary forest, sometimes on ultrabasic and limestone. Alt.: sea level to 1700 m.

VERNACULAR NAMES. Peninsular Malaysia: Pasak Lingga. Philippine Islands: Lambunau (Tagbanua). Sulawesi: Poerisihoela.

Representative specimens. THAILAND. fl., Sept. 1914, *Miss Collins* 402 (AAU!). Kaw Tao [10°06′ 99°50′], fl., Dec. 1926, *Kerr* 11202 (AAU!).

PENINSULAR MALAYSIA. Perak, K. Kangsar, Piah F.R., fr., 13 July 1967, *Kochummen* FRI 2436 (FRI!, K!, SING!). Negri Sembilan, Pasoh F.R., fr., 2 Sep. 1976, *Cheng* FRI 21579 (FRI!). Johore, mile 14 Mawai – Jemaluang road, fl., 23 April 1938, *Corner* SFN 34939 (K!).

SARAWAK. 1st Division, Semengoh F.R., st., 21 April 1974, *Mabberley* 1606 (FHO!). Baram District, Lobang Rusa, near Sungei Melinau Paku [4°02′N 114°49′E], c. 150 m, fr., 9 Feb. 1966, *Chew Wee-Lek* CWL 1029 (AAU!).

SABAH. Keningau District, nr Kg Kalumpan, Paling-paling hills, between Lagud and Sebrang, Tenom Valley and Kg Dalit, c. 300 m, fr., Oct. 1989, *Surat & Lamb* ALFB 335/89 (K!).

KALIMANTAN. Boeloengan, Kabirau, S. Bengaloen, c. 100 m, st., 29 July 1927, *Zwaan* 263, bb 11718 (L!).

The hairs of *A. grandis* have a central rhachis from which radiate 2 – 4 whorls of arms. The arms of the hairs on the twigs and fruits are long and the indumentum is dense forming a continuous, pale brown woolly covering. The hairs also densely cover the lower surface of the leaflets, but the surface is visible between the hairs. There are as many as 36 lateral veins on each side of the midrib. The twigs, petiole, rhachis and peduncle are terete.

A distinctive form of this species, which may merit recognition as a subspecies, occurs on ultrabasic rock in the Philippines; this is *A. perfulva* of Elmer. It has smaller leaflets which have a reddish-brown indumentum on the lower surface and recurved margin; it is not formally described and named here, because no distinguishing floral or fruit character has been found in the specimens examined.

Representative specimen. PHILIPPINES. Palawan, Narra, Victoria Peaks, Trident Mining Company, 490 – 590 m, yg fl. & fr., 19 May 1984, *Ridsdale* SMHI 1784 (FHO!, L!).

18. Aglaia ramotricha C.M. Pannell spec. nova *Aglaiae grandi* Miq. similis, cuius pilos habet rhachide centrali praeditos unde complures verticilli brachiorum radiant; sed ab ea indumento paginae inferioris foliolorum multo minus denso et reticulatione subprominente paginarum ambarum differt. Holotype: Malaysia, Sabah, Kota Belud District, canyon of Penataran River S. of Melangkap Tamis on N.W. side of Mt Kinabalu [6°08′N 116°30′E], 450 – 500 m, fl., 8 April 1984, *Beaman* 9308 (FHO!, 3 sheets).

Tree 6 – 15(– 20) m. Bole up to 20 cm in diameter, with small buttresses upwards to 60 cm and outwards to 30 cm. Outer bark brown; inner bark pale brown or pink; sapwood yellow or brown. Twigs stout, 2 – 4 cm in diameter, greyish-brown, longitudinally wrinkled, densely covered with compact reddish-brown stellate hairs or brown hairs which have a central rhachis and 2 – 4 whorls of arms radiating from it.

Leaves imparipinnate, up to 155 cm long and 100 cm wide; peduncle up to 46 cm, the base of the peduncle up to 2.5 cm across on the abaxial side and up to 1.5 cm across on the adaxial side, the adaxial side deeply channelled, the peduncle, rhachis and petiolules densely covered with hairs like those on the twigs or densely covered with pale brown stellate hairs or scales. Leaflets 15 – 21, the laterals subopposite or alternate, all 8.5 – 68 cm long, 3 – 22 cm wide, usually oblong, sometimes elliptical, broadly acuminate at apex, the obtuse or acute acumen 0.5 – 2 cm long, rounded or sometimes subcordate or cuneate at base, with hairs like those on the twigs numerous on the lower surface of the leaflets when young, with reddish-brown stellate hairs or pale brown or nearly white stellate hairs and scales numerous on the midrib and veins on the lower surface and few on the rest of that surface when mature; veins 16 – 32 on each side of the midrib, ascending, curved upwards near the margin and anastomosing, the midrib prominent and the lateral veins subprominent on the lower surface, the reticulation subprominent on both surfaces; petiolules up to 1 cm.

Inflorescence up to 32 cm long and 32 cm wide; peduncle up to 5.5 cm, the peduncle, rhachis and branches densely covered with hairs like those on the twigs or the distal branches densely covered with golden brown stellate hairs. Flowers (not known whether male or female) 1.6 – 3.5 mm long, 1.6 – 2.5 mm wide, ellipsoid or subglobose, sessile and clumped around the distal branches of the sparsely branched inflorescence; bracteoles 2, 12 – 14 mm long and 9 – 12 mm wide, ovate; the bracteoles and calyx with few pale brown stellate scales. Calyx 1 – 1.5 mm long, divided almost to the base into 5 subrotund, overlapping lobes, aestivation quincuncial, margins ciliate. Corolla 1.3 – 3 mm long, 1.4 – 2.5 mm wide, white, divided to half way or almost to the base into 5 broad ovate lobes; aestivation quincuncial. Staminal tube 1 – 2.5 mm long, 1.1 – 1.8 mm wide, yellow, ellipsoid or obovoid, aperture 0.4 – 0.6 mm in diameter, entire; anthers 5, 0.7 – 1.2 mm long, 0.5 – 0.8 mm wide, ellipsoid or ovoid, occupying the middle or upper part of the tube and either included or just protruding through the aperture. Ovary 0.1 – 0.5 mm long, 0.6 – 0.7 mm wide, depressed globose, densely covered with pale stellate scales or hairs; loculi 2 (or 3); stigma 0.5 – 1.1 mm long, 0.3 – 0.5 mm wide, the apex expanded to 0.4 – 0.5 mm wide, columnar, shiny dark brown, with 5 – 10 longitudinal ribs and 3 – 5 apical lobes, the apex usually occupying the aperture but sometimes shorter than the tube and not reaching the aperture.

Infructescence up to 45 cm long and 50 cm wide, peduncle 12 – 18 cm, the peduncle, rhachis and branches densely covered with hairs like those on the twigs and with some pale brown stellate hairs interspersed. Fruits reddish-yellow, c. 3.5 cm in diameter, subglobose or ellipsoid, with numerous to densely packed compact reddish-brown hairs or pale brown stellate hairs, indehiscent; loculi 2 or 3, each containing 1 seed; seeds up to 2 cm wide and 1.2 cm thick.

DISTRIBUTION. Borneo only. Fig. 25.

ECOLOGY. Understorey tree in primary forest on limestone. Alt.: 35 – 1600 m.

VERNACULAR NAMES. Borneo: Lantupak (Dusun Kinabatangan).

Representative specimens. SABAH. Mount Kinabalu: Dallas, 900 m, fl., 30 Oct. 1931, *J. & M.S. Clemens* 26870 (K!); Dallas, 900 m, fr., 6 Nov. – 16 Dec. 1932, *J. & M.S. Clemens* 26968 (K!); Tenompok, 1525 m, fr., March 1932, *J. & M.S. Clemens* 28816 (K!); Penibukan, 1200 – 1525 m, fl., 7 Feb. 1933, *J. & M.S. Clemens* 31468 (K!).

SARAWAK. 1st Division, Kuching District: Tiang Bekap, Mount Mentawa, below 75 m, fl., 25 Aug. 1963, *Chew* CWL 651 (K!); Bt Mentawa, Mile 34, Padawan road, yg fr., 24 Feb. 1981, *Mamit* S 41066 (K!). Kapit, Upper Rejang River, st., 1929, *J. & M.S. Clemens* 21046 (K!).

Aglaia ramotricha resembles *A. grandis* in its hairs which have a central rhachis and several whorls of arms radiating from it, but it differs from *A. grandis* in having a much less dense indumentum on the lower surfaces of the leaflets and subprominent reticulation on both surfaces.

19. pachyphylla

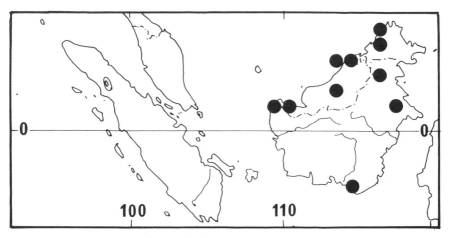

Fig. 25. Distribution of *A. ramotricha*.

19. Aglaia pachyphylla Miq., Ann. Mus. Bot. Lugd. Bat. 4: 57 (1868). Lectotype (designated here): Sumatra [occidentali], yg fl., [*Korthals*] s.n. (U!; isolectotype: L!); C. de Candolle in A. & C. de Candolle, Monog. Phan. 1: 617 (1878).

Aglaia barbatula Koord. & Valet. in Meded. 'S Lands Plantent. 16: 167 (1896). Lectotype (designated here): Java, Bantam-Tjimara-Oedjoeng, ± 10 – 200 m, st., 18 July 1892, *Koorders* 4706β (BO!; isolectotypes: G!, L!,); Koorders & Valeton, Atlas der Baumarten von Java 1: t. 153 (1913); Backer and Bakhuizen, Fl. Java 2: 126 (1965); Pannell in Ng, Tree Flora of Malaya 4: 213 (1989).
? *Aglaia clarkii* Merrill in Philipp. Gov. Lab. Bur. Bull. 29: 21 (1905). Syntypes: Philippine Islands, *Whitford* 477 (PNH†); *Whitford* 104 (PNH†); *Whitford* 215 (PNH†); *Merrill* 3724 (PNH†); For. Bur. 1195 (PNH†).
Aglaia megistocarpa Merrill in Univ. Calif. Publ. Bot. 15: 130 (1929). Lectotype (designated here): British North Borneo [Sabah], Tawao, Elphinstone Province, fr., Oct. 1922 – March 1923, *Elmer* 20930 (UC!; isolectotypes: A!, BM!, BO!, G!, GH!, L!, M, NY!, P!, SING!, UC!).

Tree up to 43 m, sometimes flowering at 1.8 m, with a few ascending branches which have terminal clusters of leaves forming a small, open crown. Bole up to 20 m, up to 65 cm in diameter, sometimes with buttresses upwards up to 3 m high, outwards to 1 m and up to 22 cm thick. Outer bark brown, greyish-brown or greenish-grey, with large corky lenticels or with pits and regularly longitudinal fissures, the fissures narrow, deep and c. 15 cm long, the intervals flat and somewhat scaly; inner bark dark brown or pale yellowish-brown, thick, firm and finely fibrous, sapwood pinkish- brown, pale brown or pale yellow; heartwood brown; latex white, when present. Twigs stout, up to 2.5 cm in diameter, dark brown, with longitudinally wavy ridges, large prominent leaf scars and densely covered with reddish-brown stellate hairs or hairs which have a long central rhachis and many arms.

Leaves up to at least 135 cm long and 60 cm wide, oblong in outline, imparipinnate, in spirals particularly towards the ends of the twigs where they are very close together, the petiole bases crowded together; petiole up to 30 cm, with a groove on the adaxial side, the petiole, rhachis and petiolules angular, with longitudinal channels and densely covered with hairs like those on the twigs; latex white. Leaflets 13 – 23, the laterals subopposite, 10 – 52 cm long, 3 – 11 cm wide, upper surface shiny, usually oblong, sometimes lanceolate or oblanceolate, acuminate at apex with the acute or obtuse acumen up to 15 mm long, usually rounded and subcordate but sometimes cuneate, at the asymmetrical base, upper surface with numerous minute pits, lower surface densely covered with pale reddish-brown hairs which have a central rhachis and 2 – 4 whorls of arms radiating from it, the surface of the leaflet barely visible between the hairs or densely covered with pale or dark brown stellate hairs or scales which sometimes have scattered darker hairs in between, sometimes densely covering the upper surface in young leaves, sometimes deciduous on lower surface of old leaves; veins (15 –)20 – 45 on each side of the midrib, ascending, markedly curved upwards near the margin and usually anastomosing, the midrib and lateral veins depressed on upper surface but prominent on lower surface, secondary veins subprominent or visible on both surfaces; petiolules up to 10 mm on lateral leaflets, up to 25 mm on terminal leaflet.

Inflorescence up to 45 cm long and 60 cm wide, flowers sessile on the final branches and often clumped together; peduncle up to 10 cm, the èpeduncle, rhachis, and branches angular, channelled and densely covered with hairs like those on the twigs. Flowers subglobose, up to 2 mm in diameter, sessile. Calyx almost as long as the corolla densely covered with pale brown stellate hairs on the outer surface, with 5 rounded lobes. Petals 5, rotund or elliptical, aestivation quincuncial. Staminal tube subglobose, thick and fleshy, deeply 5-lobed; anthers 5, as long as the tube, ovoid and curved with the tube, with tufts of white stellate hairs at the apex and at the base of each locule, those at the apex filling the aperture of the tube. Ovary depressed globose; stigma c. $^2/_3$ the length of the staminal tube, cylindrical, truncate at the apex.

Infructescence with 1 – 15 fruits, up to 20 cm long and 15 cm wide; fruit- stalks up to 15 mm. Fruit up to 8 cm in diameter, obovoid or subglobose, greyish-green when young, brown when mature, densely covered with hairs like those on the twigs, glabrescent; the pericarp 3 – 5 mm thick, with white latex. Loculi 2(– 4), each containing 1 or 2 seeds; seed completely surrounded by a fleshy, translucent aril.

DISTRIBUTION. Thailand, Peninsular Malaysia, Sumatra, Borneo, Philippine Islands, Java, Sulawesi. Fig. 26.

ECOLOGY. Found in primary forest and secondary swamp forest, and along margins of forest. Present on limestone, clay and sandstone. Alt.: sea level to 1350 m. Frequent along rivers to very common. USES. Wood is hard and said to be durable; it is used for planks and temporary construction.

VERNACULAR NAMES. Peninsular Malaysia: Semeliang. Borneo:

19. pachyphylla

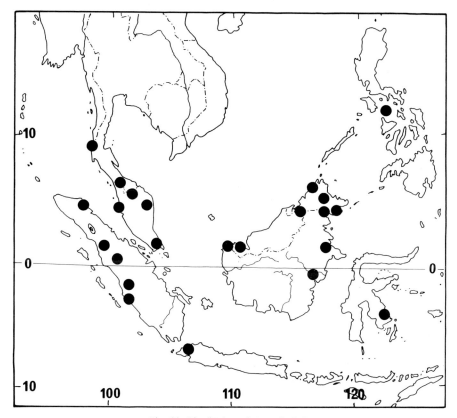

Fig. 26. Distribution of *A. pachyphylla*.

Bindjai Babi (Bilajang), Mamprarangan (Basak-Dyak), Singkok (Kutai Malay).

Representative specimens. THAILAND. Kampuam, Ranawng, c. 50 m, fr., 5 Feb 1929, *Kerr* 17016 (C!, K!).

PENINSULAR MALAYSIA. Kedah, C., Bt. Enggang, F.R., c. 60 m, fl., 25 Sep. 1903, *Pennington* 7838 (FHO!, FRI!). Perak, Ulu Sg. Perak, Chegar Daaras, nr Fort Tapong, c. 275 m, yg fr., 15 Jan. 1971, *Whitmore* FRI 15748 (FRI!, K!).

SUMATRA. Sumatra Barat: eastern foothills of Barisan Mountains, Sungaidareh, Sg Mimpi, Ulu Batang Hari, P.T. Pasar Besar logging concession [1°01:8:S 101°27:8:E], 150 m, st., 15 June 1983, *Pannell* 1902 (FHO!); 30 km along road from Tapan to Bengkulu, S. of Lunang [2°15:8:S 101°10:8:E], c. 50 m, st., 2 Oct. 1983, *Pannell* 2076 (FHO!).

SARAWAK. 1st Division, foot of Gunong Lundu, c. 60 m, fl. & fr., 19 Nov. 1963, *Pennington* 7992 (FHO!).

SABAH. Sepilok F.R. along boundary (E.W.) between compts 4 & 5, c. 15 m, fr., 22 Oct. 1963, *Pennington* 7916 (FHO!).

PHILIPPINES ISLANDS. Catanduanes, Mt Tagmasuso, c. 900 m, fr., 11 Aug. 1928, *Ramos & Edaño* 75147 (NY!). Island of Burias, fl., June 1904, *Clark* 1717 (US!).

This species resembles *A. grandis*, but it is usually a larger tree with an open crown like that of *A. rubiginosa*. The hairs are shorter than in *A. grandis* and cover the lower surface of the leaflets so densely that the lower surface is not or barely visible. There are usually more lateral veins on each side of the midrib than in *A. grandis*. The twigs, petiole, rhachis and peduncle are angular and deeply channelled. When the hairs are short, it is sometimes similar in appearance to *A. eximia*, but it may be distinguished by the shape of the leaflet, greater number of veins and the prominence of the reticulation on the dark, shiny upper surface. When the indumentum is dark, it may resemble *A. rubiginosa*, but the flowers are pentamerous in *A. pachyphylla* and trimerous in *A. rubiginosa*.

20. Aglaia bourdillonii Gamble in Kew Bull. 1915: 346 (1915). Lectotype (designated here): India, Chimunjie, 4000 ft [c. 1200 m], fl., 5 April 1895, *Bourdillon* 565 (K!). Illustration: Beddome, Fl. Sylvatica t. 130B (1871) as *A. roxburghiana;* Bombay N.H. 78: 425 (1981).

Aglaia elaeagnoidea (A. Juss.) Benth. var. *bourdillonii* (Gamble) K.K.N. Nair in Jour. Bombay Natural History Society 78: 426 (1981).

Tree of 'medium size'. Twigs densely covered with large orange-brown scales which have a fimbriate margin. Leaves 9 – 22 cm long, 12 – 20 cm wide; petiole 2.5 – 8 cm, the petiole, rhachis, and petiolules longitudinally wrinkled and densely covered with scales like those on the twigs. Leaflets 5 – 7, 3.5 – 10 cm long, 1 – 3.5 cm wide, coriaceous, recurved at the margin, obovate or elliptical, acuminate at the apex, the acumen obtuse and 2 – 5 mm long, cuneate at the asymmetrical base, with pale orange-brown peltate scales which have a longer fimbriate margin than those on the twigs numerous on the upper surface, glabrescent, leaving numerous dark reddish-brown pits, with scales like those on the twigs densely covering the lower surface, but occasionally deciduous; veins 7 – 12 on each side of the midrib, ascending, curved upwards near the margin but not anastomosing, midrib and lateral veins depressed above, the midrib prominent and lateral veins subprominent below; petiolules 3 – 10 mm on lateral leaflets, up to 20 mm on the terminal leaflet.

Inflorescence 11 – 28 cm long, 9 – 22 cm wide; peduncle 1 – 5.5 cm, the peduncle, rhachis and petiolules with numerous to densely covered with reddish-brown scales which have a fimbriate margin and sometimes have a dark blackish-brown centre like those on the twigs. Flowers 1.5 – 1.8 mm long, 1.5 – 1.8 mm wide, subglobose; pedicel 1 – 1.5 mm. Calyx 0.3 mm long, cup-shaped, divided into 5 broad shallow rounded lobes which have ciliate margins, with few scales like those on the twigs. Petals 5, aestivation quincuncial. Staminal tube 1 – 1.3 mm long, 1.3 – 1.5 mm wide, cup-shaped with an aperture 1 – 1.2 mm across; anthers 5, 0.7 – 0.5 mm long, 0.4 – 0.5 mm wide, inserted just inside or half way down the tube, with a few simple hairs and sometimes with pale yellow margins. Ovary 0.3 mm high, 0.3 mm wide, depressed-globose, densely covered with peltate scales; locule 1, containing 1 ovule; stigma 0.3 – 0.5 mm long, 0.3 – 0.4 mm wide, ovoid, densely covered with minute pale brown papillae.

Infructescence 6 – 8 cm long, 4 – 6 cm wide with c. 6 fruits; peduncle 1 – 3 cm. Fruits 2 – 2.2 cm long and 1.5 – 1.7 cm wide, obovoid, narrowed at the base to a short broad stipe up to 4 mm long, densely covered with scales like those on the twigs on the outside; fruit-stalk up to 5 mm. Loculi 2, each containing 1 seed.

DISTRIBUTION. S. India only.
ECOLOGY. Alt.: up to 1200 m.
Representative specimens. INDIA. S., Neelgherry [according to Gamble (1915), this locality may be in error for Attraymallay Ghats, Travancore, Tinevelly], fl., 1873, *Beddome* s.n. (K!). Travancore, fl., March 1872, *Beddome* 246 (K!); Travancore, April 1896, *Beddome* 811 (K!). Travancore hills, c. 1200 m, 135 (BO!); Travancore, Mutthu Kuli Vagal, c. 1000 m, fl. & fr., 24 March 1896, *Bourdillon* 811 (K!). fl., April 1896, *Bourdillon* 790 (K!). Madras, Malabar, fl., 1872, *Beddome* s.n. (K!). Madras Presidency, *Beddome* 359 (BO!);

Beddome appears to have confused this species with his *Aglaia minutiflora* var. *travancorica* (= *A. tomentosa)*, to which it bears a superficial resemblance in its dense orange-brown indumentum. However, the indumentum in *Aglaia minutiflora* var. *travancorica* is of stellate hairs and that in *A. bourdillonii* is of peltate scales. In his Fl. Sylvatica t. 130B (1871), Beddome refers to this species as *A. roxburghiana,* which also has peltate scales but they are less numerous on the lower leaflet surface than in *A. bourdillonii.*

The lower surface of the leaflets of *A. bourdillonii* is densely covered with peltate scales which have a longer fringe and paler centre than those of *A. apiocarpa;* the veins of *A. bourdillonii* may also be more prominent and the leaflet margins are recurved. See also the note under *A. apiocarpa.*

21. Aglaia eximia Miq., Fl. Ind. Bat. Suppl. 1: 197, 506 (1861). Lectotype (designated here): [Sumatra, Palembang], Oegan-oeloe, [*Teysmann*] HB 4049 (U!; isolectotypes: K!, L!, U!); Pannell in Ng, Tree Flora of Malaya 4: 215 (1989).

? *Aglaia ancolana* Miq., Fl. Ind. Bat. Suppl. 1: 506 (1861). Type: Sumatra, [bor. in prov. Angkolae sup. regione sylvatica, 1 – 3000 ft, (J)] (non vidi).

Aglaia argentea var. *eximia* (Miq.) Miq., Ann. Mus. Bot. Lugd. Bat. 4: 55 (1868); King in Jour. As. Soc. Bengal 64 : 70 (1895); Ridley, Fl. Malay Penins. 1: 405 (1922).

Aglaia argentea Blume var. *latifolia* Miq., Ann. Mus. Bot. Lugd. Bat. 4: 55 (1868). Lectotype (designated here): Sumatra, Doekoe, *Korthals* (L!; isolectotype: K!);

Aglaia argentea Blume var. *stellatipilosa* [*stellati-pilosa*] Adelb., Blumea 6: 321 (1 June 1948). Holotype: Java, Noesa Kambangan, Limoes Boentoe, fl., [18 Nov. 1907], *Amdjah* 210 (L!).

Aglaia argentea sensu Corner in Gard. Bull. Suppl. 1: 131 (1978), et mult. auctt. non Blume (1825).

Tree up to 15 m, with a rounded crown. Bole up to 10 m. Outer bark smooth, greyish-brown with some lenticels in longitudinal rows;

inner bark brown or pale brown; sapwood pale yellow or pale brown, sometimes tinged with pink; latex white. Twigs stout, dark brown with shallow longitudinal wavy ridges and densely covered with brown stellate scales.

Leaves imparipinnate, at least up to 135 cm long and 50 cm wide, narrowly elliptical in outline; petiole up to 25 cm long, the petiole, rhachis and petiolules with scales like the twigs. Leaflets 11 – 21, the laterals subopposite, 6 – 42 cm long and 2 – 11 cm wide, the terminal leaflet up to 58 cm long and up to 16 cm wide, all usually oblong, sometimes elliptical, acuminate at apex with the obtuse or acute acumen up to 15 mm long, usually rounded at the asymmetrical subcordate base but the terminal leaflet usually cuneate, lower surface thickly covered with white and pale brown stellate scales, often with some brown stellate hairs interspersed; veins 10 – 22(– 29) on each side of the midrib, ascending and markedly curved upwards near the margin, midrib and lateral veins prominent on lower surface; petiolules up to 5 mm on lateral leaflets, up to 15(– 50) mm on terminal leaflet.

Inflorescence up to 40 cm long and 28 cm wide, peduncle up to 10 cm, peduncle, rhachis and branches with surface and indumentum like the twigs. Flowers sessile, subglobose, up to 2 mm in diameter. Calyx c. $\frac{1}{2}$ the length of the corolla, densely covered with brown stellate hairs, deeply divided into 5 rounded lobes. Petals 5, white or yellow, subrotund or elliptical, aestivation quincuncial. Staminal tube sometimes shallowly cup-shaped with the rim incurved, usually ellipsoid with the aperture up to 0.5 mm in diameter and obscurely 5-lobed; anthers almost as long as the tube, ovoid, inserted near the bottom of the tube, usually included but sometimes protruding through the pore. Ovary depressed-globose; stigma c. $\frac{3}{4}$ the length of the staminal tube, cylindrical or ovoid, truncate at apex.

Infructescence up to 32 cm long and 16 cm wide; peduncle up to 10 cm. Fruits up to 3 cm long, 2.2 cm wide, ellipsoid or subglobose, grey or brown; pericarp thin, densely covered with reddish-brown stellate scales or peltate scales which have a fimbriate margin on the outside, containing white latex. Loculi 1 – 3, each containing 1 seed which has a thin aril; the aril white, pale yellow or pink. Fig. 27.

DISTRIBUTION. Thailand, Peninsula Malaysia, Sumatra, Philippine Islands, Java, Sulawesi, Maluku. Fig. 28.

ECOLOGY. Found in primary forest, secondary forest, lowland and montane forest, sometimes on limestone. Small black ants found inside hollow petioles in Sumatra. Alt.: sea level – 2000 m. Often common.

USES. Aril edible when taken in moderation.

VERNACULAR NAMES. Peninsular Malaysia: Mehabok (Temuan); Memberas, Pasak. Sumatra: Bantana Boeloeh, Baur, Kajoe Sepanas.

Representative specimens. BURMA. Tavoy District, Zimba Valley, fr., 22 Nov. 1924, *Parker* 2256 (K!).

THAILAND. Kwae Noi Basin, near Koeng Chadavill in Ran Ti River Valley, c. 200 m ♂ fl., 30 May 1946, *Kostermans* Exp. no. 775 (L!). Peninsular, Khao Chong [7°32'N 99°45'E], fl., 10 Oct. 1970, *Charoenphol, Kai Larsen & Warncke* 3568 (AAU!).

PENINSULAR MALAYSIA. Kedah, Gunong Jerai F.R., Compt. 14, c.

Fig. 27. *A. argentea* (top left). Leaf x¹/₂. Inflorescence x¹/₂. Half flower x7. Fruits x¹/₂. Peltate scale x60. *A. eximia* (bottom right). Part of leaf and part of inflorescence x¹/₂. Half flower x12. Stellate scale x70. (Reproduced from Pannell (1989) with permission from the Forest Research Institute Malaysia).

Section Aglaia

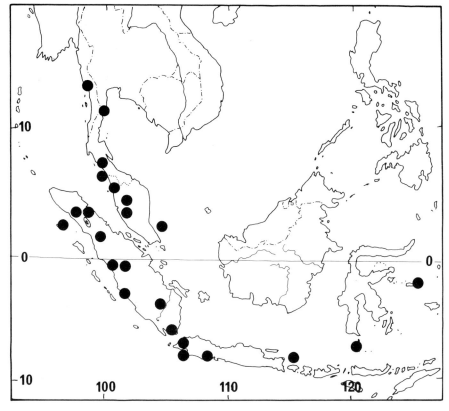

Fig. 28. Distribution of A. *eximia*.

90 m, fr., 27 Sep. 1963, *Pennington* 7850 (FHO!). Kelantan, Sg. Lebir, Eastern bank opposite Jentah, 60 m, fl., 28 April 1976, *Stone & Sidek* 12578 (FRI!). Selangor, Ulu Gombak F.R., c. 450 m, fl., 12 Sep. 1963, *Pennington* 7811 (FHO!).

SUMATRA. Aceh, Simeulue Barat, Sg. Laouloh [2°50′N 95°52′E], ± sea level, fl., 30 July 1983, *Pannell* 1998 (FHO!).

SULAWESI. Saleier Island [Salajar]: c. 200 m., fl., 20 May 1913, *Docters van Leeuwen* 1667 (BO! U!); yg fr., *Teysmann* HB 13854 (U!). Menado, Subdivision Poso, between bivouac I and Borone, top of watershed 1.5 m, st., 8 Sep. 1938, *Eyma* 3716 (L! U!).

MALUKU. Soela Mangoli, Mangoli, (Soela Eil.), Kampong Mangoli, Oostelijke oever van de Waj Mana, 30 m, st., 21 Sep. 1939, *Bloembergen* 4643 (L!, NY!).

The lower surface of the leaflets is densely covered with white or pale brown stellate scales. In the density of the indumentum, A. *eximia* resembles A. *argentea*, but in A. *argentea* the indumentum is of peltate scales. A. *eximia* sometimes flowers as a small tree.

22. Aglaia argentea Blume, Bijdr.: 170 (1825). Lectotype (designated here): Indonesia. Java, [Mt Salak], *Herb. Blume* s.n. (L!). Miq., Fl. Ind . Bat. Suppl. 1: 543 (1861). Miq., Ann. Mus. Bot. Lugd. Bat. 4: 54 (1868); King in Jour. As. Soc. Bengal 64: 70 (1895); Koorders & Valeton, Atlas der Baumarten von Java, 1: t. 151 (1913); Ridley, Fl. Malay Penins. 1: 405 (1922); C. de Candolle in A. & C. de Candolle, Monog. Phan. 1: 618 (1878); I.H. Burkill, Dictionary of Economic Products of the Malay Peninsula 1: 73 (1935) Backer & Bakhuizen, Fl. Java 2: 129 (1965); Pannell in Ng, Tree Flora of Malaya 4: 211 (1989).

[? *Milnea argentea* Reinw. in Gewassen Hort. Buitenz. 71 (1823), nom. nud.]
Aglaia hypoleuca Miq., Fl. Ind. Bat. Suppl. 1: 197, 507 (1861). Lectotype (designated here): [Sumatra] Priaman, [*Diepenhorst*] 699 HB (U!).
Aglaia speciosa Teijsm. & Binn., Cat. Hort. Bogor.: 211 (1866), non Blume (1825).
Aglaia argentea Blume var. *angustata* Miq., Ann. Mus. Bot Lugd. Bat. 4: 55 (1868). Lectotype (designated here): Java, [W.], Pengalengan [mountain], 4000 ft [c. 1200 m], fr., Dec., *Junghuhn* 44 (L!).
Aglaia argentea Blume var *borneensis* Miq., Ann. Mus. Bot. Lugd Bat . 4: 55 (1868). Lectotype (designated here): Borneo, G. Bahay, *Korthals* s.n. (L!).
? *Aglaia argentea* Blume var. *microphylla* Miq., Ann. Mus. Bot. Lugd . Bat. 4: 55 (1868). Type: Java, Cult. Hort. Bogor (non vidi).
Aglaia argentea Blume var. *superba* Miq., Ann. Mus. Bot Lugd. Bat. 4: 55 (1868). Lectotype (designated here): New Guinea, *Zippelius* 181 (L!; isolectotypes: BO!, U!).
Aglaia argentea Blume var. *hypoleuca* (Miq.) Miq., Ann. Mus. Bot. Lugd. Bat. 4: 55 (1868).
Aglaia argentea Blume var. *curtisii* King in Jour. As. Soc. Beng. **64**: 71 (1895). Syntypes: Malay Peninsula, Penang, *Curtis* 2287 (K!); Perak, *King's Coll.* 8239 (G!).
Aglaia argentea Blume var. *cordulata* C. de Candolle in A. & C. de Candolle, Monog. Phan. 1: 618 (1878). Holotype: Java, *de Vriese* (K!).
Aglaia argentea Blume var. *multijuga* Koord. et Val. in Meded. 'S Lands Plantent. 16: 165 (1896). Syntypes: Java: *Koorders* 4777β (fr. only) (G!, K!) & *Koorders* 4711β (K!, L!) & *Koorders* 4770β (L!) & *Koorders* 4767β (L!).
Aglaia argentea Blume var. *splendens* Koord. & Val. in Meded. 'S Lands Plantent. 16: 166 (1896). Lectotype (designated here): Cult. Hort. Bog. III-B-34 (L!).
Aglaia splendens (Koord. & Valet.) Koord. & Valet. in Ic. Bogor. t. 14 (1897). [*Aglaia javanica* Koord. & Valet. ex Koord. in Meded. 'S Lands Plantent. 19: 381, (1898) nom. nud.]
Aglaia bauerleni C. de Candolle in Bull. Herb. Boiss. Sér. II. 3: 175 (1903). Holotype: New Guinea, Strickland River, fr., *Bauerlen* 36 in h. Inst. Phyt. Melb. (MEL; isotype: G!).
Aglaia multifoliola Merrill in Philipp. Jour. Sci., Bot. 1914, 9: 534 (1915). Syntypes: Philippine Islands, Mindanao, District of Zamboanga, Port Banga, Jan. 1908, *Whitford & Hutchinson* 9208 (PNH†); Basilan, Jan.

1904 *Hallier* s.n. (PNH†); Basilan, 30 Sep. 1912, *Miranda* For. Bur. 18964 (BM!).

Aglaia discolor Merrill in Univ. Calif. Publ. Bot. 15: 130 (1929). Lectotype (designated here): British North Borneo, Elphinstone Province, Tawao, Oct. 1922 – Oct. 1923, *Elmer* 21029 (UC!; isolectotypes: A!, BM!, BO!, G!, GH!, L!, MICH!, NY, SING!, U!).

Tree up to 30 m, sometimes flowering and fruiting at 4.5 m, with a dense rounded crown. Bole up to 18 m, up to 60 cm in diameter; with buttresses upwards up to 1 m and outwards up to 1 m, c. 4 cm thick. Bark smooth, brown or greyish-green, longitudinally lenticellate; inner bark white, yellow or brown; sapwood pale brown, brown or reddish-brown; latex white, when present. Twigs stout, surface pale brown with longitudinal wavy ridges, thickly covered with pale brown or white peltate scales which have a darker centre.

Leaves in spirals particularly towards the ends of the twigs where they are close together, imparipinnate, 17 – 112 cm long and 14 – 75 cm wide, obovate in outline; petiole 4 – 41 cm, the petiole, rachis and petiolules ridged and with indumentum like the twigs. Leaflets 9 – 19, the laterals subopposite, all 4.5 – 30(– 33) cm long, 1.5 – 11(– 14) cm wide, often shiny above, subcoriaceous, elliptical or oblong, acuminate at apex with the acute acumen up to 10 mm long, usually rounded to an asymmetrical subcordate base, sometimes cuneate, especially on the terminal leaflet, both surfaces covered with peltate scales when young, when mature upper surface glabrous, the lower surface thickly covered with white peltate scales with few to numerous brown peltate scales interspersed, both types of scales often have a shortly fimbriate margin; veins 11 – 25 on each side of the midrib, ascending and becoming indistinct towards the margin, midrib and lateral veins depressed on upper surface, midrib prominent and lateral veins subprominent on lower surface; petiolules 5 – 10 mm on lateral leaflets, 10 – 120 mm on terminal leaflets.

Inflorescence up to 60 cm long and 60 cm wide, in the axils of the leaves near the apex of the shoot; peduncle up to 15 cm, the peduncle, rachis and branches with surface and indumentum like the twigs. Male and female flowers similar, 2 – 4 mm long and 1.5 – 2.5 mm wide, ellipsoid, sessile or occasionally with pedicels up to 2.5 mm long. Calyx c. $^1/_2$ the length of the corolla, thickly covered with scales like those on the leaves, deeply divided into 5 rounded lobes. Petals 5, white or yellow, elliptical with some scales on the outside when young, aestivation quincuncial. Staminal tube shorter than the corolla, obovoid, aperture 0.5 – 0.6 mm across with an entire or shallowly 5-lobed margin, anthers 5, c. 0.8 mm long and 0.6 mm wide, ovoid, included or just protruding. Ovary c. 0.5 mm long and 0.5 mm wide, ellipsoid or ovoid, densely covered with peltate or stellate scales; locules 2 or 3, each containing one ovule; stigma c. 0.5 mm long and 0.4 mm wide, subglobose or ovoid (see Plates 9 and 10), longitudinally ridged and with 2 or 3 apical lobes; ovary and stigma together up to $^3/_4$ the length of the staminal tube.

Infructescence 20 – 50 cm long, 25 – 35 cm wide; peduncle c. 14 cm, the peduncle, rachis and branches with indumentum like the twigs. Fruits 3 – 3.5 cm long and 2 – 3 cm wide, ovoid or obovoid, with a short stipe up to 0.5 cm long, pericarp yellow or brown, densely covered with scales like those on the twigs, sometimes glabrescent; latex white. Loculi 2 (or 3) each containing 0 or 1 seed; seeds completely surrounded with a soft, white, sweet or sweet-sour aril. Fig. 27.

DISTRIBUTION. Nicobar Islands, Thailand, Peninsular Malaysia, Sumatra, Borneo, Philippine Islands, Java, Nusa Tenggara, Sulawesi, Maluku, New Guinea, Solomon Islands, Australia (Cape York Peninsula). Fig. 29.

ECOLOGY. Found in primary, secondary and riverine forest, evergreen or semi-evergreen, on granite, basalt, sandstone, coral sand, clay or limestone. Scattered to locally rather common. Alt.: sea level up to 1200 m. Aril eaten by monkeys, hornbills, children.

VERNACULAR NAMES. Sumatra: Bajur (Sunda); Balek Balek, Bantanabalah, Boloh, Suluh, Surian. Java: Ki-siloewar (Sunda); Bangsal, Doerenan, Selang. Borneo: Punyai (Kutai), Lantupak (Dusun-Kinabatangan), Segara (Iban), Durian-durian. Sulawesi: Alo (Besoa), Lasa (Doudri), Mawowisrintek. Nusa Tenggara: Laweludja, Soerenan. Maluku: Olukalukama Manauru (Tobaro), Luka-lukam. New Guinea: Bokrar, Niebwabie (Kebar), Gaigihap (Kaigorin), Gila (Nunumai dialect, Daga), Herrib (Manikiong), Kamoenares (Mooi), Kumho (Kesewai), Massapenkabaran (Madang), Meay (Sepik), Ngraro (Aiome), Sarumo (Bembi), Segudasseken (Kemtuk), Sira (Burumbu), Waair (Kiwai). Solomon Islands: Saebala, Ulukwalo Bala (Kwara'ae).

Representative specimens: THAILAND. Peninsula: Trang, Khao Chong, fr., 3 April 1969, *Phusomsaeng* 102 (C!, E!, FHO!); Ranong, Hard Hin Dam, near the sea [9°20'N 98°25'E], 25 – 50 m, fr., 25 April 1974, *K. Larsen & S.S. Larsen* 33371 (AAU!, NY!).

PENINSULAR MALAYSIA. Penang, Waterfall, yg fr., April 1890, *Curtis* 2287 (K!). Kedah, C., Sungkup F.R., Compt. 12, 75 m, fl., 26 Sep. 1963, *Pennington* 7842 (FHO!). Johore, Mersing F.R., 8 miles W. of Jemaluang, 5 miles from the road, 60 m, fl., 19 Dec. 1963, *Pennington* 8030 (FHO!).

SUMATRA. Simaloer Island, st., 8 Nov. 1918, *Achmad* 723 (U!) & fl., 29 Oct. 1918, *Achmad* 706 (U!). Atjeh [Aceh]: Gunung Leuser Nature Reserves, southern part of the reserves, Alas River valley, near the mouth of the Renun River, c. 50 km S. of Kutacane [c. 3°N 97°50'E], 50 – 125 m, fr., 21 July 1979, *de Wilde & de Wilde-Duyfjes* 18972 (L!) & 250 m, fl. & fr. , 20 July 1979, *de Wilde & de Wilde-Duyfjes* 18914 (L!). North Sumatra, Sibolangit, c. 400 m fl., 29 Nov. 1917, *Lörzing* 5384 (L!). West Sumatra, Padang, Ad Ayer mancior, c. 360 m, ♀ fl. bud & yg fruit, Aug. 1878, *Beccari* 539 (L!).

SABAH. Sandakan, Lamag District, along the Tungkabir River Kinabatangan, ± 100 m, fl., 11 Aug. 1965, *Sinanggul* SAN 51320 (FHO!).

SARAWAK. 3rd Division, Balingian, Ulu Sg. Arip, Bt Iju, fl., 29 July 1965, *Sibat ak Luang* S 23660 (FHO!).

PHILIPPINE ISLANDS. Luzon, Rizal Province, Mt Tokduanbanoy, 1200 m, ♀ fl. & fr., 26 Nov. 1926, *Ramos & Edaño* 48635 (NY!, UC!).

JAVA. W., Ujong Kulon Nature Reserve, Pulau Peucang, seedling, 23 June 1972, *de Vogel* 1394 (L!); Besoeki: Riogodjampoi, fr., 3 Sep. 1897, *Koorders* 28991β (FHO!); Tjoeramanis, fr., 4 Nov. 1895, *Koorders* 19 979β (FHO!). BALI. *van Balgooy* 5253 (L!).

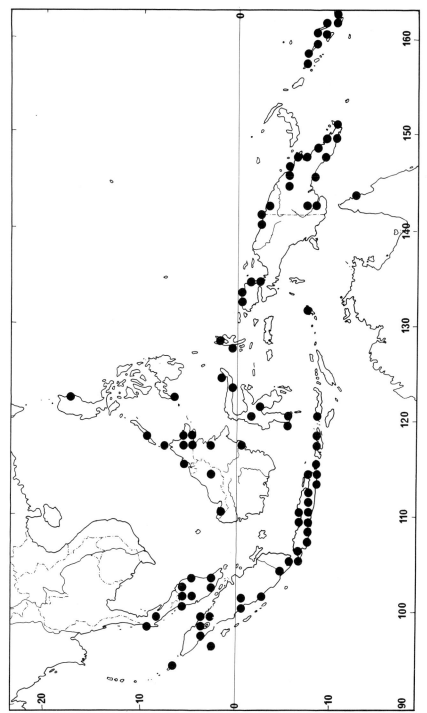

Fig. 29. Distribution of *A. argentea*.

MALUKU. Weda, Tiloppe, c. 25 m, st., 14 March 1938, Neth. Ind. For. Service bb 24842 (L!, NY!).
PAPUA NEW GUINEA. Sambio village, Morobe District, Mumeng Subdistrict [7°05'S 146°37'S], c. 670 m, fr., 11 July 1969, *Streimann & Kairo* NGF 44029 (K!, L!).
AUSTRALIA. Queensland, Cape York Peninsula, Claudie River, fr., Oct. 1968, *Webb & Tracey* 8311 (BRI!, CANB!, K!); 'Steelwire bridge', East Claudie River, 40 m, 9 Dec. 1981, yg fr., 23 June 1948, *Tucker* 278 (QRS!); Iron Range, *Brass* 19302 (BRI!, CANB!, K!); Table Range, Dead Horse Creek, *Dockrill* 778 (QRS!).
SOLOMON ISLANDS. N.E. Malaita, Faufo Area, 25 m, ♂ fl., 20 Nov. 1968, *Mauriasi & Collectors* BSIP 13405 (K!, L!). Uluwa Island, nr Olisu'u anchorage, 1 – 3 m, ♀ fl., 4 Feb. 1965, BSIP 6214 (K!, L!). Namunga, Star Harbour, 40 m, ♀ fl., & fr., 2 Oct. 1968, *Runikera & Collectors* BSIP 10955 (K!, L!).

The lower surface of the leaflets has a dense indumentum of white or brown peltate scales, the former gives them a silvery appearance, hence the specific epithet. There are at least two forms of this species, linked by intermediates; one has a darker indumentum, the lateral veins on the lower leaflet surface are prominent and the leaflet has a subcordate base; the other is less robust, the indumentum is white, the lateral veins are not prominent and the leaflet base is cuneate. Both are often found in secondary and regenerating forest and forest edge. The form with the darker indumentum is, for example, conspicuous by its copper crown in the ravine along which the Padang to Bukittinggi road runs in West Sumatra. In addition to these two forms, some collections from Thailand have small, coriaceous leaflets which are shiny dark brown above when dry.

23. – 25. Aglaia squamulosa *group*
The three species of *Aglaia*, *A. squamulosa*, *Aglaia subcuprea* and *Aglaia densisquama*, are characterised by their large, entire, peltate scales, 0.1 – 0.25 mm across. These scales thickly covering the twigs and numerous to densely covering on the lower surface of the leaflets. They differ in the size and number of the leaflets, in the extent to which the leaflet surfaces are rugose and in the colour of the scales. All three species have pale brown stellate or simple hairs on the anthers and inside the staminal tube. The fruits of *A. squamulosa* are sometimes spindle-shaped , in *A. densisquama* they are always so and in *A. subcuprea* they are obovoid.

23. Aglaia squamulosa King in Jour. As. Soc. Bengal 64: 68 (1895). Syntypes: Malay Peninsula, Perak, near Ulu Kerling, 400 – 600 ft [c. 135 – 200 m], fl., April 1886, *King's Coll.* 8805 (K!, SING!); Perak, Ulu Bubon g, 500 – 700 ft [c. 150 – 210 m], yg fr., Sep. 1886, *King's Coll.* 11013 (BM!, K!); Perak, Ulu Bubong, 400 – 600 ft [c. 120 – 180 m], fl., June 1886, *King's Coll.* 10145 (BM!, K!, SING!); Ridley, Fl. Malay Penins. 1: 407 (1922); Pannell in Ng, Tree Flora of Malaya 4: 225 (1989).

Aglaia cuprea Elmer in Leafl. Philipp. Bot. 9: 3287 (1937), sine diagn. lat.. Type no.: *Elmer* 16058, Philippine Islands, Luzon, Province of

Sorsogon, Irosin, Mt Bulusan, fr., May 1916 (A!, BM!, BO!, G!, GH!, K!, L!, NY!, P!, U!, UC!, US!, W!).

Tree up to 20 m, usually with a broad rounded crown. Bole up to 15 m, up to 120 cm in circumference, with watery exudate when cut, sometimes with L-shaped buttresses upwards up to 55 cm and outwards up to 36 cm. Outer bark brown, pale green, pale orange-brown, pinkish-brown, pale brownish- grey or grey, sometimes with transverse and longitudinal striations or rows of lenticels; inner bark yellowish-brown, orange or green; sapwood brown or pale yellowish-pink, pale brown or orange; heartwood magenta, sometimes with white latex. Twigs stout, pale grey or greyish-brown, with longitudinal wavy ridges, thickly covered with peltate scales which have a brown centre and pale brown shortly fimbriate margin.

Leaves in spirals which become dense towards the ends of the twigs, imparipinnate, up to 90 cm long and 60 cm wide, elliptical or obovate in outline; petiole up to 20 cm, the petiole, rhachis and petiolules ridged and with indumentum like the twigs, but the scales sometimes brown throughout. Leaflets 9 – 15, the laterals usually alternate, sometimes subopposite, all 4 – 30 cm long, 2 – 10 cm wide, dark yellowish-green on upper surface, paler on lower surface, coriaceous, oblong, elliptical or elliptical-oblong, recurved at the margin, acuminate at apex with the obtuse to acute acumen up to 15 mm, rounded or cuneate at the asymmetrical base, when young both surfaces densely covered with pale brown or colourless, shiny, peltate scales, when mature upper surface rugose and with a few scattered scales, lower surface with numerous to densely covered with scales, those on the main veins reddish-brown; veins 4 – 13 on each side of the midrib, ascending, and curved upwards near the margin, midrib prominent and lateral veins subprominent on lower surface; petiolules 3 – 20 mm.

Male inflorescence up to 39 cm long and 20 cm wide; peduncle up to 9 cm long, the peduncle, rhachis branches and pedicels thickly covered with scales like those on the twigs. Flowers up to 2 mm in diameter, subglobose, fragrant; pedicels up to 1 mm long. Calyx cup-shaped, c. $1/3$ the length of the corolla, wrinkled and densely covered with peltate scales near the base, divided almost to the base into 5 (or 6) subrotund, obtuse lobes which have fewer scales. Petals 5, yellow, obovate, aestivation quincuncial. Staminal tube c. 1 mm long and 1 mm wide, obovoid, yellow, the aperture c. 0.5 mm across, shallowly 5-lobed, with white stellate hairs in dense clumps on the inner surface of the tube near the apex and few on the outer surface; anthers ovoid, c. 0.5 mm long and 0.3 mm wide, ovoid, reaching or protruding just beyond the aperture, densely covered with white stellate hairs. Ovary depressed-globose, densely covered with brown peltate scales; stigma ovoid with two small apical lobes, dark brown and shiny; ovary and stigma together c. $2/3$ the length of the staminal tube. Female flowers similar to the male but up to 5 mm long, obovoid; sessile or with pedicels up to 1 mm. Calyx up to $2/3$ the length of the corolla; anthers c. $1/3$ the length of the tube and inserted in the upper $1/2$ or $1/3$.

Fruits narrowly obovoid when young, up to 5 cm long, 3.5 cm wide and subglobose, often with a short beak up to 5 mm and a short stipe up to 5

Fig. 30. *A. squamulosa*. Leaf x 1/2. Habit with inflorescence x 1/2. Half flower x25. Infructescences with immature fruits and (behind) ripe fruits x 1/2. Peltate scale x80.

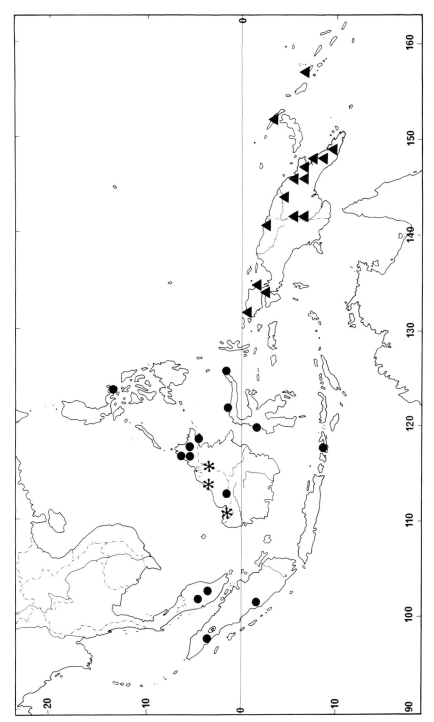

Fig. 31. Distribution of *A. squamulosa* ●, *A. densisquama* * and *A. subcuprea* ▲.

mm when mature, brown or yellow, densely covered with brown peltate scales. Loculus 1, containing one seed; aril translucent, white. Fig. 30.

DISTRIBUTION. Peninsular Malaysia, Sumatra, Borneo, Philippine Islands, Nusa Tenggara (Sumbawa), Sulawesi. Fig. 31.

ECOLOGY. Found in primary and secondary forest on sandstone, sand, clay, loam. Alt.: 17 to 2000 m.

VERNACULAR NAMES. Peninsular Malaysia: Bekak; Hati Cura (Orang Asli). Borneo: Lantupak (Dusun Kinabatangan); Segera (Iban).

Representative specimens. PENINSULAR MALAYSIA. Pahang, Taman Negara, path from K. Kenyam to K. Perkai, 9 March 1978: *Pannell* 1088 (FHO!) & *Pannell* 1091 (FHO!).

SUMATRA. Sumatra Barat [West Sumatra], Eastern foothills of Barisan mountain range, Ulu Batang Hari, logging road S.W. from Sungaidareh, Sg. Mimpi, P.T. Pasar Besar logging concession [1°12′S 101°20′E], 300 m: fl., 17 June 1983, *Pannell* 1909 (FHO!) & fr., 17 June 1983, *Pannell* 1919 (FHO!).

SABAH. Tawau District, Sandakan, Mt Wullersdolf, fl., 26 June 1969, *Ahmad Talip* SAN 65861 (FHO!). nr Sandakan, Sepilok Forest Reserve: c. 30 m, fl., 18 Oct. 1963, *Pennington* 7900 (FHO!); Jalan Kabili, fl., 8 May 1974, *Mabberley* 1665 (FHO!).

PHILIPPINE ISLANDS. Luzon: Albay Province, Mayon Volcano [13°04′N 123°42′E], fl., May – June 1953, *Mendoza* PNH 18308 (K!, L!); Province of Sorsogon, Irosin (Mt Bulusan), fr., May 1916, *Elmer* 16058 (K!); Montalban, fl., 18 April 1905, *Loher* 5698 (K!).

SULAWESI. N.E. Minahasa, N. slope of Mt Klabat [1°28′N 125°02′E], 600 m, fl., 28 June 1956, *Forman* 268 (L!).

A. squamulosa is readily distinguished by the markedly rugose upper surface of the leaves and the numerous peltate scales on the lower surface from which the specific epithet is derived. *Aglaia densisquama*, which is known only from Sarawak, appears to be an extreme but distinct variant of *A. squamulosa* and is treated here as a separate and new species. It has spindle-shaped fruits, which is also rarely found in *A. squamulosa* in Borneo (e.g. in *Chai* S 36154).

24. Aglaia densisquama C.M. Pannell spec. nova *Aglaiae squamulosae* King similis sed indumento obscuriore, margine folioli valde recurvo diversa. In *Aglaia densisquama* indumentum densum squamarum peltatarum nitentium, obscure badiarum in pagina inferiore foliolorum colore discrepat paginae superioris pallide flavovirenti. Costa nervique laterales foliolorum supra valde depressi; nervi laterales anastomosantes brochides conspicuas prope marginem efformant. Foliola apice longo caudato; fructus fusiformis, rostro longo angusto. Holotype: Malaysia, Sarawak, Kuching, 12th mile, Penrissen Road, Semengoh F.R., Arboretum, tree no. 1516, ♂ fl., 28 Sep. 1966, *Banyeng ak Nudong & Benang ak Bubong* S 25209 (FHO!; isotype: L!).

Tree up to 20 m. Bole up to 1 m in circumference. Bark smooth, with longitudinal cracks. Wood pink, becoming darker towards the centre; without latex. Twigs grey, longitudinally wrinkled, densely covered with peltate scales which are 0.3 – 0.4 mm in diameter and dark reddish-brown with a pale margin.

Leaves imparipinnate, up to 48 cm long and 44 cm wide, obovate in outline; petiole 8 – 16 cm, the petiole rhachis and petiolules longitudinally wrinkled and densely covered with scales like those on the twigs. Leaflets 9 – 11, the laterals usually alternate, sometimes subopposite, all 7 – 22 cm long, 4 – 8 cm wide, coriaceous, pale yellowish-green on upper surface when dry, elliptical or ovate, recurved at margin, caudate-acuminate at apex, the acumen obtuse, narrow, parallel-sided and up to 20 mm long, rounded or cuneate at the markedly asymmetrical base, the upper surface of the leaflet rugulose and pitted, the midrib and veins on the upper surface and the entire lower surface densely covered with scales like those on the twigs; veins 7 – 17 on each side of the midrib, ascending, curved upwards near the margin and anastomosing; the midrib and lateral veins markedly depressed on the upper surface and subprominent on the lower; petiolules up to 15 mm.

Male inflorescence c. 42 cm long and 42 cm wide; peduncle 7.5 – 11 cm, the peduncle, rhachis and branches longitudinally wrinkled and densely covered with scales like those on the twigs. Flowers c. 3 mm long and 3 mm wide; pedicels c. 1.5 mm, the pedicels and calyx densely covered with scales like those on the twigs. Calyx c. 1.5 mm long, thick and fleshy, divided into 5 sub-rotund lobes. Petals 5, thick and fleshy, with a few scales like those on the twigs on the outer surface, aestivation quincuncial. Staminal tube c. 1.7 mm long and 2.2 mm wide, the aperture c. 0.7 mm in diameter; anthers 5, c. 0.8 mm long and 0.4 mm wide, ovoid, inserted about two thirds of the way up the staminal tube and just protruding through the aperture, the staminal tube thick and fleshy below and between the anthers. Ovary 0.5 mm long and 0.6 mm wide, ovoid; stigma c. 0.3 mm long and 0.3 mm wide, ovoid.

Female inflorescence c. 28 cm long and 4 cm wide; peduncle up to 8 cm. Flowers c. 3 mm long and 4 mm wide; pedicels c. 1.5 mm, the pedicels and calyx densely covered with scales like those on the twigs. Calyx c. 2 mm long, thick and fleshy, divided into 5 sub-rotund lobes. Petals 5, thick and fleshy, with a few scales like those on the twigs on the outer surface; aestivation quincuncial. Staminal tube c. 1.5 mm long and 2.5 mm wide, depressed-globose, the aperture c. 0.7 mm in diameter; anthers 5, c. 0.5 mm long and 0.5 mm wide, ovoid, inserted c. $^2/_3$ of the way up the staminal tube and just protruding through the aperture, the staminal tube thick and fleshy below and between the anthers. Ovary c. 0.8 mm long and 1 mm wide, ovoid, densely covered with peltate scales; locules 2, each containing 1 ovule; stigma c. 0.2 mm long and 0.3 mm wide, ovoid, with two apical lobes.

Infructescence 30 – 40 cm long, 18 – 20 cm wide; peduncle 2 – 8 cm, the peduncle, rhachis and branches longitudinally wrinkled and densely covered with scales like those on the twigs. Fruits up to 5 cm long and 2 cm wide, narrowly ellipsoid, ovoid or obovoid with a long narrow beak up to 1.5 cm long and a short broad stipe up to 5 mm long; the pericarp densely covered with scales like those on the twigs and without latex. Locules 2, each containing one seed which has an entire, translucent, white aril.

DISTRIBUTION. Borneo (Sarawak) only. Fig. 31.

ECOLOGY. Found in primary and riverine forest; on alluvial sand with some clay and on clayey loam. Alt.: 15 to 1600 m.

VERNACULAR NAMES. Borneo: Segera (Iban).

Representative specimens. SARAWAK. 1st Division, Kuching, Semengoh Forest Reserve, fr., 19 April 1974, *Mabberley* 1579 (FHO!, L!); 4th Division, Bintulu, Ulu Segan, ♀ fl., 28 Aug. 1968, *Wright* S 27951 (FHO!, L!).

Aglaia densisquama resembles *A. squamulosa* but the indumentum is darker in colour and the leaflet margin is strongly recurved. In *A. densisquama*, the dense indumentum of bright, shiny, dark reddish-brown peltate scales on the lower surface of the leaflet contrasts with the pale yellowish-green upper surface of the leaflet. The midrib and lateral veins are markedly depressed on the upper surface and the lateral veins anastomose and appear as conspicuous loops near the margins of the leaflets. The leaflets have a long, caudate apex and the fruits are spindle-shaped with a long, narrow beak.

25. Aglaia subcuprea Merrill & Perry in Jour. Arn. Arb. 1940, 21: 324 (1940). Lectotype (designated here): Nordöstliches Neu-Guinea [N.E. New Guinea], Morobe-Distrikt [District], Yunzaing, 4493 ft [c. 1350 m], ♀ fl. & fr., 23 June 1936, Clemens 3416 (A!; isolectotypes: K!, L!, NY!, SING!, UC!).

Aglaia versteeghii Merrill & Perry in Jour. Arn. Arb. 1940, 21: 323 (1940).
Lectotype (designated here): New Guinea, 6 km southwest of Bernhard Camp, Idenburg River, 1230 m, fl., Feb. 1939, *Brass & Versteegh* 12531 (A!; isolectotypes: BO!, L!).

Aglaia boanana Harms in Engl. Bot. Jahrb. 72: 163 (1942). Lectotype (designated here): New Guinea, Morobe Province, Mt above Boana, 4000 ft [c. 1300 m], 12 July 1938, *Clemens* 8460 (B!; isolectotypes: BO!, L!).

Tree 4 – 30 m, diameter up to 1.6 m. Outer bark fairly smooth, brown or black, mottled with grey and fawn in large patches, sometimes flaking in small irregular scales; inner bark pale brown to reddish-brown. Sapwood cream, pink, red or dark red, latex white. Twigs slender to fairly stout, longitudinally wrinkled and thickly covered with pale brown, reddish-brown or pale orange-brown peltate scales 0.1 – 0.25 mm diameter, which have a dark centre and a paler fimbriate margin.

Leaves up to 67 cm long and 45 cm wide; petiole up to 17 cm, the petiole, rhachis and petiolules densely covered with scales like those on the twigs. Leaflets (3 –)7 – 9(– 11), the laterals subopposite, all 5 – 24 cm long, 2.5 – 8 cm wide, usually obovate or oblong, occasionally elliptical, coriaceous or subcoriaceous, sometimes recurved at the margin, cuneate at the asymmetrical base, usually with a broad rounded apex, but sometimes acuminate at apex, the obtuse acumen up to 5 mm long; with scales like those on the twigs densely covering both surfaces when young, glabrescent on upper surface leaving it wrinkled or pitted, sometimes glabrescent on lower surface; veins 6 – 15 on each side of the midrib, midrib prominent and lateral veins subprominent on lower surface; petiolule up to 10 mm on lateral leaflets, up to 20 mm on terminal leaflet.

Inflorescence up to 30 cm long and 20 cm wide the peduncle, rhachis and branches thickly covered with scales like those on the twigs; sometimes with lanceolate bracts 3 – 5 mm long and 1 – 2 mm wide. Male flowers 2 – 4 mm long, 2 – 3 mm wide, female flowers 4.5 – 5 mm long, c. 3.5 mm wide, ellipsoid or broadly ellipsoid or subglobose, fragrant; pedicel 1 – 5 mm. Calyx about half the length of the flower, cup-shaped, deeply divided into 5 obtuse ovate lobes, densely covered with scales like those on the twigs on the outside. Petals 5, white, pale yellow or yellow, aestivation quincuncial. Staminal tube 1.3 – 3 mm long, 1.2 – 2.5 mm wide, obovoid, aperture 1 – 1.4 mm across and shallowly 5(or 6)-lobed; anthers 5(– 10), 0.6 – 1.5 mm long, 0.3 – 0.5 mm wide, ellipsoid or ovoid, inserted $^1/_3$ to $^2/_3$ of the way up the staminal tube and usually protruding through the aperture; with few to numerous stellate hairs or peltate scales on the anthers and the inside of the staminal tube. Ovary c. 0.6 mm high and 0.7 mm wide, subglobose; loculi 1 – 2, each containing 1 ovule; stigma 0.7 – 1.5 mm long, 0.3 – 0.4 mm wide, ovoid to fusiform with 2 small apical lobes.

Infructescence 12 – 23 cm long, up to 14 cm wide; peduncle up to 5.5 cm, the peduncle, rhachis, branches and pericarp densely covered with scales like those on the twigs. Fruits c. 2.7 cm long and 1.5 cm wide, obovoid, sometimes with numerous wart-like bumps on the outside. Loculi 2, each with 0 or 1 seed, seeds with a pale yellow aril.

DISTRIBUTION. New Guinea, New Ireland, ? Solomon Islands. Fig. 31.

ECOLOGY. Found in periodically inundated primary forest, secondary forest, lowland and montane forest, riverine forest; on sandy clay, granite. Alt.: 5 to 2570 m. Rare to rather common.

VERNACULAR NAMES. Sulawesi: Pisek. New Guinea: Herrib* (Manikiong); Keeke (Enga); Kuni (Akuna); Lowele, Semali (Mooi); Milomb (Waskuk); Mornt (Chimbu); Noboeabi (Amberbaken); Tabuwe (Wagu).

USES. Wood is used for bullock-carts in Sulau, Minahasa (Sulawesi).

Representative specimens. PAPUA NEW GUINEA. Morobe District, Menyamya Subdistrict, Angabena Ridge [7°20'S 146°10'E], 1620 m, fl., 10 Jan. 1972, *Streimann & Stevens* LAE 54853 (K!). Sepik District, Ambunti Subdistrict, eastern ridge of Sumset (Mt Hunstein), 1200 m, fr., 8 Aug. 1966, *Hoogland & Craven* 10846 (K!). Cycloop mountains, path Dozai to Dafonsero, N. of Dozai, Baimungun Creek, 500 m, fl., 2 Aug. 1961, *van Royen & Sleumer* 6385 (K!). McAdam Memorial Park, Wau, 1200 m, fr., 11 Jan. 1964, *Pennington* 8039 (FHO!). Eastern Highlands District, Goroka Subdistrict, Asaro-Mairifutica divide, 0.5 mile S. of Daulo camp, 2400 m, fl., 3 Aug. 1957, *Pullen* 451 (L!).

The following specimens either belong here or to a new, undescribed species:

SOLOMON ISLANDS. S.W. Choiseul, W. side of Kolombangara River, c. 45 m, fl., 30 Jan. 1970, *Gafui et al* BSIP 18922 (K!, L!). N.W. Choiseul, Nanan go area, c. 45 m, fl., 4 Dec. 1969, *Gafui et al* BSIP 18580 (K!, L!). Three Sisters, Malaupaina Island, c. 15 m, fl., 8 Dec. 1969, *Mauriasi et al* BSIP 17985 (K!, L!).

The lower surface of the leaflets of *A. subcuprea* is densely covered with peltate scales which vary in colour from pale orange to dark reddish-brown. The size and leatheriness of the leaflets is also variable. A variant which has small, coriaceous leaflets with a recurved margin includes the type of *A. versteeghii* and may represent a subspecies; this impression is based solely on leaf characters, supporting flower or fruit characters and information on ecological or distributional differences from the rest of *A. subcuprea* are lacking.

26. Aglaia lancilimba Merrill in Philipp. Jour. Sci., Bot., 13: 294 (1918). Lectotype (designated here): Philippine Islands, Luzon, Province of Camarines, fl., 1916, *De Mesa & Magistrado* For. Bur. 26509 (US!; isolectotypes: K!, P!, PNH†).

Tree 1 – 15 m. Twigs yellowish-brown, densely covered with large shiny reddish-brown peltate scales which are entire or have a short fimbriate margin and often have a dark centre and a dark ring near the margin, usually 0.3 – 0.6 mm in diameter, sometimes smaller.

Leaves 18 – 50 cm long, 14 – 34 cm wide, petiole 2.5 – 12 cm, the petiole, rhachis and petiolules densely covered with scales like those on the twigs. Leaflets (7 –)9 – 13(– 17), the laterals subopposite or alternate, all 5 – 18 cm long, 1.5 – 5.5 cm wide, elliptical, pale yellowish-brown and usually with the margin recurved or undulate when dry, apex acuminate, the obtuse acumen 2 – 25 mm, cuneate or rounded at the asymmetrical base; with scales like those on the twigs thickly covering the midrib on the lower surface, sometimes with a few paler and more fimbriate scales interspersed, occasional on the rest of that surface, the lower surface with numerous pits, veins 6 – 17 on each side of the midrib, ascending and curved upwards near the margin and nearly or quite anastomosing, midrib prominent, lateral veins barely subprominent; petiolules 5 – 15 mm on lateral leaflets, up to 20 mm on the terminal leaflet.

Inflorescence up to 28 cm long and 23 cm wide; peduncle 5 – 15 mm, the peduncle rhachis and branches densely covered with scales like those on the twigs or with scales which have a longer fimbriate margin or are stellate. Flowers 1.5 – 3 mm long, 1.5 – 3.5 mm wide, pedicel 0.5 – 2 mm. Calyx 0.75 – 1 mm long, cup-shaped and divided into 5 lobes, densely covered with scales like those on the twigs on the outside. Petals 5, yellow, aestivation quincuncial. Staminal tube up to c. 1 mm long, 1 – 2 mm wide, shallowly cup-shaped, thickened below and between the anthers, the aperture 0.6 – 1.3 mm across and with the margin shallowly lobed; anthers 5, 0.4 – 0.5 mm long and 0.4 – 0.5 mm wide, dark blackish-brown when dry with a pale yellow margin, inserted inside the margin of the tube and protruding. Stigma 0.2 – 0.3 mm long and 0.3 – 0.4 mm wide, ovoid, truncate at the apex; ovary c. 0.4 – 1 mm high, 0.6 – 1 mm wide, depressed globose densely covered with reddish-brown peltate scales which have a fimbriate margin; loculi 2, each containing 2 ovules.

Infructescence up to 19 cm long and 5.5 cm wide, peduncle 2.5 cm; the peduncle, rhachis and branches densely covered with scales like those on the twigs. Fruits 2.2 – 3 cm long and 2.2 – 3 cm wide, subglobose, brown or yellow when ripe; the pericarp thin and brittle when dry, densely covered with scales like those on the twigs on the outside; loculi 2 (or 3), each containing one seed, the seed covered with a white aril.

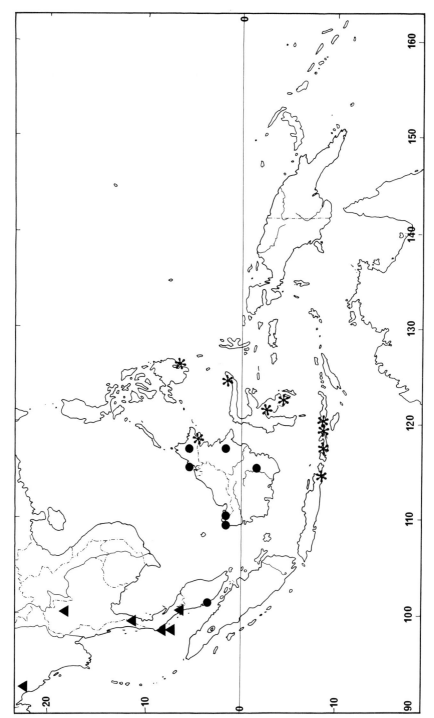

Fig. 32. Distribution of A. *chittagonga* ▲, A. *scortechinii* ● and A. *lancilimba* *.

DISTRIBUTION. Philippine Islands, E. Borneo (Bumbun Is.), Bali, Nusa Tenggara (Sumbawa & Flores), Sulawesi. Fig. 32.

ECOLOGY. Found in primary forest, swamp forest, montane forest, semi-wet forest. Alt.: 1 to 1400 m. Common.

VERNACULAR NAMES. Philippine Islands: Lowas, Pagaspas (Ma). Sulawesi: Woearaoe.

Representative specimens. SABAH. Lahad Datu, Semporna, Bumbun Is., fl., 29 April 1939, *Valera* BNB 10279 (L!, SING!).

PHILIPPINE ISLANDS. Davao, Mount Mansamuga, summit, 1030 m, fl., 4 Sep. 1949, *Edaño* (field no. 1851) PNH 11104 (L!).

NUSA TENGGARA. Bali, West Mountains above Bungbungan village, fl., 22 June 1976, *Meijer* 10605B (L!). Flores: W., Mt Satalibo, fr., 27 Oct. 1961, *Muchtar* 24 (L!); Rahong (Ruteng), 1400 m, ♀ fl., 8 Feb. 1972, *Verheijen* 3046 (L!). Sumbawa: W., Mt Batulanteh, trail from Batudulang to Pusu, 900 – 1000 m, fr., 24 Oct. 1961, *Kostermans* 19091 (K!, L!); W., Mt Batulanteh, Sampar Olat and Batu Burung ridges, N. of Batudulang, ♀ fl., 1 May 1961, *Kostermans* 18627 (L!).

SULAWESI. Malili Oesoe: fl., 20 Oct. 1931, *Boschproefst.* no. Cel./III.26 (L!) & yg fr., 9 Dec. 1931, *Boschproefst.* no. Cel./III.26 (L!) & yg fr., 6 Feb. 1932, *Boschproefst.* no. Cel./III.26 (L!). Sulawesi Tengara, S.E., around Opa swamp, Mt Makalio [122°E 4°05′S], 20 – 250 m, fr., 13 Nov. 1978, *Prawiroatmodjo & Soewoko* 1705 (L!)

Aglaia lancilimba has relatively narrow leaflets, pale green below when dry, with large peltate scales which often densely cover and are conspicuous on the midrib of the lower surface of the leaflets but are few on the rest of that surface.

27. Aglaia lepiorrhachis Harms in Engl. Bot. Jahrb. 72: 165 (1942). Lectotype (designated here): Nordöstliches Neu-Guinea, Morobe-Distrikt, Wareo, 2000 ft [c. 600 m], 25 Dec. 1935, *Clemens* 1348 (L!; isolectotypes : B†, G!).

Tree 4 – 7 m tall, about 8 cm in diameter. Outer bark smooth, pale grey or greyish-brown; inner bark pale yellowish-brown or pale brown. Twigs rugulose, somewhat longitudinally ridged towards the apex; densely covered with large shiny pale brown, peltate scales which have a fimbriate margin.

Leaves imparipinnate, up to 80 cm long, dark green above, yellowish-green below, obovate in outline; petiole up to 16 cm long, the petiole, rhachis and petiolules densely covered with scales like those on the twigs.

Leaflets 7 – 9, the laterals alternate, all 14 – 30 cm long, 4 – 10 cm wide, usually ovate or obovate, sometimes elliptical, the apex acuminate, with the obtuse acumen 2 – 10 mm long, cuneate at the slightly asymmetrical base, upper surface rugulose and pitted, lower surface with numerous dark, shallow pits, with scales like those on the twigs, densely covering the midrib and scattered on the lower surface; veins 13 – 19 on each side of the midrib, shallowly ascending and curved upwards near the margin, with some shorter lateral veins in between; midrib prominent and longitudinally wrinkled on the lower surface, depressed on the upper surface.

Male inflorescence about 30 cm long and 30 cm wide, triangular in outline; peduncle up to 5 cm, the peduncle rhachis and branches densely covered with scales like those on the twigs. Male flowers up to 2.5 mm long and 2 mm wide, globose to ellipsoid; pedicels up to 2 mm, densely covered with scales like those on the twigs. Calyx about $^1/_3$ the length of the corolla, divided into 4 – broadly ovate, obtuse lobes, with few or no scales on the outside. Petals 4 – 5 (– 7) obovate, white, with a ciliate margin, aestivation quincuncial. Staminal tube about $^2/_3$ the length of the petals, cup-shaped or obovoid, the apical margin shallowly lobed and with simple hairs; anthers 4 – 5, about $^1/_2$ the length of the tube, inserted $^2/_3 - ^3/_4$ up the tube and protruding for half the anther length, with a few simple or stellate hairs. Ovary depressed globose with peltate scales which have a fimbriate margin; stigma ovoid $^1/_3 - ^2/_3$ the length of the staminal tube longitudinally ridged where adjacent to anthers.

Fruits solitary, c. 3.5 cm long and 3.5 cm wide; pedicel 0.5 cm, with numerous reddish-brown peltate scales which have a fimbriate margin on the pedicel and calyx and few paler scales on the outside of the pericarp. Loculi 2, each containing one seed which is surrounded by a gelatinous aril.

DISTRIBUTION. New Guinea only.
ECOLOGY. Lowland and hill forest. Alt.: up to 600 m.
Representative specimens. PAPUA NEW GUINEA. Morobe District: Lae Subdistrict, Sankwep logging area [6°35'S 147°05'E], 250 m, fl., 6 Jan. 1975, *Barker et al.* LAE 66770 (K!); Lae Subdistrict, Sankwep Road, Gwambari Ex-Service Station [6°35'S 146°55'E], 65 m, fr., 19 April 1971, *Katik* NGF 46767 (K!); Burep River, about 15 miles N.E. of Lae, c. 15 m, fl., 13 May 1963, *Hartley* 11847 (K!).

A. lepiorrhachis resembles *A. lancilimba* in that it has large, shiny, brown, peltate scales densely covering the midrib on the lower surface of the leaflets. It differs from *A. lancilimba* in the paler colour of the scales and in its usually larger leaflets which have more numerous lateral veins. *A. lancilimba* is not found east of Sulawesi and *A. lepiorrhachis* is confined to the island of New Guinea.

28. Aglaia chittagonga Miq., Ann. Mus. Bot. Lugd. Bat. 4: 44 (1868). Lectotype (designated here): [Bangladesh], Chittagong [22°N 92°E], tropical region, 0 – 1000 ft [0 – c. 300 m], fr., '13 *Milnea*' in herb. Ind. Or., *Hooker fil.* et *Thomson* s.n. (U!; isolectotypes: BM!, CG E!, G-BOIS!, G-DC!, K!, L!, NY!, P!, W!).

Amoora chittagonga (Miq.) Hiern in Hooker fil., Fl. Brit. Ind. 1: 559 (1875), quoad specim. Chittagong, '13 *Milnea*' in herb. Ind. Or., *Hooker fil.* et *Thomson* s.n. (U!), non Bhotan & Upper Assam, *Griffith* s.n. (K!), nec Pegu, *Kurz* s.n. (K!); as *Meliacea*, Griffith, Itin. Notes: 168 (1848); C. de Candolle in A. & C. de Candolle, Monog. Phan. 1: 587 (1878).

Tree up to 7 (– 10) m. Twigs grey, longitudinally wrinkled, with numerous to densely covered with large (up to 0.3 mm in diameter) very pale orange brown peltate scales which have an irregular, occasionally shortly fimbriate, margin, the margin is paler than the darker orange centre and there is a small dark brown bump in the centre.

Leaves imparipinnate, up to 30 cm long and 35 cm wide, obovate in outline; petiole 3.5 – 10.5 cm, the petiole, rhachis and petiolules with numerous to densely covered with scales like those on the twigs. Leaflets 5 – 7, the laterals subopposite, all 8.5 – 20 cm long, 3.5 – 7.5 cm wide, oblong, ovate or elliptical, rarely obovate rounded or cuneate at the slightly asymmetrical base, often with the terminal 3 leaflets cuneate and the remainder rounded at the base, apex acuminate, the acumen acute or occasionally obtuse and parallel-sided, up to 10 mm long, the upper leaflet surface with few to numerous pits, with scales like those on the twigs but without a fimbriate margin scattered on the upper leaflet surface, numerous and evenly distributed on the lower leaflet surface; veins 10 – 22 on each side of the midrib, ascending, curved upwards near the margin and sometimes anastomosing, often with shorter lateral veins in between, the midrib depressed and the lateral veins and reticulation slightly prominent on the upper leaflet surface, the midrib prominent lateral veins and reticulation subprominent and reticulation sometimes visible, but not subprominent on the lower leaflet surface; petiolules up to 10 mm on lateral leaflets, up to 20 mm on the terminal leaflet.

Inflorescence 7 – 9.5 cm long, c. 8 cm wide; peduncle 2 – 10 mm. Flowers 2 – 3 mm long, 1.5 – 3 mm wide; pedicel 0.5 – 2 mm, the pedicel and calyx with numerous to densely covered with scales like those on the twigs, sometimes with a short fimbriate margin; calyx 1 – 1.5 mm long, cup- shaped with 5 short acute lobes which have a fimbriate margin; petals (4 or) 5, white; staminal tube 1.3 – 2.5 mm long, 2 – 2.5 mm wide, obovoid, the aperture 0.7 – 1.2 mm wide and lobed; anthers 7 – 10, 0.8 – 1.5 mm long, 0.4 – 0.6 mm wide, inserted $^1/_4$ to $^1/_3$ up the tube and just protruding through the aperture; ovary 0.3 – 0.5 mm high, ovoid, densely covered with pale peltate scales on the outside which have an irregular or shortly fimbriate margin, loculi 2 (or 3); stigma 0.2 – 0.5 mm high, ovoid, shiny dark brown, with 2 or 3 apical lobes.

Infructescence up to 8 cm long and 6 cm wide; peduncle 5 – 10 mm, the peduncle, rhachis and branches densely covered with scales like those on the twigs. Fruits c. 2.5 cm long and 1.5 cm wide ellipsoid, indehiscent, densely covered with very pale orange-brown scales, which may have a fimbriate margin, on the outside. Loculi 2, each containing 0 or 1 seed which is surrounded by an edible aril. Fig. 33.

DISTRIBUTION. Bangladesh (Chittagong), Thailand. Fig. 32.

ECOLOGY. Found in forest along rivers and beach, sometimes on limestone. Alt.: 50 to 200 m.

Representative specimens. BANGLADESH. see type.

THAILAND. Kao Kalakiri, Pattani, c. 200 m, fl. 29 March 1928 *Kerr* 14877 (C!, E!, K!, L!). Kaw Pipi, Krabi: fl. 11 April 1930 *Kerr* 18931 (C!, K!, L!) & c. 50 m, fl. 9 April 1930, *Kerr* 18911 (C!, K!, L!). Prachuap [11°50'N 99°49'E], Huoy Tap Bakae, 40 m, fl. 11 July 1921, *Khoon Winit* 619 (K!). Hui Me K'Mi [18°21'N 100°19'], 270 m, fr. 14 Feb. 1912, *Kerr* 2373 (E!, K!).

The lower leaflet surface of *A. chittagonga* has numerous, large, evenly distributed, peltate scales which have an entire undulate margin and dark centre; this characteristic distribution of large pale scales is diagnostic and is visible with the naked eye if the specimen is held so that light is reflected off the scales. Specimens where the scales on the lower leaflet surface scales have a fimbriate margin or they are more numerous towards the

Fig. 33. *A. chittagonga*. Habit with infructescence x$^{1}/_{2}$. Habit with inflorescence. Half flower x10. Peltate scale x50.

midrib belong to *A. lawii*. The flower of *A. chittagonga* usually has 5 calyx lobes, 5 petals and 2 loculi; the fruit is ellipsoid or obovoid, indehiscent and the seed has an edible aril. *A. lawii* usually has 4 calyx lobes, 4 petals and nearly always has 3 (or 4) loculi; the fruit is dehiscent.

In vegetative characters, *A. chittagonga* closely resembles *A. lawii* and some sheets appear to be a mixture of these two species. The fruit in *A. chittagonga* is usually 2-celled and always indehiscent and the species therefore belongs to section *Aglaia*. In *A. lawii*, the fruit is usually 3- celled and always dehiscent and the species therefore belongs to sect. *Amoora*.

In the type collection of *A. chittagonga*, the fruits are unlike those seen in other collections of this species, they are obovoid and markedly longitudinally ridged. One of these failed to soften on boiling in water and could only be dissected after immersion in sodium hypochlorite for one month. It had two loculi each of which contained a flat undeveloped seed, the seed coats were reduced to hard brittle layers and it was not possible to tell whether one of these had been a fleshy aril when fresh; the pericarp appeared to contain numerous longitudinal canals (possibly for latex). The longitudinal ridges on the dried fruits of the type collection may have been produced by the pericarp shrinking into the incompletely filled loculi during drying.

29. Aglaia elaeagnoidea (A. Juss.) Benth., Fl. Austral. 1: 383 (1863), quoad specim.: Australia. Entrance Island, Endeavour Straits, *Leichardt* (K!, P!); Koorders & Valeton, Atlas der Baumarten von Java 1: t. 154 (1913); C. de Candolle in A. & C. de Candolle, Monog. Phan. 1: 611 (1878); Backer and Bakhuizen, Fl. Java 2: 128 (1965); Mabberley in Flore de la Nouvelle-Calédonie et Dépendances 15: 75 (1988).

Nemedra elaeagnoidea A. Juss., Bull. Sci. Nat. Géol. 23: 239 (1830); A. Juss. in Mém. Mus. Nat. Hist. Par. 19: 223, 269, t. 14 ['1830'] (1832) (date uncertain, see Stafleu & Cowan, 1979: 476 & Mabberley, Taxon 31: 66, 1982). Lectotype (Mabberley, 1988): Australia, *Leschenault in Baudin* s.n. Australie (P!; isolectotypes: BM! 'Iter H. Francois', G!, K! fragment).
[*Walsura lanceolata* Wall. Cat. n. 4886 (1831 – 2) nom. nud.]
Milnea roxburghiana Wight & Arn., Prod.: 119 (1834). Lectotype (designated here): India, Herb. *Wight* 311 (K!; isolectotypes: BM!, CGE!, G!, K-W!).
[*Sapindus lepidotus* Wallich, Cat. 8036 (1847) nom. nud.]
[*Aglaia grata* Wall. ex Voigt, Hort. Suburb. Calc.: 136 (1845) nom. nud.]
[*Aglaia midnaporensis* Carey ex Voigt, Hort. Suburb. Calc.: 136 (1845) nom. nud.]
Aglaia odoratissima sensu Benth. in Hooker, Lond. Jour. Bot. 2: 213 (1843), non Blume, quoad specim. New Guinea, *Hind* 1841 (K!) & Tobie Island, *Barclay* (K!).
Aglaia lepidota Miq., Fl. Ind. Bat. Suppl. 1: 197 (nomen), 507 (1861). Lectotype (designated here): [Sumatra, S.], Lamp [Lampong Province], Poeloe Leboekoe *Anon.* 4485 HB (U!).
Aglaia lepidota var. *paupercula* Miq., Fl. Ind. Bat. Suppl. 1: 507 (1861). Lectotype (designated here): [Java] Poeloe Sangian [island in the Sunda Straits], *Teysmann* 2965 HB (U!; isolectotypes: BO!, L!).
Aglaia roxburghiana (Wight & Arn.) Miq., Ann. Mus. Bot. Lugd. Bat. 4: 41

(1868). Koorders & Valeton, Atlas der Baumarten von Java 1: t. 161 (1913); Hiern in Hooker fil., Fl. Brit. India 1: 555 (1875); Kurz in Journ. As. Sci. Beng. 44: 147 (1875); C. de Candolle in A. & C. de Candolle, Monog. Phan. 1: 604 (1878); Pierre, Fl. Forest. Cochinch. Fasc. 21, ante t. 336 (1 July 1895).

[*Aglaia spanoghei* Blume ex Miq., Ann. Mus Bot. Lugd. Bat. 4: 41 (1868) nom. in syn.]

Aglaia roxburghiana (Wight & Arn.) Miq. var. *angustata* Miq., Ann. Mus. Bot. Lugd. Bat. 4: 42 (1868). Lectotype (designated here): Java, Japara [Province], Poeloe Kellor, fl., (U!; isolectotypes: BO!, L!, U!).

Aglaia roxburghiana (Wight & Arn.) Miq. var. *balica* Miq., Ann. Mus. Bot. Lugd. Bat. 4: 42 (1868). Lectotype (designated here): Balie [Bali], *Anon.* s.n. (U!; isolectotype: L!).

Aglaia roxburghiana (Wight & Arn.) Miq. var. *paupercula* (Miq.) Miq ., Ann. Mus. Bot. Lugd. Bat. 4: 42 (1868).

Aglaia wallichii Hiern in Hooker fil., Fl. Brit. Ind. 1: 555 (1875). Lectotype (designated here): India, E. Himalayas, Silhet, *Anon.* in *Herb. E.I.C.* 8036 (K-W! as *Sapindus lepidotus*; isolectotypes: BM!, FI!, K!) ; C. de Candolle in A. & C. de Candolle, Monog. Phan. 1: 606 (1878).

Aglaia roxburghiana (Wight & Arn.) Miq. var. *obtusa* C. de Candolle in A. & C. de Candolle, Monog. Phan. 1: 605 (1878). Lectotype (designated here): Ceylon, *Anon.* in *Thwaites* C.P. 1148 (G-DC!).

Aglaia wallichii Hiern var. *brachystachya* C. de Candolle in A. & C . de Candolle, Monog. Phan. 1: 606 (1878). Holotype: *Griffith* 1045 (K!).

Aglaia hoanensis Pierre, Fl. Forest. Cochinch. Fasc. 21, ante t. 336 (1 July 1895). Lectotype (designated here): Dao Chiang on prov. Bien hoa, austro Cochinchina, Sep. 1869, *Pierre* 2779 (P!; isolectotypes: BM!, K!).

Aglaia canariifolia Koord. in Meded. 'S Lands Plantent. 19: 380, 633 (1898). Lectotype (designated here): Celebes [Sulawesi], Minahassa, Menado, 900 m, fl., 10 Jan. 1895, *Koorders* 17899β (BO!; isolectotype: L!).

? *Aglaia littoralis* Talbot, Syst. List Trees Bombay, ed 2: 76 (1902), non Miquel (1868) (= *Aglaia lawii*). Syntypes: Mysore State [N. Kanara District], Kumpta, 25 Nov. 1882, *Talbot* 2955 (BSI); Mysore State, 11 Aug. 1896, *Talbot* s.n. (BSI); Sundararaghavan in Bull. Bot. Surv. India 2: 18 4 (1969).

Aglaia elaeagnoidea (A. Juss.) Benth. var. *glabrescens* Valeton in Hochreutiner, Plant. Bog. Exs. Nov. Vel. Minus Cognitae (1905). Type: Cult. HB, *Teijsmann* 323 (K!, L!).

Aglaia elaeagnoidea (A. Juss.) Benth. var. *formosana* Hayata ex Matsumura & Hayata, Enum. Pl. Formos.: 78 (1906). Type: Taiwan, near Chokachiraisha, 2 March 1898, *Owatari* (TI?, K!).

? *Aglaia poulocondorensis* Pellegrin in Lecomte, Not. Syst. 1: 290 (1910). Lectotype (designated here): Indo-Chine, Iles de Poulo-Condor, (Cochinchine française), fl., Aug. 1876, *Harmand* 748 (P!).

Aglaia formosana (Hayata ex Matsumura & Hayata) Hayata, Ic. Pl. Formos. 3: 52 (1913).

Aglaia roxburghiana (Wight & Arn.) Miq. var. *beddomei* Gamble, Flora of the Presidency of Madras 1: 180 (1915). Lectotype (designated here): Annamallays (Tamil Nadu), 5000 ft [c. 1500 m], *Beddome* s.n. (BM!). Illustration: Beddome, Fl. Sylvatica: t. 130A (1871).

Aglaia roxburghiana (Wight & Arn.) Miq. var. *courtallensis* Gamble, Flora of the Presidency of Madras 1: 180 (1915). Lectotype (designated here): S. India, [W. Ghats, in the Hills of Tinnevelly, Courtallum, fr., 17 July 1907, *Barber* 8388 (K!).

Aglaia parvifolia Merrill in Philipp. Gov. Lab. Bur. Bull. 29: 21 (1905). Lectotype (designated here): Philippine Islands, Island of Burias, June 1904, *Clark* For. Bur. 986 [968] (NY!; isolectotypes: BM!, G!, K!, PNH†, US!).

Aglaia elaeagnoidea var. *pallens* Merrill in Philipp. Jour. Sci., Bot. 3: 413 (1908). Lectotype: Philippine Islands, Camiguin Island, Babuyanes, fl., June / July 1907, *Fenix* Bur. Sci. 4122 (NY!; isolectotype: PNH†).

Aglaia pallens (Merrill) Merrill in Philipp. Jour. Sci., Bot. 13: 297 (1918).

Aglaia cupreolepidota [*cupreo-lepidota*] Merrill in Philipp. Jour. Sci., Bot. 20: 393 (1922). Lectotype (designated here): Philippine Islands, Paluan, Mindoro, fl., April 1912, *Ramos* Bur. Sci. 39579 (A!; isolectotypes: BM!, BO!, GH!, K!, PNH†, US!).

? *Amoora poulocondorensis* (Pellegrin) Harms in Engl. & Prantl, Nat. Pflanzenfam., ed. 2, 19b1: 128, 176 (1940). *Aglaia poilanei* Pellegrin in Bull. Soc. Bot. France, 91: 179 (1944). Holotype: [Vietnam], Annam, Bù Khang pro: Vinh, fl., 11 Aug. 1929, *Poilane* 16713 (P!).

? *Aglaia talbotii* Sundararaghavan in Bull. Bot. Surv. India 2: 184 (1969), nom. nov. pro *Aglaia littoralis* Talbot, non Miq. (1868) (= *Aglaia lawii*).

Aglaia elaeagnoidea (A. Juss.) Benth. var. *beddomei* (Gamble) K.K.N. Nair in Jour. Bombay Natural History Society 78: 426 (1981).

Aglaia elaeagnoidea (A. Juss.) Benth. var. *courtallensis* (Gamble) K.K.N. Nair in Jour. Bombay Natural History Society 78: 426 (1981).

Aglaia abbreviata C.Y. Wu, Flora Yunnanica 1: 240 (1977). Type: [China, Yunnan, Meng-la (24°N 100°E)], 1400 m, fr., 4 Dec. 1953, [*P.I. Mao*] 3262 (KUN!).

Small tree or shrub 5 – 10(– 20) m; bole up to 25(– 50) cm in diameter, sometimes with small buttresses. Outer bark brown, greyish-brown or yellowish-grey, with lenticels and narrow vertical fissures, flaking in thin, irregular, stiff scroll-like scales; inner bark pink or reddish- brown; sapwood yellow; heartwood red. Twigs slender, grey or pale brown, densely covered with pale brown or pale orange-brown peltate scales which have a short fimbriate margin, sometimes in India with stellate scales interspersed at the apex.

Leaves imparipinnate, 6 – 29 cm long, 7 – 21 cm wide, obovate in outline; petiole 2.5 – 10.5 cm, the petiole, rhachis and petiolules densely covered with peltate scales like those on the twigs, sometimes almost stellate in western part of range. Leaflets (1 –)3 – 7, the laterals subopposite, all 2 – 13(– 16) cm long, 1 – 5(– 6) cm wide, subcoriaceous usually narrowly or broadly elliptical, sometimes obovate, the apex rounded or acuminate with the obtuse acumen 2 – 5(– 20) mm long, cuneate at the asymmetrical base upper surface shiny, with scales like those on the twigs, densely covering both surfaces of the leaflets when young, numerous on or densely covering the midrib and sparse to numerous elsewhere when mature, with numerous faint or conspicuous pits on both surfaces, lateral veins 5 – 10 on each side of the midrib, curved upwards and anastomosing with shorter lateral veins in between, midrib prominent, lateral veins and sometimes the reticulation subprominent on both surfaces; petiolules 5 – 15 mm on lateral leaflets, up to 20 mm on terminal leaflet.

Male inflorescence (3 –)9 – 34 cm long and (1 –)2.5 – 25 cm wide; female inflorescence up to 12.5 cm long and 10 cm wide; peduncle up to 6 cm, the peduncle, rhachis and branches with indumentum like the twigs or occasionally with the scales almost stellate. Male flowers up to 2 mm long and 2 mm in diameter, subglobose or depressed globose, sweetly fragrant; pedicel 0.5 – 1.5 mm, densely covered with scales like those on the twigs; calyx up to $\frac{1}{2}$ the length of the corolla, shallowly divided into 5 broadly ovate obtuse lobes, densely covered with scales like those on the twigs, or sometimes with a longer fimbriate margin, on the outside. Petals usually 5, white or yellow, aestivation quincuncial. Staminal tube nearly as long as the corolla, subglobose, yellow, the aperture 0.3 – 0.6 mm across with a dentate margin; anthers 5 about half the length of the tube, inserted half way up the tube with their apices usually just protruding through the aperture. Ovary subglobose, densely covered with stellate scales, with 2 loculi each containing 1 or 2 ovules; stigma ovoid with two small apical lobes; the ovary and stigma together about $\frac{2}{3}$ the length of the staminal tube. Female flowers 2 – 3 mm long, 2.5 – 3 mm wide; the aperture of the staminal tube up to 0.7 mm in diameter; otherwise like the male flowers.

Infructescence up to 12 cm long and 10 cm wide; peduncle up to 5 cm, the peduncle, rhachis, branches and fruit-stalks densely covered with peltate scales like those on the twigs. Fruits 1.1 – 2 cm long, 1.3 – 1.5 cm in diameter, subglobose, ellipsoid or obovoid, orange, brown or red, indehiscent, densely covered with scales like those on the twigs, sometimes glabrescent; pericarp thin, soft. Loculi 2, each with 0 – 1 seed; seed c. 10 mm long, 6 mm wide and 3 mm thick, usually completely covered with a thin, white, gelatinous, sweet aril. In Australia, the pericarp is red, with few scales, and the aril is a small piece of tissue attached to the raphe. Fig. 34.

DISTRIBUTION. India, Sri Lanka, Taiwan (southern cape), Vietnam, Cambodia, Thailand, Malay Peninsula, Sumatra (Bangka & Belitung), Borneo, Philippine Islands, Java, Bali, Sulawesi, Maluku, New Guinea, Australia, Vanuatu, New Caledonia, Samoan Islands. Fig. 35.

ECOLOGY. Found in secondary forest, deciduous forest, along beach, river banks, and in *Barringtonia* formation on sand, granite, coral and limestone. Rather common. Alt.: sea level to 1000 m.

USES. Aril edible, sweetish, tasty.

VERNACULAR NAMES. Vietnam: Cay Gi. Thailand: Khanghao. Java: Nanaykaan (Djasilin), Kemoebang, Kidang, Pandjal Kidang, Petatjara Prentil. Bali: Mata-mata (Bajau).

Representative specimens. INDIA. Bombay Presidency, region East of Goa boundary [15° – 16°N], fl, 26 Nov. 1950, *Fernandes* 1898 (K!). Courtallum, S. Tamil Nadu, rather dry forest above village in river valley, c. 800 m, fr., 5 June 1976, *Kostermans* 26000 (K!). Ram Ghat, fl., Nov., *Ritchie* 1658 (E!, K!).

SRI LANKA. Eastern Province, Trincomalee District, Uppuveli, in disturbed scrub close to beach near Blue Lagoon, sea level, ♂ fl., 28 March 1977, *Cramer* 4942 (K!). Ratnapura District, Belihuloya, 200 m,fl.,

Kostermans 23463 (AAU!, K!, L!). Opanake, 180 m, fr., 17 Oct., 1947, *Worthington*, 3208 (K!). Anuradhapura, just south of Akirikande, between M.P. 94 & 95, fr., 9 Dec. 1970, *Fosberg* with *Balakrishnan* 53463 (K!).

TAIWAN. Oulanpi [21°54'N 120°53'E], fr., 15 July 1967, *Yang et Liao* 10805 (L!). Heng Chun Branch [22°03'N 120°04'E], ♂ fl., 16 May 1966, *Jih-Ching Liao* 10576 (L!).

VIETNAM. Cochinchina, Trangbom, 60 m, fl., *Poilane* 184 (K!).

THAILAND. Chonburi Province: S.E., 60 km east of Sriracha [13°20'N 101°05'E], low, fl., 19 Nov. 1969, *van Beusekom & Smitinand* 2280 (AAU!, C!); Sriricha District, Kew Kiee, 200 m, fl., 28 Jan. 1975, *Maxwell* 75-41 (AAU!); Bahn Beung District, Kow Kieo [13°16'N 101°05'E], 400 m, fl., 25 Sep. 1976, *Maxwell* 76-651 (AAU!). Saraburi Province, Muang District, Sahm Lahn forest, 150 m, fl., 24 Aug. 1974, *Maxwell* 74-838 (AAU!). Chanthaburi Province, Bong Nam Rawn District, North Soy Dow Mountain, 200 m., fr., 6 May 1975, *Maxwell* 75-504 (AAU!). Prachuap, Hui Yang, fl ., 7 Oct. 1930, *Put* 3238A (C!). E., Nakhon Ratchasima Province: Khao Kieo, Khao Yai National Park, [14°45'N 102°E], 500 – 600 m, fl., 31 Oct. 1969, *van Beusekom & Charoenpol* 1951 (AAU!, C!, E!); Eastern part of the Khao Yai National Park, 80 km at the Korat-Sattahip highway, 400 m, fl., 11 Aug. 1968, *K. Larsen, Santisuk & Warncke* 3304 (AAU!, C!, E!).

PENINSULAR MALAYSIA. Perlis, Kaki Bukit, fr., 12 April 1938, *Salleh* SFN 35260 (K!).

SUMATRA. Lampong, Poeloe Seboekoe, fl., *Teysmann* 7524 (L!). Bangka Island: Soendei, fr., *Teysmann* 65 (L!); Soengei Liat, fr., *Teysmann* 7521 (L!).

SARAWAK. Miri (Niah), Gg Subis, c. 120 m, partially exposed situation on dry limestone cliff, yg fl. 5 June 1962, *Anderson* S 16032 (K!). SABAH. Pantan-pantan, seashore, fl., 18 June 1936, *Rahman* 6294 (K!). N. , Selangan Is., rocky coral shore, yg fr., 17 July 1938, *Symington* 35357 (K!).

PHILIPPINE ISLANDS. Basilan Island, fl., *Hutchinson* 6112 (NY!). Mindanao , Mt Galintan, Davao Province, c. 180 m, ♀ fl., 3 June 1927, *Ramos & Edaño* 48920 (NY!, UC!). Palawan, Ursula Island, 0 – 5 m, fr., 2 Oct. 1961, *Olsen* 482 (C!). JAVA. S.W., Tandjung Lajar, Udjung Kulon Reserve, hillside, c. 10 m, fl., 17 Aug. 1960, *Kostermans & Kuswata* 50 (K!). yg. fr., *Horsefield* 529 (K!).

BALI. fl., *Teysmann* 2769 HB (K!). Banjupoh, 200 m, old secondary forest, fl., 5 April 1964, *Dilmy* 1078 (K!).

NUSA TENGGARA. Flores, fr., 6 Dec. 1973, *Schmutz* 3596 (L!).

SULAWESI. N., Gorontalo, fl., *Riedel* s.n. (K!). C., hills above Donggala [0°30' – 1°30'S 119°30' – 120°30'E], 100 m, fl., 22 April 1979, *van Balgooy* 2975 (K!).

MALUKU. Morotai, former P.O.W. camp, 30 m, fl., *Kostermans* 1567 (K!). Dawalore, *Riedel* s.n. (K!). Little Kei Island, fl. & fr., Sept. 1874, *Moseley, Challenger Expedition s.n.* (K!).

IRIAN JAYA. N.W., Sorong [131°15'E 0°53'S], Misool, near Waima, fl., 11 Sep. 1948, *Pleyte* 785 (K!).

PAPUA NEW GUINEA. Morobe District: coast, fl. & fr., 3 Jan. 1948, *Womersley* 2970 (K!, LAE!); Lae Subdistrict, Buso Musik Island, [7°30'S 147°15'E], sea level, fl. & fr., *Streimann & Kairo* NGF 39426 (K!).

AUSTRALIA. Western Australia, 13.5 km north east of Crystal Head, S.W. Osborn Island, Kimberley Coast, fr., 26 Jan. 1989, *Keneally &*

Fig. 34. *A. elaeagnoidea*. Habit with inflorescence x¹/₂. Half flower x12. Habit with infructescence x¹/₂. Peltate scale x60.

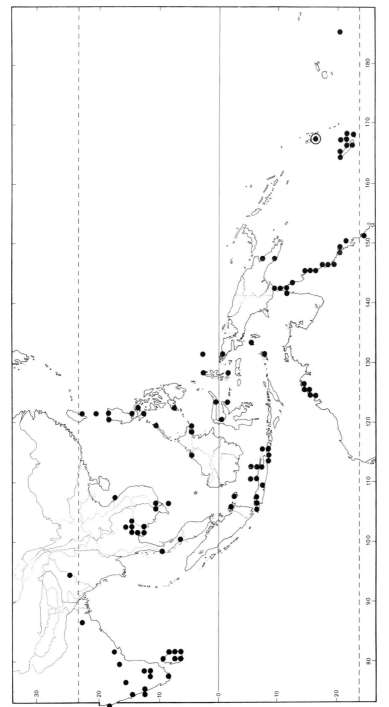

Fig. 35. Distribution of *A. elaeagnoidea*.

Hyland 108 53 (FHO!). Queensland: Mapoon, S. of Port Musgrave, fl., 27 Dec. 1980, *Morton* AM-1006 (BRI); Clump Point, fl., 4 Nov. 1951, *L.S. Smith* 4916 (BRI!, CANB!, L!); Magnetic Island in Horseshoe Bay, North Kennedy District, fr., 22 April 1980, *Sandercoe* 662 (BRI).

VANUATU. Aneityum Island, Aname, 150 m, fl., Sep. 1929 *Wilson* 987 (K!). Enomanga Island: fl., Aug. 1829, *Bennet* s.n. (K!); Dillon Bay, sea level, fl., 5 June 1928, *Kajewski* 357 (K!).

NEW CALEDONIA. Nouméa: fl., Oct. 1923, *White* 2102 (K!); Anse Vata, South, Pacific Base Command Headquarters, fl., 20 Nov. 1947, *Bucholz* 1402 (K!); Anse Vata, 5 m, fr., 30 April 1964, *McKee* 11475 (K!).

SAMOAN ISLANDS. Tobie Island, fl., *Barclay* s.n. (K!).

Aglaia elaeagnoidea is a widespread and variable species which occurs throughout the range of the genus except for the Solomon Islands and it extends beyond the range of all other species in the genus in Western Australia. It is most frequent in the non-equatorial parts of the range of *Aglaia*, which have a more seasonal climate. *A. elaeagnoidea* is usually coastal, especially in the east of the range, but it is also found inland in India, Sri Lanka and Thailand, on dry sandy soils and limestone. The large pale orange or almost white peltate scales are characteristic, but they vary in colour, in the presence and extent of the fimbriate margin and in their density on the plant. In the eastern part of the range, the scales are larger, paler in colour and more frequently entire than in the west, the density is greater on the leaves, but the fruits may have fewer scales than in the west. Spirit material and information on the fruit, received from K. Keneally, shows that, in Kimberley, Western Australia, the aril is vestigial and the pericarp ingested by birds, especially Torres Straits Pigeons, which swallow the whole fruits.

In India, *A. elaeagnoidea* is sometimes a larger tree than in the rest of the range of the species. Large-leaved specimens can be almost indistinguishable from *A. edulis*, unless fruits are present; these are much larger and are usually 3-locular in *A. edulis*; the upper surface of the leaflets in *A. elaeagnoidea* is slightly shiny when dry and is not so in *A. edulis*.

In the flowers of two unnumbered collections from India, which have fruits on the same shoot as the flowers, the dehisced anthers contain pollen grains. The pollen from both of these specimens is irregular in shape and therefore probably infertile; those from a collection by Stocks stained with cotton blue, those from one by Dalzell did not.

30. Aglaia smithii Koord. in Meded. 'S Lands Plantent. 19: 383, 635 (1898). Lectotype (designated here): Celebes [Sulawesi], Provincia Minahassa, Ratatotok, bivak Totok, fl., 25 March 1895, *Koorders* 17917B (BO!; isolectotype: L!).

Aglaia dysoxylonoides Koorders quoad specim.: Celebes, Minahassa, Menado, yg fr., 8 April 1895, *Koorders* 17893 (L!).
Aglaia bicolor Merrill in Philipp. Jour. Sci. 4: 270 (1909). Lectotype (designated here): Philippine Islands, Luzon, Province of Cagayan, fl., June 1906, *Klemme* For. Bur. 4288 (US!; isolectotypes: G!, K!, PNH†).
Aglaia badia Merrill in Philipp. Jour. Sci. 4: 270 (1909). Lectotype

(designated here): Philippine Islands, Luzon, Cagayan Prov., fl., May 1907, *Klemme* For. Bur. 7082 (US!; isolectotype: G!, PNH†).

Aglaia ramosii Quisumb. in Philipp. Jour. Sci. 41: 326, t. 6 (1930). Lectotype (designated here): Philippine Islands, Luzon, Rizal Province, Mt Irid, 2400 ft [c. 730 m], fl., 17 Nov. 1926, *Ramos & Edaño* Bur. Sci. 484 26 (UC!; isolectotypes: NY!, PNH†).

Tree up to 10 m. Twigs densely covered with dark reddish-brown peltate scales which are depressed and very dark in the centre and have a paler, irregular margin.

Leaves 27 – 47 cm long; petiole 6 – 8 cm long, the petiole, rhachis and petiolules densely covered with scales like those on the twigs. Leaflets (? 5 –)9 – 15, the laterals subopposite, all 7 – 17(– 19) cm long, 2 – 6(– 8) cm wide, usually obovate, sometimes ovate and markedly asymmetrical, acuminate at apex, the obtuse acumen 5 – 10 mm, usually cuneate, sometimes rounded at the very asymmetrical base, with numerous pits or scales like those on the twigs on the upper surface, densely covered with scales like those on the twigs and a few larger and darker ones interspersed on the lower surface; veins 9 – 13 on each side of the midrib, the midrib prominent, lateral veins subprominent on the lower surface, usually curved upwards, sometimes ascending and curved upwards near the margin; petiolule 5 – 15 mm, up to 20 mm on the terminal leaflet.

Inflorescences in the axils of several leaves near the apex of the shoot, up to 28 cm long; peduncle c. 6 cm, the peduncle, rhachis and branches densely covered with scales like those on the twigs. Flower c. 2.5 mm long and 3.5 mm wide; pedicels c. 3 mm, densely covered with orange-brown scales which have a paler margin. Calyx c. 0.8 mm long, cup-shaped, divided into 5 rounded lobes. Petals 5, pale yellow, aestivation quincuncial. Staminal tube c. 2 mm long and 3 mm wide with 10 triangular lobes, with stellate yellow hairs on the margin; anthers 5, c. 0.7 mm long and 0.4 mm wide, inserted below the margin, with yellow stellate hairs, the staminal tube thickened below and between the anthers. Ovary c. 0.6 mm high and 0.8 mm wide, depressed-globose, densely covered with large peltate scales up to 0.3 mm in diameter; loculi 2; stigma c. 0.6 mm long and 0.7 mm wide, depressed-globose.

Infructescence c. 18 cm long and 16 cm wide; peduncle 3 – 4 cm, the peduncle, rhachis and branches densely covered with scales like those on the twigs. Unripe fruits c. 1 cm long, obovoid, with numerous bumps and densely covered with scales on the outer surface.

DISTRIBUTION. Philippine Islands (Palawan and Basilan), Nusa Tenggara, Maluku (Tanimbar Island), New Guinea (Adi Island). Fig. 36.

ECOLOGY. Found in primary and coastal forest on clay. Alt.: sea level to 40 m. Rather scarce to common. USES. Wood is used for poles (Irian Jaya, Fak-fak, Kembala).

VERNACULAR NAMES. Sulawesi: Monjowojan. Maluku: Alawe. New Guinea: Doeren, Mansaambree (Biak); Timtimser (Keras).

Representative specimens. PHILIPPINE ISLANDS. Basilan, fl., Aug. 1912, *Miranda* 18888 (L!).

NUSA TENGGARA. W. Sumbawa, Batu Lanteh Mountain, trail from

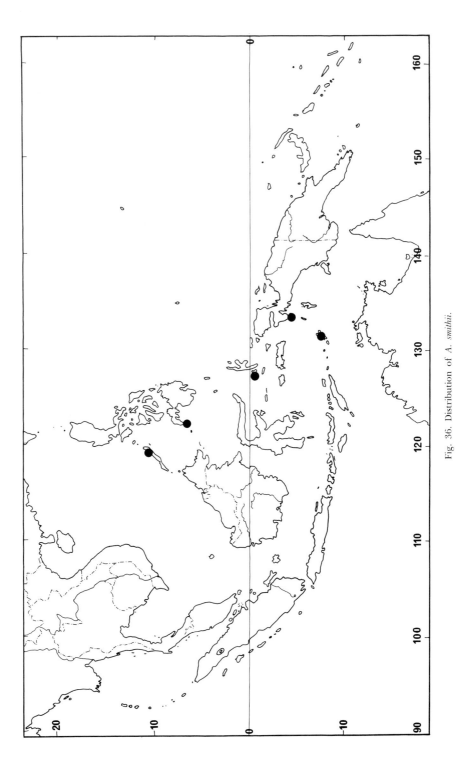

Fig. 36. Distribution of A. *smithii*.

Batudulang to summit, 700 – 800 m, fl., 13 April 1961, *Kostermans* 18114 (NY!).

MALUKU. Tenimber [Tanimbar] Island (Timor Laoet), P. Jamdena, Otimmer, low, fl., 28 March 1938, *Buwalda* 4477 (L!, NY!).

IRIAN JAYA. Kaimana District, Adi Island, fr., 8 March 1961, *Moll* BW 9892 (L!).

The indumentum in *Aglaia smithii* is similar to that of *A. korthalsii* and *A. speciosa*, but it is usually much more dense. *A. smithii* occurs only in Wallacea.

31. Aglaia variisquama C.M. Pannell spec. nova *Aglaiae squamulosae* King similis sed ramunculis crassioribus, foliis magis coreaceis, squamis fuscioribus nec dense paginam inferiorem folioli obtegentibus. Squamae aequaliter distributae obscure aurantiaco-brunneae squamis pallidis intermixtae referunt indumentum paginarum inferiorum foliolorum *Aglaiae rimosae* (Blanco) Merrill. Holotype: Sarawak, Sabal Tapang Forest Reserve, 70th mile, Serian road, c. 250 m, ♀ fl., 12 May 1974, *Tong* S 34271 (FHO!, isotype: L!).

Tree up to 20 m, with few branches and a small, compact crown. Bole up to 15 m, up to 1 m in circumference. Outer bark smooth, green, brown, reddish-grey or black; inner bark pale yellow or red; cambium white, sapwood pale yellow. Twigs greenish-brown or pinkish-brown, longitudinally wrinkled, densely covered with dark orange-brown peltate scales which are up to 0.3 mm in diameter and have an irregular margin, with some scales which have a dark orange-brown centre and paler margin interspersed.

Leaves imparipinnate, obovate in outline, up to 79 cm long and 58 cm wide; petiole 7 – 15 cm, the petiole, rhachis and petiolules densely covered with scales like those on the twigs. Leaflets 9 – 13, the laterals subopposite, all 7.5 – 30 cm long, 4 – 13 cm wide, coriaceous, obovate, recurved at the margin, acuminate at the apex, the obtuse acumen up to 5 mm long, cuneate at the base, the lower surface of the leaflet with numerous scales like those on the twigs, the scales evenly distributed and visible to the naked eye as tiny dark and pale brown spots; veins 10 – 17 on each side of the midrib, ascending and curved upwards near the margin, the midrib depressed on the upper surface and prominent on the lower, the lateral veins subprominent on the lower surface; petiolules 5 – 19 mm.

Inflorescence up to 32 cm long and 22 cm wide, in the axils of c. 3 leaves near the apex of the shoot; peduncle 1.5 – 10 cm, the peduncle, rhachis and branches densely covered with scales like those on the twigs. Male flowers c. 1.5 mm long and 2 mm wide, depressed-globose; pedicels up to 1.5 mm, the calyx and pedicels with numerous pale peltate scales which have a fimbriate margin. Calyx c. 0.5 mm long, divided into 5 sub-rotund lobes. Petals 5, yellow or white; aestivation quincuncial. Staminal tube c. 1.2 mm long and 1.5 mm wide, depressed globose, the aperture c. 0.5 mm in diameter; anthers 5, c. 0.6 mm long and 0.4 mm wide, inserted about half way up the staminal tube and just protruding through the aperture, with a few simple hairs at the base and apex of each anther and on the inner surface of the staminal tube, the staminal tube with longitudinal, thickened ribs, one below each anther and one between each pair of adjacent anthers. Ovary c. 0.2 mm high and 0.6 mm wide, depressed-globose, densely covered with pale orange-brown peltate scales which have a fimbriate margin; locule 1, containing 2 ovules; stigma c. 0.7 mm long and 0.8 mm wide, ovoid.

Female flowers c. 3 mm long and 4 mm wide, depressed-globose; sessile. Calyx divided into 5 sub-rotund lobes densely covered with scales like those on the twigs. Petals 6, thick and fleshy, with few orange-brown or pale orange-brown peltate scales which have a fimbriate margin and a dark central spot on the outer surface. Staminal tube c. 1.8 mm long, 1.7 – 2 mm wide with a few pale brown stellate scales on the outer surface, the aperture 0.5 mm in diameter; anthers 5, c. 0.6 mm long and 0.4 mm wide, with hairs and with the staminal tube thickened as in the male; ovary c. 0.6 mm long and 0.9 mm wide, depressed-globose, densely covered with pale orange-brown peltate scales which have a fimbriate margin; loculi 2; stigma c. 0.7 mm long and 0.8 mm wide, ovoid and longitudinally ridged.

Infructescence c. 16 cm long and 10 cm wide. Fruits c. 4 cm long and 4 cm wide, subglobose, sometimes with a small beak, yellowish-brown, indehiscent, densely covered with pale orange-brown peltate scales on the outer surface; latex white. Seed 1, with a fleshy translucent aril; testa dark brown.

DISTRIBUTION. Peninsular Malaysia, Borneo. Fig. 37.

ECOLOGY. Found in secondary forest, primary forest, kerangas, swamps and along rivers; on sand. Alt.: 200 to 430 m. Scattered.

VERNACULAR NAMES. Peninsular Malaysia: Bekak, Kaum Buka, Kedongong. Borneo: Lantupak (Dusun Kinabatangan); Segera (Iban).

Representative specimens. PENINSULAR MALAYSIA. Perak: tributary of the Sg Trong, W. of the Bubu Massif, E. of Trong, 200 m, fr., 27 Feb. 1970, *Everett* FRI 13982 (L!); W. side of Gg Bubu massif, E. of Trong, 300 m, fl., 26 Feb. 1970, *Everett* FRI 13975 (L!). Johore, 5.5 mile Kota Tinggi – Mawai road, low, fr., 11 May 1935, *Corner* SFN 29303 (L!).

SARAWAK. 1st Division: Gg Gaharu, 70th mile, Serian – Simanggang road, Simunjan, 400 m, fr., 7 Oct. 1974, *Ilias & Azahari* S 35640 (L!); Sabal Sawmill, Sabal Forest Reserve, 70th mile Serian road, fl., 14 May 1974, *Tong et al* S 34303 (FHO!, L!).

SABAH. Sandakan, Segaliud Lokan Forest Reserve, mile $33^{1}/_{2}$, fr., 18 Jan. 1975, *Gibot* SAN 80967 (L!). Lahad Datu, Mostyn mile 16 Kalumpang Forest Reserve, 100 m, fl., 18 Mar. 1964, *Agam & Aban* SAN 40869 (L!). Tawau, Ulu Umas Umas, 100 m, fl., 21 Oct. 1965, *Nordin* SAN 46062 (L!). Kinabatangan, Tamegang Timber Camp near Kampong Pangkaian, st., 23 Nov. 1968, *Kokawa & Hotta* 1501 (L!).

KALIMANTAN. E., Balikpapan District, Mentawir River region, yg fr., 24 July 1954, *Kostermans* 9811 (L!).

Aglaia variisquama resembles *A. squamulosa* but the twigs are stouter, the leaves more coriaceous and the scales darker and not densely covering the lower leaflet surface. The evenly distributed dark orange-brown scales with paler scales interspersed is similar to the indumentum on the lower surfaces of the leaflets of *A. rimosa*.

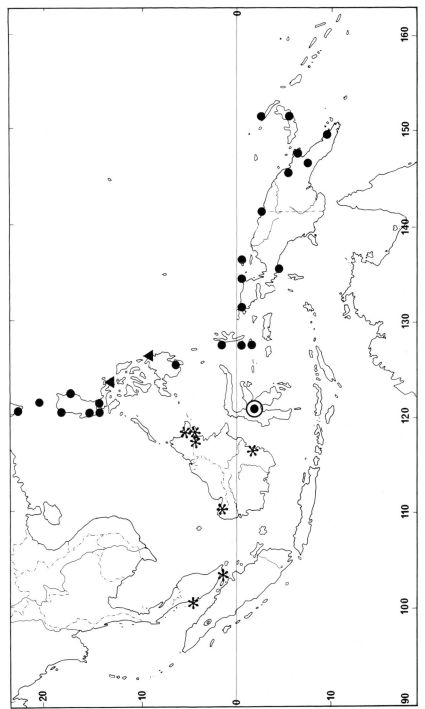

Fig. 37. Distribution of A. *variisquama**, A. *rimosa* ● and A. *costata* ▲

32. Aglaia rimosa (Blanco) Merrill, Sp. Blancoanae, 212 (1918).

Portesia rimosa Blanco, Fl. Filip.: 297 (1837). Neotype: Philippine Islands, Luzon, Batangas Province, San José, Feb. 1915, *Merrill Sp. Blancoanae* 765 (PNH†; isoneotypes: BM!, BO!, K!, P!); see also Veldkamp, Flora Malesiana Bulletin 10(2): 143 – 145 (1989).

Trichilia rimosa (Blanco) Blanco, op. cit. ed. 2: 250 (1845), ed. 3, 2: 99 (1878).

Aglaia [Aglaja] denticulata Turcz. in Bull. Soc. Imp. Nat. Mosc. 31: 410 (1858). Syntypes: Philippine Islands, Luzon, Prov. Tayabas, fl., *Cuming* 761 (BM!, CGE!, E!, FI!, G!, G-DC!, K!, L!, LE!, NY!, OXF!, P!); C. de Candolle in A. & C. de Candolle, Monog. Phan. 1: 612 (1878).

Aglaia [Aglaja] hexandra Turcz. in Bull. Soc. Imp. Nat. Mosc. 31: 410 (1858). Syntypes: Philippine Islands, [Luzon, Nuova Euja], fl., *Cuming* 1410 (BM!, CGE!, FI!, G!, G-BOIS!, K!, LE!, OXF!, W!); Mindora, *Cuming* 1583 (BM!, CGE!, FI!, G-BOIS!, G-DC!, K!, L!, LE!, OXF!, W!); C. de Candolle in A. & C. de Candolle, Monog. Phan. 1: 614 (1878).

Aglaia [Aglaja] macrobotrys Turcz. in Bull. Soc. Imp. Nat. Mosc. 31: 409 (1858). Syntypes: Philippine Islands, [Luzon, Prov. Albay], fl., *Cuming* 902 (BM!, CGE!, E!, FI!, G!, G-BOIS!, K!, LE!, OXF!, W!); C. de Candolle in A. & C. de Candolle, Monog. Phan. 1: 609 (1878).

Aglaia batjanica Miq., Ann. Mus. Bot. Lugd. Bat. 4: 46 (1868). Lectotype (designated here): [Maluku], Batjan, *Anon.* 5620 (U!); C. de Candolle in A. & C. de Candolle, Monog. Phan. 1: 625 (1878).

Aglaia llanosiana C. de Candolle in A. & C. de Candolle, Monog. Phan. 1: 621 (1878). Holotype: Philippine Islands, *Llanos* (G-DC!).

Aglaia goebeliana Warb. in Bot. Jahrb. 13: 345 (1891). Syntypes: Dutch New Guinea, Sattelberg, 3000 ft [c. 900 m], *Warburg* 20103 (B!, BM!).

Aglaia elliptifolia Merrill in Philipp. Jour. Sci. 3: 413 (1909). Lectotype (designated here): Philippine Islands, Batanes Island, Sabtan, fl., June 1907, *Fenix* Bur. Sci. 3733 (G!; isolectotypes: PNH†, NY!); Huang, Pollen Flora of Taiwan: 166, t. 106: 42 – 44 (1972).

Aglaia lanceolata Merrill in Philipp. Jour. Sci. 5: 184 (1910). Lectotype (designated here): Philippine Islands, Luzon, Nueva Vizcaya, fl., May 1909, *Ramos* Bur. Sci. 8141 (G!; isolectotypes: PNH†, US!).

Aglaia loheri Merrill in Philipp. Jour. Sci., Bot. 1914, 9: 533 (1915). Lectotype (designated here): Philippine Islands, [Luzon], Rizal, fl., *Loher* 5682 (US!).

Aglaia diffusiflora Merrill in Philipp. Jour. Sci., 14: 410 (1919). Lectotype (designated here): Philippine Islands, Panay, Capiz Province, Jamindan, fl., April – May 1918, *Ramos & Edaño* Bur. Sci. 31098 (A!; isolectotypes: K!, P!, PNH†, US!).

Aglaia bulusanensis Elmer ex Merrill, Enum. Philipp. Fl. Pl. 2: 373 (1923), in obs., pro syn.: *A. diffusiflora*. Type no.: *Elmer* 16596, Philippine Islands, Island of Luzon, Province of Sorsogon, Irosin (Mt Bulusan), fl., July 1916 (BM!, BO!, G!, GH!, L!, NY!, UC!, W!).

Aglaia reticulata Elmer ex Merrill, Enum. Philipp. Fl. Pl. 2: 377 (1923), in obs., pro syn.; Elmer in Leafl. Philipp. Bot. 9: 3306 (1937), sine diagn. lat. Type no.: *Elmer* 10735, Philippine Islands, Island of Mindanao,

District of Davao, Todaya (Mt Apo), fr., May 1909 (A!, BM!, BO!, G!, GH!, K!, NY!, P!, UC!, US!).

Aglaia subviridis Elmer ex Merrill, Enum. Philipp. Fl. Pl. 2: 373 (1923) , in obs., pro syn.: Elmer in Leafl. Philipp. Bot. 9: 3312 (1937), sine diagn. lat.. Type no.: *Elmer* 13609, Philippine Islands, Island of Mindanao, Province of Agusan, Cabadbaran (Mt Urdaneta), fl., Aug. 1912 (A!, BM!, G!, GH!, K!, L!, NY!, P!, SING!, UC!, US!, W!).

Shrub or tree, 2 – 30 m high. Bole 5 – 52 cm in diameter with buttresses upwards up to 50 cm, outwards up to 1 m, and up to 3 cm thick. Outer bark dark brown to greenish-grey, scaly, with small lenticels, c. 2 mm thick; middle bark green; inner bark pink to dull red, with white latex. Sapwood white to orange-yellow; heartwood red-brown. Twigs pale greyish-brown or orange brown, densely covered with peltate scales which are radiate and have a dark orange brown or dark reddish-brown centre with a paler margin, the margin irregular, entire or ragged (often fimbriate in New Guinea); the scales are variable in size on any one plant, 0.15 – 0.3 mm in diameter, the largest being on the apical bud.

Leaves 30 – 71 cm long, 22 – 50 cm wide, petiole 7 – 26 cm; the petiole, rhachis and petiolules with few to densely covered with scales like those on the twigs. Leaflets (3 –)9 – 11(– 15), the laterals usually subopposite, sometimes alternate, all 5.5 – 23 cm long, 1.5 – 8 cm (– 10 cm in New Guinea) wide, usually obovate, sometimes ovate, acuminate at apex, sometimes tapering into an obtuse acumen up to 1.5 cm long, the acumen usually narrow, acute and up to 1 cm long, sometimes obtuse and up to 4 mm long, rounded or cuneate at the asymmetrical base; dark glossy green on upper surface and dull light green on lower surface when fresh, orange, orange-brown, orange green or pale green when dry, sometimes pitted on upper and lower surfaces, often rugulose on upper surface, with scales like those on the twigs scattered or numerous on the lower leaflet surface, with a few darker scales interspersed, numerous on to densely covering the midrib below, numerous on upper leaflet surface of young leaves, deciduous; veins 7 – 17(– 20 in New Guinea) on each side of the midrib, ascending and curved upwards near the margin, not anastomosing, sometimes with shorter laterals veins in between, midrib prominent, lateral veins usually subprominent, sometimes barely so, reticulation rarely subprominent; petiolules 0.5 – 2 mm on lateral leaflets, up to 2.5 mm on terminal leaflet.

Male inflorescence 13 – 36 cm long, 10 – 33 cm wide, peduncle c. 1.5 cm. Female inflorescence 6 cm long, 5 cm wide, peduncle c. 1 mm. The inflorescence branches with few to densely covered with peltate scales like those on the twigs. Flowers (1.1 –)1.5 – 2.2(– 2.5) mm long, (1.1 –)1.7 – 2.2 mm wide, subglobose; pedicel (0.3 –)0.5 – 0.7 mm to articulation, subtending branchlet 1 – 1.5 mm, with few to densely covered with peltate scales and with a small ovate bract at the base. Calyx 0.7 – 1 mm long, divided into (4 or) 5 rounded lobes which have entire or slightly ciliate margins, usually densely covered on the outside with scales like those on the twigs, sometimes with few or no scales, or calyx pale yellowish-brown in colour and with scales only near the junction with the pedicel. Petals 4 or 5, white to red, pale yellowish-brown when dry. Staminal tube (0.6 –)0.8 – 1.2(– 1.5) mm long, (0.8 –)1 – 1.6 mm wide, obovoid or cup-shaped, aperture 0.7 – 1 mm, shallowly lobed, 5 (or 6) anthers, 0.4 – 0.5 mm long, 0.2 – 0.4 mm wide, narrowly ovoid, curved inwards, with pale yellow margins and dark brown centre, usually inserted about ⅔ up the tube but sometimes lower down, protruding and with the apices pointing towards the centre of the flower, staminal tube thickened (not in New Guinea) below and between the anthers so that each anther occupies a depression, with simple hairs often densely covering the lower part of the staminal tube and between the anthers and along the margins of the anthers. Ovary 0.2 – 0.5 mm high, 0.3 – 0.5(– 0.9) mm wide, depressed globose or ovoid, densely covered with scales like those on the twigs; stigma (0.3 –)0.4 – 0.5 mm long, (0.3 –)0.4 – 0.6 mm wide ovoid or subglobose with two small lobes at the apex, black when dry; loculi 2, each containing ? 1 ovule.

Infructescence 8 – 36 cm long, 13 cm wide, peduncle up to 8 cm, densely covered with scales like those on the twigs. Fruits 1.3 cm (– 1.5 cm in New Guinea) diameter, dull orange to brown, obovoid or sometimes ellipsoid with a beak up to 5 mm long and narrowed to a stipe up to 5 mm long, the pericarp thin, rigid and brittle when dry, densely covered with scales on the outside, green on the inside, with latex; loculi 2, each containing 0 or 1 seed; seed surrounded by a translucent yellow aril.

DISTRIBUTION. Taiwan, Philippine Islands, Sulawesi, Maluku, New Guinea, (including New Britain and New Ireland). Fig. 37.

ECOLOGY. Found in secondary forest along river and along coast on limestone and sandy clay. Alt.: sea level to 1350 m. Rather scattered up to common.

VERNACULAR NAMES. Philippine Islands: Busilac. Maluku: Hitang Mararu (Sahu); Lalasa Daare, Orie. New Guinea: Bowie, Herrig (Manikiong); Mansaambra (Biak); Ihoe, Nanegne. USES. Wood is used for house construction (Halmaheira, Maluku).

Representative specimens. TAIWAN. Mt Hiiran, Heng Chun [22°03′N 120°45′E], fr., 3 Oct. 1966, *Jih-Ching Liao* 10642 (L!).

PHILIPPINE ISLANDS. E. Cagayan, N.E. Luzon, Bagio Cove, coastal sand, fl., 20 June 1981, PNH 150011 (L!). Alabat Island. fr., Sep. – Oct. 1926, *Ramos & Edaño* 48052 (NY!, UC!, US!). Island of Mindanao, District of Davao, Todaya (Mt Apo), fl., July 1909, *Elmer* 11784 (U!).

MALUKU. Halmaheira, W. Tobelo, fr., 18 Dec. 1922, *Beguin* 2296 (L!). Exp. Obi, fl., 1899, *Atasrip* 65 (L!).

PAPUA NEW GUINEA. McAdam Memorial Park, Wau, 1200 m, fr., 11 Jan. 1964, *Pennington* 8038 (FH0!, L!). Milne Bay District, Baniara Subdistrict, E. of Nowata airstrip [9°59′S 149°44′E], fl., 500 m, 2 July 1969, *Kanis* 1126 (K!). Sepik District, Sereni Creek [3°00′S 141°20′E], 30 m, fr., 26 March 1964, *C.D. Sayers* NGF 19541 (K!).

A. rimosa has a characteristic indumentum on the vegetative parts of the plant. This indumentum and the woolly white indumentum or white deposit in the staminal tube bring together a wide diversity in leaflet size and shape. However, there appears to be no consistent characters upon which the group can be subdivided. The species as defined here is recognisable with the naked eye because of the shiny peltate scales which look like evenly distributed orange dots on the lower surfaces of the leaflets.

33. Aglaia costata Merrill in Philipp. Jour. Sci., Bot. 3: 146 (1908). Lectotype (designated here): Philippine Islands, Mindanao, Lake Lanao, Camp Keithley, June 1907, *M.S. Clemens* s.n. (BO!; isolectotype: PNH†).

Aglaia umbrina Elmer ex Merrill, Enum. Philipp. Fl. Pl. 2: 372 (1923) in obs. pro syn.; Elmer in Leafl. Philipp. Bot. 9: 3317 (1937), sine diagn. lat. Type no.: *Elmer* 13770, Philippine Islands, Island of Mindanao, Province of Agusan, Cabadbaran, Mt Urdaneta, fl., Sep. 1912 (A!, BM!, G!, GH!, K!, L!, NY!, P!, U!, UC!, US!, W!).

Tree c. 6 m high, bole c. 10 cm in diameter. Twigs brown or pale pinkish-brown, densely covered with reddish-brown peltate scales which have a fimbriate margin.

Leaves up to 55 cm long and 46 cm wide; petiole up to 22 cm, the petiole, rhachis and petiolules densely covered with scales like those on the twigs. Leaflets 9 – 11, the laterals subopposite, all 6 – 23 cm long and 2.5 – 10 cm wide, ovate or elliptical, acuminate at apex, the broad obtuse acumen 3 – 7 mm, rounded at the slightly asymmetrical base; veins 8 – 12 ascending with pale orange brown peltate scales which have a fimbriate margin and a dark central spot numerous on the lower leaflet surface, with some darker orange brown scales which have an almost entire margin interspersed; petiolules 1 – 2 cm.

Inflorescence c. 38 cm long and 28 cm wide; peduncle 8 cm, the peduncle, rhachis and petiolules densely covered with scales like those on the twigs. Flower c. 2.5 mm long and 2.5 mm wide; pedicel 1 – 3.5 mm, the pedicel and calyx densely covered with reddish-brown peltate scales which have a fimbriate margin. Calyx cup-shaped and divided into 5 rounded lobes. Petals 5, aestivation quincuncial. Staminal tube cup-shaped 0.6 mm high, 1.6 mm wide; anthers 5, inserted just inside the margin of the tube, pointing towards the centre of the flower and occupying the entire aperture. Ovary 0.3 mm high, 0.6 mm wide; loculi ? 5; stigma 0.5 mm long, 0.5 mm wide, ovoid.

Infructescence c. 17 cm long and 8 cm wide; peduncle c. 5.5 cm, the peduncle, rhachis and pedicels densely covered with scales like those on the twigs. Fruits (1 –)1.2 – 2 cm long and (1.5 –)2 cm wide, subglobose or depressed-globose with 10 deep longitudinal grooves so that the fruit has 10 longitudinal lobes, densely covered with scales like those on the twigs; fruit-stalks up to 2.5 cm. Loculi 10, each containing 1 small seed.

DISTRIBUTION. Philippine Islands only. Fig. 37.

Representative specimens. PHILIPPINE ISLANDS. Samar: Mt Bohaton, April 1970, *Madulid et al* 1021, PNH 117996 (L!) & April 1914, *Ramos* Bur. Sci. 1658 (L!).

This species resembles *A. rimosa* except in details of the trichome structure (in the Philippines, the scales of *A. rimosa* are entire, although they are fimbriate in New Guinea) and pigmentation (the outer two thirds of the scale is paler in *A. rimosa*) and in the distinctive fruits which have 10 longitudinal lobes and 10 locules, each of which contains 1 seed; the locule number in the flowers dissected was not clearly visible, but appeared to be 5. This fruit structure and locule number is found in no other species of *Aglaia*.

34. Aglaia agglomerata Merrill & Perry in Journ. Arn. Arb. 21: 322 (1940). Lectotype (designated here): Nordöstliches NeuGuinea [Papua New Guinea, N.E.], Yunzaing, 4500 ft [c. 1350 m], fl., 21 June 1937, *Clemens* 6432 (A!).

Aglaia leeuwenii Harms in Engl. Bot. Jahrb. 72: 172 (1942). Lectotype (designated here): New Guinea [Irian Jaya], Rouffaer River, 175 m, fl., Aug. 1926, *Docters van Leeuwen* 9913 (L!; isolectotypes: B†, BO!).
Aglaia doctersiana Harms in Engl. Bot. Jahrb. 72: 170 (1942). Lectotype (designated here): New Guinea [Irian Jaya], Rouffaer River, 175 m, fl., Aug. 1926, *Docters van Leeuwen* 9962 (L!; isolectotype: B†).

Tree 3 – 32 m, sometimes bearing fruit at 3 – 5 m; bole up to 19 m, 5 – 92 cm in diameter with buttresses upwards up to 2.5 m, outwards up to 3 m and 20 cm thick. Outer bark pale grey, greyish-brown or brown, smooth, with vertical lines of brown lenticels, inner surface yellow; middle bark greenish-yellow to dark red; inner bark yellow or reddish-yellow; sapwood yellow or pink; heartwood dark red or brown, aromatic; latex white. Twigs greyish-brown with numerous to densely covered with, small, very dark reddish-brown peltate scales which have a paler and slightly irregular or fimbriate margin.

Leaves imparipinnate, 20 – 50 cm long, 22 – 40 cm wide, obovate in outline; petiole 3 – 13 cm, the petiole, rhachis and petiolules with numerous to densely covered with scales like those on the twigs. Leaflets (7 –)9 – 13, all similar, the laterals opposite or subopposite, all 8 – 20 cm long 2.5 – 7.5 cm wide, ovate or elliptical, apex acuminate, the obtuse acumen up to 10 mm long, rounded or cuneate at the slightly asymmetrical base, glossy green above, paler below, the upper surface with numerous pits and sometimes rugulose, with numerous to densely covered with scales like those on the twigs on the midrib and numerous on the surface below, with some pits interspersed; veins 11 – 19 on each side of the midrib, ascending, curved upwards near the margin and not quite anastomosing, with some shorter lateral veins in between, midrib prominent below, lateral veins subprominent; petiolules 5 – 10 mm.

Male inflorescence 21 cm long, 10 cm wide, peduncle 4 cm, the peduncle, rhachis, branches, pedicels and calyx densely covered with reddish-brown peltate scales which have a fimbriate margin, or stellate scales. Flower c. 1.5 mm long and 2 mm wide, depressed globose, fragrant; pedicels 0.5 – 3 mm. Calyx 0.5 mm long divided into 5 broad rounded lobes. Petals 5, yellow, aestivation quincuncial. Staminal tube 0.7 mm long, 1 mm wide, cup-shaped, aperture 1 mm across, the margin shallowly lobed; anthers 5, ovoid, inserted $3/4$ up the staminal tube and protruding for $2/3$ of their length. Ovary depressed globose densely covered with, pale brown peltate scales which have a fimbriate margin. Stigma ovoid, longitudinally ridged; the ovary and stigma together 1 mm long or longer and the stigma filling the space within the anthers and the staminal tube.

Female inflorescence c. 11 cm long and 11 cm wide; peduncle 2 – 3 cm, the peduncle, rhachis and branches densely covered with scales like those on the twigs. Flowers 2 – 2.5 mm long, 1.5 – 2.5 mm wide, subglobose; pedicels 0.5 – 1 mm, calyx about one third as long as the corolla divided into 5 deep broad lobes, with a few peltate scales which have a fimbriate margin or stellate hairs on the outside, petals 5, staminal tube 1.2 – 1.5

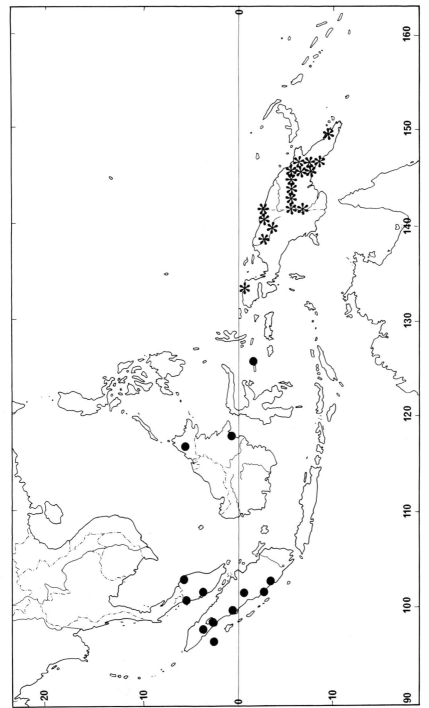

Fig. 38. Distribution of *A. speciosa* ● and *A. agglomerata* *.

mm long, 1.2 – 1.5 mm wide, obovoid or subglobose, the aperture c. 0.7 mm wide, shallowly lobed or dentate, anthers 5, $^{1}/_{2} - ^{2}/_{3}$ as long as the tube, sometimes with a few simple hairs, just protruding through the aperture; the ovary and stigma together nearly as long as the staminal tube, the stigma ovoid, longitudinally ridged with the impression of the anthers, ovary about one third the length of the tube, densely covered with scales on the outside, with 1 or 2 loculi, each containing 1 ovule.

Infructescence either with 1 or 2 large fruits or 8 – 10 cm long and 7 – 13 cm wide with up to 36 small fruits; peduncle 0.5 – 2.5 cm long. The large fruits up to 6.5 cm long and 4 cm wide obovoid or ellipsoid, reddish-brown or dull yellow, indehiscent, the pericarp thick, woody and wrinkled, densely covered with pale or reddish-brown peltate scales which have a fimbriate margin on the outside and yellow inside. Seed 1, with a complete pale yellow aril. The small fruits 1.7 cm long, 1.4 cm wide, obovoid, pericarp dull orange, 1 – 2 mm thick and soft when fresh, brittle when dry, indehiscent; loculi 2, each containing one seed; seeds c. 1.4 cm long, 1 cm across and 0.6 cm thick, with a complete translucent orange or white aril less than 1 mm thick and black testa.

DISTRIBUTION. Found only on the island of New Guinea. Fig. 38.

ECOLOGY. Found in primary lowland to midmontane forest, secondary forest, river banks on clayey and volcanic soil, sometimes on limestone. Rare to sometimes relatively common, scattered. Alt.: 10 to 1800 m.

VERNACULAR NAMES. Sila, Wangimbu (Naho language); Herrab, Herrib, Oensommie, Seraka, Serierieribga (Manikiong); Milom (Waskuk); Bani (Wagu); Teble-tiri (Wigote dial., Wapi); Arla (in E. Highlands).

Representative specimens. PAPUA NEW GUINEA. Morobe District, Bulolo, Crooked Creek, [7°12′S 146°39′E], c. 730 m, Aug. 1962, *Crompton NGF 21023* (K!). Eastern Highlands, 1 mile W. of Ikana [6°22′S 145°56′E], c. 1500 m, fr., 24 July, 1963, *Hartley* TGH 12099 (K!, L!).

A. agglomerata has numerous peltate scales on the lower leaflet surface. It resembles *A. crassinervia*, but the leaflets are more coriaceous and they have a recurved margin. It is not clear whether the two forms of the fruits described reflect different stages of maturity or whether the plants with infructescences which have many small fruits belong to a variant distinct from that which has few large fruits.

35. & 36. Aglaia speciosa *group*

The *Aglaia speciosa* group clearly consists of more than two taxa, but with the material available, it is only possible to separate *Aglaia speciosa* and *Aglaia korthalsii*, of which *A. korthalsii* is more widespread and variable.

The leaflet shape, including the apex, shows no pattern which can be correlated with the variation observed in other organs. All the scales have more or less the same morphology, showing only subtle differences in density and coloration, which does not correlate with what is known of the variation in the fruits. The leaves have from 3 to 11 leaflets and these vary in shape from elliptical to ovate; the base is cuneate or rounded and the obtuse acumen may be quite short and blunt or up to 15 mm long. The indumentum consists of small, reddish brown peltate scales, varying in density from sparse to almost densely covering the lower surface. In some specimens, these scales are uniformly coloured throughout, and in others they have a dark centre and paler margin; the margin is usually slightly fimbriate. In some specimens, pigmented scales are interspersed with others which are almost colourless. *A. speciosa* sensu stricto appears distinct because of the greater density of scales on the lower surface of the leaflets and usually higher number of leaflets, but in practise the distinction between these specimens and the rest of the aggregate cannot always be made; for instance some specimens of *A. korthalsii* (defined by leaflet number) have equally numerous scales. The terminal leaflet of the cultivated form of *A. korthalsii* in Kelantan is usually shorter than the uppermost pair of lateral leaflets.

The flower structure is fairly constant, with 5 (or 6) calyx lobes and the same number of petals, a cup-shaped staminal tube and five anthers inserted inside the rim and pointing towards the centre of the flower; the subglobose ovary is densely covered with reddish-brown stellate scales and the stigma is either narrow, cylindrical and longitudinally ridged (only in *A. speciosa*?) or ovoid and minutely bilobed at the apex. Fig. 42.

Few collections include mature fruits, but among those available, three different fruit types may be distinguished. All are indehiscent, with a single line of weakness in the pericarp which encircles the fruit longitudinally and loculicidally. The pericarp is orange in colour and densely covered with peltate scales on the outside, while the inner surface is without scales and shiny. The one or two, rarely three, seeds are surrounded by a thin, translucent, orange or yellow, gelatinous aril. The three types differ from each other in the following ways. The fruit which appears to belong to *A. speciosa* sensu stricto measures up to 3 cm long by 2.7 cm wide and has a thin (c. 1 mm), rather inflexible pericarp which becomes brittle when dried or preserved in alcohol. That which vegetatively most closely resembles the type specimen of *A. korthalsii* has a larger fruit, up to 4 cm long and 3.5 cm or more wide, with a thicker, more flexible and shallowly longitudinally wrinkled pericarp, which if ripe becomes squashed when pressed and dried, but remains flexible when preserved in alcohol. The third, which closely resembles *A. aquatica* Pierre, and includes trees cultivated in Kelantan, N.E. Peninsular Malaysia, has smaller fruits (2 – 2.5 cm long and 1 – 1.5 cm wide), with a thin, flexible pericarp which may be peeled away both in fresh and alcohol-preserved fruits; it is included in *A. korthalsii* in the present treatment. The aril of the first two is orange in colour and has a characteristic sweet taste, while that of the last is yellow and somewhat astringent. Biochemical analysis of the arils of the first two, collected in Ketambe, N. Sumatra, reveal dry weight values of 15.5% sugars, 5.8% starch, 11% lipids, 2.1% protein and 65.6% residue for *A. speciosa* and 24.5% sugars, 5.2% starch, 12.8% lipids, 4.8% protein and 52.7% residue for *A. korthalsii* (Pannell & Kozioł, 1987, where they are assigned to the two species: *A. speciosa* and *A. korthalsii*). The fruits of both species are eaten by *Pongo pygmaeus* at Ketambe, in the Gunong Leuser National Park in Sumatra (Rijksen, 1978). In his ecological study of this ape, Rijksen treats the two species of *Aglaia* recognised here as a single species, *A. speciosa*. I have observed *Hylobates lar* ingesting the aril of *A. korthalsii* at the same site (Ketambe, Aceh, Sumatra). The aril of the cultivated form of *A. korthalsii* found in Kelantan, Peninsular Malaysia is eaten by man. Some specimens of *A. korthalsii* from Borneo as well as *A. cauliflora* (here treated as a synonym of *A. korthalsii* and in which the fruits are borne on the older wood of the twigs) from Sulawesi have small fruits which appear, in the dried state, to most closely resemble the Kelantan plant. I have not seen fruits of *A. aquatica*, but on vegetative and floral characters, it falls within the range of variation here recognised for *A. korthalsii*. Occasionally a specimen which is identified as *A. korthalsii* from its leaves has the fruits of *A. speciosa* (e.g. *Chivers* DCS 073 from Kuala Lompat, Peninsular Malaysia). The two species at Ketambe were fruiting at the same time in July 1983 and it would be interesting to discover whether there are differences in dispersal biology between the two, such as preference for one or other by different primate species. Even if mature fruits were available with all collections from female trees, the absence of reliable distinguishing features in the leaves and flowers for trees bearing the different forms of fruits might mean that male trees could not be placed, if the group was further subdivided.

More complete material for the group throughout its range might result in the detection of patterns of variation which would allow the delimitation of reliably distinguishable species or subspecies. Similarly, more information about the animals which eat the arils and disperse the seeds and about the biochemistry of the aril might, however, result in an improved understanding of the biology of the different forms and the functional significance of the morphological variation. *A. speciosa* is here delimited on its higher number of leaflets, their narrow, usually elliptical, shape and the density of the indumentum on the lower surfaces of the leaflets, and in fruit type. *A. korthalsii* usually has fewer than 7 leaflets, which are broader and have fewer scales than *A. speciosa*; it includes a wide range of fruit forms.

35. Aglaia speciosa Blume, Bijdr.: 171 (1825). Lectotype (designated here): Java, [at the foot of Salak], fl. [Oct. – Nov.], *Anon.* 627 (L!); Miq. Fl. Ind. Bat. 1: 543 (1859); Miq., *Ann. Mus. Bot. Lugd. Bat.* 4: 46 (1868); C. de Candolle in A. & C. de Candolle, Monog. Phan. 1: 614 (1878); Koorders & Valeton, Atlas der Baumarten von Java, 1: t. 162 (1913); Backer and Bakhuizen, Fl. Java 2: 127 (1965).

? *Aglaia speciosa* Blume var. *macrophylla* C. de Candolle in A. & C. de Candolle, Monog. Phan. 1: 614 (1878). Holotype: Java, *Zollinger* 802 (G-DC!).

Tree up to 35 m, bole up to 20 m, diameter at breast height up to 60 cm; buttresses outwards up to 1 m, upwards up to 5 m, the trunk fluted above. Bark reddish-brown, with small orange lenticels, inner bark magenta or pink, sapwood yellow, latex white. Twigs greyish-brown with pale pinkish- orange lenticels scattered or in longitudinal rows and densely covered with peltate scales which have a dark reddish-brown centre and pale margin, barely fimbriate or with a short irregular fimbriate margin. Leaves 25 – 30 cm long, 22 – 24 cm wide, obovate in outline; petiole 4.5 – 10 cm, the petiole, rhachis and petiolules densely covered with scales like those on the twigs; leaflets (5 –)7 – 11, the laterals subopposite, all 7 – 16 cm long, 2.5 – 4 cm wide, elliptical or narrowly obovate, apex acuminate with the obtuse acumen often parallel-sided and up to 10 mm long, base rounded or cuneate, with numerous scales like those on the twigs on the lower leaflet surface, sometimes deciduous and leaving that surface with numerous pits; veins 12 – 15 on each side of the midrib, the midrib depressed above and prominent below, lateral veins barely prominent below; petiolules 5 – 10 mm.

Inflorescence up to 22 cm long, 20 cm wide; peduncle 0.5 – 4 cm, the peduncle, rhachis and branches densely covered with scales like those on the twigs. Male flower 1 – 1.5 mm long, 1 – 2 mm wide, pedicels 0.5 – 1 mm; calyx 0.5 – 1 mm long, the pedicel and calyx densely covered with peltate scales which have a fimbriate margin; staminal tube 0.5 – 1 mm long, 0.8 – 1 mm wide, cup-shaped or obovoid, anthers 5, inserted just inside the rim of tube and pointing towards the centre of the flower; ovary 0.2 – 0.3 mm long, 0.2 – 0.3 mm wide, ovoid, densely covered with scales like those on the twigs; stigma 0.2 – 0.5 mm long, 0.2 – 0.3 mm wide, columnar, bearing the longitudinal impressions of the anthers, the apex truncate. Female flower 1.5 – 1.75 mm wide, 1.5 mm long, pedicel 0.5 – 1.5 mm; calyx 0.5 – 0.8 mm long, the pedicel and calyx densely covered with peltate scales which have a fimbriate margin, staminal tube 0.8 – 1.2 mm long, 1.5 mm wide, anthers 0.3 – 0.5 mm long, 0.8 – 0.3 mm wide, inserted inside the rim of the staminal tube. Ovary c. 0.4 mm long and 0.5 mm wide, densely covered with shiny peltate scales; stigma c. 0.5 mm long and 0.5 mm wide, columnar or obovoid with a depression in the centre at the apex.

Infructescence up to 21 cm long, 20 cm wide, peduncle 3 – 5 cm. Fruits 2.3 – 3 cm long, 1.7 – 2.7 cm wide, obovoid, orange or brown, indehiscent; fruit-stalks up to 5 mm, pericarp c. 1 mm thick, brittle, without dehiscence lines, densely covered with scales either like those on the twigs or paler and more fimbriate; loculi (1 or) 2, each containing 0 – 1 arillate seed; the aril c. 2 mm thick, translucent, yellow or orange, edible, firmly adhering to the testa; seed without the aril c. 1.6 cm long, 1.3 cm wide and 0.8 cm through, the seed coat with branched venation. Fig. 39 & 42.

DISTRIBUTION. Peninsular Malaysia, Sumatra, Borneo, Sulawesi. Fig. 38.

ECOLOGY. Found in primary forest, secondary forest, marsh forest and forest edges on loam with lime. Alt.: 5 to 2200 m.

VERNACULAR NAMES. Peninsular Malaysia: Bekak, Memberas. Sumatra: Ganggo, Ganggo Oedang, Setur Padi. Borneo: Kopeng.

Representative specimens. PENINSULAR MALAYSIA. Pahang, Krau Game Reserve, Kuala Lompat [3°40′N 102°20′E], fr., *Chivers* DCS 073 (FHO!); Pahang, Raub, Sg. Sempam, 600 m, fr., 15 April 1970, *Soepadmo* 627 (FRI!).

SUMATRA. Atjeh [Aceh]: Gunong Leuser Nature Reserve, Ketambe, 200 – 400 m, fr., 6 Feb. 1975, *de Wilde & de Wilde-Duyfjes* 14583 (K!, L!); Gajoland, Goempang to Kongke, newly cut forest near Alas river, fl., 12 March 1937, *van Steenis* 9779 (K!).

SARAWAK. 4th Division, Miri District, Niah, South side of Gunong Subis, Sekaloh river [4°00′N 113°87′E], fr., 29 Nov. 1966, *Anderson, Tan & Wright* S 27569 (FHO!).

SABAH. lower slopes of Mount Kinabalu, between mile 34 & 40, Kundasang-Ranau road, c. 1280 m, fl., 27 Oct. 1963, *Pennington* 7926 (FHO!, L!).

Fig. 39. *A. speciosa*. Habit with infructescence x1/2. Seed x1/2. Inflorescence x1/2. Half flower x20.

A. speciosa usually has 7 – 11 leaflets, whereas *A. korthalsii* usually has 5 leaflets. In Borneo some, but not all leaves on a specimen may have fewer than 7 leaflets, but then the dense indumentum of small peltate scales which have a distinctly dark centre and clearly defined pale margin which is closely adherent to the leaflet surface distinguish it from *A. korthalsii*. The (10 –)13 – 15 veins together with the indumentum of peltate scales only, distinguish *A. speciosa* from *A. odoratissima*, which has 5 – 9(– 11) veins and a mixture of peltate and stellate scales. The stigma is often longer and narrower than in *A. korthalsii*. The fruit is similar to, but larger than, that of *A. odoratissima*, and contains one or two seeds.

36. Aglaia korthalsii Miq., Ann. Mus. Bot. Lugd. Bat. 4: 42 (1868). Lectotype (designated here): Sumatra, [occidentali], Doekoe, fl., *Korthals* 899a (L!); C. de Candolle in A. & C. de Candolle, Monog. Phan. 1: 611 (1878); non Pellegrin in Lecomte, Fl. Gén. Indo-Chine, 1: 771 (1911); Corner, Wayside Trees of Malaya 1: 457, fig. 151 (1940), ed 3, 2: 496, fig. 154 (1988) as *Aglaia* sp.; Pannell in Ng, Tree Flora of Malaya 4: 228 (1989) as *Aglaia* sp. 4.

Hearnia sarawakana C. de Candolle in A. & C. de Candolle, Monog. Phan. 1 : 632 (1878). Lectotype (designated here): Borneo [Sarawak], fl., *Beccari* 3902 (G!) (isolectotypes: FI!, K!).
Hearnia aquatica Pierre, Fl. Forest. Cochinch. Fasc. 21, ante t. 333b (1 July 1895). Lectotype (designated here): ad Bao-Chiang, austro Cochinchine, fl., July 1877, *Pierre* 446 (P!; isolectotypes: K!, L!, NY!) .
Aglaia aquatica (Pierre) Harms in Engl. & Prantl, Pflanzenf. 3(4): 300 (1896).
Aglaia cauliflora Koord. in Meded. 'S. Lands Plantent. 19: 381 nomen, 633 (1898). Lectotype (designated here): Celebes. Oerwoud bij bivak Pinamorongan, naby Kajoewatoe, Boschnummer 1861, fr., 28 Feb. 1895, *Koorders* 17907β (BO!).
Aglaia dysoxylifolia Koord. in Meded. 'S Lands Plantent 19: 634 (1898). Lectotype (designated here): Celebes, Minahassa, Menado, 500 m, st., 4 Mar. 1895, *Koorders* 17923β (L!); excl. *Koorders* 17893β (L!) (= *A. smithii* Koord.).
[*Aglaia dysoxylonoides* Koord. in Meded. 'S. Lands Plantent. 19: 380 (1898), nom. nud. (= *Aglaia dysoxylifolia*)].
Aglaia celebica Koord. in Meded. 'S Lands Plantent. 19: 634 (1898). Lectotype (designated here): Celebes, bivak Tatok naby Ratatotok, 200m, fl., 25 March 1895, *Koorders* 17909β (BO!; isolectotype: L!).
Aglaia longipetiolulata Baker fil. in Jour. Bot., Lond. 62: Suppl. 20 (1924). Lectotype (designated here): Sumatra [S., Lampong, Penanggoengan, 600 ft], 1880, *Forbes* 1647 (BM! excl. carpological collection (= ?)).
Aglaia confertiflora Merrill in Univ. Calif. Publ. Bot. 15: 125 (1929). Lectotype (designated here): British North Borneo, Elphinstone Province, Tawao, fl., Oct. 1922 – March 1923, *Elmer* 20874 (UC!; isolectotypes: A!, BM! BO!, G!, GH!, K!, L!, NY!, P!, SING!, U!).

Tree up to 26 m. Bole up to 16 m; up to 70 cm in diameter; with triangular buttresses outwards up to 1.5 m with exposed roots beyond, upwards up to 2 m, bole fluted above; branches patent or ascending.

Bark pale to dark reddish-brown or pinkish-brown or orange brown, sometimes with coarse longitudinal fissures or rows of lenticels, flaking in large irregular roundish scales, exposing orange or yellowish-green bark beneath; inner bark pink, dark greenish-pink or purplish-pink, fibrous; sapwood pale pink, pale yellow, yellow or white; latex white. Twigs pale pinkish- or greyish-brown, longitudinally wrinkled, with pale pink lenticels and numerous to densely covered with shiny reddish-brown peltate scales which have a dark centre, becoming paler towards the margin or are pale throughout, the margin irregular or shortly fimbriate, latex white. Leaves up to 40 cm long and 48 cm wide, obovate in outline; petiole 7 – 12 cm, the petiole, rhachis and petiolules with numerous scales like those on the twigs. Leaflets (3 –)5(– 7), the laterals subopposite, all 8 – 27(– 36) cm long, (2.5 –)3 – 8 cm wide lateral leaflets opposite or subopposite, elliptical, ovate or sometimes obovate, sometimes subcoriaceous in Thailand, pale green when young turning dark green and slightly glossy above when mature, apex acuminate with the usually obtuse but sometimes acute acumen up to 10(– 15) mm long, usually rounded but sometimes cuneate at the asymmetrical base; with scales like those on the twigs few on the upper surface and scattered or numerous on the lower surface at fairly regular intervals, sometimes with faint reddish-brown pits; veins 10 – 25 on each side of the midrib, ascending and curved upwards near the margin, the midrib prominent and the lateral veins subprominent below; petiolules usually 5 – 10 mm, rarely up to 30 mm.

Inflorescences borne in leaf axils or on old wood of the twigs. Male inflorescence up to 30 cm long and 30 cm wide, ovate in outline, sturdy; peduncle 0.5 – 2 cm, the peduncle rhachis and branches with numerous to densely covered with scales like those on the twigs. Female inflorescence like the male but often much smaller, with fewer branches and flowers. Flower c. 1.5 – 2 mm long and 1.6 – 2.5 mm wide, subglobose or depressed- globose, sweetly fragrant; pedicel 0.8 – 3 mm; calyx $^1/_4$ to $^1/_2$ the length of the corolla, with few to numerous peltate scales like those on the twigs, divided almost to the base into 5 rounded lobes which have fimbriate margins. Petals 5, white or yellow, obovate, aestivation quincuncial. Staminal tube 0.8 mm long, 1.9 mm wide, obovoid or cup-shaped with the apical margin incurved and shallowly 5-lobed; anthers 5, c. 0.4 mm long, ovoid, inserted inside the rim of the tube, protruding and pointing towards the centre of the flower; ovary depressed-globose densely covered with peltate scales; stigma 0.5 mm long, 0.7 mm wide, depressed globose or obovoid, often with a central depression at the apex or with two small lobes, narrowed at junction with the ovary, the ovary and stigma together c. $^1/_2$ the length of the staminal tube.

Infructescence up to 17 cm long and 14 cm wide, with up to 15 fruits; peduncle 1 – 2 cm, the peduncle, rhachis and branches with numerous scales like those on the twigs. Fruits 2 – 4 cm long, 1 – 3.5(– 5) cm wide, ellipsoid or subglobose, orange, densely covered with orange-brown peltate scales which have a fimbriate margin on the outside and with small longitudinal wrinkles, the pericarp indehiscent with a dehiscence line running longitudinally around the fruit along which the ripe fruit breaks open when pressure is applied, the pericarp 1 – 10 mm thick, fibrous and flexible, with some white latex, the inner surface, without hairs or scales, shiny orange; fruit-stalks 1 – 2 cm. Loculi 2 (or 3), each containing 0 – 1 seed, septum persistent, up to 0.5 mm thick, membranous. Seeds 1.5 – 2 cm long, 1 – 1.5 cm wide, 0.8 mm through, ellipsoid, with inner surfaces flattened; aril c. 2 mm thick, translucent yellow or pale orange, juicy or gelatinous, edible, sweet or rather bitter tasting, firmly attached to the testa especially at the hilum and main antiraphe vascular bundle, usually not quite complete on the antiraphe side, the seed coat with branched venation. Cotyledons transverse; radicle enclosed, pointing towards the middle of the raphe side and the micropyle which is beneath the hilum; the plumule covered with pale yellow peltate scales. Fig. 41 & 42.

DISTRIBUTION. N.E. India (Assam), Bhutan, Nicobar Islands, Burma, Vietnam, S. Thailand, Peninsular Malaysia, Sumatra, Borneo, Philippine Islands, Nusa Tenggara (Sumbawa & Flores), Sulawesi. Fig. 40.

ECOLOGY. Found in primary forest, secondary forest, riverine forest, peat swamp forest, along rivers and in villages; on clay, loam, limestone, sandstone, sand. Cultivated in Kelantan (N.E. Malay Peninsula). Alt.: 5 to 1700 m. Rare to common.

USES. Edible fruits. Timber used for house poles.

VERNACULAR NAMES. Thailand: Gee Ya. Peninsular Malaysia: Bekok, Duku Hutan, Ganggo, Langsat, Saeloe Pajo. Borneo: Belajang Merah (Banjar Malay); Langsat Munchit (Sungei); Lantupak, Lantupak Burong (Dyak Kinabatangan); Segera (Iban); Bunjau, Langsat, Mula, Mulah, Mulak, Sagiar. Sulawesi: Buno (Uma).

Representative specimens. BHUTAN. Mirichana Timpu, 900 m, yg fl., 20 April 1915, *Cooper for Bulley* 3747 (E!); Saruchi District, Khagra valley near Gokti, [26°49′, 89°12′], 550 m, 2 March 1982, *Grierson & Long* 3400 (E!).   BURMA. Tavoy District, Thit Kataung, Kalein Aung Res., 75 m, fl., 30 Jan. 1930, ? *Sukoy* 10873 (K!).

THAILAND. Ranawng, Luang Ko, Kao Pawta, c. 700 m, fl., 31 Jan. 1929, *Kerr* 16913 (C!).

PENINSULAR MALAYSIA. Kelantan, Kota Bahru, Kg. Dermit: ♀ fl. & fr., 7 Sep. 1979, *Pannell* 1452 (FHO!) & ♂ fl., 7 Sep. 1979, *Pannell* 1454 (FHO!). Pahang: ? Danau, fr., 4 June 1923, *Bain* FMS 6002 (FRI!); Tembeling , 27 May 1939; fl., 27 May 1931, *Henderson* SFN 24534, (K!, FHO!); Temerloh, Sungei Lameh, fl., 13 Dec. 1919, *Hamid* FMS 5153 (FRI!).

SUMATRA. Aceh, Gunong Leuser Nature Reserve, Ketambe [3°40′N, 97°40′E], 300 – 350 m: fl., 12 Feb. 1975, *de Wilde & de Wilde-Duyfjes* 14823 (K!, L!) & fr., 27 July 1979, *de Wilde & de Wilde-Duyfjes* 19198 (K!, L!). North Sumatra, Sibolangit, 350 – 500 m, fl., 4 Dec. 1927, *Lorzing* 12649 (L!).

KALIMANTAN. East Kutei Reserve: 120 m, st., 21 Aug. 1979, *Leighton* 1009 (FHO!) & fr., 17 Feb. 1979, *Leighton* 516 (FHO!). W. Kutei [Koetei], Djembajan (Sei Koentap), c. 6 m, st., 28 June 1938, *Neth. Ind. For. Serv.* bb 25137 (L!).

PHILIPPINE ISLANDS. Mindanao, Province Lanao, yg fr., 22 Oct. 1938, *Zwickey* 287 (NY!). SULAWESI. N. Peninsula, between Palu and Parigi, 35 km from Palu, fl., 17 April 1975 *Meijer* 9320 (L!). Manado, Donggala, Wombo, 800 m, st., 28 Jul y 1934, *Bisch* 159, bb 18659 (BO!).

A. korthalsii is separated from *A. speciosa* by its usually 5, usually larger, leaflets; less dense indumentum; and its fruits. The fruit of *A. korthalsii*

Section Aglaia

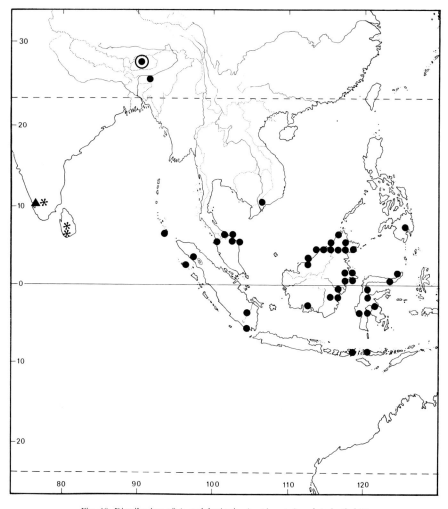

Fig. 40. Distribution of *A. malabarica* ▲, *A. apiocarpa** and *A. korthalsii* ●.

is usually similar to that of *A. elliptica* in that it is indehiscent and has a longitudinal ridge along which the ripe pericarp splits in two if pressure is applied, but the indumentum is of peltate scales in *A. korthalsii*, whereas it is of stellate scales in *A. elliptica*. In Sulawesi, *A. korthalsii* is sometimes ramiflorous. In Sumatra, the arillate seeds are eaten by the primates *Pongo pygmaeus* (Orang Utan) and *Hylobates lar* (White-handed Gibbon); the aril is digested and the seeds voided or discarded intact (Rijksen, 1978; Pannell & Kozioł, 1987).

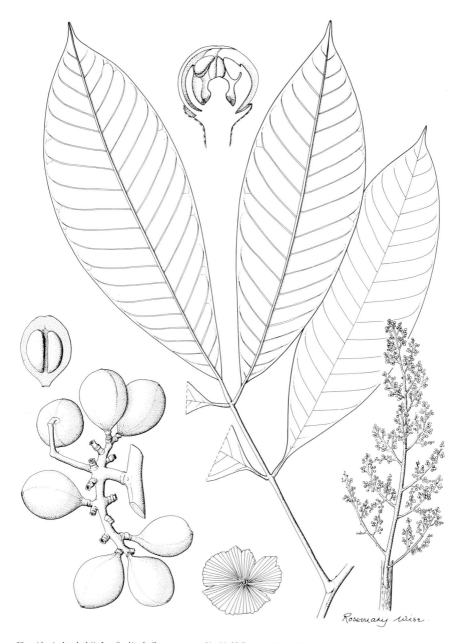

Fig. 41. *A. korthalsii*. Leaf x1/2. Inflorescence x1/2. Half flower x20. Infructescence x1/2. Peltate scale x100.

Fig. 42. *A. korthalsii* and *A. speciosa* (A, G & H). Half flowers x15. Peltate scales x50.

37. Aglaia apiocarpa (Thwaites) Hiern in Hooker fil., Fl. Brit. Ind. 1: 555. 1875. C. de Candolle in A. & C. de Candolle, Monog. Phan. 1: 620. 1878; Trimen, Hand-book to the Flora of Ceylon 1: 245. 1893; Alston in Trimen, Hand-book to the Flora of Ceylon 6: 45 (1931).

Milnea apiocarpa Thwaites, Enum. Pl. Zeyl.: 60 (1858). Lectotype (Kostermans, 1982): sine loc., *Anon.* in *Thwaites* C.P. 405 (PDA!).
Aglaia congylos Kostermans in Acta Botanica Neerlandica 31: 130 (1982). Holotype: Ceylon, Adam's Peak, fl., *Gardner* s.n. in *Thwaites* C.P. 405 (PDA!).

Tree up to 20 m, sometimes bearing fruits at 1 m, diameter up to 40 cm; buttresses upwards up to 2 m, outwards up to 1 m. Bark pale pinkish- brown, reddish-brown or grey, flaking in squarish scales; inner bark red or pink; sapwood white or red; latex white. Twigs greyish-brown or reddish-brown, longitudinally wrinkled and with numerous orange lenticels, densely covered with dark reddish-brown peltate scales which usually have a darker central spot and a dark ring near the fimbriate margin. Leaves up to 42 cm long and 26 cm wide; petiole 15 cm, green, pitted, flattened on the adaxial side, densely covered with scales like those on the twigs, glabrescent. Leaflets, (3 –)5 – 9, the laterals usually subopposite, rarely alternate, 4 – 20.5 cm long, 1 – 7 cm wide, elliptical, sometimes coriaceous and with the margin recurved, the apex acuminate, with the obtuse acumen sometimes parallel-sided and 5 – 25 mm, cuneate at the base, either with dark reddish-brown and orange peltate scales like those on the twigs numerous on the midrib of both surfaces and scattered on the rest of the lower surface, or with peltate scales which have a dark centre and paler irregular or fimbriate margin numerous on the upper surface of leaflets when young and usually densely covering, or sometimes numerous on, the lower surface and with some paler scales which have a longer fimbriate margin interspersed; with numerous pits on both surfaces; veins 7 – 10 on each side of midrib, curved upwards, midrib prominent below, lateral veins visible but barely prominent; petiolules 1.5 – 12 mm on lateral leaflets, up to 16 mm on terminal leaflet, sometimes flattened on the adaxial side, green and with numerous pits.

Male inflorescence 9 – 26 cm long, and 5 – 25 cm wide; female inflorescence up to 5.5 cm long and 3.5 cm wide; peduncle 0.2 – 7 cm, the peduncle, rhachis and branches densely covered with scales like those on the twigs. Flowers c. 2 mm long, 2.5 mm wide; pedicels 2 – 3 mm, the pedicel and calyx densely covered with very dark reddish-brown peltate scales which have an irregular margin. Calyx about half the length of the corolla, cup shaped and divided into 5 subrotund lobes which have fimbriate margins. Petals 5, yellow, aestivation quincuncial. Staminal tube 1.2 mm long, 1.7 mm wide, depressed globose, thick and fleshy, the aperture 0.4 mm across, lobed; anthers 0.8 mm long, 0.6 mm wide, broadly ovoid, inserted c. $1/3$ of the way up the tube, just protruding and filling the aperture. Ovary 0.5 mm long, 0.9 mm wide at the base, ovoid, densely covered with scales like those on the calyx, stigma 0.5 mm long, 0.5 mm wide, ovoid with two small lobes at the aperture; locules 2, each containing one ovule.

Infructescence 6 – 18 cm long, 6 – 7 cm wide, with up to 12 fruits;

Fig. 43. *A. apiocarpa*. Habit with infructescence x¹/₂. B. Habit of the variant which has a denser indumentum (right) x¹/₂. Male inflorescence x¹/₂. Half flower, male x7. Female inflorescence x¹/₂. Half flower, female x7. Details of lower leaflet surfaces showing indumentum x7. Peltate scale x90.

peduncle 1 – 6 cm, the peduncle and branches densely covered with scales like those on the twigs. Fruits up to 4 cm long and 2.3 cm wide, obovoid or subglobose with a small beak and tapering to a short stipe c. 5 mm long, brown, reddish-orange or yellow; pericarp 1 – 2 mm thick, brittle, densely covered with scales like those on the twigs on the outside, fruit- stalks 6 – 10 mm. Loculi 2, one of which contains one seed, the seed 15 – 21 mm long, 11 – 13 mm wide and 8 – 10 mm through; the aril translucent and sweet. Fig. 43.

DISTRIBUTION. South India and Sri Lanka. Fig. 40.

ECOLOGY. Wet evergreen forest, primary and secondary forest. Alt.: 140 to 2100 m.

Representative specimens. INDIA. S., Kerala, near Devicolam, Sholas Devicolam to Periyar Road, mostly beyond the gap, Cardamom Hills, fl., 25 March 1980, *Ridsdale* 687 (FHO!, L!), same locality, fl., 24 March 1980, *Ridsdale* 671 (FHO!, L!).

SRI LANKA. Kandy District, Corbet's Gap, fl., 10 Oct. 1973, *Waas* 206 (K! , PDA). Central Province, Merisketiya, c. 1070 m, fr., 1 Feb. 1975, *Waas* 1053 (K!, PDA). Nuwaraeliya District, Road to Maskeliya near Doublecutting, [c. 6°45'N 80°35'E], 900 m, yg fl., Nov. 1978, *Kostermans* 27085 (G!, K!, PDA). Badulla District, Dothulagalle Estate above Haputale, [c. 6°45'N 80°50'E], 1600 m, yg fl., 7 May 1969, *Kostermans* 23369 (G!, K!, PDA). Kalutara District, Weligala, 120 m, fr., 29 June 1975, *Waas* 1306 (K!, L!).

A. apiocarpa appears to occur in two forms and it is not clear whether two closely related species should be recognised. One form has larger leaflets with scales mainly on and near the midrib. The other frequently has one more pair of leaflets, the leaflets are smaller and leathery and the scales densely cover the lower leaflet surface; Kostermans described this as a new species, *A. congylos*. Since intermediates between the two extremes exist and cannot be placed with certainty in one or other of the two species and since there is little or no difference in flower and ripe fruit structure, I have treated them as a single species. Both collections cited by Thwaites in the original description of *Milnea apiocarpa* are mixtures of these two forms.

Alston (1931) noted the similarity between some specimens of *A. apiocarpa* and *A. bourdillonii* Gamble from S. India, but the latter differs in its consistently dense indumentum of orange-brown scales which differ in colour and have a longer fimbriate margin than the scales in *A. apiocarpa* and in its lateral veins, which are subprominent on the lower leaflet surface, whereas these are barely visible in *A. apiocarpa*. He also mentions *A. odoratissima*, but this species is quite distinct and does not occur in Sri Lanka nor in S. India, the nearest record being from Pegu in Burma.

38. Aglaia scortechinii King in Jour. As. Soc. Beng. 64: 64 (1895). Syntypes: [Peninsular Malaysia], Perak, *Scortechini* 722 (G!, L!, SING!).

Tree up to 22(– 30) m; bole 9 m, diameter 25 cm. Bark smooth, green, greyish-brown or reddish-brown, inner bark yellow, pale brown or pink, wood white; latex white. Twigs densely covered with dark reddish-brown peltate scales which have a fimbriate margin.

Leaves 45 cm long and 38 cm wide; petiole 7 cm, the petiole, rhachis and petiolules densely covered with scales like those on the twigs. Leaflets (7 –)9 – 13(– 15), 5 – 18.5 cm long, 1.5 – 4.5 cm wide, yellowish-brown when dry, elliptical or ovate, acuminate at apex, the obtuse acumen parallel-sided and up to 2 cm long, base cuneate, with peltate scales numerous on the midrib below and scattered elsewhere on the lower leaflet surface; veins 6 – 15, ascending, curved upwards near the margin and nearly or quite anastomosing, midrib prominent and lateral veins subprominent below; petiolules 5 – 10 mm.

Inflorescences in the axils of up to 6 leaves near the shoot apex, c. 13 cm long and 19 cm wide; peduncle up to 1 cm, the peduncles, rhachis and branches with numerous to densely covered with scales like those on the twigs. Flowers c. 1.5 mm long and 1.5 mm wide, fragrant; pedicel 0.5 – 1 mm. Calyx c. $^1/_3$ the length of the flower, cup-shaped and divided into 5 lobes densely on the outside covered with peltate scales which have a fimbriate margin. Petals 5, yellow or orange. Staminal tube c. 0.8 mm high, thickened below and between the anthers; anthers 5, usually inserted inside the margin and pointing towards the centre of the flower, sometimes enclosed within the tube. Stigma ovoid with two apical lobes; ovary with 2 (or 3) locules, each containing 1 ovule.

Infructescence up to 16 cm long and 8 cm wide, peduncle c. 1 cm, the peduncle, rhachis and branches densely covered with scales like those on the twigs. Fruits 1.5 – 3 cm long and 2 – 3.5 cm wide, subglobose, red when ripe; the pericarp thin and brittle when dry, densely covered with scales like those on the twigs on the outside; loculi 2 (or 3), each containing one seed. Seed c. 1.3 cm long, 1.1 cm wide and 0.9 cm thick, completely covered with a white or yellow aril.

DISTRIBUTION. Peninsular Malaysia, Borneo. Fig. 32.
ECOLOGY. Found in primary forest on limestone. Alt.: 100 to 1000 m.
VERNACULAR NAMES. Borneo: Lantupak (Dusun Kinabatangan).
Representative specimens. PENINSULAR MALAYSIA. Selangor: 6th mile from Ginting Hotel, fr. 26 March 1973, *Kochummen* FRI 16680 (L!); Ulu Langat, fr., 7 May 1969, *Suppiah* FRI 11273 (L!).
SABAH. Sandakan District, Kinabatangan, Tidok, 50 m, fl., 9 March 1960, *Brand* SAN 20814 (L!). Sipitang District, Ulu Moyah, 17 miles E.S.E. of Sipitang, 700 m, fl., 17 Aug. 1955, *Wood* SAN 16569 (L!). Tawau, Apas road, Mile 10, 15 m, fr., 21 Oct. 1937, *Tandom* 8811 (SING!).
SARAWAK. 1st Division: Bidi at Bau, on foot of Gg Meraja, 100 m, fl., 30 March 1977, *Othman et al* S 37513 (L!); Tiang Bekap limestone, yg fl., 29 April 1974 *Mabberley* 1634 (FHO!, L!); Kuching District, Tiang Bekap, fl., 5 June 1960, *Anderson* 12558 (L!)
KALIMANTAN. E., Berau, Mt Njapa on Kelai River, 1000 m, fr., 26 Oct. 1963, *Kostermans* 21523 (L!). Kalimantan Selatan Province, Batu Kampai, 3 km from Djaro, N.E. of Muara Uja [1°50'S 115°40'E], 100 m, fr., 20 Nov. 1971, *de Vogel* 932 (L!).

This species sometimes resembles *A. edulis*, but differs in its indumentum and fruits. It resembles *A. korthalsii*, but the leaflets are usually more numerous, they are paler when dry and the scales are almost confined to the midrib. The fruits are subglobose, with a thin brittle pericarp.

39. Aglaia glabrata Teijsm. & Binn. in Nat. Tijdschr. Ned. Ind. 27: 42 (1864). Type: Cult. Hort. Bogor, origin: Bangka, vernacular: name Bawang (no specimen seen which pre-dates publication of the name); illustrative specimen: Cult. Hort. Bogor III-B-60 (origin: Bangka, vernacular name: Bawang), fl., 28 May 1892 (BO!); Miquel, Ann. Mus. Bot. Lugd. Bat 4: 58 (1868).

Tree up to 20 m. Bole up to 10 m, up to 30 cm in diameter, with steep buttresses upwards to 1.2 m. Outer bark greyish-brown or reddish-brown with numerous lenticels, sometimes flaking in irregular scales; inner bark pink or pale yellow; sapwood red or pale yellowish-brown, latex white. Twigs greyish-brown, densely covered with dark purplish-brown peltate scales which have a fimbriate margin, or sometimes with stellate scales. Leaves imparipinnate, up to 22 cm long and 20 cm wide, obovate in outline; petiole 3.5 – 7 cm long, the petiole, rhachis and petiolules densely covered with scales like those on the twigs. Leaflets 5 – 9, the laterals usually subopposite, sometimes alternate, all 4 – 13 cm long, 1.5 – 5 cm wide, often pale green when dry, usually elliptical, rarely ovate, apex acuminate with the obtuse acumen often parallel-sided and up to 15 cm long, cuneate at the slightly asymmetrical base, the upper and lower surfaces sometimes with numerous pits, densely covered with scales like those on the twigs on the midrib of the lower surface and nearly always absent from the rest of the leaflet surfaces; veins 6 – 13 on each side of the midrib, ascending and curved upwards near the margin and nearly or quite anastomosing, midrib prominent, lateral veins subprominent and the reticulation often subprominent on both surfaces; petiolules 3.5 – 7 mm.

Inflorescence up to 12 cm long and 12 cm wide; peduncle 1 – 2 cm, either with scales like those on the twigs or densely covered with reddish-brown or orange-brown scales on the peduncle, rhachis and branches. Flowers (not known whether male or female) 1 – 1.5 mm long, 1 – 1.5 mm wide, subglobose; pedicels 0.5 – 1 mm. Calyx about half the length of the flower, cup-shaped and shallowly divided into 5 obtuse lobes, densely covered with scales like those on the rest of the inflorescence on the outer surface. Petals 5; aestivation quincuncial. Staminal tube c. 1 mm long and 1 mm wide, obovoid, thickened and appressed against the ovary below the anthers, the aperture 0.5 – 0.8 mm in diameter and shallowly lobed; anthers 5, 0.5 – 0.7 mm long, 0.3 – 0.5 mm wide, ovoid, inserted below the margin of the tube and protruding to fill or almost fill the aperture. Ovary 0.3 – 0.5 mm long, 0.3 – 0.5 mm wide, subglobose, densely covered with reddish-brown stellate hairs or scales; loculi 2; stigma 0.2 – 0.5 mm long, 0.2 – 0.5 mm wide, ovoid with two apical lobes.

Infructescence c. 5 cm long; peduncle c. 1.5 cm; the peduncle, rhachis and branches densely covered with scales like those on the inflorescence. Fruits c. 1.5 cm long and 1.4 cm wide, subglobose, either densely covered with scales like those on the inflorescence, or with scales which have a dark grey central spot.

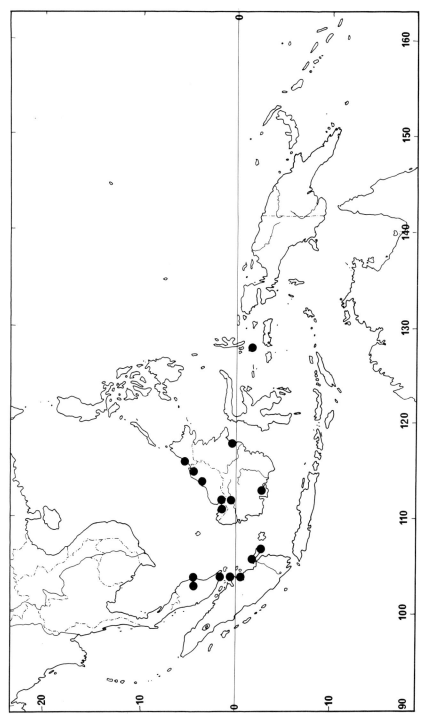

Fig. 44. Distribution of *A. glabrata*.

DISTRIBUTION. Peninsular Malaysia, Sumatra, Borneo, Maluku. Fig. 44.

ECOLOGY. Found in primary forest, secondary forest, riverine forest and in kerangas; on granite, sandy loam. Alt.: 10 to 800 m. Rather common.

VERNACULAR NAMES. Peninsular Malaysia: Pasak. Sumatra: Kaju Pasak, Membalo Burung, Parek. Borneo: Kori (Dyak); Bunjau, Empiras.

Representative specimens. PENINSULAR MALAYSIA. Pahang, Taman Negara, Lata Bakoh, 250 m, yg fr., 29 1975, *Ang Khoon Cheng* FRI 23416 (L!). 10th mile Dungun – Bukit Besi road, Compt. 12B, Bukit Bauk F.R., fl., 18 June 1967, *Kochummen* FRI 2373 (L!).

SUMATRA. Bangka, Lobok-besar: 23 m, fr., 16 Oct. 1949, *Kostermans & Anta* 1175 (NY!); Gg Pading, 50 m, fr., 29 Sep. 1949, *Kostermans & Anta* 951 (NY!); Gg Pading, 20 m, fr., 29 Sep. 1949, *Kostermans & Anta* 960 (L!).

SARAWAK. Baram District, Melinau Gorge (Mulu formation), 300 m, yg fl., 8 July 1961, *Anderson* 4326 (L!). SABAH. Papar district, Bangawan F.R., peat swamp, yg fr., 6 Feb. 1976, *Dewol & Talib* SAN 80347 (L!).

KALIMANTAN. E., west of Samarinda, Loa Djanan River, fl., 6 Sep. 1954, *Kostermans* 9965 (L!).

MALUKU. Ternate, Tidore en Batjan, Masoeroeng eil. Batjan, c. 500 m, fl., 12 Aug. 1937, Neth. Ind. For. Serv. bb. 23124 (L!).

CULTIVATED. Hort. Bogor, *Kostermans* 11206 (L!); Cult. Hort. Bogor III-B- 60, '360' (K!, L!); Hort. Bogor III-D-15, ♂ fl., 25 May 1957, *Sutrisno* 15 (L!).

A. glabrata is characterised by having dark purplish-brown peltate scales with a fimbriate margin confined to the midrib on the lower surface of the leaflets. *Aglaia glabrata* resembles *A. oligophylla*, but the indumentum of the latter is of stellate hairs, not peltate scales. The distal branches of the inflorescence and infructescence in herbarium specimens of this species are frequently malformed and appear to be galled. *A. glabrata* appears to have been described from a living plant in the Bogor Botanic Garden; no type specimen has been seen. Some of the specimens collected on Bangka Island (from where the cultivated material originated) and labelled *A. glabrata* belong to *A. malaccensis* and others belong to *Dysoxylum halmaheirae* (Miq.) C. de Candolle. This may explain why Miquel included *A. glabrata* in 'species incertae', at the end of his treatment of *Aglaia*.

40. Aglaia flavescens C. de Candolle in Lorentz, Nova Guinea 8: 426 (1910). Lectotype (designated here): Nova Guinea neerlandica meridionalis, fluv. 'Lorentz', fl., 11 July 1907, *Versteeg* 1417 (BO!; isolectotype: K!, L!, U!).

Aglaia lauterbachiana Harms in Engl. Bot. Jahrb. 72: 164 (1942). Lectotype (designated here): Neu-Guinea, Schumann Flusse, fl., Dec. 1901, *Schlechter* 13815 (BO!; isolectotype: B†).

Small tree up to 4 m. Twigs greyish-brown, densely covered with shiny orange-brown peltate scales which have a fimbriate margin.

Leaves 25 – 56 cm long, 20 – 50 cm wide; petiole 8 – 16 cm. Leaflets 5 – 9, the laterals subopposite, all 9 – 25 cm long, 3 – 11 cm wide usually obovate, sometimes elliptical, pale greenish-yellow when dry, acuminate at apex, the obtuse acumen up to 10 mm long, cuneate at the base, with scales like those on the twigs numerous on the midrib below and scattered on the rest of that surface, with numerous minute reddish-brown pits on the upper and lower surfaces; veins 9 – 20 on each side of the midrib, ascending, curved upwards at the margin and usually anastomosing, the midrib prominent, lateral veins subprominent and reticulation visible below; petiolule 0.3 – 10 mm on the lateral leaflets, up to 15 mm on terminal leaflet.

Male inflorescence 5 – 6 cm long, 8 – 12 cm wide, sessile or with a peduncle up to 0.5 cm, the peduncle, rhachis and branches densely covered with scales like those on the twigs, the scales more fimbriate on the distal branches. Flowers c.1 mm long and 1 mm in diameter, subglobose, pedicel 0.5 mm, the pedicel and calyx with few orange-brown peltate scales which have a fimbriate margin. Calyx spreading, divided almost to the base into 5 overlapping lobes. Petals 5. Staminal tube 0.3 – 0.4 mm high, 0.4 – 0.5 mm wide, cup-shaped, sometimes with the margin incurved; anthers 5, c. 0.3 mm long and 0.2 mm wide, ovoid, inserted inside the tube near the margin, protruding and filling the aperture. Ovary c. 0.2 mm long and 0.2 mm wide, depressed globose, densely covered with scales like those on the inflorescence branches; stigma c. 0.2 mm long and 0.2 mm wide, ovoid.

Infructescence c. 4 cm long and 2 cm wide; sessile, the rhachis and branches densely covered with scales like those on the twigs. Fruits 2 cm in diameter, subglobose, densely covered with scales like those on the twigs on the outside, indehiscent, the pericarp thin and brittle when dry; fruit-stalks c. 0.3 cm.

DISTRIBUTION. New Guinea only.

Representative specimens. IRIAN JAYA. Mamberamo, Albatros Bivak: c. 30 m, fl., May 1926, *Docters van Leeuwen* 9128 (BO!) & c. 60 m, fl., Nov. 1926, *Docters van Leeuwen* 11327 (BO!, L!).

PAPUA NEW GUINEA. Madang District, near Amiaba River, Usino Road, Dumpu, fl., 18 Dec. 1969, *Henty & Foreman* NGF 42878 (K!).

A. flavescens is recognised by its shiny orange peltate scales which have a ragged margin and, when dry, its pale yellowish-green leaves. The flowers are small and the fruit has a thin pericarp, which is brittle when dry.

41. Aglaia rubrivenia Merrill & Perry in Jour. Arn. Arb. 22: 318 (1940). Lectotype (designated here) : Solomon Islands, Malaita Island, Quoimonapu, fl., 11 Dec. 1930, *Kajewski* 2337 (A!; isolectotypes: BM!, BO!, BRI: photo SING!).

Tree up to 14 m, diameter up to 90 cm, without buttresses. Bark smooth, brown; inner bark pale yellowish-brown; sapwood pale yellowish-brown or pale reddish-brown. Twigs pale yellowish-brown densely covered with reddish-brown peltate scales which have a darker centre and a slightly fimbriate margin.

Leaves 17 – 70 cm long, 14 – 64 cm wide; petiole 4 – 17 cm, the petiole, rhachis and branches densely covered with scales like those on the twigs. Leaflets 5 (or 6), the laterals subopposite, all 7 – 35 cm

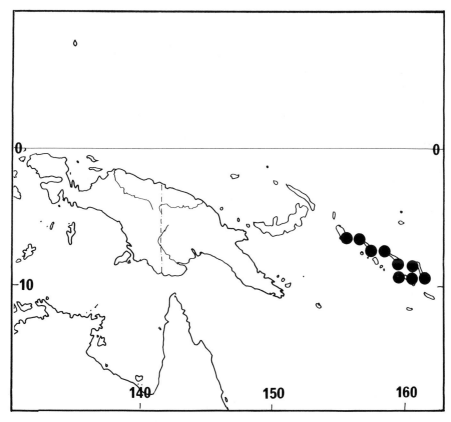

Fig. 45. Distribution of *A. rubrivenia*.

long, 2.5 – 10.5 cm wide, ovate, obovate or elliptical, acuminate at apex, with the obtuse acumen up to 10 mm long, rounded or cuneate at the slightly or markedly asymmetrical base, with scales like those on the twigs densely covering the midrib above and below and numerous on to densely covering the lateral veins above and below; veins 8 – 18 on each side of the midrib, ascending and curved upwards near the margin, not anastomosing, midrib prominent, lateral veins subprominent; petiolule 8 – 15 mm.

Inflorescence 7 – 21 cm long, 8 – 14 cm wide; peduncle 1.5 – 4 cm; the peduncle, rhachis and branches densely covered with scales like those on the twigs. Flower 2 – 3 mm long and 2 – 3 mm wide, pedicel 1 – 4 mm, the calyx and pedicel densely covered with scales like those on the twigs. Calyx cup-shaped and divided into 5 sub-rotund lobes. Petals 5, yellow, densely covered with peltate scales on the outside, aestivation quincuncial. Staminal tube 1.3 – 1.5 mm long, 1 – 1.2 mm wide, aperture 0.9 mm, obovoid with a few peltate scales on the outside; anthers 5, 0.6 mm long and 0.4 mm wide, inserted just inside the margin, pointing towards the centre of the flower and filling the aperture. Stigma 0.4 – 0.8 mm long, 0.2 – 0.5 mm wide, narrowly ovoid, longitudinally ridged and densely covered with peltate scales except at the apex.

Infructescence c. 23 cm long and 9 cm wide; peduncle c. 3 cm, the peduncle, rhachis and branches densely covered with scales like those on the twigs. Fruits 1.5 – 2 cm long, 1 – 1.2 cm wide, ellipsoid or obovoid with a small beak, red or orange when mature; fruit-stalks 4 – 10 mm.

DISTRIBUTION. Bougainville and Solomon Islands only. Fig. 45.
ECOLOGY. Found in primary and montane forest. Alt.: sea level to 970m. Common.
USES. Wood is used for bows (Solomon Islands: Malaita Island).
VERNACULAR NAMES. Solomon Islands: Ulukwalo (Kwara'ae); Quoi Mon Apu, Una-quala.
Representative specimens. PAPUA NEW GUINEA. Bougainville, c. 23 km N. of Buin, lower southern slopes of Lake Loloru crater, fr., 13 Aug. 1964, *Craven (& Schodde)* 253 (K!).
SOLOMON ISLANDS. N.W. Santa Ysabel: Binusa, fl., 17 Jan. 1966, *Beer's Collectors* BSIP 6765 (K!); S.E. of Paehena Point, fr., 1 Dec. 1965, *Beer's Collectors* BSIP 7088 (K!). W. Choiseul, E. Mbirambira, 15 m, fl., 17 Jan. 1978, *Gafui & Collectors* BSIP 18856 (K!). N.W. Guadalcanal, Mt Mambulu, c. 475 m, fr., 17 Nov. 1964, *Kere's collectors* BSIP 5592 (K!).

A. rubrivenia resembles *A. euryanthera* and *A. vitiensis*, but differs from both in its flower structure and the conspicuous scales on the lateral veins.

42. Aglaia samoensis Asa Gray, United States Exploring Expedition. 1838 – 1842. Botany Phanerogamia 1: 236 (1854). Lectotype (designated here): one of the Samoan or Navigator's Islands, Tutuila, *Anon.* s.n. (GH!, K!, NY!, P!); C. de Candolle in A. & C. de Candolle, Monog. Phan. 1: 616 (1878).

Aglaia edulis sensu A. Gray, United States Exploring Expedition. 1838 – 1842. Botany Phanerogamia 1: 237 (1854), quoad specim.: Samoa, *Whitmee* 1 12 (received 1867 & 1877) (BM!), *Whitmee* s.n. (received 1867 & 1877) (CGE!), non (Roxb.) Wall..
Aglaia betchei C. de Candolle in Bull. Herb. Boiss. Sér. II 3: 179 (1903). Lectotype (designated here): Ins. Samoa, *Betche* in H. Cand. (G!).
Aglaia whitmeei C. de Candolle in Bull. Herb. Boiss. Sér. II. 3: 178 (1903). Lectotype (designated here): Samoa, *Whitmee* 22 (G!).
Aglaia psilopetala A.C. Smith in Contrib. U.S. Nat. Herb. 30: 482 (1952). Lectotype (designated here): Polynesia, Wallis Islands, fl., *Burrows* W19 (US!; isolectotype: BISH: photo NY!).

Forest tree up to 7 m. Twigs densely covered with dark reddish-brown peltate scales which have a short fimbriate margin and some orange-brown scales which have a longer fimbriate margin interspersed.

Leaves 19 – 55 cm long; petiole 4.5 – 11 cm, the petiole, rhachis and petiolules densely covered with scales like those on the twigs. Leaflets (5 –)7 – 9(– 13), the laterals subopposite, all 5.5 – 22 (– 29) cm long, 2 – 8 cm wide, ovate or elliptical, acuminate or almost caudate at the apex, the acute or obtuse acumen often parallel sided and 10 – 20 mm long, rounded or cuneate at the slightly asymmetrical base, with few to numerous scales like those on the twigs on the midrib and scattered elsewhere and the younger leaflets with numerous pale almost stellate scales on the lower surface, usually glabrescent, but sometimes a few persist on the older leaves; the lower surface with numerous small dark reddish-brown pits and the same but more faintly on the upper leaflet surface; veins 8 – 12 on each side of the midrib, curving upwards and not quite anastomosing; petiolules up to 5 mm.

Inflorescence up to 18 cm long and 22 cm wide; peduncle 0.5 – 2 cm, the peduncle, rhachis and branches with scales like those on the twigs, but with a greater portion of those with a longer fimbriate margin. Flowers 1 – 1.5 mm in diameter, subglobose, fragrant; the pedicel 1 – 1.5 mm. Calyx 0.5 – 0.7 mm, deeply divided into 5 broad rounded lobes; the pedicel and calyx densely covered with fimbriate peltate scales. Petals 5 (or 6), pale yellow or pink, without hairs or scales, aestivation quincuncial. Staminal tube 0.75 – 1 mm long, obovoid, with few to numerous simple (occasionally stellate) white hairs on the inside, especially in the upper $^2/_3$ of the tube; anthers about 0.4 mm long and 0.3 mm wide, inserted inside the margin of the tube and protruding for c. $^1/_2$ their length, with some simple white hairs. Ovary subglobose densely covered with scales like those on the inflorescence; loculi 2, each containing 1 ovule; stigma ovoid, minutely bilobed at the apex; the ovary and stigma together about 0.7 mm long, the apex of the stigma reaching to just above the base of the anthers.

Infructescence up to 9.5 cm long and 8 cm wide, peduncle about 2.5 cm, the peduncle, rhachis and branches with indumentum like the inflorescence. Fruits 2 – 3.2 cm long and 1.5 – 2.5 cm wide, indehiscent, brown, red, orange or yellow, ellipsoid or subglobose when ripe, often asymmetrical when only one seed develops, obovoid when young; the pericarp thin, densely covered with reddish-brown peltate scales which have a fimbriate margin (sometimes with numerous pale peltate scales which have a long fimbriate margin, e.g. BSIP 8058). Loculi 2, each containing 0 or 1 seed which has a thin translucent aril. Fig. 46.

DISTRIBUTION. Papua New Guinea (including New Britain, New Ireland and Bougainville), Solomon Islands, Santa Cruz Islands, Vanuatu, Wallis Islands, Samoan Islands. Fig. 47.

ECOLOGY. Found in poorly to usually well-drained primary forest, secondary forest, montane forest and on coral terraces; on limestone and clay. Alt.: sea-level to 830 m. Common. USES. Red timber-wood (Bougainville).

VERNACULAR NAMES. New Guinea: Kokeqolo, Kukupiriki. Solomon Islands: Kokengolo (Roviana); Makiliau, Ulukwalo, Vulukwalo (Kwara'ae); Nganalba (Vanikoro); Rauapu (Longu); Chu Chu. Isles de Hornes: Langakali.

Representative specimens. PAPUA NEW GUINEA. Bougainville, Siwai, fl., Oct. 1930, *Waterhouse* 366 (K!).

SOLOMON ISLANDS. S.E. Kolombangara, Shoulder Hill area, 275 m, fl., 17 June 1968, *Mauriasi & Collectors* BSIP 11503 (K!). S.E. New Georgia, Segiline 83, fr., 8 Sep. 1966, *Burn-Murdoch's collectors* BSIP 6877 (K!, L!). WALLIS ISLANDS. Uvea, Utuleve, 5 m, fl., 2 Nov. 1968, *Mackee* 19906 (K!).

Fig. 46. *A. samoensis*. Habit with infructescence x½. Inflorescence x½. Half flower, with detail of the outside of the margin of the staminal tube x25. Stellate scale x90.

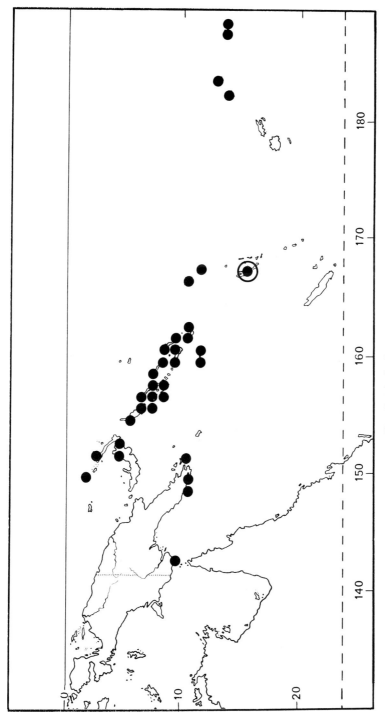

Fig. 47. Distribution of *A. samoensis*.

SAMOAN ISLANDS. Savaii, above Matavanu, 1030 m, fr., 27 July 1931, *Christophersen & Hume* 2195 (K!, NY!, UC!).

Some specimens from the Solomon Islands are intermediate between *A. subminutiflora* and *A. samoensis* and these are cited and mapped under *A. subminutiflora*.

In the following specimens, from Isles de Horn and Isles de Wallis, the trichomes are peltate at the base, but free arms of varying lengths project from the centre of the peltate scale to form a stellate hair. These have the distinctive dark reddish-brown colour of the majority of the scales of typical *A. samoensis* and similar leaf shape, size and colour when dry to the rest of the species, so that they are classified with it.

ISLES DE HORN. Futuna, Sausau, 2 m, fl., 25 Oct. 1968, *Mackee* 19765 (K!) ; Futuna, Mt Falanaise pente N. Vers 350 m, fr., 11 June 1982, *Veillon* 504 8 (K!, L!).

ISLES DE WALLIS. Cratère du Lalolalo, fr., 20 Oct. 1982, *Morat* 7150 (L!) .

A. samoensis has an indumentum of dark reddish-brown peltate scales which have a short fimbriate margin, with some orange-brown scales which have a longer fimbriate margin interspersed. These are few to numerous on the midrib on the lower surface of the leaflets and are sparse or absent elsewhere, so that the midrib has a distinctive dark reddish-brown appearance. The lower surface of the leaflets have numerous small dark reddish-brown pits. The staminal tube has few to numerous simple (occasionally stellate) white hairs on the inside, especially in the upper two thirds of the tube.

43. Aglaia gracilis A.C. Smith in Contributions from the U.S. National Herbarium 30: 489 (1952). Holotype: Fiji, Viti Levu, Mba, Western slopes of Mt Nanggaranambulata, east of Nandarivatu, 850 – 1000 m, fl., 19 June – 2 Oct 1947, *Smith* 6325 (A!; isotypes: K!, US!); A.C. Smith, Flora Vitiensis Nova 3: 544 (1985).

Dysoxylum obliquum Gillespie in Bishop Mus. Bull. 83: 13 (excl. fig. 15) (1931). Holotype: Fiji, Viti Levu, Tholo-north Province, vicinity of Nadarivatu, 900 m, 14 Dec. 1927, *Gillespie* 4316 (BISH; isotype: GH!).

Didymocheton obliquum (Gillespie) Harms in Engler & Prantl, Nat. Pflanzenfam., ed 2, 19b1: 157 (1940).

Slender, sometimes unbranched tree 2 – 5 m. Twigs densely covered with very dark reddish-brown or purplish-brown peltate scales which may or may not have a fimbriate margin.

Leaves 7.5 – 63 cm long; petiole 1.5 – 14 cm long; the petiole, rhachis and petiolules densely covered with scales like those on the twigs. Leaflets (1 –)3 – 11, the laterals subopposite, all 4 – 22.5 cm long, 1.5 – 6.5 cm wide, elliptical or obovate, sometimes ovate; rounded or cuneate at the slightly asymmetrical base; rounded at the apex or with a broad short rounded acumen, with few to numerous scales like those on the twigs on the midrib below and few on the surface in between; veins 9 – 20, ascending at 60 – 70° to the midrib, curved upwards near the margin and nearly anastomosing, with some shorter intermediate lateral veins; petiolules 3 – 15 mm.

Fig. 48. *A. gracilis*. Habit with immature infructescences x½. Habit with inflorescence x½. Mature fruit x½. Stellate scale x150.

Inflorescence 2 cm long, 1 cm wide, often ramiflorous, sometimes in the axil of an existing leaf, the peduncle, rhachis and branches with numerous or densely covered with scales like those on the twigs. Flowers 1.2 – 1.5 mm in diameter, up to 2 mm in female, subglobose, pedicel 1 – 1.5 mm, up to 2.5 mm in female, with few or no scales, the subtending branchlet c. 1 mm long and densely covered with reddish-brown peltate scales which have a fimbriate margin. Calyx 0.5 – 1 mm long, divided into 5 broad rounded lobes, with no scales. Petals 5, aestivation quincuncial. Staminal tube short, lobed and forming a shallow ring; anthers 5, large, c. 0.6 mm long, 0.3 mm wide, inserted on the margin, curving over the stigma and meeting at the centre of the flower. Ovary depressed globose with a ring of stellate or peltate scales, loculi 2, each containing 1 ovule; stigma 0.3 – 0.5 mm, ovoid, longitudinally ridged where adpressed up to the anthers, with two lobes at the apex. Fruits 2 cm long, 1.5 cm wide, ellipsoid or longitudinally half-ovoid, indehiscent; fruit-stalks up to 1.5 cm; the pericarp thin with numerous scales with a fimbriate margin like those on the twigs. Loculi 1 or 2, each containing 0 or 1 seed. Fig. 48.

DISTRIBUTION. Fiji Islands only.
ECOLOGY. Forest. Alt.: 50 to 1200 m.
Representative specimens. FIJI. Viti Levu, Mba, ridge between Mt Nanggaranambulutu and Mt Namama, east of Nandarivatu, 1050 – 1120 m, fr., 30 June – 18 Aug. 1947, *A.C. Smith* 4991 (A!); Viti Levu, Tholo North Province, Thol-i-Nandarivatu, 1150 m, fl., 19 Nov. 1927, *Gillespie* 3951 (GH!).

A. gracilis resembles *A. samoensis* except that it is sometimes ramiflorous and it has a rounded leaflet apex and widely spreading veins at 60 – 70° to the midrib. It occurs only in Fiji, where *A. samoensis* is not found.

44. Aglaia vitiensis A.C. Smith in Bulletin of the Bernice Bishop Museum, Honolulu, 141: 80 (1936). Holotype: Fiji, Koro, eastern slope of main ridge, 300 – 500 m, fl., 29 Jan. – 5 Feb. 1934, *Smith* 981 (BISH; isotypes: BO! GH!, K!, NY, P!, UC!); A.C. Smith, Flora Vitiensis Nova 3: 540 (1985).

Aglaia axillaris A.C. Smith in Sargentia 1: 43 (1942). Holotype: Fiji, Viti Levu, Tholo North, Mount Matomba, Nandala, vicinity of Nandarivatu, 750 – 900 m, fr., 15 – 18 Feb. 1941, *Degener* 14505 (A!; isotypes: K!, L!, MICH!, NY!, US!).
Aglaia vitiensis A.C. Smith var. *vitiensis* A.C. Smith, Contrib. U. S. Nat. Herb. 30: 487 (1952). A.C. Smith, Flora Vitiensis Nova 3: 543 (1985).
Aglaia vitiensis A.C. Smith var. *minor* A.C. Smith, Contrib. U.S. Nat. Herb. 30: 488 (1952). Holotype: Fiji Islands, *A.C. Smith* 1788 (US 1674996; isotypes: GH!, K!, NY!); A.C. Smith, Flora Vitiensis Nova 3: 543 (1985).

Tree 3 – 23 m high; bole up to 7 m, up to 30 cm in diameter. Branches ascending, arching, then descending. Bark smooth, grey, with small white lenticels; inner bark pale brown; sapwood pale yellow; heartwood pink and scented; latex white. Twigs grey, densely covered with dark reddish-brown (sometimes paler) peltate scales with or without a short fimbriate margin.

Leaves imparipinnate up to 47 cm long and 32 cm wide, obovate in outline; petiole up to 11 cm, the petiole, rhachis and petiolules densely covered with scales like those on the twigs. Leaflets 5 – 9, the laterals subopposite, all 3.5 – 22 cm long, 1.5 – 8 cm wide, shining mid-green above, paler below, elliptical or obovate, apex rounded or slightly obtuse-acuminate, cuneate or rounded at the asymmetrical base, with scales like those on the twigs densely covering the midrib above and below and few on the rest of the lower surface; both surfaces with numerous reddish-brown pits; veins 8 – 13(– 15) on each side of the midrib, shallowly ascending at an angle of 60 – 65° to the midrib, curved upwards near the margin, not or only faintly anastomosing; some lateral veins between these not reaching the margin; midrib prominent and lateral veins subprominent or not raised on the lower surface; petiolule up to 2(– 2.5) cm.

Male inflorescence up to 12 cm long and 7 cm wide. Female inflorescence up to 8 cm long and 2.5 cm wide. The peduncle, rhachis and branches densely covered with scales like those on the twigs. Male flowers 2.5 mm long, subglobose; pedicels up to 1.5 (– 4.5) mm, densely covered with scales like those on the twigs. Calyx up to 1 mm, divided into 5 broadly ovate, obtuse lobes, densely covered with scales on the outside. Corolla aestivation quincuncial, petals 5, obovate, densely covered with scales like those on the twigs on the outside; staminal tube 0.5 mm long, shallowly cup-shaped, usually divided to the base into 5 free lobes each with one anther inserted at the apex; anthers 0.8 mm long, 1 mm wide, inserted on the margin or occasionally inside the tube. Ovary depressed globose, 0.5 mm long, densely covered with peltate scales; stigma ovoid, 0.5 mm long, 0.7 mm wide with longitudinal ridges where adjacent to the anthers, apex of stigma just below the aperture between the anthers. Female flowers up to 3 (– 5) mm long and 5 mm across, subglobose; pedicel up to 2 mm; staminal tube up to 2 mm long; otherwise like the male flowers.

Infructescence 6 cm long. Fruits 3.5 cm long, 1 cm wide, ellipsoid, longitudinally wrinkled, pale orange or orange-brown, densely covered with peltate scales like those on the twigs. Loculi 2, one with a seed and the other with only the aril. Fig. 49.

DISTRIBUTION. Fiji islands only.
ECOLOGY. Found in secondary and primary forest. Alt.: 30 to 1150m.
VERNACULAR NAMES. Fiji: Lindiyango, Mbaumbulu, Thawarn, Nggiliyango, Sasawira.

Representative specimens. FIJI. Viti Levu, Nandronga & Navosa, northern portion of Rairaimatuku Plateau, between Nandrau and Nanga, 725 – 825 m, fl., 4 – 7 Aug. 1947, *A.C. Smith* 5556 (L!); Ngau, hills east of Herald Bay, inland from Sawaieke, 300 – 450 m, fr., 15 – 18 June 1953, *A.C. Smith* 7762 (L!).

A. vitiensis has dark peltate scales, which are either entire or have a short fimbriate margin, densely covering the stems and the midrib of the lower surface of the leaflet. It has the spreading lateral veins and rounded leaflet apex which are characteristic of many of the endemic species of Fiji; these characters and the absence of a fimbriate margin on many of the scales separate *A. vitiensis* from *A. samoensis*. *Aglaia vitiensis* is distinguished from *A. basiphylla* by its indumentum which consists of

Fig. 49. *A. vitiensis*. Habit with inflorescence x$1/2$. Small-leaved variant with inflorescence x$1/2$. Half flower, with detail of the staminal tube from the outside x10. Young infructescences, mature fruit and seed x$1/2$. Peltate scale x70.

peltate scales only; *A. basiphylla* varies in its indumentum from stellate scales interspersed with some scales like those of *A. vitiensis* to mainly stellate hairs which have long arms. The leaflets of *A. vitiensis* vary in size and number and the staminal tube may be lobed or divided to the base. This variation might form the basis for subspecific division of the species.

45. Aglaia unifolia P.T. Li et X.M. Chen in Acta Phytotaxonomica Sinica 22: 495 (1984), nom. nov. pro *Aglaia haplophylla* A.C. Smith.

Aglaia haplophylla A.C. Smith in Contrib. U.S. Nat. Herb. 30: 496 (1952), non Harms (1941) (= *A. simplicifolia*). Holotype: Fiji, Viti Levu, Mba

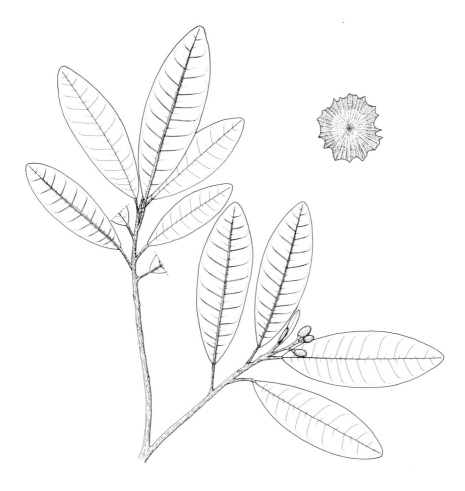

Fig. 50. *A. unifolia*. Habit with young fruits x$^1/_2$. Peltate scale x70.

[Tholo North], ridge between Mt Nanggaranambuluta [Lomalangi] and Mt Namama, east of Nandarivatu, 1050 – 1120 m, yg fr., 30 June – 18 Aug. 1947, *Smith* 5683 (A!; isotype: K!, L!, NY!, P!, US!); A.C. Smith, Flora Vitiensis Nova 3: 539 (1985).

Small tree about 7 m high. Twigs greyish-brown, densely covered with large purplish-brown and reddish-brown peltate scales. Leaves simple, 3.5 – 13.5 cm long, 1 – 4 cm wide, elliptical or obovate; cuneate at the base, rounded at the apex; veins 6 – 12 on each side of the midrib, ascending at an angle of 65 – 70° to the midrib, curving upwards near the margin and nearly or quite anastomosing, with some shorter lateral veins in between; with peltate scales like those on the twigs densely covering both surfaces of the young leaves, numerous on to densely covering the midrib below of the mature leaves and few on the rest of the lower surface when mature, with numerous pits on the lower surface; petiole 0.5 – 2.5 cm, densely covered with scales like those on the twigs. Flowers not seen. Infructescence with up to 3 fruits, c. 3.5 cm long; the peduncle, rhachis and branches densely covered with scales like those on the twigs. Young fruits c. 1 cm long and 0.5 cm wide, obovoid, densely covered with scales like those on the twigs. Fig. 50.

DISTRIBUTION. Fiji Islands, known only from the type collection.
ECOLOGY. Forest. Alt.: 1050 to 1120 m.

A. unifolia resembles *Aglaia vitiensis*, but has simple rather than imparipinnate leaves.

46. Aglaia leucoclada C. de Candolle in Bull. Herb. Boiss. Sér. II 3: 172 (1903). Holotype: S.E. New Guinea, fl., *Forbes* 52 (G!).

Small tree up to 5(– 25) m tall. Bark smooth light greyish-brown, inner bark pale yellowish brown. Twigs yellowish-brown, densely covered with small reddish-brown peltate scales, sometimes with a fimbriate margin. Leaves simple, 10 – 28 cm long, 2.5 – 7.5 cm wide, elliptical or obovate, yellowish-green, pale green or pale brown when dry, apex acuminate, the obtuse acumen parallel-sided, often curved and 1.5 -2.5 cm long; cuneate at the slightly asymmetrical base; with a few scales like those on the twigs on the midrib below and scattered on the rest of that surface; veins 10 – 13 on each side of the midrib, ascending, often divided near the margin and either with both branches curving upwards or one branch curving upwards and the other downwards, anastomosing; midrib prominent, veins subprominent on lower surface; petiole 1.5 – 3.5 cm with few to numerous scales like those on the twigs.

Inflorescence (3 –)13 – 20 cm long and 13 – 16 cm wide, with few spreading branches, sessile or with a peduncle up to 5 mm, the peduncle rhachis and branches with few to numerous peltate scales like those on the twigs and some stellate scales interspersed. Flowers 0.8 – 1.5 mm long, 1.8 – 0.8 mm wide, subglobose or obovoid, white or pale yellow; pedicel 0.3 – 0.8 mm, with a point of weakness between this and the rest of the final branch of the inflorescence, with few to numerous pale brown stellate scales. Calyx 0.3 – 0.5 mm, divided into 5 ovate acute or blunt lobes with few to numerous pale brown stellate scales on the outside. Petals 5, pale yellow or yellow; aestivation quincuncial. Staminal

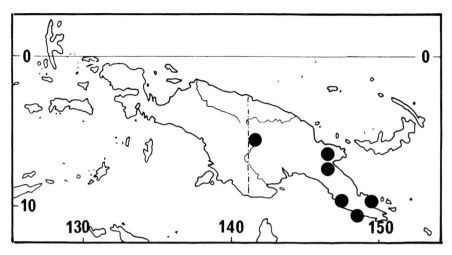

Fig. 51. Distribution of *A. leucoclada*.

tube 0.3 – 1.5 mm long, 0.7 – 1 mm wide, cup-shaped and lobed, anthers 5, 0.2 – 0.4 mm long, c. 0.2 mm wide, ovoid, inserted between the lobes just inside the margin of the tube and pointing towards the centre of the flower. Ovary 0.1 – 0.3 mm high, 0.2 – 0.5 mm wide, depressed-globose, with numerous pale brown stellate scales; loculus 1; stigma 0.1 – 0.2 mm high, 0.1 – 0.4 mm wide, depressed-globose.

Infructescence up to 3 cm, sessile. Fruits obovoid, up to 1.8 cm long and 1 cm wide, yellow or orange, the pericarp thin and brittle when dry and with few to densely covered with peltate scales which have a fimbriate margin on the outside; fruit-stalks up to 10 mm. Ripe fruit 3.5 cm in diameter, subglobose, with 2 locules each containing 1 arillate seed; aril gelatinous; testa brown.

DISTRIBUTION. Papua New Guinea only. Fig. 51.
ECOLOGY. Found in primary forest, montane forest and in secondary regrowths. Understorey tree. Alt.: 5 to 870 m.
VERNACULAR NAMES. New Guinea: Gun (Daga).
Representative specimens. PAPUA NEW GUINEA. Central District, Abau subdistrict, Bamguina river [10°10'S 148°25'E], fl., 19 April 1964, *Sayers* NGF 19622 (K!, L!); Western section, 11½ miles E. of Subitana [9°25'S 147°32'E], c. 600 m, fr., 24 Sep. 1962, *Hartley* TGH 10793 (K!); Western District, Kuenga Subdistrict, 50 km N.E. of Ningerum, 600 m, fl., HYN 251 (LAE!).

Aglaia leucoclada has simple leaves which are pale green or yellowish-green when dry; the indumentum is inconspicuous and the inflorescence is small and delicate.

47. Aglaia silvestris (M. Roemer) Merrill, Interpr. Rumph. Herb. Amboin. 210 (1917). Neotype: [Maluku], Amboina, [Hitoe Lama], July – Nov. 1913

[11 Oct 1913], in forests alt c. 150 m, *Robinson Plantae Rumphianae Amboinense* 490 (PNH†; isoneotypes: BM!, BO!, K!, L!, NY, P!).

[*Lansium silvestre* Rumph., Herb. Amb. 1: 153, t.55 (1741)].
Lansium silvestre M. Roemer, Fam. Nat. Syn. Monogr. 1: 99 (1846).
Aglaia ganggo Miq., Fl. Ind. Bat. Suppl. 1: 506 (1861). Lectotype (designated here): Sumatra, Bonjol, fr., [*Teijsmann*] 577 HB (U!; isolectotype: L!); Miq., Ann. Mus. Bot. Lugd. Bat. 4: 47 (1868); C. de Candolle in A. & C. de Candolle, Monog. Phan. 1: 627 (1878); King in Jour. As. Soc. Bengal 64: 65 (1895); Koorders & Valeton, Atlas der Baumarten von Java 1: t. 156 (1913); Backer & Bakhuizen, Fl. Java 2: 129 (1965); Pannell in Ng, Tree Flora of Malaya 4: 216 (1989).
Aglaia pyrrholepis Miq., Ann. Mus. Bot. Lugd. Bat. 4: 47 (1868). Lectotype (designated here): Java, fl., 1859 – 1860, *Teijsmann & de Vriese* s.n. (U!; isolectotype: L!); C. de Candolle in A. & C. de Candolle, Monog. Phan. 1: 628 (1878).
Aglaia forstenii Miq., Ann. Mus. Bot. Lugd. Bat. 4: 46 (1868). Lectotype (designated here): Moluccas [Maluku], Amboina, Laha, fl., *Forsten* s.n. (L!; isolectotypes: BO!, K!, U!); C. de Candolle in A. & C. de Candolle, Monog. Phan. 1: 627 (1878).
Amoora ganggo (Miq.) Kurz in Jour. As. Soc. Beng. 45: 123 (1876).
Aglaia pyramidata Hance in Jour. Bot. (N.S.) 6: 331 (1877). Lectotype (designated here): *Pierre* in herb. *Hance* 20062 (BM!).
Epicharis bailloni Pierre in Bull. Soc. Linn. Soc. Paris: 292 (1881). Lectotype (designated here): [Cambodia], in montibus Cam Chay ad Kamput Province,, yg fr., Aug. 1874, *Pierre* 1470 (P!; isolectotype: K!).
Lepiaglaia pyramidata (Hance) Pierre, Fl. Forest. Cochinch. Fasc. 21, ante t. 334 (1 July 1895).
[*Aglaia cochinchinensis* Pierre, Fl. Forest. Cochinch. Fasc. 21, ante t. 334A (1 July 1895) nom. in syn.]
Lepiaglaia baill(i)oni (Pierre) Pierre, Fl. Forest Cochinch. Fasc. 22, ante t. 352 (1 July 1896).
Amoora mannii King ex Brandis, Indian Trees: 142 (1906). Syntypes: S. Andaman, North Bay, fl., 5 Aug. 1893, *King's Coll.* (K!); Man Gunj, fr., Jan. 1894, *King's Coll.* s.n. (K!).
Aglaia baillonii (Pierre) Pellegrin in Lecomte, Fl. Gén. L'Indo-Chine 1: 774 (1911).
Aglaia acuminata Merrill in Philipp. Jour. Sci., Bot. 1914, 9: 531 (1915). Lectotype (designated here). Philippine Islands, Palawan, Taytay, [c. 15 m], fl., May 1913, *Merrill* 9306 (NY!; isolectotypes: A!, BO!, PNH†, US!).
Aglaia obliqua C.T. White & Francis in Proc. Roy. Soc. Queensl. (1926), 38: 236 (1927). Type: Papua [New Guinea], 8 miles West of Buna, Northern Division, on the Ambogi River, fl., *Lane-Poole* 132 (K!).
Aglaia micropora Merrill in Univ. Calif. Publ. Bot. 15: 129 (1929). Lectotype (designated here): British North Borneo, [Sabah], Elphinstone Province, Tawao, fl., Oct 1922 – March 1923, *Elmer* 21866 (UC!; isolectotypes: A!, BM!, BO!, G!, K!, M, NY, P!, U!).
Aglaia copelandii Elmer in Leafl. Philipp. Bot. 9: 3286 (1937), sine diagn. lat. Type no.: *Elmer* 14070, Philippine Islands, Island of Mindanao,

Cabadbaran, Mt Urdaneta, Oct. 1912 (A!, BM!, BO!, G!, GH, K!, L!, P!, U!, W!).
Aglaia cedreloides Harms in Engl. Bot. Jahrb. 72: 164 (1942). Lectotype (designated here): New Guinea, [N.E.], Morobe Dist., Wareo, 3200 ft [c. 1000 m], fl., 8 Jan. 1936, *Clemens* 1563 (B!; isolectotypes: G!, L!)
Aglaia mannii (King ex Brandis) Jain & Gaur, Jour. Econ. Tax. Bot. (1985) 7: 466 (1986).

Tree up to 30 m (– 50 m), with a broad rounded crown. Bole up to 13 m, up to 113 cm in circumference, with up to 7 L-shaped buttresses upwards up to 120 cm and outwards up to 215 cm. Outer bark pale greyish-brown or reddish-brown, longitudinally split at wide intervals, with longitudinal rows of lenticels; inner bark reddish-brown, or dark orange-brown; sapwood paler than inner bark; heartwood pale yellowish-brown, almost white; latex white. Branches ascending or arching. Twigs slender, brown, densely covered with peltate scales which have a dark brown centre and pale margin.

Leaves imparipinnate, 19 – 65 cm long, 14 – 35 cm wide, broadly oblong in outline; petiole 10 – 20 cm, the petiole, rhachis and petiolules densely covered with scales like those on twigs. Leaflets (5 –)13 – 19, the laterals alternate, all 8 – 24 cm long, 1.7 – 6.5 cm wide, yellowish-green when young, turning dark green above when mature, subcoriaceous, usually oblong, sometimes lanceolate, or elliptical, acuminate at apex with the obtuse acumen up to 18 mm long, rounded or shortly cuneate at the sometimes asymmetrical base, with scales like those on the twigs densely covering the upper surface when young but deciduous before maturity and sparse on to densely covering the lower surface; veins 12 – 21 on each side of the midrib, ascending and curved upwards near the margin, midrib ± prominent on lower surface; petiolules 5 – 10(– 20) mm.

Inflorescence up to 30 cm long and 20 cm wide; peduncle up to 15 cm, the peduncle, rhachis, branches and pedicels clothed like the twigs. Flowers up to 3.5 mm long and 2.5 mm wide, obovoid, fragrant; pedicel up to 2 mm long with up to 5 ovate or elliptical bracts up to 0.8 mm long, pedicel and bracts densely covered with scales like those on the twigs. Calyx up to $1/2$ the length of the corolla, cup-shaped, densely covered with peltate scales on the outside, deeply divided into 5 broadly ovate lobes which have ciliate margins. Petals 5 (or 6), yellow, elliptical or ovate, aestivation quincuncial. Staminal tube longer than the corolla, obovoid, with a minute pore 0.2 – 0.3 mm across which is entire at the margin; anthers 5, $1/3$ to $1/2$ the length of the tube, ovoid, inserted near the base and included in the tube. Ovary depressed-globose, densely covered with peltate scales; stigma ovoid, with two small lobes, longitudinally ridged; ovary and stigma together c. $1/3$ the length of the staminal tube.

Infructescence up to 30 cm long and 25 cm wide, with c. 50 fruits; peduncle up to 15 cm, the peduncle, rhachis, branches and fruit-stalks with indumentum like twigs. Fruit up to 2 cm long and 2 cm wide, usually obreniform in outline, flattened, wrinkled when dry, brown, red, orange or yellow, densely covered with scales like those on the twigs, indehiscent, sometimes subglobose (Australasia); fruit-stalks up to 1 cm. Loculi 1 – 2 (or 3), each containing 1 seed; aril thin, brown, translucent and sweet.

DISTRIBUTION. Andaman Islands, Nicobar Islands, Vietnam, Cambodia, Thailand, Peninsular Malaysia, Sumatra, Borneo, Philippine Islands, Java, Sulawesi, Maluku, New Guinea, New Britain, Solomon Islands. Fig. 52.

ECOLOGY. Found in primary forest, swamps, savannah, kerengas, monsoon forest, moss forest, along roads, along rivers on clayey loam, sandstone, stand, limestone. Alt.: sea level to 2100 m. Few and scattered to common.

FIELD OBSERVATIONS. Kuala Lompat, Pahang, Peninsular Malaysia: when young, the leaflets have a dense covering of bronze-coloured peltate scales which give the flushing leaves a pinkish colour. In April – June 1978 all the trees of this species were flushing new leaves and therefore were quite conspicuous in the forest. This species does not flower annually at Kuala Lompat.

USES. Buttresses for house-building (Solomon Islands: San Cristobal); wood used for spear shafts (Philippine Islands: Luzon, Mountain prov.), axe-handles (Solomon Islands: Guadalcanal), as board in houses (Halmaheira: Weda).

VERNACULAR NAMES. Peninsular Malaysia: Bekak. Sumatra: Balam Kembang Tandjoeng, Kajoe Pasak, Kakambrang, Patjar Kidang, Toeba Oelat. Java: Hentjit, Langir, Ki (Sundanese). Borneo: Buniau, Bunya, Segera (Iban); Bunyo, Gayan (Kayan); Jalungan Sasak, Langsat-Langsat (Malay); Lantupak (Dusun Kinabatangan); Lepuniau (Kenyah); Bilajang Langsap, Djaring Burung. Sulawesi: Galatri. Moluccas: Moloa, Motoa (Weda); Parawunut'a. New Guinea: Boor, Bowor (Kebar); Deng (Moejoe); Dumpahop (Dumpu); Gun (Daga); Hargifa (Okapa); Herrib (Manikiong); Kamoenem, Lowele (Mooi); Kawi (Amberbaken); Kawiij (Sidei); Kini (Bilia); Mansaambree, Mansrodoewane (Biak); Mavoa (Biaru); Mouge (Faita); Ojakobo (Kapau-koe); Rihjoh (Nemo); Sahote (Amele); Sengsa (Kemtoek); Thorer. Solomon Islands: Pira, Ulukwalo, Vulukwalo (Kwara'ae).

Representative specimens. NICOBAR ISLANDS. fl., Feb. 1875, *Kurz* 25951 (K!).

ANDAMAN ISLANDS. *King's Coll.* 421 (L!).

BURMA. Tavoy District, Kalainaung Reserve, 75 m, fl., 31 Jan. 1930, *Suksy* 10875 (K!); Tavoy Range, Taungbaung, 75 m, fl., 7 Jan. 1930, *Suksy* 10829 (K!).

VIETNAM. Annam, Ile Donai, near Blao, *Poilane* 19883 (L!).

THAILAND. Chantaburi, Kao Soi Dao [13°N 102°10'E], fl., Dec. 1924, *Kerr* 9681 (C!, K!).

PENINSULAR MALAYSIA. Kedah: Sg. Petani, Conservator of Forest, FMS 9005 (K!); Rayor F.R., fl., 1933 *Babjee* KEP 32917 (K!); Gg Jerai (Kedah Peak), Pannell 1337 (FHO!). Pahang: Taman Negara, *Ang* FRI 23306 (FRI!); Krau Game Reserve, K. Lompat, ♂ fl., 30 Nov. 1978, *Pannell* 1321 (FHO!); Jenderak Estate, *Kadim & Mahmood* 82 (K!). Selangor: mile 3 Ginting Highlands road, *Kochummen* FRI 16218 (K!); mile 17 Ulu Gombak, *Nur* SFN 34243 (K!, SING!); Ulu Gombak F.R., 250 m, *Ahmad* KEP 94458 (L!); Ulu Gombak F.R., *Pannell* 1029 (FHO!).

SUMATRA. Bandjol, *Anon.* 577 HB (U!). Atjeh [Aceh], Gunung Leuser Nature Reserves, Southern part of reserves, Alas River Valley, near the mouth of the Renun River, c. 50 km, S. of Kutacane [c. 3°N 97°50'E], 50–125 m, fr., 20 July 1979, *de Wilde & de Wilde-Duyfjes* 18929 (L!).

SARAWAK. N Amau, Ulu Mujong, Balleh [2°15'N 113°60'E], c. 300 m, fr.,

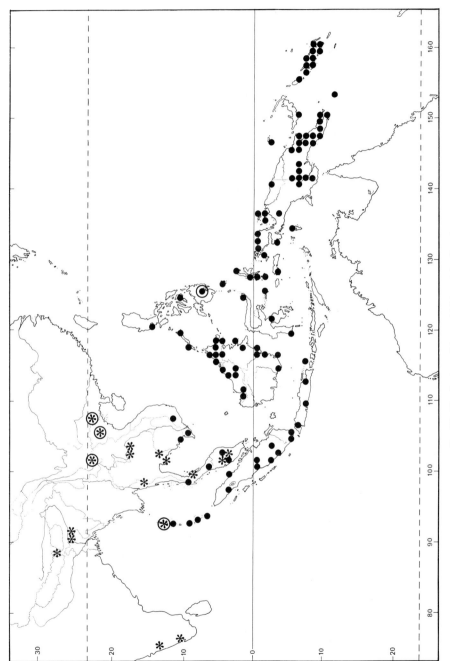

Fig. 52. Distribution of *A. perviridis** and *A. silvestris* ●.

17 Apr. 1964, *Othman b. Haron* S 19940 (FHO!).

SABAH. nr Sandakan, Sepilok F.R., Sample plot 3, compt. 4, Plot H, N.T. 861, fl., 23 Oct. 1963, *Pennington* 7920 (FHO!). Mount Kinabalu, Penibukan, 1350–1650 m, *J. & M.S. Clemens*, 31239 s.n. (BM!). Elphinstone Province, Tawao: *Elmer* 21819 (U!) & *Elmer* 21866 (U!) & *Elmer* 21884 (SING!, U!).

SULAWESI. N., Minahassa, Bolaan Mongondo, Kec. Lolak, between Pinogaluman and Pindol [0°49'N 124°02'E], 70 m, fr., 17 Oct. 1973, *de Vogel* 2522 (FHO!).

PAPUA NEW GUINEA. Morobe District: hillside above Dengaloo in Bulolo, c. 1220 m, fl. 19 Jan. 1964, *Pennington* 8070 (FHO!); near Bulolo, Crooked creek, c. 760 m, fr., 21 Jan. 1964, *Pennington* 8081 (FHO!). Eastern Highlands District, Kainantu Subdistrict, Kassam Pass [6°12'S 146°02'E], c. 400 m, fl., 9 Jan. 1968, *Henty & Coode* NGF 29233 (K!). Milne Bay District, Cameron Plateau, headwaters of Gumini River [10°20'S 150°10'E], fr., 11 June 1964, *Womersley* NGF 19300 (K!). Bougainville, Siwai, fr., June 1931, *Waterhouse* 469-B (K!).

SOLOMON ISLANDS. S.E. New Georgia, Tita River, ♂ fl., 4 April 1966, *Burn- Murdoch's Collectors* BSIP 7160 (K!, L!). Guadalcanal, Monitor Creek, below Mt Gallego, 600 m, fl., 6 July 1965, *Whitmore* BSIP 6036 (K!). N.W. Kolombangara, Rei Area, 30 m, fr., 1 July 1968, *Mauriasi & collectors* BSIP 11620 (K!, L!).

In Western Malesia, the leaves of *A. silvestris* have 13 – 19 leaflets and when the leaflets are mature, the scales are conspicuous only on the lower surface; the scales are typically bronze in colour, with a paler ragged margin and blackish central spot; the flowers have an obovoid staminal tube with a minute apical pore; the flattened obreniform fruit which has a conspicuously wrinkled surface when dry is peculiar to this species of *Aglaia*.

In the eastern part of the range, e.g. New Guinea, the fruits are larger (c. 2.5 cm long and 1.7 cm wide) and obovoid rather than small and obreniform, but they often still have the characteristic wrinkled surface when dry. The venation is often reddish, resembling that of *A. perviridis*. The leaflets are more variable in size (including smaller and larger than elsewhere) and may be fewer in number (e.g. 5 or 7), they are still asymmetrical in shape and the asymmetry is sometimes more pronounced. The inside of the staminal tube and the anthers are sometimes hairy. In some specimens, the peltate scales are pale brown, but the asymmetrical leaflets and the minute aperture of the staminal tube distinguish these specimens as *A. silvestris*.

A. samoensis resembles *A. silvestris*, but the peltate scales of *A. samoensis* lack the very dark centre which is usually found in *A. silvestris* and peltate scales with a long fimbriate margin or stellate scales are more frequent in *A. samoensis*, especially on the inflorescenc e; the flowers of *A. samoensis* are subglobose and have an obovoid staminal tube with a wide aperture, whereas, in *A. silvestris*, the flower is ovoid or ellipsoid and the staminal tube is a similar shape, with a minute aperture.

48. Aglaia perviridis Hiern in Hooker fil., Fl. Brit. Ind. 1: 556 (1875). Lectotype (designated here): India, Khasia, subtrop., 4000 ft [c. 1200 m], fl., '8 *Milnea*', *J.D. Hooker & T. Thomson* s.n. (CGE!; isolectotypes: G-DC!,

K!, L!, W!); C. de Candolle, Monog. Phan.: 610 (1878); C.Y. Wu, Flora Yunnanica 1: 239 (1977).

Aglaia perviridis var. *sikkimiana* C. de Candolle in A. & C. de Candolle, Monog. Phan. 1: 610 (1878). Holotype: Sikkim, below Karsung, fr., *J.D. Hooker* s.n. (K!).

Aglaia maiae Bourdillon in Jour. Bombay Nat. Hist. Soc. 12: 350, t.3 (1899). Type: India, Travancore, Ariyaukum, 1000 ft [c. 300 m], fl., 10 June 1895, *Bourdillon* 625 (K!);

Aglaia kingiana Ridley in Jour. As. Soc. Straits. 82: 175 (1920). Lectotype (designated here): Malay Peninsula, Perak, Gopeng, Trinta C.P., 500 – 800 ft (c. 150 – 240 m], fl., July 1883, *King's Coll.* (*Kunstler*) 4606 (SING!; isolectotypes: CGE!, E!, K!, L!, W!); I.H. Burkill, Dictionary of Economic Products of the Malay Peninsula 1: 74 (1935).

Aglaia canarensis Gamble in Kew Bull. 1915: 347 (1915). Lectotype (designated here): India, [Madras Presidency], South Canara, fl., 1873, *Beddome* s.n. (K!).

Tree up to 12(– 25) m. Twigs greyish-brown or pale yellowish or greenish-brown, longitudinally wrinkled and often with numerous very pale orange brown yellowish-brown lenticels, with numerous to densely covered with small reddish-brown peltate scales which have a dark central spot and an irregular or fimbriate margin, occasionally with stellate scales. Leaves up to 54 cm long; peduncle up to 10 cm, the petiole, rhachis and petiolules with few to numerous scales like those on the twigs. Leaflets 11 – 13, the laterals subopposite, all 7.5 – 23 cm long, 2 – 6 cm wide, blackish-brown or reddish-brown when dry, usually markedly ovate, sometimes (China) lanceolate or narrowly elliptical, rounded on one side and cuneate on the other at the markedly asymmetrical base, tapering to a caudate apex, the obtuse acumen usually parallel sided and up to 2 cm long; with few to numerous scales like those on the twigs, and sometimes in China with some pale stellate scales, on the midrib below and few on the rest of that surface, usually pitted and sometimes rugulose on both surfaces, sometimes with occasional scales on the upper surface; veins 12 – 18 on each side of the midrib, often reddish when dry, ascending and curving upwards near the margin, not anastomosing, often with shorter lateral veins in between, the midrib prominent, the lateral veins usually and the reticulation sometimes subprominent below; petiolule up to 10 mm, up to 20 mm on the terminal leaflet.

Inflorescence up to 35 cm long and 24 cm wide; peduncle up to 12 cm, the peduncle, rhachis and branches with few to numerous scales like those on the twigs, but usually with a fimbriate margin. Flowers with few or no scales on the calyx and pedicel, the pedicel articulated with 3 small ovate bracteoles below the articulation, the calyx lobes and bract with ciliate margins. Flower 1.2 – 2.3 mm long, 1.2 – 1.8 mm wide, ellipsoid, pedicel 0.5 – 1 mm, articulated with branchlet up to 1 mm long, which has few to numerous reddish-brown peltate scales with a fimbriate margin, the pedicel and calyx without or with occasional or no hairs or scales. Calyx about $1/3$ the length of the flower, cup-shaped, divided to c. $1/2$ way into 5 broad ovate lobes which have a fimbriate margin. Petals 5, elliptical, yellow, aestivation quincuncial. Staminal tube 1 – 2 mm long, 0.8 – 1.8 mm wide, narrower at the base below the insertion of the anthers, aperture 0.4 – 1 mm across, the margin shallowly lobed, anthers 5, 0.4 – 0.8 mm long, 0.2 – 0.5 mm wide, inserted near the base or about half way up the tube, included or just protruding through the aperture. Ovary 0.2 – 0.4 mm long, 0.3 – 0.4 mm wide, depressed globose; loculus 1, containing 1 ovule; stigma 0.3 – 0.4 mm long, 0.2 – 0.4 mm wide, ovoid.

Infructescence up to 26 cm long and 20 cm wide, pendulous; peduncle up to 6 cm, the peduncle, rhachis and branches with few to numerous scales like those on the twigs. Fruits few, up to 3 cm long and 1.7 cm wide, asymmetrically ellipsoid with one side flat or slightly concave, yellow or brown, with a thin brittle pericarp which is densely covered with scales like those on the twigs or paler scales or occasionally reddish-brown stellate scales on the outside, inner surface smooth and shiny; loculus 1, containing one seed; the seed up to 2.7 cm long and 1.5 cm wide, completely surrounded by an aril, the aril thin, papery and dark reddish-brown when dry and with a network of veins, the shrunken seed within completely separate from the aril and up to 1.6 cm long and 1 cm wide.

DISTRIBUTION. India, Bangladesh, Bhutan, Andaman Islands, China, Thailand, Peninsular Malaysia. Fig. 52.

ECOLOGY. Found in evergreen forest, primary forest, secondary forest; on limestone. Alt.: 100 to 1330 m. Common.

Representative specimens. INDIA. Mysore, Hassan District, Attahalla, Bisle Ghat [13°N 76°E], fl., 10 April 1970, *Saldanha* 16772 (E!), fl., 14 March 1969, *Saldanha* 13058 (E!). E. Himalaya, fr., *Ribu & Rhomoo* s.n. (E!).

BHUTAN. Mirichana Timpu, 900 m, fr., 20 April 1915, *Cooper* for *Bulley* 3776 (BM!, E!). Tongsa District, between Pertimi & Tintibi Bridge, Mangde Chu [27°12'N 90°40'E], 1090 m, fr., 5 April 1982, *Grierson & Long* 4349 (E!, K!). Saruchi District, Tamangdhanra Forest, Saruchi [26°54'N 89°06'E], 700 m, 28 Feb. 1982, *Grierson & Long* 3314 (E!, K!). Sarbhang District, 2.5 km below Getchu, on Chirang road [26°57'N 90°14'E], 1480 m, fr., 12 March 1982, *Grierson & Long* 3676 (E!, K!).

BANGLADESH. Chittagong Hill tracts, Rangamattia, st., March 1880, *Gamble* 7777 (K!).

ANDAMAN ISLANDS. S. Wandur, road-sides, sandy soil, sea level, yg fl., 3 July 1974, *Balakrishnan* 1667 (E!). fl., 1854, *King's coll.* 421 (L!).

BURMA. Kachin State, Sumprabum Sub-Division, Eastern approaches from Sumprabum to Kumon Range [26°40'N 97°20'E], between Ning W'Krok and Kanang, on the eastern aspects of Gwe-Kya-Kat-Bum. 1200 – 1525 m, 20 Jan. 1962, *Keenan, Tun Aung & Tha Hla* 3353 (E!). Tenasserim division, Tavoy District, area with a radius of 12 miles from Paungdaw, south of Paungdaw Power Station [14°N 98°30'E], 425 m, fl., Sep. 1961, *Keenan, Tun Aung & Rule* 1437 (E!).

VIETNAM. Tonkin, Taai Wong Mo Shan & Vicinity, Chuk-phai, Ha-coi, fr., 16 – 22 Oct. 1936 *Tsang* 26992 (E!). CHINA. Kwangsi, S.E. Shang-sze District, Kwangtung Border, Shap Man Taai Shan: Tang Lung village, yg fr., 21 Aug. 1934 *Tsang* 24088 (NY!) & Iu Shan village, fr., 19 May 1934 *Tsang* 22333 (BM!). Yunnan, Szemeo [22°46'N 101°03'E], Yulo, 1200 m, *Henry* 12996 (E!, K!, NY!).

THAILAND. N., Phitsanulok, Tung Salaeng Luang, 600 m, fr., 25

July 1966, *Larsen, Smitinand & Warncke* 898 (AAU!). S.W., Kanchanaburi, Khao Yai, E. of Sangkhla, c. 850 – 950 m, fr., 10 April 1968, *Beusekom & Phengkhlai* 415 (AAU!, K!, L!).

Aglaia perviridis has 11 – 13 usually markedly ovate leaflets, the upper and lower surfaces of which are pitted and the midrib and lateral veins are reddish-brown when dry. The fruit is 1-locular and of a characteristic asymmetrically ovoid shape. *A. perviridis* resembles *A. leptantha* and *A. silvestris*. The ovary and fruit of *A. leptantha* have two loculi; the pericarp appears to be softer and it dries moulded around the seeds, whereas in *A. perviridis* the dry seed contracts within the pericarp, which is brittle and retains its shape. *A. silvestris* differs from *A. perviridis* in leaflet shape in having a staminal tube with narrow, entire aperture and in its fruit which is frequently obreniform in shape and has 1, 2 or 3 loculi. Specimens of *A. perviridis* from China have leaflets which are not ovate in shape, but lanceolate or narrowly elliptical, and which often have some stellate scales on the lower surface. In China, the fruit is said to be edible.

49. Aglaia leptantha Miq., Ann. Mus. Bot. Lugd. Bat. 4: 51 (1868). Lectotype (designated here): Sumatra, [W.], *Korthals* s.n. (L!; isolectotype: BO!, K!); C. de Candolle in A. & C. de Candolle, Monog. Phan. 1: 604 (1878).

Aglaia glabriflora Hiern in Hooker fil., Fl. Brit. India 1: 555 (1875). Lectotype (designated here): Malacca [Peninsular Malaysia], Mt Ophir, *Griffith* 1042 (K!); C. de Candolle in A. & C. de Candolle, Monog. Phan. 1: 608 (1878); King in Jour. As. Soc. Bengal 64: 63 (1895); Ridley, Fl. Malay Penins. 1: 404 (1922); I.H. Burkill, Dictionary of Economic Products of the Malay Peninsula 1: 73, 137 (1935); Pannell in Ng, Tree Flora of Malaya 4: 217 (1989).

? *Aglaia leptantha* [*lepantha*] Miq. var. *borneensis* C. de Candolle in A. & C. de Candolle, Monog. Phan. 1: 604 (1878). Type: Borneo (U).

Aglaia laevigata Merrill in Philipp. Gov. Lab. Bur. Bull. 35: 31 (1906). Lectotype (designated here): Philippine Islands, Luzon, Province of Rizal, Bosoboso, fl., July 1908, *Merrill* 2818 (NY!; isolectotypes: G!, K!, P!, US!).

Aglaia multiflora Merrill in Philipp. Jour. Sci. 1, Suppl.: 73 (1906). Lectotype: Philippines, Luzon, Province of Bataan, Lamao River, Mt Mariveles, fl., July – Aug. 1904, *Ahern's Coll. For. Bur.* 1420 (K!; isolectotypes: G!, PNH†).

Aglaia glabrifolia Merrill in Univ. Calif. Publ. Bot. 15: 129 (1929). Lectotype (designated here): British North Borneo [Sabah], Elphinstone Province, Tawao, sea level, fl., Oct. 1922 – March 1923, *Elmer* 21394 (UC!; isolectotypes: A!, BM!, BO!, G!, GH!, K!, MICH!, NY!, P!, SING!, U!, UC!).

Aglaia gamopetala Merrill in Univ. Calif. Publ. Bot. 15: 126 (1929). Lectotype (designated here): British North Borneo [Sabah], Elphinstone Province, Tawao, fl., Oct. 1922 – March 1923, *Elmer* 21748 (UC!; isolectotypes: A!, BM!, BO!, G!, GH!, L!, MICH!, NY!, P!, U!); Pannell in Ng, Tree Flora of Malaya 4: 216 (1989).

? *Aglaia annamensis* Pellegrin in Bull. Soc. Bot. France 91: 179 (1944).

Lectotype (designated here): [Vietnam], Annam, Quang-Tri [Province], Dent du tigre, *Poilane* 10249 (P!).

Tree up to 30 (– 40) m, sometimes flowering and fruiting at 4 m, with an irregularly rounded or narrowly conical crown. Bole up to 16 m, up to 132 cm in circumference, fluted near the base or with small buttresses upwards up to 30 cm and root-like buttresses outwards up to 135 cm, branches patent or ascending. Outer bark pale grey, greenish-brown, yellowish-brown, or reddish-brown with greyish-green patches, with longitudinal and transverse cracks and round lenticels; surface of inner bark green and pale yellow; inner bark green, pale orange-brown or reddish-brown; sapwood pale brown, pale yellowish-brown, dark orange-brown or reddish-brown, soft; latex white or exudate pinkish-orange. Twigs slender, grey, with longitudinal wavy ridges, densely covered with reddish-brown, pale brown or grey peltate scales which have a fimbriate margin, when pale, the scales have a dark grey central spot.

Leaves imparipinnate 30 – 83 cm long, 20 – 68 cm wide, obovate in outline; petiole 6.5 – 16 cm, the petiole rhachis and petiolules with ridges and with few to densely covered with scales like those on the twigs. Leaflets 7 – 11 (or 12), the laterals alternate, all 3.5 – 34 cm long, 1.5 – 14 cm wide, dark glossy green above, paler beneath, young leaflets brownish- green when dry, older leaflets black or blackish-green when dry, elliptical, ovate, oblong or oblanceolate-oblong, rarely obovate, sometimes asymmetrical, acuminate-caudate at apex, the obtuse acumen often narrow and parallel-sided and up to 20 mm long, broadly cuneate or rounded to a sometimes subcordate asymmetrical base, upper surface occasionally rugulose or with a few scales on the veins, lower surface with scales like those on the twigs few to numerous on the midrib, few on the veins and rarely on the surface in between, sometimes in Borneo with numerous orange brown stellate hairs, glabrescent, but the scales present on older parts like those on the twigs, sometimes in Borneo with numerous pits on upper and lower leaflet surfaces; veins 4 – 14 on each side of the midrib, black or reddish-brown when dry, ascending and markedly curved upwards near the margin, midrib and lateral veins subprominent, secondary veins slightly prominent on upper surface, midrib prominent, lateral veins subprominent and secondary veins slightly prominent on lower surface; petiolules 2 – 10 mm on lateral leaflets, up to 25 mm on terminal leaflet.

Inflorescence up to 40 cm long and 40 cm wide; peduncle up to 10 cm, the peduncle, rhachis and branches with indumentum like the twigs. Flowers 1.8 – 3 mm long, 1.7 – 3 mm wide, obovoid or ellipsoid, fragrant of citronella, pedicels 0.8 – 2 mm, densely covered with scales like those on the twigs. Calyx $1/3 - 1/2$ the length of the corolla without or densely covered with scales on the outside, divided into 5 shallow obtuse lobes which have fimbriate margins. Petals 5, white or lemon yellow, elliptical, aestivation quincuncial. Staminal tube slightly shorter or longer than the corolla, pale lemon yellow, obovoid, either with a shallowly 5-lobed aperture c. 0.5 mm in diameter or with a minute apical pore at the apex c. 0.3 mm across, occasionally with a few stellate scales on both surfaces or simple hairs inside; anthers 5, 0.9 – 1 mm long, 0.4 – 0.5 mm wide, narrowly ovoid inserted c. $1/6 - 1/3$ up the tube and included or just protruding, with a few pale brown stellate scales, ovary depressed globose densely covered with stellate hairs or scales, loculi 1 or 2, each containing 1 ovule; stigma 0.4 – 0.5 mm long and c. 0.3 mm wide, ovoid with two small apical lobes, brown and shiny when dry.

Infructescence up to 10 cm long and 5 cm wide with few fruits; peduncle up to 6 cm with surface like the twigs and few to numerous hairs or scales like those on the twigs. Fruits 1.5 – 3.2 cm long and 1.5 – 3 cm wide, ellipsoid or subglobose, sometimes laterally compressed and then c. 2.3 cm thick, with a longitudinal ridge encircling the fruit; the pericarp either thin or hard and woody and up to 5 mm thick, brown, orange, yellow, white or green, densely covered with pale brown stellate scales on the outside; fruit-stalks c. 2 mm. Loculi 1 – 2, each containing 1 seed. Seed up to 2.3 cm long, 1.4 cm wide and 1 cm thick, ovoid, the inner surface flattened; with a complete gelatinous, translucent, sweet-sour edible aril. Fig. 53.

DISTRIBUTION. ? Vietnam (Annam), Thailand, Peninsular Malaysia, Singapore, Sumatra, Bunguran Island, Borneo, Philippine Islands, Nusa Tenggara (Flores). Fig. 54.

ECOLOGY. Found in primary forest, seasonal swamp, ridge forest, montane forest and in kerangas; on sandstone, limestone, sand, granitic sand, clay and podzolic sand. Alt.: 20 to 1700 m. Rare to common.

VERNACULAR NAMES. Peninsular Malaysia: Memberas. Sumatra: Bomberang, Koedo-koedo Delok, Lampat, Langadai Delok, Langadai Pojo, Langsat, Sare- sare Boeloeh, Sare-sare Delok. Borneo: Hahi (Kayan); Kela Buno (Kelabit); Langsat-langsat (Malay); Lantupak (Dusun Kinabatangan); Segera (Iban); Kaju Lilin.

Representative specimens. THAILAND. Surat, Kao Nawng, c. 800 m, fl., 9 Aug. 1927, *Kerr* 13251 (C!). N. Sritamerat, Kao Luang, c. 1100 m, fr., 2 9 April 1928, *Kerr* 15476 (C!).

PENINSULAR MALAYSIA. Kedah, Sungkop V.J.R., c. 110 m, fl., 6 Aug. 1974, *Chan* FRI 021701 (FRI!). Perak: Ulu Slim, *King's Coll.* 10724 (SING!), fl., *Scortechini* 482 (CGE!). Pahang: Tahan Woods, S. Teku, c. 180 m, fr., 21 Feb. 1968, *Whitmore* FRI 4783 (FRI!); Krau Game Reserve, K. Lompat, ♂ fl. , 29 Sep. 1978, *Pannell* 1335 (FHO!). Malacca: Merlimau, *Derry* 56 (SI NG!); *Griffith* 1041 (K!); *Alvins* 2059 (SING)!. *Maingay* 1380 (Kew Dist. 336) (K!), *Maingay* 1442 (Kew Dist. 360) (K!).

SINGAPORE. Tangkin, *Ridley* 3890 (K!), Changi, *Ridley* 1891 (SING!). Changi 1812 (BM!, SING!). Botanic Gardens Jungle, *Sinclair* SFN 40958 (K!). Reservoir Woods, *Ridley* 12668 (SING!), *Ridley* 2800 (SING!).

SUMATRA. Atjeh [Aceh]: c. 35 km N.W. of Kutacane, Gunung Leuser Nature Reserve, Ketambe, Valley of Lau Alas, near tributary of Lau Ketambe, c. 600 m, ♀ fl., 28 May 1972, *de Wilde & de Wilde-Duyfjes* 12429 (L!); c. 50 km S. of Kutacane, southern part of Gunung Leuser Nature Reserves, Alas River Valley, near the mouth of the Renun River, [3°N 97°50'E], 250 m, fr., 20 July 1979, *de Wilde & de Wilde-Duyfjes* 18917 (L!).

SARAWAK. 1st Division: Bt Merayong, 2½ hrs upstream from Lundu, 17 m, fr., 18 Nov. 1963, *Pennington* 7990 (FHO!); 25th mile, Bau/Lundu Road, Sampadi F.R., path to Kg Selampit [1°60'N 110°E], c. 210 m, fr., 3 July 1968, *Ilias Paie* S 26993 (FHO!). 4th Division, Bintulu Similajau –

Fig. 53. *A. leptantha*. Leaf x$^1/_2$. Inflorescence x$^1/_2$. Half flower x10. Immature infructescence x$^1/_2$.

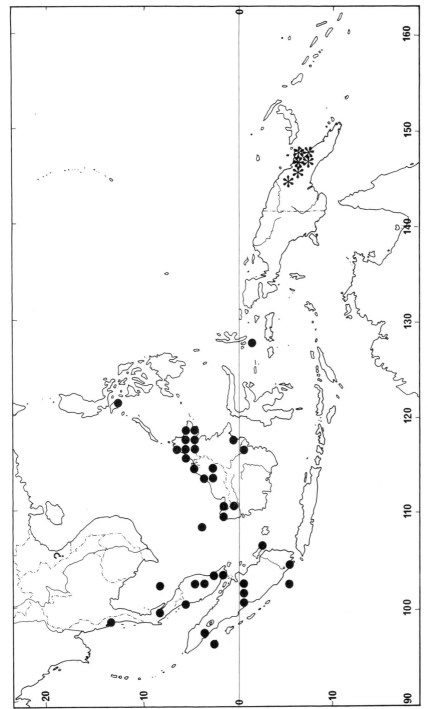

Fig. 54. Distribution of *A. leptantha* ● and *A. cremea* *.

Labang, [3°45'N 113°40'E], fl., 4 Sep. 1968, *Ilias Paie* S 28012 (FHO!). Upper Plieran, Sg. Kenaban: *Pickles* 3417 (FHO!) & 3630 (FHO!).

PHILIPPINE ISLANDS. Mindanao, Davao Province, Mati, fr. Jan. 1931, *Ondeada* For. Bur. 31346 (NY!). Negros, fl., May – Aug. 1914, *Contreras* For. Bur. 23408 (NY!).

JAVA. Megamendoeng, fl., *Anon.* s.n. (L!).

NUSA TENGGARA. Flores: southern part, Mt Ndeki, 200 m, fr., 10 Apr. 1965, *Kostermans & Wirawah* 74 (AAU!); Nunang, 1000 m, fr., 6 Nov. 1966, *Schmutz* 678 (L!).

The leaflets of *A. leptantha* have a caudate, obtuse acumen and are sometimes markedly asymmetrical at the base; the peltate scales are usually pale, with a fimbriate margin and a dark central spot; the secondary venation is visible or subprominent.

A. glabriflora is a form of *A. leptantha* but one which cannot be clearly differentiated from it and is included as a synonym. At Kuala Lompat in Pahang, Peninsular Malaysia, *A. glabriflora* is found in closed forest and *A. gamopetala* in partially cleared forest where there is more light. At this site they can be distinguished, but appear to be closely allied. *A. gamopetala* has larger leaves, paler indumentum and larger flowers in which the staminal tube extends beyond the corolla and has a minute apical pore; in *A. glabriflora*, the inflorescence is without scales and the aperture of the staminal tube is wider. However, intermediates between these two are found in other parts of their range. *A. perviridis* is also similar to *A. leptantha*; the differences are described under *A. perviridis*.

50. Aglaia cremea Merrill & Perry in Jour. Arn. Arb. 21: 319 (1940). Lectotype (designated here): Northeastern New Guinea, Morobe District, Sattelberg, 3500 ft [c. 1050 m], fr., 15 Feb. 1936, *Clemens* 1855 (A!; isolectotype: G!).

Tree 4 – 13 m high. Bole 7.5 – 10 cm in diameter. Outer bark light brown; inner bark yellow or greenish-yellow, latex white. Sapwood pale yellow; heartwood pinkish-yellow. Twigs densely covered with dark reddish-brown peltate scales which have a paler fimbriate margin.

Leaves 17 – 40 cm long, 13 – 30 cm wide, petiole 2.5 – 6 cm; the petiole, rhachis and petiolules with numerous scales like those on the twigs. Leaflets 5 – 7(– 9), the laterals usually subopposite, sometimes alternate, all 4.5 – 15 cm long, 2 – 5.5 cm wide, elliptical or obovate, cuneate at base, acuminate at apex, the parallel-sided acumen obtuse 10 – 15 mm long, dull dark green above, green below, with few to numerous pale peltate scales like those on the twigs on the lower leaflet surface with paler scales scattered in between; veins 6 – 8 on each side of the midrib, curved upwards, not anastomosing; petiolules 3 – 5 mm on lateral leaflets, up to 15 mm on terminal leaflet.

Inflorescence 2.5 – 16 cm long, 4 – 13 cm wide; the peduncle, rhachis and branches with few pale brown peltate scales which have a fimbriate margin. Flower 1.5 – 2 mm long, 1.5 – 2 mm wide, depressed globose, pedicels 0.5 – 1 mm, without or with a few stellate hairs. Calyx 0.5 mm long, with 5 broad spreading lobes which are without or with numerous to densely covered with stellate scales. Petals 5, white or yellow, aestivation quincuncial. Staminal tube 0.5 – 1 mm long, 1.2 – 2 mm wide, cup-shaped with the margin incurred, aperture 0.8 – 1.3 mm across with a lobed margin, anthers 0.4 – 0.5 mm long, 0.7 mm wide, inserted just inside the margin or about a third of the way down the tube, protruding and pointing towards the centre of the flower. Ovary 0.3 – 0.5 mm long, 0.4 – 0.5 mm wide, depressed globose, densely covered with pale peltate scales which have a fimbriate margin; stigma c. 0.5 mm long, 0.4 – 0.5 mm wide depressed globose or subglobose, the stigma visible through the aperture of the staminal tube; loculi 1 – 2 each containing 1 ovule.

Infructescence c. 6 cm long and 4 cm wide, peduncle c. 1 cm; the peduncle, rhachis and branches with few pale brown peltate scales which have a fimbriate margin. Fruits 1 – 2.5 cm long, obovoid, brownish-green, densely covered with reddish-brown peltate scales which have a short fimbriate margin; loculus 1, seed 1, ? with a thin gelatinous aril.

DISTRIBUTION. Papua New Guinea only. Fig. 54.
ECOLOGY. Found in secondary and in hill forest. Alt.: 200 to 1670 m.
USES. Locally used for tool handles.
Representative specimens. PAPUA NEW GUINEA. Eastern Highlands District: Kassam Pass, Rupert Haviland Memorial Lookout, N.E. Hang, 1200 m, fl., 22 Jan. 1964, *Stauffer & Sayers* 5566 (K!, LAE!, UC!); Kainantu Subdistrict, Kassam Pass [6°12′S 146°02′E], 1200 m, fl., 25 Jan. 1968, *Henty & Vandenberg* NGF 29350 (K!).

A. cremea resembles the less robust forms of *A. leptantha*, but it is a smaller tree and the staminal tube is short and cup-shaped with a wide aperture. This species is found only in Papua New Guinea, where *A. leptantha* does not occur.

51. Aglaia forbesii King in Jour. As. Soc. Bengal 64: 68 (1895). Syntypes: Malay Peninsula, Perak, Ulu Kali, 1000 – 1500 ft [c. 300 – 450 m], fl., Aug. 1886, *King's Coll.* 10787 (BM!, G!, K!, L!); Malay Peninsula, Perak, Larut, 500 – 800 ft [c. 150 – 250 m], fl., Aug. 1883, *King's Coll.* 4762 (BM!, CGE!, G!); Pangkore, Sg. Bruas, *Curtis* 1631 (SING!), Perak, Kota, Larut, yg fr., Oct. 1888, *Wray Jr.* 3265 (SING!); Sumatra, Mt Napalhitju, 2500 ft [c. 760 m] 1881, *Forbes* 3179 (BM!, K!, L!); Ridley, Fl. Malay Penins. 1: 406 (1922); Pannell in Ng, Tree Flora of Malaya 4: 215 (1989).

Aglaia humilis King in Jour. As. Soc. Bengal 64: 69 (1895). Syntypes: Malay Peninsula, Perak, nr Ulu Selangor, 400 – 600 ft [c. 135 – 200 m], fr., March 1886, *King's Coll.* 8619 (BM!, G!, K!, SING!); Upper Perak, 10 00 ft [c. 300 m], July 1889, *Wray Jr.* 3763 (K!); Ridley, Fl. Malay Penins. 1: 407 (1922); Pannell in Ng, Tree Flora of Malaya 4: 218 (1989).

Tree up to 35 m, with a dense rounded crown. Bole up to 60 cm in diameter. Outer bark often smooth, brown, pale brown, greenish-brown or greyish- brown, with vertical cracks and lenticels in longitudinal rows, sometimes flaking in irregular scales; inner bark brown, yellowish-brown or reddish- brown; sapwood orange-brown, pale brown, pale yellowish-brown or sometimes reddish-brown; latex white. Twigs usually slender, sometimes stout, greenish-brown or brown, almost smooth or with longitudinal wavy ridges, densely covered with dark brown, reddish-brown or greyish-brown stellate scales or hairs.

Leaves imparipinnate, up to 100 cm long and 60 cm wide, elliptical in outline, petiole up to 35 cm, the petiole, rhachis and petiolules clothed like the twigs. Leaflets 9 – 15, the lateral usually alternate, sometimes subopposite, all 8.5 – 30 cm long, 2 – 10 cm wide, subcoriaceous, green or blackish-green on upper surface, brown or greenish-brown on lower surface when dry, usually elliptical or oblong, sometimes ovate, acuminate at apex, the acute and tapering acumen up to 12 mm long, rounded or cuneate at the asymmetrical base, upper surface rugulose and pitted, lower surface granular, sometimes rugulose and pitted, with white, pale brown or reddish-brown stellate hairs or scales few to numerous on the midrib, few on the lateral veins and scattered on the surface in between; veins 9 – 24 on each side of the midrib, ascending and markedly curved upwards near the margin, midrib and lateral veins slightly depressed on upper surface, midrib prominent and lateral veins subprominent on lower surface, both often nearly black when dry; secondary veins usually barely visible; petiolules up to 15 mm on lateral leaflets, up to 20 mm on terminal leaflet.

Inflorescence up to 35 cm long and 25 cm wide; peduncle up to 6 cm, peduncle and rhachis often flattened, peduncle, rhachis, branches and petiolules with longitudinal wavy ridges and with numerous to densely covered with white, pale brown or reddish-brown stellate scales or hairs. Flowers up to 2 mm long, obovoid, fragrant; pedicels up to 1.5 mm. Calyx cup-shaped, c. $1/3$ the length of the corolla, densely covered with stellate scales or hairs on the outside, divided up to c. $1/2$ way into 5 rounded lobes which have fimbriate margins. Petals 5 – 6, up to 1.7 mm long and 1 mm wide, obovate, white, yellow or orange-yellow, aestivation quincuncial or imbricate. Staminal tube shorter than the corolla, up to 1.5 mm long, cup-shaped or subglobose, slightly incurved and shallowly 5- lobed at the apical margin with the aperture up to 1.3 mm wide or with a pin-prick apical pore; anthers 5, $1/3$ to nearly the length of the tube, obovoid and just protruding through the aperture. Ovary subglobose, densely covered with white stellate hairs; stigma black and shiny, ovoid, with two small apical lobes, longitudinally ridged; the ovary and stigma together c. $1/2$ the length of the staminal tube.

Infructescence with c. 20 fruits, up to 35 cm long; peduncle c. 8 cm; peduncle and branches with surface and indumentum like the twigs. Fruits up to 4 cm long and 3.8 cm wide, ellipsoid or subglobose, sometimes with a small beak at the apex and sometimes narrowed at the base to a short stipe 3 mm long, sometimes with a longitudinal ridge encircling the fruit; pericarp up to 4 mm thick, soft, fibrous and flexible, white, yellow, orange, grey or greenish-grey, longitudinally wrinkled when dry, densely covered with white or yellowish-grey stellate scales or peltate scales which have a fimbriate margin on the outside, shiny inside, with white latex. Loculi 2, each containing 0 or 1 seed. Seed with aril 1.5 – 3 cm long, 2 – 2.2 cm wide, c. 1.5 cm thick, with a complete translucent, gelatinous, yellow or pink aril up to 3 mm thick; the aril sweet-sour or with a flavour like that of *Lansium domesticum* Corr. Fig. 55.

Fig. 55. *A. forbesii*. Habit with male inflorescence x1/2. Half flower, male x10. Female inflorescence x1/2. Half flower, female x10. Infrutescence x1/2.

Section Aglaia

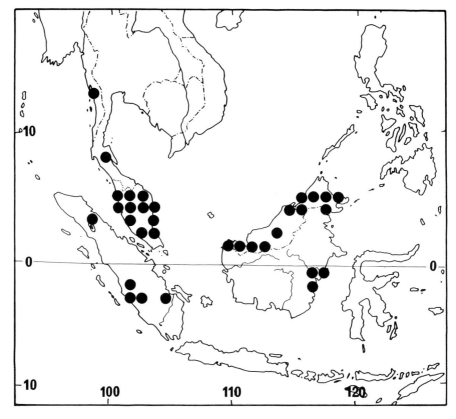

Fig. 56. Distribution of *A. forbesii*.

DISTRIBUTION. S. Burma, S. Thailand, Peninsular Malaysia, Sumatra, Borneo. Fig. 56.

ECOLOGY. Found in evergreen forest, primary forest, secondary forest, moss forest, along rivers, in belukar on sand, clay, sandy loam, sandstone, ultrabasic. Alt.: 30 to 1000 m.

VERNACULAR NAMES. Peninsular Malaysia: Chualing (Temuan); Bekak, Langsat, Memberas. Borneo: Langsat Burung (Malay); Lantupak (Dusun Kinabatangan); Segera (Iban); Suloh (Selakau melanan); Engkuang, Merlangsat.

Representative specimens. BURMA. Tenasserim Division, Tavoy District, hills S. of Paungdaw Power Station, 425 m, fl., Sep. 1961, *Keenan, Tun Aung & Rule* 1437 (K!).

THAILAND. Peninsular District, Nakhon Si Thammarat, Khao Luang area, near Khiri Wong village, c. 500 – 600 m, fr., 16 May 1968, *van Beusekom & Phengkhlai* 796 (AAU!, C!).

PENINSULAR MALAYSIA. Ulu Sg. Trengganu, ½ mile upstream from Kg Petang, c. 275 m, ♀ fl. & fr., 4 June 1968, *Cockburn* FRI 8449

(FRI!). Selangor: Kanching F.R., c. 180 m, fr., 14 Sep. 1963, *Pennington* 7817 (FHO!); Ulu Gombak F.R., mile 15 logging track, c. 210 m, fl., 7 Oct. 1963, *Pennington* 7862 (FHO!). Negri Sembilan, Jelebu F.R., c. 900 m, fl., 4 July 1969, *Suppiah* 11305 (K!).

SARAWAK. 1st Division, Santubong Peak, c. 300 m: ♀ fl. & fr., 24 April 1974, *Mabberley* 1617 (FHO!) & fl., 24 April 1974, *Mabberley* 1618 (FHO!).

The leaflets of *A. forbesii* are sometimes coriaceous, the midrib is prominent and lateral veins sub-prominent on the lower surface, both are often black or red when dry. The tiny stellate scales scattered on the lower surface are inconspicuous, but stellate hairs are sometimes also present on immature leaves.

It is sometimes difficult to separate the leaves from those of *A. leptantha*, but the leaflets of that species are thinner and the secondary venation is visible, all the veins are black (or sometimes red in Borneo) when dry; the leaf apex is obtuse and parallel-sided. *A. leptantha* usually has peltate scales, with a fimbriate margin and a dark central spot.

52. Aglaia foveolata C.M. Pannell spec. nova *Aglaiae forbesii* King interdum similis, a qua foliolis angustioribus et indumento numquam non stellato differt; atque *Aglaiae perviridi* Hiern etiam similis, cui tamen foliola pauciora grandiora, venatione et indumento dissimilia. *Aglaia foveolata* C.M. Pannell praeterea foliis imitatur *Aglaiam multinervem* C.M. Pannell, sed differt nervis paucioribus, foliolis vulgo utrumque foveolatis, supra nitidis in sicco, floribus 5 petalis, 5 antheris; *Aglaiae multinervis* paginae foliorum opacae sunt in sicco et flores 3 petalis, 3 antheris. Holotype: Malaya, Kaula Trengganu, Jerangau Forest Reserve, Compt. 8, 250 ft [c. 75 m], ♂ fl., 5 Sep. 1966, *Kochummen* FRI 2112 (K!); Pannell in Ng, Tree Flora of Malaya 4: 230 (1989), as *Aglaia* sp. 7.

Tree up to 20(−25) m with a dense feathery crown. Bole up to 2 m in circumference, with buttresses upwards to 75 cm and outwards to 30 cm. Outer bark smooth, reddish-brown or greyish-brown, with shallow longitudinal fissures; inner bark pale brown or reddish-brown; sapwood yellowish-brown; heartwood brown or reddish-brown; latex white. Twigs slender, greyish-brown or reddish-brown, longitudinally wrinkled, densely covered with reddish-brown stellate hairs or scales near the apex. Leaves imparipinnate, up to 42 cm long and 22 cm wide, elliptical in outline; petiole up to 8 cm, the petiole, rhachis and petiolules densely covered with stellate hairs or scales like those on the twigs. Leaflets (11 −)13 − 17(− 27), the laterals subopposite, all 5 − 10.5 cm long, 1.3 − 2.2 cm wide, narrowly elliptical, narrowly oblong or lanceolate, acuminate-caudate at apex with the obtuse acumen narrow, parallel-sided and up to 15 mm long, usually cuneate at the asymmetrical base but sometimes rounded on one side, with scales or hairs like those on the twigs few on or densely covering the midrib on the upper and lower surfaces and absent to numerous on the rest of the leaflet; the upper surface usually shiny and with numerous pits, the lower surface rugulose and often with numerous pits; veins 9 − 15(− 24) on each side of the midrib, either curved upwards or ascending and curved upwards at the margin, sometimes with shorter lateral veins in between, midrib prominent, lateral veins subprominent or depressed, secondary

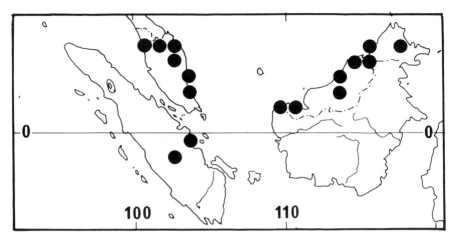

Fig. 57. Distribution of A. foveolata.

veins sometimes subprominent or visible on the lower surface; petiolules up to 8 mm.

Inflorescence up to 22 cm long and 22 cm wide, ovate in outline; peduncle up to 5 cm long; the peduncle, rhachis and branches clothed like the twigs. Flowers (not known whether male or female) up to 1.5 mm in diameter, subglobose, fragrant; pedicels up to 1.5 mm, with indumentum like the twigs. Calyx a third to half the length of the corolla, deeply divided into 5, ovate, obtuse lobes, with few to densely covered with stellate hairs or scales on the outer surface. Petals 5, obovate, yellow, aestivation quincuncial. Staminal tube two thirds as long to nearly as long as the corolla, subglobose, the aperture 0.3 – 0.5 mm in diameter which has a wavy margin; anthers 5, ellipsoid, about two thirds as long as the staminal tube, inserted near its base or about half way up and usually just protruding. Ovary subglobose or depressed-globose, densely covered with stellate scales or hairs; stigma narrowly ovoid or ellipsoid with longitudinal ridges and two apical lobes; the ovary and stigma together about two thirds the length of the staminal tube.

Infructescence c. 35 cm long and 20 cm wide; peduncle up to 8 cm, the peduncle, rhachis and branches with indumentum like the twigs. Fruits up to 2.5 cm long, subglobose or broadly ellipsoid, purple, brown, orange or yellow, indehiscent, densely covered with stellate hairs or scales; locules 1 (or 2), each containing one seed; aril translucent, sweet.

DISTRIBUTION. Peninsular Malaysia, Sumatra, Borneo. Fig. 57.

ECOLOGY. Found in primary forest, secondary forest, swamps, riverine forest and in ridge forest; on sand, silty clay, clay. Alt.: 6 – 1000 m.

VERNACULAR NAMES. Peninsular Malaysia: Bekak, Memberas. Borneo: Segera (Iban).

Representative specimens. PENINSULAR MALAYSIA. Pahang, Balok F.R., fl., 3 Sep. 1966, *Kochummen* 2092 (SING!). Johore, Kluang F.R., 45 m, fl., 31 Jan. 1966, Jan. *Ng* Kep 97964 (FRI!, K!).

SARAWAK. Kuching District, 12th Mile Penrissen Road [1°50'N 110°35'N], fr., 21 Sep. 1966, *Banyengak Nudong & Benang ak Bubong* S 25483 (FHO!). Kuching, Semengoh Arboretum [1°45'N 110°35'E], c. 10 m, fl. 23 May 1963, *Banyingan* S 13955 (FHO!). 3rd Division, Bukit Raya, 2½ hrs upstream from Kapit, 210 m, fl., 25 Nov. 1963, *Pennington* 7997 (FHO!). 4th Division, Bintulu, Similajau / Labang [3°40'N 113°35'E], 90 m, fr., 5 Sep. 1968, *Wright* S 27979 (FHO!).

Aglaia foveolata sometimes resembles *A. forbesii*, but it has narrower leaflets and the indumentum is always stellate. It also resembles *A. perviridis* Hiern, but the leaves of *A. perviridis* have fewer leaflets which are larger and differ in their venation and indumentum. The leaves of *Aglaia foveolata* resemble those of *Aglaia multinervis*, but in *Aglaia* sp. L, the veins are fewer in number and the leaflets are usually pitted on the upper and lower surfaces, the upper surface is shiny when dry, the flowers have five petals and 5 anthers; *Aglaia multinervis* has dull leaflet surfaces when dry and its flowers have 3 petals and 6 anthers. The leaves of *Aglaia foveolata* differ from those of *Aglaia multinervis* as described above, but in some specimens, not included in the description above of *A. foveolata*, the number of lateral veins is similar to that in *Aglaia multinervis* and vegetative distinction is based on the indumentum, which is peltate in *A. multinervis* and stellate in *A. foveolata*.

53. Aglaia crassinervia Kurz ex Hiern in Hooker fil., Fl. Brit. Ind. 1: 556 (1875). Lectotype (designated here): [Burma], Tenasserim, fl., 1838, *Helfer* 1609 (Kew Dist. 1038) (K!, isolectotypes: L!, W!); Kurz, Preliminary report on the forest and other vegetation of Pegu: 32, no. 233 (1875). C. de Candolle, Monog. 1: 615 (1878); Pannell in Ng, Tree Flora of Malaya 4: 229 (1989) as *Aglaia* sp. 6.

Aglaia cinerea King in Jour. As. Soc. Bengal 64: 66 (1895). Syntypes: 'Burma and Malaya', *Griffith* 1047 (K!). Malay Peninsula, Perak, 3000 – 3500 ft [900 – 1000 m], Dec. 1883, *King's Coll.* 5285 (K!); *King's Coll.* 2730 (BM!, G!); [Peninsular Malaysia], Perak, *Scortechini* 347 (CGE!, SING!); Ridley, Fl. Malay Penins. 1: 404 (1922).

Chisocheton sumatranus Baker fil. in Jour. Bot., Lond. 62 Suppl.: 18 (1924). Types: Indonesia, Sumatra, Paoe, Palembang, Mt Dempe, 500 ft [c. 150 m], fr., 1881, *Forbes* 2278 (BM, K!, L!).

Aglaia pyricarpa Baker fil. in Jour. Bot., Lond. 62 Suppl.: 20 (1924). Lectotype (designated here): Sumatra [Lampongs], fr., 1880, *Forbes* 1977a (BM!).

Tree up to 22 m, flowering at 6 m. Bole up to 15 m, up to 50 cm in diameter; buttresses upwards up to 2 m, outwards up to 30 cm, up to 20 cm thick. Bark smooth, greyish-brown or greyish-green, lenticellate, sometimes with longitudinal cracks or flaking in small scales; inner bark pale yellow or dark reddish-brown or pink with white latex from cambium region; sapwood yellow, orange or pale brown, sometimes turning purple near the heart; latex white. Twigs fairly stout, greyish-green or brown with numerous yellowish-brown or orange brown, peltate scales which are entire or have a fimbriate margin, with white latex.

Leaves imparipinnate up to 1 m long and 70 cm wide, obovate in outline; petiole up to 20 cm long, the petiole, rhachis and petiolules with indumentum like the twigs. Leaflets 11 – 15(– 17), the laterals alternate or subopposite, all 7 – 35 cm long, 4 – 12 cm wide, dull dark green above, paler below, often greyish-green when dry, elliptical or occasionally ovate, acuminate-caudate at the apex with the acute acumen up to 1.5 cm long, rounded or cuneate at the slightly asymmetrical base, upper surface with numerous almost white scales when young, glabrescent, usually rugulose and with numerous pits and lower surface usually pitted, with numerous scales like those on the twigs on the midrib, veins and surface in between on the lower surface; veins 7 – 19 on each side of the midrib, ascending and markedly curved upwards near the margin, midrib prominent and lateral veins subprominent on the lower surface; petiolules up to 1.5 cm.

Male inflorescence up to 50 cm long and 60 cm wide. Female inflorescence up to 30 cm long, peduncle, rhachis and branches with indumentum like the twigs. Flowers subglobose, c. 1.5 mm in diameter, fragrant; pedicels up to 1.5 mm. Calyx $1/3 - 1/2$ the length of the corolla, usually densely covered with scales like those on the twigs, deeply divided into 5 rounded lobes. Petals 5, bright yellow in the male, dark yellow or dull pale yellow in the female, obovate or elliptical, glabrous and without scales, aestivation quincuncial. Staminal tube subglobose $2/3 - 3/4$ the length of the corolla, with a wide aperture 0.4 – 0.5 mm across, entire; anthers 5, ovoid, $1/2 - 2/3$ length of the staminal tube, inserted $1/3 - 1/2$ the way up the tube and protruding through the aperture. Ovary small, depressed-globose, densely covered with scales like those on the twigs; stigma c. $1/2$ the length of the staminal tube, ovoid, dark brown, shiny, glabrous.

Infructescence up to 30 cm long and 20 cm wide, with up to 50 fruits. Fruits 5.5 – 6 cm long and 3.5 – 4 cm in diameter, subglobose or pyriform, dull green, grey, brown or blackish-purple, becoming yellow or orange when ripe; pericarp woody, 1 mm thick, densely covered with scales like those on the twigs on the outside; stalk stout, up to 2 cm long and 1 cm wide. Loculi 1 or 2, each containing 0 or 1 arillate seed; seed with aril 2.5 – 3 cm long, 1.5 – 1.7 cm wide and 1 – 1.8 cm thick; aril transparent, reddish-brown slightly sour. Fig. 58.

DISTRIBUTION. Nicobar Islands, Burma, Thailand, Peninsular Malaysia, Sumatra, Borneo, Philippine Islands. Fig. 59.

ECOLOGY. Found in evergreen, primary and secondary forest and in seasonal swamp on acidic rock, over basalt, sandstone, sand, sandy loam, clay. Alt.: sea level to 1570 m. The aril is eaten by monkeys.

VERNACULAR NAMES. Peninsular Malaysia: Bekak. Sumatra: Lasoen (Fatoeh); Balam; Balek Balek Bolon, Selabai Angin Putih, Seng-seng Koeda. Borneo: Balim (Kadayan); Langsat-langsat (Malay); Lantupak (Dusun-Kinabatangan); Segera (Iban); Sigirah (Dyak); Boko Boko Utan.

Representative specimens. BURMA. see type of *A. crassinervia*.

THAILAND. Yala, fr., 5 Feb. 1973, *Sangkachard et al.* 1602 (C!). Nakhon Si Thammarat Province, Peninsular District, Khao Luang, c. 750 – 800 m, fr., 21 May 1968, *van Beusekom & Phengkhlai* 913 (AAU!, C!)

PENINSULAR MALAYSIA. Perak, Gunong Bubu [4°40'N 100°50'E], 600 m, ?♀ fl., 16 Aug. 1966, *Chew Wee-lek* 1214 (FRl!). Kedah, Gunong Jerai F.R. Compt. 14, c. 90 m: ?♀ fl., 8 Nov. 1963, *Pennington* 7854

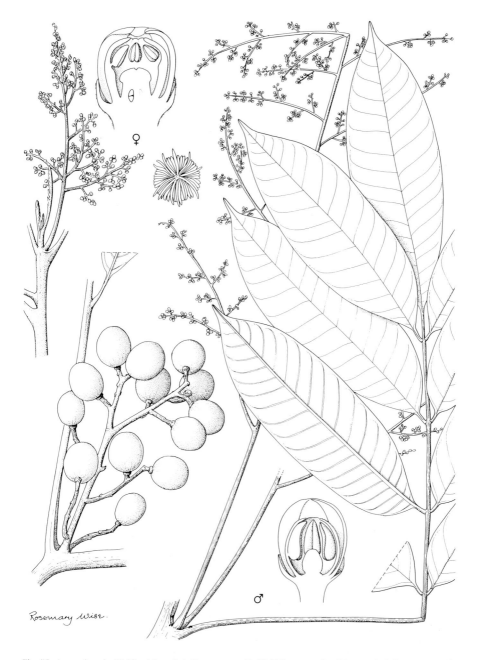

Fig. 58. *A. crassinervia*. Habit with male inflorescence x½. Half flower, male x20. Female inflorescence x½. Half flower, female x20. Infructescence x½. Peltate scale x70.

Section Aglaia

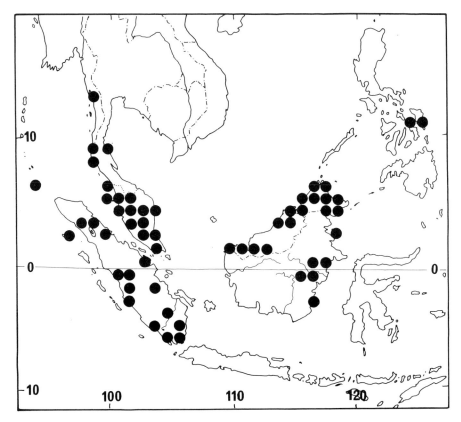

Fig. 59. Distribution of *A. crassinervia*.

(FHO!) & fl., 28 Sep. 1963, *Pennington* 7855 (FHO!). Pahang: Raub, Sg. Sempam, 600 – 900 m, fr., 16 April 1970, *Soepadmo* 671 (FRI!); near Temerloh, Kemasul F.R., Compt. 43, 60 m, ?♂ fl., 8 Oct. 1963, *Pennington* 7864 (FHO!). Selangor , Ulu Gombak Forest Reserve, 210 m, fl., 13 Sep. 1960, *Ahmad* Kep 94457 (FRI!, K!).

SUMATRA. Simaloer [Simalur] Island: fl., 3 Feb. 1919, *Achmad* 900 (U!); fl., 14 Feb. 1918, *Achmad* 242 (U!). Atjeh [Aceh], Gunung Leuser Nature Reserves, Southern part of the reserves, Alas River Valley, near the mouth of the Bengkong River, c. 50 km S. of Kutacane [c. 3°N 97°50'E], 50 – 125 m, fr., 16 July 1979, *de Wilde & de Wilde-Duyfjes* 18765 (L!).

SARAWAK. 3rd Division, Bukit Raya 2½ hours upstream from Kapit, c. 20 m, fl., 27 Nov. 1963, *Pennington* 7999 (FHO!)

SABAH. Sandakan, Sepilok Forest Reserve, Jalan Kabili, ?♀ fl., 8 May 1974, *Mabberley* 1667 (FHO!). Labuk Road, mile 80 west of Sandakan, ?♀ fl., 19 May 1974, *Mabberley* 1703 (SAN 75485) (FHO!). Elphinstone Province, Tawao, fr., Oct. 1922 – Mar. 1923, *Elmer* 20602 (U!).

The leaves of *A. crassinervia* resemble *A. rufinervis*, but there are fewer leaflets and the indumentum is of peltate scales with a fimbriate margin rather than stellate hairs or scales. The length of the fimbriate margin on the peltate scales varies in length. When the scales have a long fimbriate margin, it is sometimes difficult to distinguish *A. crassinervia* from *A. aspera*. *A. aspera* has numerous stellate scales on the lower leaflet surface and some compact stellate hairs interspersed; the latter are absent in *A. crassinervia*.

This species sometimes resembles *A. edulis*, but differs in its indumentum and fruits. It resembles *A. korthalsii*, but the leaflets are usually more numerous, they are paler when dry and the scales are almost confined to the midrib. The fruits are subglobose, with a thin brittle pericarp.

54. Aglaia aspera Teijsm. & Binn. in Nat. Tijdschr. Ned. Ind. 27: 42 (1864). Type: Cult. Hort. Bogor, origin: Java, Mt Salak; vernacular: name Saliem (no specimen seen which pre-dates publication of the name); illustrative specimen: Cult. in Hort. Bog. III-B-36a (origin: Java, Mt Salak) fr., July 1889 [same illustrative specimen as for *Aglaia acuminatissima*] (BO!). Miq., Ann. Mus. Bot. Lugd. Bat. 4: 52 (1868) Koorders & Valeton, Atlas Baumart. Java, t. 152 (1913); C. de Candolle in A. & C. de Candolle, Monog. Phan. 1: 626 (1878); Backer and Bakhuizen, Fl. Java 2: 127 (1965); Pannell in Ng, Tree Flora of Malaya 4: 211 (1989), as *Aglaia* sp. 5.

Aglaia sexipetala Griffith, Notulae ad Plantas Asiaticas 4: 505 (1854). Holotype: 'Birma and Malay Peninsula', Ching, Nhinghuk, fl., 27 Jan. 1845, *Griffith* 1036 (K!).

Aglaia acuminatissima Teijsm. & Binn. in Nat. Tijdschr. Ned. Ind. 27: 42 (1864). Type: Cult. Hort. Bogor, origin: Java, Mt Salak; vernacular: name Saliem (no type specimen seen) [same illustrative specimen as for *Aglaia aspera*].

Aglaia polyphylla Miq., Ann. Mus. Bot. Lugd. Bat. 4: 56 (1868). Lectotype (designated here): Java, *Junghuhn* 35 (U!; isolectotype: L!).

Aglaia minutiflora Bedd. var. *macrophylla* C. de Candolle in A. & C. de Candolle, Monog. Phan. 1: 616 (1878) superfl. nom. illegit. pro *Aglaia sexipetala* Griffith.

Aglaia myristicifolia C. de Candolle in Bull. Herb. Boiss. Sér. II 3: 17 6 (1903). Holotype: S.E. New Guinea, fr., *Forbes* 67 (G!).

Aglaia aspera var. *sumatrana* Baker fil. in Jour. Bot., Lond. 62, Suppl.: 20 (1924). Types: Sumatra, S., Lampongs, Perangoengan, *Forbes* 1668 (BM?, L!).

Aglaia calelanensis Elmer in Leafl. Philipp. Bot. 9: 3283 (1937), sine diagn. lat.. Type no.: *Elmer* 11804, Philippine Islands, Island of Mindanao, District of Davao, Todaya, Mt Apo, fr., July 1909, (A!, BM!, BO!, G!, K!, L!, NY!, P!, SING!, U!, UC!, US!, W!).

Tree up to 29 m with ascending branches and a rounded crown. Bole up to 12 m, up to 40 cm in diameter, sometimes with L-shaped buttresses upwards up to 1.8 m, outwards up to 2 m and up to 20 cm thick. Bark smooth, greyish-brown, yellowish-brown or reddish-brown with grey, green and pale brown patches, inner bark pink or brown;

sapwood yellowish-brown, orange, pinkish-yellow or reddish-brown; latex white. Twigs slender, almost smooth, dark brown or grey, densely covered with reddish-brown stellate scales.

Leaves imparipinnate, up to 72(–100) cm long and 50(–75) cm wide, obovate in outline; petiole up to 17 cm, the petiole, rhachis and petiolules with indumentum like the twigs. Leaflets 7–13(–17), the laterals subopposite, all 7–22(–45) cm long, 1.5–8.5(–15) cm wide, dark green above and silvery-green or brownish-green below when dry, elliptical, lanceolate-oblong, oblong, ovate or obovate, usually asymmetrical, acuminate at apex with the obtuse or acute acumen up to 10(–15) mm long, rounded, subcordate or cuneate at the asymmetrical base, with a few stellate scales and numerous pits on the upper leaflet surface and with numerous to densely covered with scales like those on the twigs or peltate scales with a long fimbriate margin on the midrib below, numerous on the rest of the lower surface, with brown stellate hairs which have many short arms interspersed; veins 7–14(–29) on each side of the midrib, either curved upwards or ascending slightly and markedly curved upwards near the margin, midrib prominent, and lateral veins subprominent on lower surface; petiolules up to 12 mm on lateral leaflets, up to 18(–35) mm on terminal leaflet.

Male inflorescence up to 40 cm long and 40(–50) cm wide with indumentum like the twigs; peduncle c. 5 mm. Flowers up to 1.5 mm in diameter, depressed-globose; pedicels 1–2 mm long. Calyx c. $^1/_3$–$^1/_2$ the length of the corolla, with numerous stellate scales on the outside, divided almost to the base into 5 subrotund lobes which are obtuse at the apex. Petals 5, elliptical or subrotund, yellow, aestivation quincuncial. Staminal tube 0.5–1 mm, pale yellow or yellow, either shallowly cup-shaped with the apical margin incurved and shallowly 5-lobed or ovoid and with the aperture less than 1 mm in diameter; anthers 5, c. $^1/_2$ to longer than the length of the tube, inserted near the base or just below the margin of the tube and protruding through the aperture, pointing towards the centre of the flower. Ovary subglobose densely covered with stellate hairs or scales; stigma ovoid with two small apical lobes; ovary and stigma together $^1/_2$ to nearly as long as the staminal tube. Female inflorescence little branched, up to 30 cm long and 30 cm wide. Flowers similar to the male but slightly larger.

Infructescence c. 20 cm long. Fruits subglobose or obovoid, up to 5 cm long and 4 cm in diameter, yellow, reddish-brown or orange-brown, the pericarp up to 2 mm thick, hard and brittle or woody, densely covered with reddish-brown stellate hairs or scales on the outside, sometimes containing white latex. Loculi 2, each containing 1 seed; seed with a complete transparent or white aril. Fig. 60.

DISTRIBUTION. Thailand, Peninsular Malaysia, Sumatra, Java, Borneo, Philippine Islands (type of *A. calelanensis* only), New Guinea. Fig. 61.

ECOLOGY. Found in primary, secondary, kerangas, monsoon forest and along banks on yellow sand, yellow to dark brown loam and silty clay. Alt.: sea level to 1600 m. Scarce to rather common. Aril 'much eaten by monkeys'.

USES. Aril edible. Timber for house construction (Papua New Guinea: Morobe).

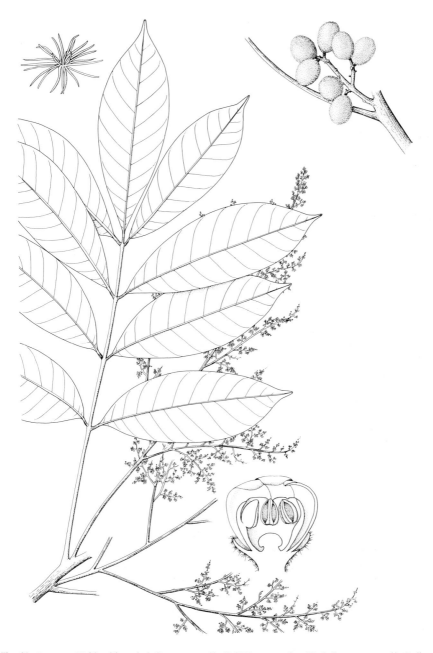

Fig. 60. *A. aspera*. Habit with male inflorescence x½. Half flower, male x30. Infructescence x½. Stellate scale x100.

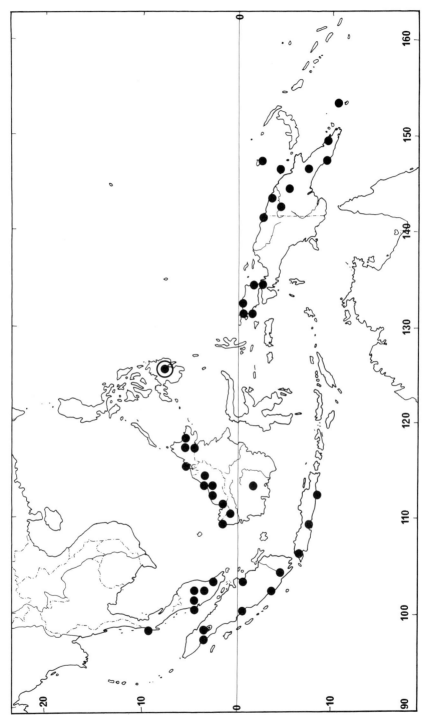

Fig. 61. Distribution of *A. aspera*.

VERNACULAR NAMES. Peninsular Malaysia: Bekak. Sumatra: Sepanas (Sunda). Java: Doekoe Leuweung, Lido. Borneo: Lantupak (Malay), Segera (Iban). New Guinea: Bowor (Kebar), Kwai (Samu dialect, Abelam), Segusemakebab (Kem- toek), Seraka (Manikiong), Bengan.

Representative specimens. THAILAND. Langsuan, Kao Nam Sao [9°57′N 98°50′E], c. 400 m, fl., 22 Feb. 1927, *Kerr* 12088 (C!).

PENINSULAR MALAYSIA. Pahang, Taman Negara, c. 5 km from Kuala Tahan H.Q., hill above plot III nr Tabing salt lick, fr., 24 April 1975, *van Balgooy* 2533 (AAU!, L!). Negri Sembilan, Ulu Pedas on Gg Angsi, c. 600 m, fl., 26 Nov. 1923, *Nur* SFN 11715 (K!).

SUMATRA. Lampong, helling G. Raté, Berenong, fl., 19 Nov. 1921, *Tcoet?* 161 (BO!).

JAVA. Batavia, Pabangbon, W.V. Buitenzorg, fr., 5 Nov. 1927, *Bakhuizen van den Brink* 6975 (L!).

IRIAN JAYA. Manokwari Reserve, Maepi II [1°10′S 134°08′E], c. 10 m, fr., 23 Oct. 1954, *Koster* BW 1096 (K!, L!).

PAPUA NEW GUINEA. Lower Inakanda Logging Area, Bulolo, Morobe District, c. 750 m, [7°12′S 146°39′E], fr., *Havel & Kairo* NGF 17093 (K!). Northern District, Bobodura Plain, Embi Lakes area, c. 150 m, fl., 16 March 1945, *Cavanagh & Fryer* NGF 2089 (K!, LAE!, L!).

The leaves of *A. aspera* have numerous stellate scales on the lower surfaces of the leaflets, sometimes interspersed with stellate hairs which have numerous short arms. The scales are sometimes peltate with a long fimbriate margin and it may then become difficult to distinguish this species from *A. crassinervia* which has numerous peltate scales with a short fimbriate margin, but in these cases the scales of *A. aspera* are interspersed with compact stellate hairs. In the Western part of the range, the flower of *A. aspera* has a shallow cup-shaped staminal tube, while in New Guinea the staminal tube is ovoid and has a narrower aperture. The absence of records of this species from Nusa Tenggara, Sulawesi and Maluku suggests that it has a disjunct distribution. Further material, especially of ripe fruits in spirit, may provide characters by which the New Guinea and Vanuatu plants may be reliably distinguished from the remaining, western parts of the range, and therefore be recognised as a separate species (*Aglaia myristicifolia*).

55. Aglaia parviflora C. de Candolle in Bull. Herb. Boiss. Sér. II 3: 176 (1903). Holotype: S.E. New Guinea, fl., *H.O. Forbes* 55 (G!).

Aglaia forbesiana C. de Candolle in Bull. Herb. Boiss. Sér. II. 3: 174 (1903). Holotype: S.E. New Guinea, [Strickland River], fl., *Forbes* 63 (G!).
Aglaia procera C. de Candolle in Denkschr. Akad. Wiss. Wien Math.- Nat., 89: 565 (1913). Lectotype (designated here): Neu-Pommern [New Britain], Gazelle Halbinsel, Bairing Gebirges, fr., Sep. 1905, *Rechinger* 3676 (W!) .
Aglaia ulawaensis Merrill & Perry in Jour. Arn. Arb. 21: 327 (1940). Lectotype (designated here): Solomon Islands, Ulawa Island, coast to 100 m, fl., 8 Oct 1932, *Brass* 2986 (A!; isolectotype: BO!).
Aglaia acariaeantha Harms in Engl. Bot. Jahrb. 72: 170 (1942). Lectotype (designated here): Papua [Papua New Guinea, N.E.], Kaiser-Wilhelmsland, Wakeak, 500 m, fl., 30 Aug. 1908, *Schlechter* 18117 (NY!; isolectotypes: B†, K!).

Tree up to 20 m high. Bole up to 16 m; up to 30 cm in diameter. Outer bark yellowish-brown or pale greyish-brown, with scattered lenticels; inner bark pinkish-brown, sapwood pale yellowish-brown, becoming pinker inwards; white latex. Twigs greyish-brown, with numerous orange brown and yellowish-brown peltate scales which have a fimbriate margin.

Leaves 28 – 95 cm long, 22 – 62 cm wide, obovate in outline; petiole 5 – 19(– 28) cm long, the petiole, rhachis and petiolules with numerous to densely covered with scales like those on the twigs. Leaflets 5 – 7(– 9), the laterals subopposite, all 8 – 37(– 45) cm long, 2.5 – 13(– 21) cm wide, obovate, or occasionally elliptical or ovate; apex acuminate, the obtuse acumen often parallel sided and 5 – 25 mm long; usually cuneate or sometimes rounded at the slightly or sometimes markedly asymmetrical base; with numerous to densely covered with scales like those on the twigs and some pale yellow stellate scales on the midrib below, few to numerous mainly stellate scales on the lateral veins and the rest of the lower surface sometimes with numerous pits; veins 7 – 17(– 26) on each side of the midrib, ascending and curving upwards near the margin, nearly anastomosing; petiolules 5 – 15 mm long, up to 45 mm on the terminal leaflet.

Inflorescence 10 – 31(– 39) cm long, 6 – 20(– 24) cm wide, peduncle 0.5 – 6 cm, the peduncle, rhachis and branches with numerous scales like those on the twigs and some stellate scales increasing in frequency distally. Male flowers 1 – 1.5 mm long, 1 – 1.5 mm wide, subglobose, fragrant; pedicel 1.2 – 2 mm sometimes with an articulation and sometimes with a few stellate scales. Calyx 0.5 mm, deeply divided into 5 ovate obtuse or acute lobes with a fimbriate margin and a few stellate scales on the outside. Petals 5, white, yellow or orange; aestivation quincuncial. Staminal tube cup-shaped, 0.5 mm – 1 mm long with a wide aperture 0.4 – 0.6 mm across, the margin sometimes ciliate, margin shallowly lobed, anthers 5, broad, c. $^1/_4$ to the same length as the tube, usually inserted just below the margin protruding and pointing towards the centre of the flower, sometimes inserted lower down in the staminal tube and included, usually with simple white hairs on the inside of the tube and on the anthers, sometimes densely clumped and visible in the aperture of the staminal tube. Ovary depressed globose or subglobose, densely covered with stellate scales; stigma depressed globose or obovoid, the apex flattened and the margin raised and lobed, loculi 2, each containing 1 ovule. Female flowers c. 2.5 mm long and 2 mm wide; calyx c. 1.5 mm long; aperture of staminal tube c. 0.3 mm in diameter; anthers c. 0.4 mm long and 0.4 mm wide; ovary 0.2 – 0.5 mm high, c. 1 mm wide, depressed-globose; loculi 2, each containing 1 ovule; stigma c. 0.1 mm high, 0.4 – 0.5 mm in diameter, flattened and with shallow marginal lobes; otherwise like the male flowers.

Fruits 2 – 2.5 cm long, 1.5 – 2 cm wide, obovoid or ellipsoid, dull brown or orange-brown, indehiscent. Pericarp c. 1 mm thick, hard, densely covered with scales like those on the twigs on the outside. Loculi 2, each containing 0 or 1 seed; seed with aril c. 1.4 cm long, 1 cm wide and 0.7 cm through; surrounded by a translucent, white, gelatinous aril. Fig. 62.

DISTRIBUTION. Maluku, New Guinea, Vanuatu (New Britain), Solomon Islands. Fig. 63.

Fig. 62. *A. parviflora*. Leaf x½. Male inflorescence x½. Half flower, male x15. Half flower, female x15. Infructescence x½. Stellate scale x40.

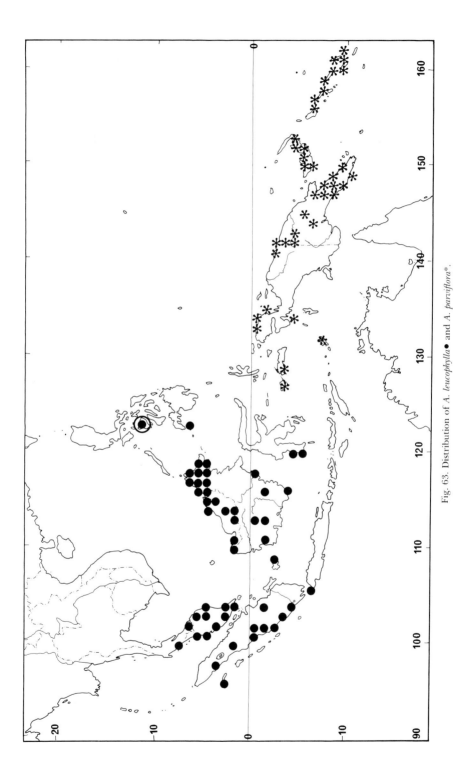

Fig. 63. Distribution of A. *leucophylla* ● and A. *parviflora* *.

ECOLOGY. Found in primary forest, ridge forest, secondary forest, riverine forest, on occasionally flooded plains, along the coast and along rivers; on sand, sandy clay, alluvial, volcanic loam, limestone. Alt.: sea level to 1700 m. Rather scarce to common.

USES. Wood is used in house construction (Papua New Guinea: Waskuk).

VERNACULAR NAMES. Maluku: Lai, Lasaba. New Guinea: Conabo (Waria); Dowa, Fieria, Tatrao (Kebar); Herrib (Manikiong); Kuton (Sko); Mansaambree (Biak); Mebutopo (Wagu); Rila (Garumaia); Sandisi (Orokaiva, Mumuni dial.); Urumiaugume (Waskuk); Peeparu, Piurai, Rapogu. Solomon Islands: Ngorisigwane, Raourbu, Ulukwalo, Valukwalo, Valukwalobula (Kwara'ae).

Representative specimens. MALUKU. Morotai: Totodoku, 30 m, fr., March 1949, *Kostermans* 7901 (L!); fl., 1949, *Kostermans* 857 bis (L!). Soela Eilanden, Eil. Mangoli, Djiko Sangatoemba, fl., *Atje* 371 (L!).

IRIAN JAYA. Vogelkop Subdistrict, Manokwari, Warsui near Ransiki, 10m., fl., 22 July 1948, *Kostermans* 2649 (K!).

PAPUA NEW GUINEA. Morobe District, Lae Subdistrict: Gizan [6°30'S 146°50'E], 900 m, fr., 19 Feb. 1977, *Conn, Masapuhafo & Kairo* 34 (LAE!); 3 m from Boana airstrip, track to Kalem village [6°25'S 146°50'E], 1200m, ♀ fl. & fr., 14 Sep. 1970, *Katik* NGF 46742 (K!). Sepik District, Vanimo Sub-district, East of Warimo village [2°40'S 141°20'E], c. 30 m, ♀ fl. and fr., 9 Aug. 1967, *Eddowes & Kumul* NGF 36009 (K!).

SOLOMON ISLANDS. N. Malaita, Takwa, Kwaerasi, fl., 6 Sep. 1968, *Runikera & collectors* BSIP 10738 (K!). Santa Ysabel: N.E., Gehe River, fr., 20 Jan. 1967, *Beer's collectors* BSIP 7730 (K!); Tatamba, 10 m, ♀ fl., 5 Oct. 1965, *Hunt* BSIP 2884 (K!).

A. parviflora sometimes resembles *A. sapindina*, but it lacks the dark purplish-brown peltate scales on the midrib on the lower surface of the leaflets. The flower of *A. parviflora* has a subglobose staminal tube, the anthers are hairy, half to three quarters the length of the tube and just protruding; the stigma is obconical, horizontally compressed, with a broad, flattened apex and lobed edge. The indumentum of *A. parviflora* is similar to that of *A. saltatorum*, but it differs in flower structure. These two species are apparently geographically separated; *A. parviflora* in not recorded from the Santa Cruz Islands, while this is the only part of the Solomon Islands where *A. saltatorum* occurs.

56. Aglaia heterotricha A.C.Smith in Contrib. U.S. Nat. Herb. 30: 486 (1952). Holotype: Tonga, Eua Island, fl., *H.E. Parks* 16305 (U.S. Nat. Herb. 1527045) (US!; isotypes: BM!, K!, NY!, UC!).

Twigs pale greenish-brown, densely covered with tiny pale orange-brown peltate scales which have a fimbriate margin. Leaves 28 – 60 cm long, 20 – 46 cm wide; petiole 8.5 – 16 cm; the petiole, rhachis and petiolules with few scales like those on the twigs. Leaflets 7, the laterals subopposite, all 10 – 27 cm long, 4.5 – 9.5 cm wide, ovate, obovate or elliptical, with peltate scales like those on the twigs numerous on the midrib; scattered and with a longer fimbriate margin on the rest of the lower leaflet surface; veins 10 – 17 on each side of the midrib, ascending at an angle of 55 – 60° to the midrib, curving upwards near the margin, midrib prominent, lateral veins prominent and reticulation subprominent on lower surface; petiolules 5 – 20 mm, enlarged at the base.

Inflorescence 5 – 10 cm long, c. 4 cm wide; sessile, the rhachis and branches densely covered with scales like those on the twigs. Flowers 1.5 – 2 mm long, 1.5 – 2 mm wide, subglobose; pedicel 1 mm, the pedicel and calyx densely covered with stellate scales. Calyx 0.5 mm long, cup-shaped and deeply divided into 5 ovate acute lobes. Petals 5, densely covered with scales on the outside except near the apex, aestivation quincuncial. Staminal tube 1 – 1.2 mm long and c. 1 mm wide, truncate-ovoid, the lower $^2/_3$ densely covered with scales on the outside, aperture 0.6 – 0.8 mm wide, with a wavy margin; anthers 5, c. 0.5 mm long, 0.5 mm wide, broad, white or pale yellow, just protruding and filling the aperture. Ovary depressed globose, surrounded by a ring of stellate scales; stigma obovoid or subglobose, black, bilobed, the stigma and ovary together 0.5 mm. Fruits not seen.

DISTRIBUTION. *A. heterotricha* is known only from the type collection from Eua Is., Tonga.

A. heterotricha resembles *A. saltatorum*, but the leaves are larger and the indumentum is of peltate scales only.

57. Aglaia leucophylla King in Jour. As. Soc. Bengal 64: 66 (1895). Syntypes: Malay Peninsula, *King's Coll.* 1874 (non vidi); Perak, Larut, <300 ft [c. 90 m], fl., April 1882, *King's Coll.* 2998 (K!); Perak, Larut, yg fr., Aug. 1884, *King's Coll.* 6494 (K!); Perak, Assam Kumbong, yg fr., 1888, *Wray Jr.* 2935 (SING!, W!); Ridley, Fl. Malay Penins. 1: 403 (1922); Pannell in Ng, Tree Flora of Malaya 4: 218 (1989).

Aglaia kunstleri King in Jour. As. Soc. Bengal 64: 69 (1895) Syntypes: Malay Peninsula, Perak, Larut, ≥300 ft [c. 90 m], fr., Dec. 1883, *King's Coll.* 5287 (BM!, G!, K!, L!, P!, SING!); Perak, Ulu Bubong, 400 – 600 ft [c. 120 – 180 m], fl., July 1886, *King's coll.* 10610 (BM!, G!, K!, L!, P!); King, Kew Bull. 1949: 166 (1949).

Aglaia heteroclita King in Jour. As. Soc. Bengal 64: 78 (1895). Syntypes: Malay Peninsula, Perak, Ulu Kal, 500 – 700 ft [150 – 210 m], *King's Coll.* 10896 (BM!, K!); Perak, Lower Camp, Gunong Batu Puteh, 3400 ft [c. 1500 m], fr., *Wray Jr* 1135 (BM!, G!, K!, L!); Perak, Blanda Mabok, *Wray Jr* 3994 (SING!); Sumatra, Lampong, Goemoeng Trang, *Forbes* 1558 (SING!), *Forbes* 1696 (non vidi); Ridley, Fl. Malay Penins. 1: 410 (1922).

? *Aglaia pallida* Merrill in Philipp. Journ. Sci., Bot. 3: 147 (1908). Lectotype (designated here): Philippine Islands, *M.S. Clemens* 1228 (K!; isolectotypes: G!, PNH†).

Aglaia mirandae Merrill in Philipp. Jour. Sci., Bot. 13: 295 (1918). Lectotype (designated here): Philippine Islands, Basilan, fl., Oct 1912, *Miranda* For. Bur. 18970 (US!; isolectotypes: G!, PNH†).

Aglaia agusanensis Elmer ex Merrill, Enum. Philipp. Fl. Pl. 2: 374 (1923), in obs., pro syn.; Elmer in Leafl. Philipp. Bot. 9: 3275 (1937), sine diagn. lat. Type no. *Elmer* 14028, Philippine Islands, Island of Mindanao , Province of Agusan, Cabadbaran, Mt Urdaneta, fr., Oct 1912 (A!, BO!, BM!, G!, GH!, K!, L!, P!, US!, W!).

Aglaia elmeri Merrill in Univ. Calif. Publ. Bot. 15: 127 (1929). Lectotype (designated here): British North Borneo, Elphinstone Province, Tawao, fl., Oct. 1922 – March 1923, *Elmer* 21756 (UC!; isolectotypes: BM!, BO!, G!, GH!, L!, M, MICH!, NY!, P!, U!).

Aglaia simplex Merrill in Univ. Calif. Publ. Bot. 15: 128 (1929). Lectotype (designated here): British North Borneo, Elphinstone Province, Tawao, fl., Oct. 1922 – March 1923, *Elmer* 21489 (UC!; isolectotypes: BO !, G!, K!, L!, SING!).

Aglaia insignis Schwartz in Mitt. Inst. Bot. Hamburg 7: 235 (1931). Type : West Borneo, Sungei Malang, 100 m, 30 Jan. 1925, *Winkler* 1385 (BO!).

Tree up to 20 m, sometimes flowering when 1.5 m high. Bole up to 65 cm in circumference; sometimes fluted at base. Outer bark smooth, grey, brown or greyish-brown; inner bark pale yellow; sapwood reddish-brown or white; latex white. Twigs slender to fairly stout, greenish-brown or yellowish-brown with shallow longitudinal wavy ridges and densely covered with golden-brown or brown stellate scales usually only near the apex. Leaves imparipinnate, up to 80 cm long and 50 cm wide, obovate in outline; petiole up to 22 cm, the petiole, rhachis and petiolules with surface and few to densely covered with scales like those on the twigs. Leaflets 9 – 15(– 17), the laterals alternate or subopposite, all 9 – 28 cm long, 3.5 – 11.5 cm wide, pale green or yellowish-green when dry, elliptical ovate or obovate, with numerous reddish-brown pits on the lower leaflet surface and few to numerous tiny golden-brown stellate scales on lower surface, sometimes with darker peltate scales or reddish-brown stellate hairs interspersed, often rugulose on upper and lower surfaces, acuminate or caudate at apex, the acute or obtuse acumen up to 2.5 cm long, rounded or cuneate at the asymmetrical base; veins 8 – 14 on each side of the midrib, ascending and markedly curved upwards near the margin, nearly or quite anastomosing, lateral veins visible on upper surface, the midrib prominent, lateral veins subprominent and secondary veins usually visible on lower surface, petiolules 1 – 20 mm long. Male inflorescence up to 60 cm long and 25 cm wide; peduncle up to 15 cm, the peduncle, rhachis, branches and petiolules longitudinally wrinkled and with few to numerous golden-brown stellate scales. Male flowers subglobose, up to 1.5 mm in diameter, fragrant; pedicels up to 1 mm. Calyx up to $^1\!/_2$ the length of the flower, with few to densely covered with golden-brown stellate hairs on the outside, deeply divided into 5 elliptical and acute or subrotund lobes. Petals 5, yellow or orange, subrotund or elliptical, aestivation quincuncial. Staminal tube up to 0.9 mm long, shorter than the corolla, usually subglobose sometimes obovoid, dull yellow when fresh, black when dry, with a pore c. 0.3 mm across which is shallowly 5-lobed; anthers 5(– 7), $^1\!/_2$ to as long as the tube, broadly ovoid, inserted near the base or in the upper half of the tube, curved and just protruding through the pore, with a few simple hairs which sometimes fill the pore. Ovary subglobose densely covered with golden- brown stellate hairs; stigma ovoid, with 2 small apical lobes, longitudinally ridged, brown and shiny; the ovary and stigma together $^1\!/_2$ to nearly as long as the staminal tube. Female inflorescence c. 8 cm long and 2.5 cm wide, with lanceolate bracts up to 9 mm long and 2 mm wide, with few branches and

Section Aglaia

c. 200 flowers; peduncle 3 – 20 mm. Flower up to 3 mm long and 3 mm wide; pedicel c. 0.5 mm. Calyx c. 0.8 mm long cup-shaped and divided into 5 lobes. Petals 5. Staminal tube up to 2 mm long and 2 mm wide, depressed globose or obovoid, with a few hairs inside; aperture c. 0.5 mm; anthers 7, c. 0.7 mm long and 0.5 mm wide with pale margins and tufts of hairs at the apices, inserted near the apex of the tube and protruding. Ovary c. 0.4 mm high and 1 mm wide, depressed globose; loculi 2 each containing 1 ovule; stigma c. 0.3 mm high and 0.8 mm wide, depressed globose.

Infructescence up to 20 cm long and 18 cm wide with 3 – 10 fruits; peduncle up to 5 cm, the peduncle branches and fruit-stalks with surface and indumentum like the twigs. Fruits up to 4.5 cm in diameter, usually pyriform, sometimes subglobose, sometimes with a beak and narrowed at the base to a stipe c. 5 mm long, usually with a thick, hard, woody pericarp, sometimes the pericarp thin and brittle, yellow or brown, densely covered with golden brown or pale brown stellate hairs or scales. Loculi 2, each containing 1 seed. Seed c. 2.3 cm long and 1 cm wide; aril white or red, edible, sweet or sour; testa brown.

DISTRIBUTION. S. Thailand, Peninsular Malaysia, Sumatra, Borneo, Philippine Islands, Sulawesi. Fig. 63.

ECOLOGY. Found in evergreen forest, primary forest, riverine forest, moss forest, but usually in secondary forest; on sand, limestone, sandy clay, loam, and alluvial. Alt.: sea level to 1300 m.

USES. Boles are used for house-poles (Borneo: Tumbang Tubus).

VERNACULAR NAMES. Peninsular Malaysia: Malay: Bekak Kedondong, Memberas, Pasak Lingga. Sumatra: Awa Toetoe Mah, Ganggo Bareh, Ganggo Falah, Lasoen Balah, Letoeng, Sibadaroeh, Soebeh. Borneo: Kayo Bunyo (Dahoi); Langsat Hutan, Langsat-langsat (Malay); Lantupak (Dusun Kinabatangan); Pangak (Kenyah); Bunyo Sahi, Mata Oelat, Perumpong Hutan.

Representative specimens. PENINSULAR MALAYSIA. Perak, Gg Bubu [4°40'N 100°50'E], 180 m, fl., 14 Aug. 1966, *Chew* 1191 (AAU!, FRI!). Trengganu, Ulu Telemong F.R., Bt. Batu Kota, ♀ fl. & fr., 16 Sep. 1969, *Suppiah* FRI 11417 (FRI!). Selangor, Kepong, Bt Lagong F.R., *Mabberley & Loh* 1559 (FHO!). Johore: Labis F.R., ♀ fl., 28 July 1964, KEP 94837 (FRI!).

SUMATRA. Simaloer [Simalur] Island, Landschap Tapah (Defajan), fl., 29 Sep. 1919, *Achmad* 1408 (L!).

SARAWAK. 6½ miles Bakam Road, Miri, c. 60 m, fr., 12 April 1966, *Sibat ak Luang* S 25190 (FHO!).

BRUNEI. Kuala Abang Road, mile 7, c. 60 m, fr., 5 Jan. 1959, *Ashton* BRU N 5095 (FHO!).

SABAH. S.W. Sandakan Bay, Sekong Kechil, fl., 24 May 1974, *Mabberley* 1719 (FHO!). mile 81 west of Sandakan, E. of wireless relay station, fr., 17 May 1974, *Mabberley* 1701 (FHO!). Sandakan District, Sandakan, mile 84 Telupid road, c. 150 m, fr., 2 Oct. 1969, *Talip & Laban* SAN 62443 (FHO!). mile 32 – 33 Ranau road, Tenompok F.R., fl. & fr., 29 Oct. 1963, *Pennington* 7944 (FHO!). Lohan F.R., nr Ranau (5 miles). fl., 28 Oct. 196 3, *Pennington* 7935 (FHO!). Ranau district, Lower Bukit Kulimpisau, c.

1300 m, yg fr. 6 Nov. 1916, *Mujin* SAN 18819 (K!). Tawau district, East Sg. Serudong, hillside, c. 90 m, fr. 6 Sep. 1961, *Baker* SAN 26158 (K!).

PHILIPPINE ISLANDS. S.E. Mindanao, Davao Province, Santa Cruz: c. 240 m, fr., 5 July 1905, *Williams* 3017 (NY!) & fr., 17 May 1905, *Williams* 2813 (NY!).

SULAWESI. Papekang, Bonthain, fl., *Teysmann* 13983 (BO!, L!).

The leaflets of *A. leucophylla* are pale green or yellowish-green when dry. The small stellate scales on the lower surface are often deciduous, leaving minute pits where they were attached. *A. leucophylla* resembles *A. edulis*, but is distinguished from it by the more numerous scales on the lower leaflet surface and the pear-shaped fruits. *A. edulis* has a subglobose, 3-locular fruit, although sometimes only one seed develops. The fruit of *A. leucophylla* is occasionally subglobose in for example, the Philippine Islands, with 2 loculi, each of which contains one seed.

58. Aglaia edulis (Roxb.) Wall., Calc. Gard. Rep.: 26 (1840). Hiern in Hooker fil., Fl. Brit. India 1: 556 (1875); C. de Candolle in A. & C. de Candolle, Monog. Phan. 1: 609 (1878) pro parte; I.H. Burkill, Dictionary of Economic Products of the Malay Peninsula 1: 76 (1935).

Milnea edulis Roxb., Hort. Beng.: 18 (1814), nom. nud.; Fl. Ind., ed. Carey & Wall., 2: 430 (1824); Fl. Ind., ed. Carey, 1: 637 (1832). Lectotype (designated here): India, Silhet [native of Garrow hills and of the Silhet district (vern. 'Genui')], Hort. Bot. Calc., *Wall. Cat.* 1279.C (K-W!).
? *Nyalelia racemosa* Dennst. in Schlüssel, Hort. Malab. 14, 23, 30 (1818), invalid name based on 'Nyalel' Rheede, Hort. Malab. 4: 37, as 'Nialel': t. 16 (1683); *Nialel* Adans., Fam. Pl. 2: 446, 582 (1763); Hiern in Hooker fil., Fl. Brit. India 1: 554 (1875); Harms in Engl. & Prantl Nat. Pflanzenf., Ed. 2., 19b1: 141 (1940); Mabberley, Taxon 26: 530 (1977), as *A. elaeagnoidea*; Nicolson & Suresh in Taxon 35: 388 (1986); Nicolson, Suresh & Manilal, An Interpretation of van Rheede's Hortus Malabaricus: 177 (1988), as *A. elaeagnoidea*.
Aglaia sulingi Blume, Bijdr. 170 (1825). Lectotype (designated here): Java, [Mount Suling, Bogor (Buitenzorg) Province], fl., *Anon.* 915 (L!; isolectotype: U!); Hasskarl, Cat. Hort. Bogor. 220 (1844). Miq., Ann. Mus. Bot. Lugd. Bat. 4: 44 (1868); C. de Candolle in A. & C. de Candolle, Monog. Phan. 1: 612 (1878); Backer and Bakhuizen, Fl. Java 2: 128 (1965).
[*Camunium bengalense* Buch.-Ham. ex Wallich, Cat. 1279 (1829) nom. nud.]
Nyalelia racemosa Dennst. ex Kostel, Allg. Med.-Pharm. Fl. 5: 2005 (1836). Type: Rheede, Hort. Malab. 4: t. 16 (1683).
Milnea racemosa (Kostel.) M. Roemer, FAM. Nat. Syn. Monogr. 1: 98 (1846).
Beddomea indica Hooker fil. in Benth. & Hooker fil., Gen. Pl. 1: 336 (1862) [n.b. type species, at end of description of new genus, *Beddomea* Hooker fil.]. Lectotype (designated here): India, [Mount] Nilgherris, *Gardner* s.n. (K!; isolectotype: BM!); C. de Candolle in A. & C. de Candolle, Monog. Phan. 1: 600 (1878).
Milnea sulingi (Blume) Teijsm. & Binn., Cat. Hort. Bog.: 211 (1866).
Aglaia undulata Miq., Ann. Mus. Bot. Lugd. Bat. 4: 44 (1868). Lectotype

(designated here): India, Cult. in Hort. Calcuttensis, J.D. Hooker & Thomson '*17 Milnea*' (K!); C. de Candolle in A. & C. de Candolle, Monog. Phan. 1: 605 (1878).

[*Milnea undulata* Wall. ex Miq., Ann. Mus. Bot. Lugd. Bat. 4: 44 (1868), nom. in syn.]

Aglaia latifolia Miq., Ann. Mus. Bot. Lugd. Bat. 4: 42 (1868). Lectotype (designated here): Java [In insula Iava ad sinum maris Tjidoeran et prope Tjiboenoer anno 1821], *van Hasselt* (L!); Koorders & Valeton, Atlas der Baumarten von Java, 1: t. 158 (1913) [but fruit drawn dehisced]; C. de Candolle in A. & C. de Candolle, Monog. Phan. 1: 604 (1878); Backer and Bakhuizen, Fl. Java 2: 129 (1965).

Aglaia khasiana Hiern in Hook. fil., Fl. Brit. Ind. 1: 554 (1875). Lectotype (designated here): India, Khasia, alt. 4 – 5000 ft [c. 1200 – 1500 m], *J.D. Hooker & T. Thomson* '16 Milnea' (CGE!; isolectotypes: BM!, CGE!, G-DC!, K!, L!, W!); C. de Candolle in A. & C. de Candolle, Monog. Phan. 1: 621 (1878).

Aglaia pirifera Hance in Jour. Bot. (N.S.) 6: 331 (1877). Lectotype (designated here): Cambodia, [at the foot of Mount Kam Chai, near the border with Thailand], May 1874, *Pierre* s.n. (BM!).

Aglaia mucronulata C. de Candolle in A. & C. de Candolle, Monog. Phan. 1 : 601 (1878). Holotype: Java, *Kuhl et Off* '48' (L!).

Milnea cambodiana Pierre, Fl. Forest. Cochinch. Fasc. 21, ante t. 334 (1 July 1895). Lectotype (designated here): [Cambodia], in montibus Krewanh, in prov. Pusath, June 1870, *Pierre* 879 (P!).

Aglaia cambodiana Pierre, Fl. Forest. Cochinch. Fasc. 21, sub t. 334 (1 July 1895); Pierre in Lecomte, Fl. Gén. L'Indo-Chine 1: 765 (1911).

Milnea rugosa Pierre, Fl. Forest. Cochinch. Fasc. 21, ante t. 335 (1 July 1895). Lectotype (designated here): in montibus Krewanh, in regio Cambodiae, in prov. Pusath, fr., June 1888, *Pierre* 880 (P!).

Aglaia rugosa Pierre, Fl. Forest. Cochinch. Fasc. 21, sub t. 335 (1 July 1895).

Milnea verrucosa Pierre, Fl. Forest. Cochinch. Fasc. 21, ante t. 335 (1 July 1895). Lectotype (designated here): Cochinchina, ad provincia Compongxoni? Cambodiarum, [L'une at l'autre rive du Mékong depuis la province de Kamput jusqu' celle d'Attopeu], fr., *Harmand* in *herb. Pierre* 5939 (P!).

Aglaia verrucosa Pierre, Fl. Forest. Cochinch. Fasc. 21, sub t. 335 (1 July 1895).

Aglaia indica (Hooker fil.) Harms in Engl. & Prantl, Pflanzenf. 3(4): 300 (1896).

? *Aglaia latifolia* Miq. var. *teysmannii* Koord. & Valet. in Meded. 'S Lands Plantent. 16: 140 (1896). Type: *Koorders* 4735β (K!); Koorders & Valeton, Atlas der Baumarten von Java 1: t. 158 (1913).

Aglaia acida Koord. & Valet. in Meded. 'S Lands Plantent. 16: 143 (1896) . Lectotype (designated here): Java, Afd. Djember, Res. Besoeki, G. Watangan bij Poeger, in Zuid-Besoeki, fl., 28 Feb. 1898, *Koorders* 30065β (BO!; isolectotypes: K!, L!); Backer and Bakhuizen, Fl. Java 2: 128 (1965).

Milnea pirifera (Hance) Pierre, Fl. Forest. Cochinch. Fasc. 22, ante t. 339 (1 July 1896).

[*Aglaia oblonga* Pierre, Fl. Forest. Cochinch. Fasc. 22, ante t. 339 (1 July 1896), nom. in syn. = *Milnea pirifera* (Hance) Pierre].
? *Lepiaglaia montrouzieri* Pierre, Fl. Forest. Cochinch. Fasc. 22, ante t . 340 (1 July 1896). Lectotype (designated here): [Vietnam], in montibus Nghê an, Tonkin Province, fl., 1876, *Montrouzier* s.n. in *herb. Pierre* 5941 (P!).
? *Aglaia montrouzieri* Pierre, Fl. Forest. Cochinch. Fasc. 22, sub t. 340 (1 July 1896). Pellegrin in Lecomte, Fl. Gén. L'Indo-Chine 1: 775 (1911).
Aglaia minahassae Koord. in Meded. 'S Lands Plantent. 19: 382, 635 (1898). Lectotype (designated here): Celebes [Sulawesi], Minahassa, Menado, 1 May 1985, *Koorders* 17918β (BO!; isolectotype: L!).
Aglaia curranii Merrill in Philipp. Jour. Sci., Bot. 7: 276 (1912). Lectotype (designated here): Philippine Islands, Luzon, Prov. Bataan, fl., Nov. 1909, *Curran* For. Bur. 17580 (NY!; isolectotypes: G!, PNH†, US!).
Aglaia diffusa Merrill in Philipp. Jour. Sci., Bot. 1912, 7: 277. Lectotype (designated here): Philippine Islands, Luzon, Prov. Tayabas, [Guinayangan], fl., 21 Nov. 1909, *Darling* For. Bur. 18684 (NY!; isolectotypes: BO!, G!, PNH†, US!).
Aglaia magnifoliola C. de Candolle in Meded. Herb. Leid. No. 22: 9 (1914). Holotype: Sunda [Nusa Tenggara], S.W. Lombok, Sepi Berg N., 300 m, yg fr., 2 July 1909, *Elbert* 2489 (L!; isotypes: G!, L!).
? *Aglaia barberi* Gamble in Kew Bull. 1915: 346 (1915). Lectotype (designated here): S. India, [Madras Presidency, Coimbatore District], Udumanparai, Anamalais [Hills], fl., 22 Nov. 1901, *C.A. Barber* 4113 (K!).
Aglaia samarensis Merrill in Philipp. Jour. Sci., Bot. 11: 186 (1916). Lectotype (designated here): Philippine Islands, Samar, Catubig River, fl., Feb. – March 1916, *Ramos* Bur. Sci. 24197 (A!; isolectotypes: K!, PNH†, US!).
Aglaia motleyana Stapf ex Ridley in Kew Bull. 1930: 368 (1930). Lectotype (designated here): Borneo, Labuan, fl., *Motley* 225 (K!).
Aglaia testicularis C.Y. Wu, Flora Yunnanica 1: 240 (1977). Type: China, Yunnan, Mar-li-po (23°15'N 104°E), Hwang-jing-in, 1400 – 1800 m, fr., 6 Nov. 1947, *K.M. Feng* 12938 (KUN!).

Tree up to 20 m, sometimes flowering at 4 m. Bole up to 50 cm diameter; buttresses upwards up to 1.5 m, outwards up to 50 cm and up to 15 cm thick. Outer bark reddish-brown, yellowish-brown or greyish-green, flaking to expose orange-brown bark beneath; inner bark pink or brown; sapwood pale brown, red or yellow; latex white. Twigs greyish-brown, reddish-brown or yellowish-brown, longitudinally wrinkled and densely covered with reddish-brown, pale brown or orange-brown stellate hairs and scales or peltate scales which have an irregular or fimbriate margin. Leaves up to 44 cm long and 40 cm wide; petiole 3.5 – 9 cm, the petiole, rhachis and petiolules with few to densely covered with hairs or scales like those on the twigs. Leaflets 5 – 9(– 11), the laterals subopposite or alternate, all (4 –)5.5 – 23 cm long, (1.8 –)2 – 9 cm wide, alternate or subopposite, often pale brown or yellowish-brown when dry, usually elliptical, sometimes ovate, often coriaceous, acuminate at apex, the obtuse acumen up to 1.5 cm, rounded or cuneate at the slightly asymmetrical base, with few to numerous hairs or scales like those on the twigs on the midrib below and occasional on the rest of that surface, often with numerous reddish-brown pits on the upper and lower surfaces; veins 5 – 16 on each side of the midrib, ascending, curved upwards near the margin, not or just anastomosing, sometimes with shorter lateral veins in between, midrib prominent, lateral veins subprominent and reticulation subprominent or visible below, petiolules 5 – 12 mm on lateral leaflets, up to 20 mm on terminal leaflet.

Inflorescences usually in the axils of the leaves, sometimes borne on the older wood of twigs. Male inflorescence up to 38 cm long and 32 cm wide, peduncle 0.5 – 5 cm, the peduncle rhachis and branches with numerous to densely covered with hairs or scales like those on the twigs. Flowers 1 – 1.5 mm (– 7 mm in India) long, 1 – 1.8 mm (– 7 mm in India) wide, yellow or orange, fragrant, pedicels 0.5 – 1.5 mm, sometimes articulated near its junction with the branchlet, densely covered with pale brown or reddish-brown stellate scales or orange-brown peltate scales. Calyx cup-shaped, about half the length of the corolla, with (4 or) 5 rounded lobes which have a ciliate margin and few to densely covered with scales like those on the pedicels. Petals (4 or) 5, yellow or orange, aestivation quincuncial or imbricate. Staminal tube cup-shaped 0.5 – 1 mm (– 5 mm in India) long, up to 1.3 mm (– 5 mm in India) wide, yellow, thickened inside below the insertion of the anthers, aperture 0.6 – 1 mm, margin lobed; anthers 5(– 6), 4 – 0.5 mm (– 2 mm in India) long, 0.2 – 0.5 mm (– 1.3 mm in India) wide, ovoid, inserted half way down the tube, either included and visible or protruding through aperture, sometimes dehiscent in the lower half only. Ovary 0.2 – 0.3 mm long, 0.3 mm wide, depressed globose, densely covered with yellowish-brown stellate scales or peltate scales which have a fimbriate margin, loculi (2 or) 3, each containing 1 or 2 ovules; stigma 0.2 – 0.4 mm (– 1 mm in India) long, 0.3 – 0.5 mm (– 0.8 mm in India) wide, either ovoid and longitudinally ridged with impressions from the anthers, the apex truncate and flattened and the margin crenulate, or depressed globose with an apical depression. Female inflorescence c. 5 cm long and 4 cm wide; peduncle up to 2.5 cm. Flowers c. 2 mm (– 4 mm in India) long and 2.2 mm (– 5 mm in India) wide, pedicels nearly 2 mm. Calyx divided into 5 ovate lobes, pedicel and calyx densely covered with stellate scales. Petals 5(– 7), aestivation quincuncial or imbricate. Staminal tube 1 mm (– 2 mm in India) long, cup-shaped, thickened inside below the insertion of the anthers, aperture c. 1 mm (– 2.5 mm in India), anthers 5, 0.5 mm (– 1.5 mm in India) long and 0.4 mm (– 1.2 mm in India) wide, included or protruding from the aperture of the staminal tube. Ovary 0.3 – 1 mm long, 0.9 – 1 mm wide, densely covered with stellate scales, loculi 3, each containing 1 or 2 ovules; stigma 0.4 mm (– 0.8 mm in India) long, 0.8 mm (– 1.3 mm in India) wide, ovoid or depressed globose, longitudinally ridged with the impressions from the anthers,.

Infructescence 7 – 12 cm long and 7 – 10 cm wide; peduncle up to 2.5 cm, the peduncle, rhachis and branches with few to numerous hairs or scales like those on the inflorescence. Fruits up to 3.2 cm long and 3.8 cm in diameter, subglobose with a central depression at the apex, grey or greenish-brown when unripe, dull orange or brown or yellow when ripe; pericarp 3 – 6 mm thick, woody or granular, often with numerous warts, with numerous on the outside to densely covered with small pale

58. edulis

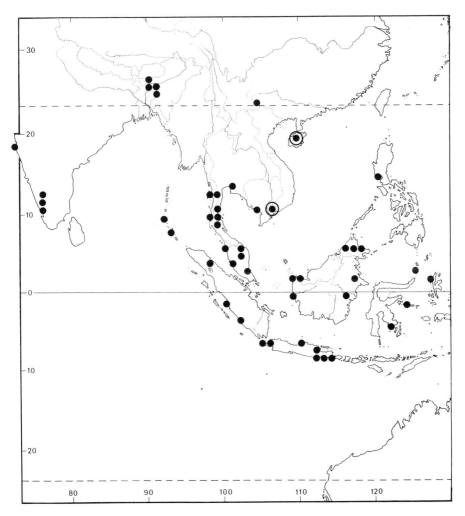

Fig. 64. Distribution of A. *edulis*.

brown or nearly white peltate scales which have a fimbriate margin pale brown and rugulose inside, sometimes with white latex. Loculi 3, each containing 0 – 1 seed; seed pale brown with a complete, thick, sour, juicy, translucent, white or orange-brown edible aril, up to 2 mm thick; seed without aril 14 – 20 mm long, 10 – 19 mm wide and 5 – 9 mm through, with the main vascular bundle running through the raphe and antiraphe, divaricately branching from the raphe over the sides of the seed; cotyledons subequal, obliquely transverse.

DISTRIBUTION. India, Bhutan, Nicobar Islands, Burma, China, Vietnam, Cambodia, Thailand, Peninsular Malaysia, Sumatra, Borneo, Philippine Islands, Java, Bali, Nusa Tenggara (Lombok), Sulawesi, Maluku. Fig. 64.

ECOLOGY. Found in evergreen and primary forest, along rocky seashore, on coral and in belukar on sandy loam with clay, sandstone, coral. Rare to scattered. Alt.: sea level to 1670 m.

VERNACULAR NAMES. Sumatra: Balik-balik, Balik-bolon, Ganggo Peram-poean, Lanting. Java: Kokosan Monjet (Djasilin); Ki Soerengede, Pantjal Kidang, Perkoso, Solokembang. Borneo: Besoliman (Sungei-Kinabatangan); Lantupak (Dusun-Kinabatangan); Segera (Iban). Java: Langsatan, Pantjal Kedang. Bali: Langsat-lotoeng. Philippine Islands: Malasaguin) Moluccas: Sala.

USES. Aril edible. Pericarp taken against diarrhoea. In China the wood is said to be red, light in weight and used in making cargo boat boards.

Representative specimens. INDIA. Assam, Garo [Garrow] hills, nr Nokrek, c. 1220 m, fr., 10 March 1950, *Koelz* 24675 (UC!, W!). Conc. Ghat, 760 m, *Gamble* 18329 (K!). Madras, Nilgiris, Devala, Nov. 1884, *Gamble* 15542 (K!). Bolampatty hills, 900 – 1200 m, fl., Dec. 1880, *Beddome* '1160' (BM!). Nilgiris, fl., *Beddome* '1163' & '1164' (BM!). Wynad, fl., *Beddome* '1162' (BM!). Tambachiring Ghat, Wynad, fl., *Beddome* '1161' (BM!).

BHUTAN. Loring Falls, 23 km above Sarbhang [26°57' 90°13'], fl., 1 June 1979, *Grierson & Long* 1562 (FHO!).

BURMA. Mergui, *Cinchona* plantation, fr., 8 Feb. 1927, *Parker* 2558 (K!).

CAMBODIA. Kamput, Cam Choy mountain, fr., June 1874, *Pierre* 5939/4264 (K!).

VIETNAM. S. Cochinchina, Baria, Dinh mountain, fl., Feb. 1896, *Pierre* 1498 (L!).

CHINA. Hainan: Shi Kai Leng, c. 1200 m, fr., 1932 – 33, *N.K. Chun & C.L. Tso* 44255 (W!); Ngai district, Yeung Ling Shan, fr., 10 June 1932 *Lau* 58 (BM!, K!, W!); Lai (Loi) area, Hung Mo Shan and vicinity, fl., 21 June 1929, *Tsang & Fung* 347 L.U. 17881 (BM!, K!).

THAILAND. Peninsula, Trang, Khao Chong: ♂ fl., 17 April 1969, *Phusomsaeng* 178 (C!, E!) & fr., 23 Apr. 1969, *Phusomsaeng* 229 (C!, E!, FHO!). Sriricha forest, Nong Kaw, 16 miles inland, c. 90 m, fr., 15 April 1915, *Collins* 388 (BM!, C!, K!).

PENINSULAR MALAYSIA. Kedah, Gunong Jerai F.R. Compt. 14, c. 90 m, fl., 28 Sep. 1963, *Pennington* 7857 (FHO!). Pahang, Taman Negara, plot 1 along Sungai Tahan trail, c. 1.5 km from Kuala Tahan H.Q., c. 90 m, fr., 18 April 1975, *van Balgooy* 2441 (AAU!). Selangor: Ulu Gombak, on slope by the lay-by directly opposite the University of Malaya Field Studies Centre, at about the $15\frac{1}{2}$ mile stone, fr., 28 April 1972, *Chin* 1825 (AAU!); Ulu Gombak F.R., 16th mile, c. 300 m, fr., 2 Oct. 1963, *Pennington* 7860 (FHO!); Ulu Langat, high hill forest, Ayer Siam, Sempada n Looi [c. 3°55'N 101°55'E], fr., 30 March 1960, *Gadoh anak Umbai* for *Millard* KL 2082 (FRI!). K. Lumpur, Weld Hills Reserve, fl., 5 March 1918, *Abdul Rohman* 2814 (FRI!).

SARAWAK. 1st Division: Kuching, off 32nd mile Padawan Road, Bt. Mentawn, fr., 14 Jan. 1973, *Mamit* S 32672 (FHO!); Tiang Bekap limestone, left side of road from Kuching, fl., 29 Apr. 1974, *Mabberley* 1638 (FHO!).

JAVA. Besuki [Besoeki] Province, Djember, Poeger, Watangan: st., 26 Oct. 1895, *Koorders* 19967β (BO!) & st., 22 Sep. 1889, *Koorders* 13089β (BO!) & st., 10 Oct. 1889, *Koorders* 4759β (BO!).

A. *edulis* is the only species in section *Aglaia* which nearly always has a 3-locular ovary. The fruit is up to 3.5 cm in diameter, subglobose, with a thick pericarp, and a seed which has a thick white edible aril in each of the three locules. However, some specimens, including the type of *Aglaia testicularis* have one large seed and a thin, brittle pericarp. Similar wide variation in fruit occurs in several species in section *Aglaia* (e.g . *A. korthalsii* and *A. lawii*) and the taxonomic status of the variants requires further investigation. As in *Aglaia lawii*, some specimens from India have much larger flowers than elsewhere.

Many of the published names from India and Indo-China are known only from the type or a small number of collections. However, material is so poor, particularly because of the absence of ripe fruits, that it is unclear whether *A. edulis* as defined here represents several species or a single variable one. Attempts to separate Thai collections into *A. pirifera*, with stellate hairs and another, unnamed, species with peltate scales failed because of the presence of intermediate specimens which have both hairs and scales.

In many Indian, Chinese and Indo-Chinese specimens, the large orange peltate scales resemble those of *A. elaeagnoidea* from India, which causes further confusion. *A. barberi* from southern India, here provisionally treated as a synonym of *A. edulis*, is an example. It has distinctive, longitudinally wrinkled fruits, but when fruits are absent, it cannot be reliably separated from *A. edulis*.

In Java, the distinction between this species and another variable species, *Aglaia lawii*, which belongs to section *Amoora*, is not always clear. The main area of confusion is in *Aglaia latifolia*. From the flower structure, Miquel's plant appears to belong to *A. edulis*; *Aglaia latifolia* var. *teysmannii* is similar to the type variety in indumentum, but it also closely resembles *A. lawii* from Java. Both *A. edulis* and *A. lawii* have 3-locular fruits in Java, the main difference lying in whether or not they dehisce and in the type of aril, neither of which can be seen in herbarium specimens of *A. latifolia* var. *teysmannii*. The fruit in the illustration of *A. latifolia* in Koorders & Valeton, Atlas der Baumarten von Java, t. 158, based on *Koorders* 4693β and 4735β, is shown as just dehiscent; this could represent either the fruit of *A. lawii*, which dehisces on the tree or the fruit of *A. edulis* which has three longitudinal ridges on the pericarp where the ripe fruit probably splits when it is over-ripe or if pressure is applied to the pericarp.

The edible aril suggests that this species may have some value as a fruit tree. According to Wallich in Roxburgh Fl. Ind., ed Carey & Wall., 1: 430 (1924), *A. edulis* is 'A native of the Garrow Hills and of the Silhet district, where it is called *Gumi* by the natives, who eat the large succulent aril which surrounds the seed under the cortex of the berry'. The resolution and full description of this complex, and the determination of the taxonomic status of the synonyms included here, would best be achieved by careful study, documentation and collection of local variants in the field throughout its range and especially of surviving populations in India and Java. It is particularly important that male and female flowers and ripe fruits are collected from each population whenever possible.

The application of the name *A. edulis* is confused, because Asa Gray (185 4) reported finding *A. edulis* (Roxb.) Wall. in Samoa and Fiji. His plant is not *A. edulis*, but *A. samoensis* A. Gray; it has 5 leaflets (rarely

7) which have a broad blunt apex. Nevertheless, Asa Gray was treated by many authors (including Hiern, 1875) as the author who transferred *Milnea edulis* to *Aglaia*, and the Fiji islands have been included in its distribution. Seeman's *A. edulis* from Fiji also does not belong here, but with *A. saltatorum*; the flowers of *Seeman* 60 differ in the structure of both the androecium and the gynoecium and the ovary has only one loculus.

59. Aglaia macrostigma King in Jour. As. Soc. Bengal 64: 78 (1895). Syntypes: Malay Peninsula, Perak, Larut, 500 – 600 ft [c. 150 – 180 m], fr., Sep. 1884, *King's Coll.* 6531 (CGE!, G!, P!, U!); *King's Coll.* 6919 (non vidi); Perak, Larut 1000 – 1500 ft [300 – 450 m], fl., May 1885, *King's Coll.* 7559 (BM!, G!, K!, L!); Ridley, Fl. Malay Penins. 1: 410 (1922); Pannell in Tree Flora of Malaya 4: 219 (1989).

Tree up to 25 m. Bole up to 125 cm in circumference. Bark smooth brown or grey; inner bark pink or red and fibrous; sapwood pink or pale pink; latex white. Branches patent or ascending. Twigs stout with longitudinal wavy ridges and densely covered with brown or orange-brown peltate or stellate scales.

Leaves imparipinnate, up to 40 cm long, obovate in outline; petiole up to 15 cm, the petiole, rhachis and petiolules with surface and indumentum like the twigs. Leaflets 3 – 9, the laterals subopposite or alternate, all 7.5 – 25 cm long, 4 – 10 cm wide, often slightly bluish-green above and orange-brown below when dry, narrowly elliptical, ovate or obovate, acuminate at apex, the obtuse acumen up to 15 mm long, rounded or cuneate at base, with few to numerous peltate or stellate scales on the midrib and lateral veins 8 – 22 on each side of the midrib, ascending and curved upwards near the margin, midrib and lateral veins slightly depressed on upper surface, midrib prominent and lateral veins subprominent on lower surface; petiolules up to 10 mm on lateral leaflets, up to 15(– 35) mm on terminal leaflet.

Inflorescence axillary or ramiflorous, up to 20 cm long and 15 cm wide; peduncle up to 1.5 cm, peduncle rhachis, branches and pedicels with longitudinal wavy ridges and densely covered with stellate hairs or scales. Flowers about 2.5 mm long, obovoid; pedicels up to 1 mm. Calyx $1/3 - 1/2$ the length of the flower, deeply divided into 5 broadly ovate, obtuse lobes, with densely covered with stellate scales on the outside. Petals 5, elliptical, concave, fleshy in the centre, glabrous, aestivation quincuncial. Staminal tube shorter than the corolla, subglobose, with a wide entire aperture; anthers 5, c. $1/2$ the length of the staminal tube, ovoid, inserted just above the centre and curved with the tube with the apices just protruding through the aperture. Ovary depressed globose densely covered with stellate hairs; stigma ovoid with 2 small lobes; the ovary and stigma together c. $2/3$ the length of the staminal tube.

Infructescence axillary or ramiflorous, up to 40 cm long with up to 25 fruits; peduncle up to 9 cm, the peduncle, branches and pedicels stout with surface and indumentum like the twigs. Fruits up to 4.5 cm long and 3 cm wide, pyriform, yellow or brown; densely covered with orange-brown stellate hairs with numerous short arms; fruit-stalks up to 1.5 cm. Loculus 1, containing 1 seed.

DISTRIBUTION. Peninsular Malaysia only. Fig. 65.

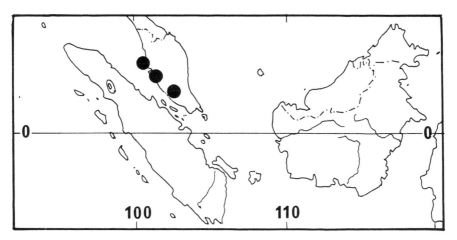

Fig. 65. Distribution of *A. macrostigma*.

ECOLOGY. Found in primary forest. Alt.: 160 to 1000 m.
VERNACULAR NAMES. Peninsular Malaysia: Bekak, Langsat Hutan, Memberas.
Representative specimens. PENINSULAR MALAYSIA. Selangor, Bt Lagong F.R., Jungle Plot 101, tree no. 227, fl., 9 June 1947, *Sow* KEP 51970 (FRI!); same tree KEP 78798 (FRI!). Negri Sembilan, Jelubu, 300 m, fr., 11 Oct. 1969, *Everett* KEP 104942 (FRI!).

A. macrostigma resembles *A. forbesii*, but the leaves of *A. macrostigma* have 3 – 9 leaflets and those of *A. forbesii* which have 9 – 15 leaflets; the leaflets of *A. macrostigma* are usually bluish-green above and orange-brown below when dry and the tiny peltate or stellate scales are inconspicuous. The inflorescences are often ramiflorous and the fruits are pyriform while in *A. forbesii* the inflorescences are axillary and the fruits subglobose or ellipsoid.

The indumentum of *A. macrostigma* resembles that of *A. edulis*, but the fruit of *A. edulis* is subglobose.

60. Aglaia odoratissima Blume, Bidjr. 171 (1825). Lectotype (designated here): Java, [Mt Salak], *Herb. Blume* s.n. (L!; isolectotypes: BO!, L!); Miq., Ann. Mus. Bot. Lugd. Bat. 4: 43 (1868); C. de Candolle in A. & C. de Candolle, Monog. Phan. 1: 602 (1878); King in Jour. As. Soc. Bengal 64: 67 (1895); Koorders & Valeton, Atlas der Baumarten von Java 1: t. 160 (1913); Ridley, Fl. Malay Penins. 1: 404 (1922); Corner, Wayside Trees of Malaya 1: 457 (1940), ed 3, 2: 496 (1988); I.H. Burkill, Dictionary of Economic Products of the Malay Peninsula 1: 75, 137 (1935); Backer & Bakhuizen, Fl. Java 2: 128 (1965); Corner in Gard. Bull. Suppl. 1: 131 (1978); Pannell in Ng, Tree Flora of Malaya 4: 221 (1989).

Aglaia diepenhorstii Miq., Fl. Ind. Bat. Suppl. 1: 507 (1861). Lectotype (designated here): [Sumatra, W.], Priaman, *Diepenhorst* 329 H.B. (U!; isolectotypes: L!, 1329 BO!); Miq., Ann. Mus. Bot. Lugd. Bat. 4: 43

(1868); C. de Candolle in A. & C. de Candolle, Monog. Phan. 1: 603 (1878).
Milnea blumei Teijsm. & Binn., Cat. Hort. Bog.: 211 (1866) nom. nud.
Aglaia paniculata Kurz, Prelim. Rep. Forest Pegu: 34 (1875), nom. nud.; Kurz in Journ. As. Soc. Beng. 44: 146 (1875), nom. nud.
Aglaia paniculata Kurz in Jour. As. Soc. Beng. 44: 199 (1876). Lectotype (designated here): Burmah, Pegu, fl., *Kurz* 2043 (K!); C. de Candolle in A. & C. de Candolle, Monog. Phan. 1: 606 (1878).
Aglaia odoratissima Blume var. *parvifolia* Koord. et Val. in Meded. 'S Lands Plantent. 16: 150 (1896). Lectotype (designated here): Java, Distr. Majang, Afd. Djember, Res. Besoeki, Boschterrein Tjoermanis, fr., 1895, *Koorders* 4677β (BO!; isolectotypes: G!, K!).
Aglaia odoratissima Blume var. *pauciflora* Koord. et Val. in Meded. 'S Lands Plantent. 16: 150 (1896). Lectotype (designated here): Java, Distr. Tjisondari, Afd. Bandoeng, Res. Preanger, Boschterrein Tjigenteng, fl., 1895, *Koorders* 4715B (BO!; isolectotypes: G!, K!, L!).
Aglaia luzoniensis (Vidal) Merrill & Rolfe var. *trifoliata* Merrill & Rolfe in Philipp. Jour. Sci. 3: 105 (1908). Lectotype (designated here): Philippine Islands, Mindanao, Province of Misamis, Mt Malindang, fl., May 1906, *Mearns & Hutchinson* For. Bur. 4724 (UC!; isolectotypes: G!, K!, PNH†).
Aglaia affinis Merrill in Philipp. Jour. Sci. 3: 235 (1908). Lectotype (designated here): Philippine Islands, Island of Balabac, fl., March 1906, *Mangubat* Bur. Sci. 446 (NY!; isolectotypes: K!, PNH†, US!).
Aglaia heterophylla Merrill in Philipp. Jour. Sci., Bot. 13: 77 (1918). Lectotype (designated here): Sarawak, Baram District, Mt Trekan, 1000 ft [c. 300 m], fl., July 1895, *Hose* 555 (A!; isolectotypes: BM!, E!, K!, L!, PNH†).
Aglaia odoratissima Blume var. *forbesii* Baker fil. in Jour. Bot., Lond. 62, Suppl.: 19 (1924). Lectotype (designated here): Sumatra, Lampongs, Kotta Djawa, 400 – 700 ft [c. 120 – 210 m], *Forbes* 1337 (BM!; isolectotype: L!).
Aglaia cuspidella Ridley in Kew Bull. 1930: 367 (1930). Lectotype (designated here): Borneo. Sarawak, nr Kuching, Mt Koum, fl., 22 Sep. 1892, *Haviland* 1706 (K!; isolectotype: SING!).
Aglaia fraseri Ridley in Kew Bull. 1930: 368 (1930). Lectotype (designated here): Borneo, British North, Kudat, *Fraser* 239 (K!).

Small tree up to 12 m (– 20 m), with a diffuse, irregular crown which is sometimes asymmetrically obovate in outline. Bole up to 5 m, up to 25 cm in circumference. Bark smooth, greenish-grey to brown, with small lenticels in longitudinal rows; inner bark green or magenta; sapwood pale pink or pale yellowish-brown or dark reddish-brown; occasionally with some white latex. Branches few, long, much divaricate, patent or ascending patent, with the ultimate branches horizontal in one plane. Twigs slender, greyish-green or brown, smooth or with longitudinal wavy ridges, densely covered with peltate scales and usually with pale yellowish-brown stellate hairs interspersed, the scales dark brown, some with a marked fimbriate margin, others much less so.

Leaves few on each twig, imparipinnate, 10 – 30 cm long, 5 – 30 cm wide, ovate or obovate in outline; petiole 1.5 – 6.5 cm, greenish brown, the petiole, rhachis and petiolules densely covered with scales and stellate hairs similar to those on twigs. Leaflets (1 –) 3 – 5 (– 7), the laterals subopposite or opposite, all 4.5 – 18(– 23.5) cm long, 2 – 6(– 8.5) cm wide, yellowish-green when young turning dark green above when mature, bluish-green above and pale brown below when dry, usually elliptical sometimes asymmetrically so, sometimes ovate or obovate, acuminate-caudate at apex with the obtuse acumen often narrow and parallel-sided and up to 20 mm long, rounded or cuneate at the usually asymmetrical base, with scales or hairs like those on the twigs occasional on midrib of upper surface, numerous on midrib of lower surface and scattered on the rest of that surface (occasionally numerous, especially in Borneo); veins 5 – 9 (– 11) on each side of the midrib, curved upwards, midrib prominent and lateral veins sub-prominent on lower surface; petiolules up to 20(– 35) mm.

Male inflorescence 7 – 35 cm long, 2 – 25 cm wide, peduncle 0 – 20 mm, green; rhachis, branches and pedicels yellow, all with numerous reddish-brown stellate scales. Flowers up to 1.5 mm in diameter, depressed globose, very fragrant of citronella; pedicels c. 1 mm. Calyx $^1/_2$ the length of the corolla with numerous reddish-brown stellate scales on the outer surface, divided almost to the base into 5 (or 6) subrotund lobes, obtuse at the apex, patent at anthesis. Petals 5, c. 0.6 mm long, lemon yellow, elliptical or obovate-rotund, glabrous, aestivation quincuncial. Staminal tube less than $^1/_2$ the length of the corolla, shallowly cup-shaped with the apical margin incurved and shallowly 5-lobed; anthers c. 0.2 mm long, c. 0.2 mm wide, yellow when immature, brown or purple at anthesis, turning black later, ovoid, inserted just below and protruding through the aperture of the tube, pointing towards the centre of the flower. Ovary subglobose densely covered with stellate hairs; stigma ovoid with two small apical lobes; ovary and stigma together c. $^1/_2$ the length of the staminal tube. Female inflorescence a narrow spike-like raceme 3.5 – 12 cm long with up to 20 brownish-yellow flowers. Flowers up to 2 mm in diameter, obovoid, like the male in structure but slightly larger; ovary and stigma together c. 0.6 mm long; loculi 1 or 2, each containing 1 ovule.

Infructescence up to 20 cm long and 15 cm wide. Fruits 1.5 – 2 cm long, 1 – 1.5 cm in diameter, ellipsoid or obovoid, rounded at apex and tapering at base, yellow, orange or orange-red, densely covered with pinkish-orange stellate scales turning brown when dry; pericarp 1 – 1.5 mm thick, fibrous and flexible, the innermost layer a detachable membrane which surrounds the seed. Loculus 1 (or 2), containing 1 seed; seed c. 1.3 cm long, 0.9 cm wide and 0.8 cm through; aril ± completely covering the seed, pale pink, translucent, gelatinous, sweet-tasting, attached along the raphe; seed coat thin, hard, dark brown; main vascular bundle running through the raphe and antiraphe, divaricately branching from the raphe over the sides of the seed. Cotyledons equal, transverse, with shiny yellow wrinkled surfaces; radicle short, enclosed covered with pale yellow stellate scales.

DISTRIBUTION. Nicobar Islands, Burma, Thailand, Peninsular Malaysia, Sumatra, Borneo, Philippine Islands, Java, ? Nusa Tenggara, Sulawesi. Fig. 66.

ECOLOGY. Found in riverine forest, evergreen forest, primary forest, secondary forest and periodically inundated swamp forest, along road-sides; on limestone, sandstone, metamorphic rock, alluvial, granitic sand, coral, clay, loam, basalt. Alt.: sea level to 1870 m. Common.

FIELD OBSERVATIONS. Kuala Lompat, Pahang, Peninsular Malaysia: the inflorescences are small, being 10 – 20 cm long in the male and 3 – 10 cm long in the female. The pale yellow male flowers, which have a strong perfume of citronella, are borne on delicate, pendulous inflorescences. The inconspicuous female inflorescences are erect and spicate with 5 – 20 flowers, which are orange-yellow and larger than the male, with little or no perfume. At anthesis the petals separate at the apex of the flower, leaving an aperture through which the anthers may be seen, resembling a neat ring of 5 teeth. There are usually 10 – 20 inflorescences on a tree. The infructescences have up to ten 1-seeded, indehiscent fruits which ripen one or two at a time, turning bright pinkish-orange, over a period of weeks.

VERNACULAR NAMES. Peninsular Malaysia: Belankas, Gigi Buntal, Kasai Hutan, Memberas, Mesenduk, Telur, Subulat Jantan, Tumilang. Sumatra: Boeroenai Silai, Ganggo, Gonggofaloh, Loeloe Pajo, Tekir, Toetoeen Oeding. Java: Kamantjing, Sajong (Sundanese, Djasilin dialect); Ki-hoera (Sundanese); Ande-andean, Kikopijan, Ramboetan Oetan.* Borneo: Anda Giwi (Murut); Kopeng (Bassap Mapulu); Lambunau (Dusun); Langsat-langsat, Langsat Munyit (Malay); Lantupak, Tanggal (Dusun Kinabatangan); Lemiyak (Sebop); Mata Kuching Munchit (Idahan); Segera (Iban); Tangiran Manok (Bajau Lobok); Tangsat Gusing (Sungei); Birajang Merah, Boeno, Lagas, Mulak. Philippine Islands: Hungo. Lesser Sunda Islands.: Lanang Soai, Lantji Tan, Melantji Tan, Muku Te'e.

Representative specimens. THAILAND. Southwestern District, Kanchanaburi Province, Khao Yai, E. of Sangkhla, c. 700 m, fl., 30 March 1968, *van Beusekom & Phengkhlai* 207 (AAU!, C!). Peninsula, Pangnga Province, Khlong nang yon, [9°15′N 98°20′E], 100 m, fr., 30 April 1973, *Geesink & Santisuk* 5065 (AAU!, C!).

PENINSULAR MALAYSIA. Penang: Penang Hill, *Pannell* 1348 (FHO!). Pahang: Taman Negara, K. Tahan, *Pannell* 1072 (FHO!); Krau Game Reserve, K. Lompat: ♀ fl., 3 June 1978, *Pannell* 1270 (FHO!) & ♂ fl., *Pannell* 1293 (FHO!) & fr., *Pannell* 1579 (FHO!). Kuantan, Balok, *Mahmud* FMS 3745 (FHO!) . Selangor: mile 16 Ulu Gombak F.R., *Pennington* 7859 (FHO!); Ulu Gombak F.R., *Kochummen* FRI 16581 (FRI!), *Kochummen* KEP 99991 (FRI!); Ulu Langat, Menuang Gasing, *Kloss* s.n. (K!). Gemas, *Burkill* SFN 3547 (K!). Malacca, between Jasin & Chaban, *Ridley* s.n. (SING!).

SINGAPORE. Botanic Gardens Jungle, *Sinclair* SFN 40656 (K!).

SUMATRA. Atjeh [Aceh]: Southern part of Gunung Leuser Nature Reserves, Atlas River Valley, near the mouth of the Bengkong River, c. 50 km S. of Kutacane [c. 3°N 97°50′E], 50 – 125 m, fl., 16 July 1979, *de Wilde & de Wilde-Duyfjes* 18750 (L!); Gunung Leuser Nature Reserves, Ketambe Research Station and vicinity, Alas River Valley, c. 35 km N.N.W. of Kutacane c. 5 km S. of Ketambe along Alas Road [c. 3°40′N 97°40′E], 300 – 350 m. fr., 9 June 1979, *de Wilde & de Wilde-Duyfjes* 18027 (L!); Gunung Leuser Nature Reserve, slopes of Gunung Mamas c. 3 km S.W. from the mouth of Lau Ketambe, c. 30 km N.W. of Kutatjane, c. 600 – 1000 m, fl., 22 March 1975, *de Wilde & de Wilde-Duyfjes* 15730 (L!). fl., 1881, *Forbes* 31509 (CGE!).

SARAWAK. 1st Division: Semengoh Forest Reserve, fl., 19 April 1974, *Mabberley* 1584 (FHO!); Gunong Gading F.R., Lundu, c. 240 m, fr., 14

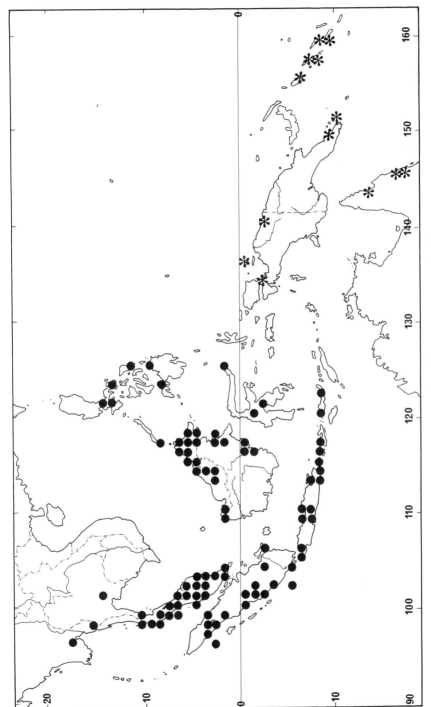

Fig. 66. Distribution of *A. odoratissima* ● and *A. brassii* *.

Nov. 1963, *Pennington* 7974 (FHO!); Gunong Gading, Lundu [1°60′N 109°85′E], 70 m, fl., 17 July 1963, *Chai* S 18454 (FHO!). 4th Division, Ulu Mayeng, Kakus, [2°55′N 113°85′E], c. 200 m, fl., 13 July 1964, *Sibat ak Luang* S 21728 (FHO!).

SABAH. Sepilok F.R., nr Sandakan, sea level, fr., 16 Oct. 1963, *Pennington* 7891 (FHO!).

PHILIPPINE ISLANDS. Luzon, Province of Sorsogon, Irosin (Mt Bulusan), fr., July 1916, *Elmer* 16814 (U!). Mindanao, Province of Agusan, Cabadbaran (Mt Urdaneta), st., Sep. 1912, *Elmer* 13945 (U!)

JAVA. Bantam, G. Karang, c. 1050 m, fl., 12 June 1892, *Koorders* 4719β (L!). Batavia: Tjampea, c. 200 – 300 m, fr., 5 July 1898, *Koorders* 30497β (L!); Goenoeng Tjibodas, Bij Tjiampea, W.V. Buitenzorg, 200 m, fl., 1 Oct. 1922, *Bakhuizen van den Brink* 5712 (U!).

The following specimen may also belong here. NUSA TENGGARA. W. Flores, S.E. part, Mbengan, 600 m, fl., 10 May 1965, *Kostermans* 22119 (AAU!).

A. odoratissima is usually a small tree with slender, more or less horizontal branches. The leaves, which rarely have more than 5 leaflets, are well spaced and borne horizontally in one plane on the slender branches. In occasional, much taller specimens, the branches are ascending, as they are in most species of *Aglaia*, and the leaves have 7 leaflets. At Kuala Lompat, in the Krau Game Reserve of Peninsular Malaysia, the small tree is common in the lower area of the forest where it may be subjected to intermittent flooding, while the taller tree with 7 leaflets is rarer and occurs in more elevated areas of the forest where flooding is unlikely. The leaflet indumentum is inconspicuous, but the peltate scales interspersed with stellate hairs which occur along the midrib may be seen with a lens. The presence of peltate scales distinguishes this species from *A. elliptica*.

61. Aglaia luzoniensis (Vidal) Merrill & Rolfe in Philipp. Jour. Sci., Bot. 3: 105 (1908).

Beddomea luzoniensis Vidal, Rev. Pl. Vasc. Filip.: 84 (1886). Lectotype (designated here): Philippine Islands, Province Tayabas (Quezon), Mauban, fl., Oct. 1881, *Vidal* 169 (A!, isolectotype: K!).
Aglaia unifoliolata Koord. in Meded. 'S Lands Plantent. 19: 383, 635 (1898). Lectotype (designated here): Celebes [Sulawesi], Minahassa (Menado), Kakas, 500 m, fl., 25 Jan. 1895, *Koorders* 17935ß (BO!; isolectotype: L!).
Aglaia monophylla Perkins in Frag. Fl. Philipp.: 33 (1904). Lectotype (designated here): Philippine Islands, Luzon, Camarines Sur, Pasacao, fr., *Ahern* 123 (US!; isolectotype: PNH†).
Aglaia brevipetiolata Merrill in Philipp. Jour. Sci., Bot. 11: 14 (1916). Lectotype (designated here): Philippine Islands, Luzon, Province of Sorsogon, fr., July – Aug. 1915, *Ramos* Bur. Sci. 23522 (A!; isolectotypes: BM!, BO!, K!, L!, NY!, P!, PNHβ, US!).
Aglaia rizalensis Merrill in Philipp. Jour. Sci., Bot. 13: 289 (19 18). Lectotype (designated here): Philippine Islands, Luzon, Province of Rizal, fl., July 1917, *Ramos et Edaño* Bur. Sci. 29640 (US!; isolectotypes: K!, PNH†).

61. luzoniensis

Tree up to 10 m, up to 15 cm in diameter. Outer bark smooth or lenticellate, brown or red, soft; inner bark red or reddish-brown, soft; cambium pink; sapwood pale brown, white or reddish-brown. Twigs greyish-brown, longitudinally wrinkled or with longitudinal cracks, the apex densely covered with orange brown or reddish-brown peltate scales which sometimes have a fimbriate margin. Leaves simple, 5 – 23 cm long, 2 – 6.5(– 8) cm wide, elliptical; apex acuminate with the obtuse, parallel-sided acumen up to 10 mm long; base cuneate; with numerous scales like those on the twigs on the midrib below and scattered or numerous on the rest of that surface and sometimes with paler brown scales in between; veins 5 – 18 on each side of the midrib, curving upwards, not or just anastomosing, the midrib prominent and the lateral veins subprominent below; petioles 1 – 3 cm, with a swelling at both ends.

Inflorescence up to 19 cm long and 9 cm wide; peduncle 3 – 7 mm, the peduncle, rhachis and branches densely covered with scales like those on the twigs. Male flowers 1 – 1.5 mm long, 1 – 1.5 mm wide, pedicel 0.3 – 1 mm to articulation with tiny triangular bracteole, subtending branchlet up to 3 mm, the pedicel and branchlet densely covered with peltate scales which have a fimbriate margin. Calyx 0.5 – 1 mm, cup-shaped with 5 ovate lobes, margins fimbriate, densely covered with scales like those on the twigs on the outside. Petals 5 (or 6), yellow, aestivation quincuncial. Staminal tube 0.5 – 0.6 mm high, 1 mm wide, cup-shaped with the margin incurved, thickened below the anthers; anthers 5, 0.2 – 0.3 mm long, 0.2 – 0.3 mm wide, pointing towards the centre of the flower. Ovary 0.2 mm high, 0.3 – 0.4 mm wide, depressed globose, densely covered with peltate scales; stigma c. 0.2 mm high, 0.3 – 0.4 mm wide, ovoid or depressed globose, sometimes with a central depression at the apex. Female inflorescence either about 2.5 cm long and 1 cm wide with few branches or c. 8 cm long and 4 cm wide and more branched; peduncle 1 – 6 mm. Female flowers sometimes with peltate or stellate scales on the outside of petals; staminal tube 0.5 – 1 mm high and c. 1 mm wide, the anthers c. 0.5 mm long and 0.5 mm wide; stigma 0.3 mm high and 0.4 – 0.5 mm wide, ovoid or depressed globose, sometimes with a central depression at the apex; otherwise like the male. Infructescence 5 – 11 cm long, 4 – 5.5 cm wide, with about 3 fruits. Fruits c. 1.8 cm long and 1.5 cm wide, dark brown, reddish-brown, pale orange or yellow, densely covered with scales like those on the twigs on the outside, indehiscent; loculi 1 or 2, each containing one arillate seed.

DISTRIBUTION. Borneo, Philippine Islands, Sulawesi. Fig. 67.

ECOLOGY. Found in primary forest, secondary forest, near mangrove; on sandstone, sandy alluvial, sand, limestone, clay, volcanic clayey sand. Alt.: 10 to 1400 m.

VERNACULAR NAMES. Borneo: Lambunan, Lambungau, Lantupak (Dusun Kina-batangan); Langsat-langsat, Lansat Munyit (Malay); Onsod Onsod (Murut); Bahko. Philippine Islands: Lambunaw Bagat (Tagbanua), Malacamote. Sulawesi: Pisek-rintek.

Representative specimens. SARAWAK. Bt Taji Buloh, Hose mountains [2°15′N 113°60′E], 1100 m, fr., 22 March 1964, *Banying ak Nyndong* S 14000 (FHO!). Baram District, Lobang Rusa, near Sungei Melinau Paku [4°02′N 114°49′E], c. 150 m, fl., 9 Feb. 1966, *Chew* CWL 1034 (AAU!).

Section Aglaia

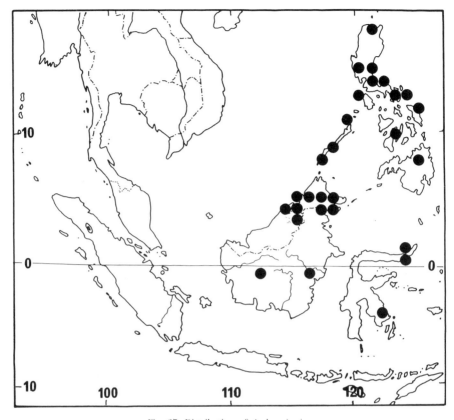

Fig. 67. Distribution of *A. luzoniensis*.

SABAH. Sepilok F.R., nr Sandakan: sample plot 3, compt. 4, Plot E, c. 15 m, fl., 23 Oct. 1963, *Pennington* 7919 (FHO!); Charcoal Creek, sea level, fl. & fr., 20 Oct. 1963, *Pennington* 7906 (FHO!).

KALIMANTAN. Bukit Raya, [112°44'E 0°40'S], c. 400 m, fl., 22 Jan. 1983, *Nooteboom* 4548 (FHO!). *A. luzoniensis* resembles A. *odoratissima* Blume, but differs in having simple leaves and more numerous peltate scales on the lower leaflet surface.

62. Aglaia yzermannii Boerl. & Koord. in Ic. Bogor.: t. 87 (1901). Lectotype (designated here): Sumatra, bij Lampatan, afd. Koeanten [Ajer Pinger], 90 m, fl., 21 Feb. 1891, *Koorders* 10377β (BO!); van Steenis, Rheophytes of the World: 287, t. 31 (1981); Pannell in Ng, Tree Flora of Malaya 4: 227 (1989).

Aglaia salicifolia Ridley in Jour. As. Soc. Straits Branch 54: 32 (1910).
    Lectotype (designated here): Malay Peninsula, Pahang, Tahan River, fl. & fr., [16 July 1891], *Ridley* 2660 (SING!; isolectotype: K!); Ridley, Fl. Malay Penins. 1: 403 (1922); I.H. Burkill, Dictionary of Economic

Products of the Malay Peninsula 1: 76 (1935); Corner, Wayside Trees of Malaya 1: 457 (1940), ed 3, 2: 496 (1988).

Small tree up to 5 m, with a broad irregular crown. Bark pale brown. Branches from near the base, usually patent and projecting horizontally from the river bank over the water, sometimes ascending. Twigs slender, pale brown, irregularly wrinkled, with numerous to densely covered with brown or yellowish-brown stellate scales, especially towards the apex.

Leaves imparipinnate, 14 – 27 cm long, 10 – 27 cm wide, obovate in outline; petiole 10 – 25 mm, the petiole, rhachis and petiolules with few to numerous scales like those on the twigs. Leaflets 3 – 5, the laterals subopposite, all 5 -17 cm long, 0.5 – 2 cm wide, slightly bluish-green and sometimes becoming brown when dry, linear, linear-lanceolate or narrowly lanceolate, slightly curved, acuminate or caudate at apex with the obtuse acumen up to 15 mm long, cuneate at the asymmetrical base, glabrous or with a few scales on lower surface, particularly along the midrib; veins 9 – 15 on each side of the midrib, curved upwards, midrib prominent on lower surface; petiolules 2 – 7 mm. Male inflorescence up to 17 cm long and 12 cm wide; peduncle up to 1 cm, the peduncle, rhachis, branches and pedicels with numerous to densely covered with scales like those on the twigs. Male flowers c. 1.5 mm in diameter, sub-globose, strongly fragrant of citronella; pedicels up to 1.5 mm. Calyx c. $\frac{1}{3}$ the length of the corolla, densely covered with stellate scales on the outside, deeply divided into 5 broadly elliptical lobes. Petals 5, yellow subrotund, aestivation quincuncial. Staminal tube $\frac{1}{2} - \frac{1}{3}$ the length of the corolla, cup-shaped with the apical margin slightly incurved and shallowly lobed; anthers 5, c. $\frac{1}{3}$ the length of the tube, broadly ovate, inserted just below the margin of the tube, protruding through the aperture and pointing towards the centre of the flower.

Ovary subglobose, densely covered with stellate scales; stigma subglobose with 2 – 3 small apical lobes; ovary and stigma together c. $\frac{1}{2}$ the length of the staminal tube. Female inflorescence and flowers similar to those of the male but the inflorescence smaller, flowers fewer, up to 2 mm in diameter, pedicels up to 2.5 mm, ovary slightly larger. Infructescence up to 15 cm long. Fruits up to 2 cm long and 2 cm wide, orange-brown or orange-red, ellipsoid or subglobose, densely covered with stellate scales like those on the twigs; inner surface of pericarp pink. Loculi 1 – 2, each containing 1 seed; aril translucent, edible, sweet. Fig. 68.

DISTRIBUTION. Peninsular Malaysia, Sumatra. Fig. 69.

ECOLOGY. Rheophyte found in riverine forest on granite. Alt.: 30 to 120m. Common.

VERNACULAR NAMES. Peninsular Malaysia: Bunga Telor Ikan, Tado Ikan, Pelir Pelandok.

Representative specimens. PENINSULAR MALAYSIA. Pahang, Taman Negara, Sg. Tahan: *Mat* s.n. (SING!); Soepadmo 86 (K!); *Sow* FMS 41007 (FRI!); 12 March 1978, ♂ fl., *Pannell* 1109 (FHO!) & 12 March 1978, ♀ fl., *Pannell* 1116 (FHO!) & 12 March 1978, ♀ fl. & fr., *Pannell* 1119 (FHO!); below Lata Berkoh, fr., 16 Aug. 1973 *Stone* 11572 (AAU!).

A rheophyte with leaflets which resemble the leaves of *Salix viminalis* L., hence the name of the synonym *A. salicifolia*. This species grows on

Fig. 68. *A. yzermannii*. Habit x¹/₂; with male inflorescence, with female inflorescence and with infructescence. Half flower, male x20. Half flower, female x15. Peltate scale x80.

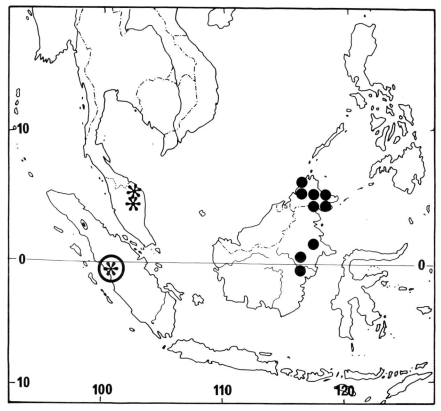

Fig. 69. Distribution of *A. yzermannii** and *A. rivularis*●

river banks with the branches projecting horizontally over the water. It is most easily found by swimming near the bank until the strong citronella perfume emitted by the male flowers is detected! Several trees are usually found growing together, the female often bearing all reproductive stages from flower buds to ripe fruits. *A. yzermannii* is found mainly in the Malay Peninsula where it seems to be restricted to the banks of relatively deep stretches of otherwise stony, fast flowing rivers. There is one gathering known from Sumatra: the type specimen of *A. yzermannii*. The simple-leaved species, *A. rivularis* Merrill, is found in Sabah and N.E. Kalimantan. The leaves are almost identical with the leaflets of *A. yzermannii*, differing only in the presence of scales scattered on the lamina as well as on the midrib. Both species are rheophytes.

63. Aglaia rivularis Merrill in Univ. Calif. Publ. Bot. 15: 125 (1929). Lectotype (designated here): British North Borneo, Elphinstone Province, Tawao, fl., Oct. 1922 – March 1923, *Elmer* 21789 (UC!; isolectotypes: A!, BM!, G!, GH!, L!, MICH!, SING!, U!, UC!); van Steenis, Rheophytes of the World: 291, t. 32 (1981).

Small tree up to 15 m. Bole up to 10 m, up to 50 cm in diameter. Branches often projecting horizontally from the river bank over the water. Bark smooth, brown or whitish-brown; inner surface of bark brown; middle bark reddish-brown; inner bark reddish-brown or pale brown, cambium pink or yellowish-brown; sapwood pink or almost white. Twigs reddish-brown or greyish-brown, longitudinally wrinkled, densely covered with reddish- brown, orange brown or pale brown peltate scales which often have a short fimbriate margin.

Leaves simple, 6.5 – 24 cm long, 1 – 4 cm wide, lanceolate, tapering to an acuminate apex, the acute acumen up to 2 cm long, tapering to a cuneate base; with numerous or densely covered with scales like those on the twigs on the midrib below, few on the rest of the lower leaflet surface, sometimes numerous when young; veins 10 – 17, curved upwards, midrib prominent, lateral veins subprominent and reticulation visible or occasionally subprominent below; petiole 1 – 2 cm, densely covered with scales like those on the twigs.

Inflorescences up to 6 in the axils of the leaves of the apical shoots, 12 – 23 cm long, 6 – 12 cm wide, usually ovoid in outline; peduncle 2.5 – 4.5 cm, the peduncle, rhachis and branches densely covered with scales like those on the twigs, or with a longer fimbriate margin, with small triangular bracts at base of branches. Male and female flowers similar, 1 – 2 mm long, 1.5 – 2.5 mm wide; pedicel 0.5 – 3.5 mm densely covered with peltate scales which have a fimbriate margin. Calyx c. $\frac{1}{3}$ the length of the flower, divided into (4 or) 5 rounded lobes which are densely covered with peltate scales and have a ciliate margin. Petals 5 (or 6), white, yellow or pale yellow, aestivation quincuncial. Staminal tube 1 – 1.3 mm long, 0.9 – 1.5 mm wide, thickened inside below the bases of the anthers, aperture 0.5 – 0.7 mm, margin lobed; anthers 5, c. 0.3 mm long, 0.3 mm wide, ovoid, inserted inside the margin of the tube, protruding through the aperture and pointing towards the centre of the flower. Ovary 0.2 – 0.4 mm long, 0.3 – 0.4 mm wide, depressed globose, densely covered with peltate or stellate scales on the outside; loculus 1, containing 1 ovule; stigma 0.2 – 0.3 mm long, 0.3 – 0.4 mm wide, ovoid or depressed globose and with a central apical depression.

Infructescence 2 – 18 cm long, with up to 20 fruits, peduncle 1.5 – 4 cm, the peduncle rhachis and branches slender and flexible densely covered with scales like those on the twigs. Fruits 1.5 cm long, 0.8 – 1 cm wide, ellipsoid, brown, reddish-brown or yellow; fruit-stalks up to 0.8 cm; pericarp reddish-brown, indehiscent; seed 1, surrounded by an edible aril.

DISTRIBUTION. E. Borneo. Fig. 69.

ECOLOGY. Rheophyte, found along river banks in riverine forest; on sand. Alt.: 3 to 500 m.

USES. Wood is used for fence posts (Borneo: Dusun Labuk).

VERNACULAR NAMES. Borneo: Lambunan (Dusun Labuk); Runu (Sungei Segaliud); Besok, Kalambunau, Rubmal.

Representative specimen.SABAH. Tawao, Elphinstone Province, fr., Oct. 1922 – March 1923, *Elmer* 21583 (L!).

A common rheophytic, riverside tree in Borneo only. *A. rivularis* resembles *A. yzermannii*, except that it has simple leaves.

64. Aglaia brassii Merrill & Perry in Jour. Arn. Arb. 21: 325 (1940). Lectotype (designated here): Solomon Islands, Ysabel, Meringe 250 m, 23 Nov. 1932, *Brass* 3189 (A!; isolectotypes: BO!, BRI: photo SING!).

Tree up to 2 – 20 m; bole diameter 7 – 15 cm, sometimes with buttresses upwards up to 1.2 m. Outer bark smooth or scaly, greyish-brown or brown; inner bark pinkish-brown; sapwood white, pinkish-yellow or reddish-brown. Twigs reddish-brown, densely covered with reddish-brown peltate scales, sometimes with paler stellate scales interspersed, sometimes with stellate hairs at the apex. Leaves 8.5 – 29 cm long, 4.5 – 29 cm wide; petiole 1 – 7 cm, the petiole, rhachis and branches densely covered with peltate and stellate scales like those on the twigs. Leaflets (1 –)3 – 7, the laterals subopposite or alternate, all 4.5 – 22 cm long, 2 – 7 cm wide, obovate or elliptical, acuminate at apex, the obtuse acumen parallel-sided, and 7 – 10 mm long, cuneate at the slightly asymmetrical base, with dark reddish- brown peltate scales and some paler stellate scales interspersed densely covering the midrib below and scattered elsewhere on the lower leaflet surface; veins 6 – 11 on each side of the midrib, ascending at an angle of 50 – 60° to the midrib and curved upwards, not anastomosing; petiolules up to 1 cm on lateral leaflets, up to 4 cm on terminal leaflet.

Inflorescence c. 13.5 cm long and 8.5 cm wide, with some ovate bracts c. 0.6 mm long and 0.6 mm wide; peduncle 1.5 – 3.5 cm, the peduncle, rhachis and branches with reddish-brown and pale brown stellate scales and peltate scales which have a fimbriate margin. Flowers 2 – 4 mm long and 2 – 4 mm wide, ellipsoid; pedicel 1 – 3 mm, the pedicel and calyx densely covered with dark or pale brown stellate scales. Calyx 0.5 – 1 mm long, cup-shaped, shallowly divided into 5 lobes. Petals 4, 5 or rarely 6, pale yellow or orange yellow, aestivation quincuncial or imbricate. Staminal tube 1.2 – 3 mm long, 1.5 – 3 mm wide, obovoid with the aperture 0.2 – 1 mm across; anthers 5, 1 – 1.3 mm long, 0.4 – 0.7 mm wide, ovoid, inserted a quarter up to half the way up the tube, either just protruding or included and visible through the aperture. Ovary 0.2 – 0.5 mm high, 0.3 – 1 mm wide, depressed-globose, densely covered with pale brown stellate hairs or scales; loculi 2, each containing 1 ovule; stigma 0.2 – 0.5 mm high, 0.3 – 0.8 mm wide, ovoid with two small apical lobes, dark blackish- brown when dry.

Infructescence either with several fruits, c. 7 cm long and 6.5 cm wide with the peduncle c. 2 cm or the fruits solitary with a peduncle up to 3 cm, the peduncle, rhachis and branches with indumentum like the inflorescence. Fruits 2–2.8 cm long, 1.3–2.5 cm wide, orange-brown or yellow, obovoid or ellipsoid, densely covered with reddish-brown and pale brown peltate scales which have a fimbriate margin, glabrescent. Loculi 2, each containing 1 seed which is enclosed in a brown, translucent aril.

DISTRIBUTION. New Guinea (including Bougainville), Solomon Islands, Australia (N. Queensland, Mount Lewis Range). Fig. 66.

ECOLOGY. Found in undergrowth of primary and secondary forest. Alt.: sea level to 500 m. Rather common.

VERNACULAR NAMES: Solomon Islands: M/Norasingwane, Ulukwalo (Kwara'ae).

Representative specimens. PAPUA NEW GUINEA. Milne Bay District: North Sagarai Valley, c. 60 m [10°22'S 150°15'E], fl., 31 May 1964, *Henty*

NGF 16878 (K!); Raba Raba Subdistrict, Junction Ugat and Mayu Rivers, near Mayu Is., 450 m [9°37'S 149°10'S], fr., 21 July 1972, *Streiman & Katik* NGF 34053 (K!). Bougainville, Buin, Karngu, fl., 30 Oct. 1930, *Kajewski* 2306 (BO!).

SOLOMON ISLANDS. Guadalcanal, Monitor Creek, 210 m, fr., 9 July 1965, *Corner* RSS 2112 (K!). N.E. Kolombangara, Kokove Area, 15 m, fl., 9 Jan. 1968, *Gafui & collectors* BSIP 7556 (K!).

AUSTRALIA. Queensland: 5 miles N. of crossing on Massey Creek, on road between Silver Plains Station and Rocky River [13°50'S 143°29'E], c. 70 m, fr., Oct. 1969, *Webb & Tracey* 9734 (BRI!); Between Massey Creek and Rocky River, fr., 20 Feb. 1980, *Hyland* 10296 (QRS!); 'Bunja' site, Mt Lewis, 1200 m, fl., 3 Oct. 1973, *Webb & Tracey* 13729 (BRI!, CANB!, QRS!); Main ridge, 1.6 km south of Mt Lewis, 70 m, fr., Oct. 1969, *Schodde* 4148 (CANB!).

A. brassii resembles *Aglaia odoratissima* except that the flowers are usually larger, and the staminal tube is obovoid with a pin-prick aperture rather than cup-shaped. *A. brassii* is confined to Australasia, where *A. odoratissima* does not occur.

65. Aglaia amplexicaulis A.C. Smith in Bulletin of the Bernice Bishop Museum, Honolulu 141: 78 (1936). Holotype: Fiji, Kandavu, in dense forest on hills above Namalata and Ngaloa Bays, 200 – 400 m, 16 Oct. 1933, *Smith* 156 (BISH; isotypes: BO!, GH!, K!, NY!, P!); A.C. Smith, Flora Vitiensis Nova 3: 544 (1985).

Tree 1 – 10 m in dense forest. Twigs densely covered with reddish-brown stellate scales.

Leaves simple, 3.5 – 25.5 cm long, 2 – 6.5 cm wide, sessile, elliptical, rounded at apex, deeply cordate and amplexicaul at base; with a few scales like those on the twigs on the midrib on the lower surface; veins (6 –)20 – 30 on each side of midrib, shallowly ascending at an angle of 70 – 80° to the midrib, curving upwards near the margin and anastomosing; some reticulation often continuing between the ends of the lateral veins and the margin of the leaf; some short intermediate lateral veins between the main laterals; midrib prominent and lateral veins subprominent on lower surface.

Flowers not seen. [According to A.C. Smith (1985) the inflorescences are borne in the axils of the leaves, are 4.5 cm long with few branches and few flowers; pedicel 2 – 3.5 mm; calyx lobes 0.8 – 1 mm, petals 2 – 2.2 mm long, 1.5 mm wide, yellow, with few scales on the outside; staminal tube c. 1 mm long, anthers c. 1 mm long, inserted on the margin of the tube].

Infructescence with up to 6 fruits, up to 7 cm long, peduncle up to 1.5 cm, Fruits up to 2 cm long and 1.5 cm wide, ellipsoid, indehiscent; pericarp thin dark reddish-brown, with a few stellate scales; loculi 2, each containing 1 seed. Fig. 70.

DISTRIBUTION. Fiji Islands only, only the type seen.
ECOLOGY. Forest. Alt. 200 to 600 m.

A. amplexicaulis is unlike any other species of *Aglaia* in that it has simple, sessile, deeply cordate leaves.

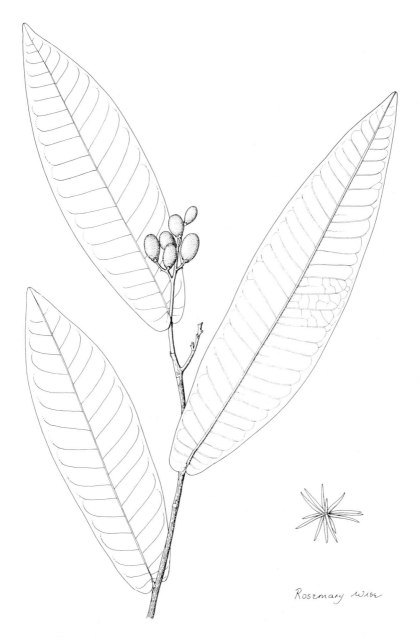

Fig. 70. *A. amplexicaulis*. Habit with infructescence x½. Stellate scale x30.

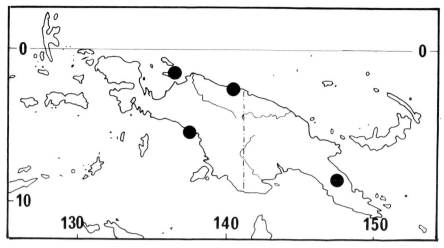

Fig. 71. Distribution of *A. puberulanthera*.

66. Aglaia puberulanthera C. de Candolle in Lorentz, Nova Guinea 8: 1013 (1914). Lectotype (designated here): Nova Guinea neerlandica septentrionalis, fluv. Tami sup., 50 m, fl., 8 July 1910, *Gjellerup* 279 (L!; isolectotypes: BO!, K!, U!).

Aglaia rubra Ridley in Hooker fil., Ic. Pl. t. 3052 (1916) et in Trans. Linn. Soc., Bot. 9: 26, t. 3052 (1916). Lectotype: Dutch New Guinea, Utakwa River to Mt Carstensz, 1912 – 13, *Boden Kloss* s.n. (K!; isolectotype: BM!).

Small tree up to 3 m. Twigs greyish-brown with numerous to densely covered with reddish-brown stellate hairs, interspersed with darker peltate scales.

Leaves simple, 5.5 – 27.5 cm long and 1.8 – 7 cm wide elliptical or obovate, or subcordate at the base, apex acuminate, the obtuse or acute acumen up to 1.8 cm long, with few to numerous stellate hairs and peltate scales on the midrib; veins 10 – 40 on each side of the midrib, ascending and curved upwards, anastomosing; almost sessile or with a petiole up to 1 cm.

Inflorescence 4 – 6.5 cm long, 2.5 – 4 cm wide; peduncle 5 – 15 mm, the peduncle, rhachis and branches densely covered with hairs and scales like those on the twigs. Flowers 2.5 mm long, 1.5 – 2.5 mm wide, obovoid, pedicel 1 – 2.5 mm. Calyx 0.5 mm, divided into 5 acute spreading lobes, with dark brown stellate scales on the outside. Corolla tube divided to about half way into 5 lobes, or petals 5, 2.5 mm long and 1.5 mm wide, elliptical, orange; aestivation quincuncial. Staminal tube c. 1.2 mm long and 1.2 mm wide, cup-shaped; anthers 5, c. 0.7 mm long, inserted on the margin of the tube, the outer surface of the anther continuous with the outer surface of the tube, densely covered with simple yellow hairs along the margins of the anthers. Ovary densely covered with stellate

scales, stigma ovoid with two apical lobes; the ovary and stigma together c. ⅓ the length of the staminal tube.
Fruits not seen.

DISTRIBUTION. New Guinea only. Fig. 71.
ECOLOGY. Found as understorey tree in primary and montane forest; on clay. Alt.: 80 to 750 m.
Representative specimen. PAPUA NEW GUINEA. Northern District, Kokoda Subdistrict, 3 km W. of Sisireta village [8°55′S 147°50′E], 180m, fl., 29 Sep. 1975, *Waikabu & Simaga* LAE 70263 (K!, L!).

A. puberulanthera has simple leaves with numerous lateral veins. The deep lobes of the staminal tube are densely covered with simple hairs, resembling those of *A. euryanthera*. There is a wide range in the number of lateral veins; in some specimens, including the type specimens of *A. rubra*, the lateral veins are numerous (17 – 40) and close together, in others, including the type specimens of *A. puberulanthera*, they are fewer (10 – 16).

67. Aglaia euryanthera Harms in Engl. Bot. Jahrb. 72: 171 (1942). Lectotype (designated here): Papua [Papua New Guinea] [Südöstliches Neu-Guinea] Lala River, 5500 ft [c. 1800 m], fl., 24 July 1935, *Carr* 13989 (BM!; isolectotypes: K!, L!, SING!).

Small tree 6 – 12 m high, with a narrowly spreading crown; bole up to 30 cm in circumference. Inner bark green; sapwood pale yellow, fibrous, with a strong odour. Twigs densely covered with dark reddish-brown or orange brown shiny peltate scales which have an irregular or fimbriate margin and dark centre and margin with a paler band in between. Leaves 15 – 49 cm long, 20 – 46 cm wide; petiole (3 –)4 – 13 cm, the petiole, rhachis and petiolules densely covered with scales like those on the twigs. Leaflets 1 – 5(– 7), the laterals subopposite, all 7 – 26 cm long, 3 – 10 cm wide, often pale green above when dry, usually elliptical, sometimes obovate, cuneate at the base, acuminate at apex, the usually obtuse, but sometimes acute, acumen often parallel-sided and 2 – 20 mm long, with numerous scales like those on the twigs on the midrib and few on the lateral veins below, with numerous dark reddish-brown peltate scales and usually some paler stellate scales in between, with numerous pits on the rest of that surface, veins 9 – 25 on each side of the midrib, ascending and curved upwards near the margin, not anastomosing, the midrib prominent and lateral veins subprominent below; petiolules 5 – 10 mm on lateral leaflets, 15 – 25 mm on terminal leaflet. Inflorescence 7 – 18 cm long and 5 – 28 cm wide; peduncle 1 – 3 cm, with scales like those on the twigs. Flower 2 – 4 mm long, 1.5 – 4 mm wide, ellipsoid; pedicels 1 – 3 mm, the pedicel and calyx with few to densely covered with scales like those on the twigs. Calyx 0.5 – 1 mm, cup-shaped with 5 rounded lobes. Petals 5 (or 6), elliptical, white or yellow. Staminal tube 1.5 – 2.5 mm long, 1 – 2 mm wide, deeply cup-shaped, anthers 5, 0.7 – 1.2 mm long and 0.5 – 0.8 mm wide, ovoid, inserted on the margin of the tube, the outer surface of the anther continuous with the outer surface of the tube, densely covered with simple yellow hairs along the margins of the anthers with pale yellow simple or stellate hairs

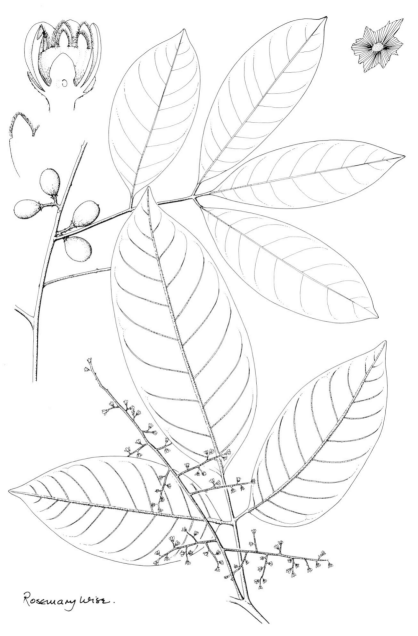

Fig. 72. *A. euryanthera*. Habit with inflorescence x1/2. Half flower, with detail of outer surface of staminal tube x10. Habit with infructescence x1/2. Peltate scale x70.

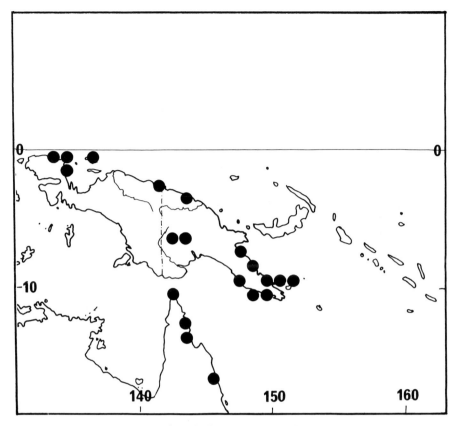

Fig. 73. Distribution of *A. euryanthera*.

on the lower part of the staminal tube inside. Ovary 0.3 – 0.6 mm long and 0.4 – 0.6 mm wide, depressed globose, densely covered with scales like those on the twigs; loculi 1 – 2, each containing one ovule; stigma 0.4 – 0.9 mm long and 0.4 – 1 mm wide, ovoid or obovoid and domed at the apex. Infructescence 3 – 11 cm long, with 1 or 2 fruits, sessile. Fruits 1.3 – 3.2 cm long, 1.4 – 2 cm wide, subglobose or narrowly ellipsoid and narrowed to a short stipe c. 5 mm long, orange, orange-brown or yellow densely covered with scales like those on the twigs on the outside; pericarp thin; fruit-stalks c. 1 cm. Loculi 2 each containing 0 – 1 seed; seed c. 2.5 cm long, 1 cm wide and 1 cm thick, surrounded by a thin dark translucent aril; testa brown or black. Fig. 72.

DISTRIBUTION. New Guinea, Australia. Fig. 73.

ECOLOGY. Found in primary forest, secondary forest, dry monsoon forest, gallery forest, forest on flood-plains, along rivers, along coast on red clay and sandy clay. Alt.: 5 to 2100 m. Found occasional to rather common. Seeds eaten by birds or rats.

VERNACULAR NAMES: New Guinea: Boewi (Hattam); Bokerar (Andjai); Kasia (Amele, Amele dialect); Mutturuma (Musa, Safia dialect); Tiri (Wapi, Miwaute dial.).

Representative specimens. IRIAN JAYA. S.E. West Irian, Bagampa, Maio River, fl., 30 April 1967, *Reksodihadjo* 217 (K!, L!).

PAPUA NEW GUINEA. Central District: Port Moresby Subdistrict, Astrolabe Range, S. of Sirinumu Dam and E. of Tupuseleia [9°33'S 147°26'E], 550 – 600 m, fr., 25 Aug. 1970, *Kanis* 1322 (K!, L!); c. 18 miles N. of Port Moresby, Rubulogo Creek, c. 40 m, fl., 6 April 1967, *Pullen* 6605 (K!, L!). Normanby Island, Waikaiuna, 20 m, fl., 14 May 1956, *Brass* 25863 (K!, L!) .

AUSTRALIA. Queensland: Lockerbie Scrub, fr., 15 Sep. 1985, *Williams* 85217 (BRI!); Iron Range, *Brass* 19048 (BRI!, CANB!, K!); Whitfield Spur, near Cairns, 250 m, fl., Oct. 1978, *Jaga* 83 (BRI!).

A. euryanthera is variable in the structure and distribution of the indumentum. The scales are usually mainly large and dark purple, but there may be some pale stellate scales interspersed and these are sometimes numerous on the lower leaflet surface. The leaves are often pale green above when dry.

The leaves and leaflet indumentum of *A. euryanthera* resembles those of *A. sapindina*, but the proportion of purple peltate scales on the midrib on the lower surface of the leaflets is usually greater in *A. euryanthera* than in *A. sapindina*, from which it is distinguished by the structure of the staminal tube. *A. sapindina* always has stellate scales on the inflorescence and infructescence whereas *A. euryanthera* has peltate scales.

68. Aglaia polyneura C. de Candolle in Lorentz, Nova Guinea 8: 1015 (1914). Lectotype (designated here): Nova Guinea, Septentr. Humboldt-Bai, 10 m, *Gjellerup* 415 (BO!; isolectotype: L!, U!).

Tree up to 10 m, shrubby or loosely branched. Twigs greyish-brown with numerous to densely covered with dark reddish-brown peltate scales which have a fimbriate margin.

Leaves 10 – 40 cm long, 12 – 42 cm wide, obovate in outline; petiole 3 – 10 cm long, the petiole rhachis and petiolules with numerous to densely covered with scales like those on the twigs. Leaflets 3 – 5, the laterals subopposite, all 8 – 26 cm long, 3 – 8 cm wide elliptical obovate or sometimes ovate; apex acuminate caudate with the obtuse acumen often parallel sided and 5 – 10 mm long, cuneate or sometimes rounded at the slightly asymmetrical base, with scales like those on the twigs, few to numerous on the midrib and scattered among numerous pale stellate scales or with numerous entire peltate scales on the surface below; veins 10 – 18 on each side of the midrib, ascending and curving upwards near the margin, nearly or quite anastomosing; midrib prominent, lateral veins subprominent on the lower surface; petiolules 2 – 6(– 9) mm. Male inflorescence up to 26 cm long and 22 cm wide, peduncle 1 – 5 cm; the peduncle, rhachis and branches with few to numerous reddish-brown peltate scales and stellate hairs. Flowers up to 1.7 m long and 2 mm wide, fragrant, depressed globose, pedicels up to 1.5 mm with occasional stellate scales. Calyx divided into 4 – 5 broad rounded lobes with fimbriate margins and with numerous stellate scales on the outside. Petals 4 or 5, yellow, aestivation imbricate or quincuncial. Staminal tube about 0.7 mm long, cup-shaped, shallowly lobed, with 5 small ovoid anthers, 0.3 mm long, inserted on the margin of the tube and pointing towards the centre of the flower. Ovary depressed globose with a ring of stellate scales; stigma ovoid; the ovary and stigma together c. $^1/_2$ the length of the staminal tube. Infructescence up to 6 cm long and 6 cm wide, sessile or with a peduncle up to 5 mm, peduncle rhachis and branches with numerous peltate and stellate scales like those on the inflorescence. Ripe fruits up to 2.5 cm long, 2 cm wide, subglobose, the pericarp soft, yellow, and with numerous stellate scales on the outside. Seeds 1 – 3, 10 – 12 mm long, 6 – 7 mm wide and c. 4 mm thick, brown, with a thin gelatinous aril.

DISTRIBUTION. New Guinea only.
ECOLOGY. Forest. Alt.: 20 to 1330 m.
VERNACULAR NAME : Dekemtom (Hattam).
Representative specimen. PAPUA NEW GUINEA. Morobe District, Bulolo, Manki Divide Range [7°10'S 146°40'E], 1200 m, fl., 10 June 1960, *Womersley & Thorne* (L!).

A. polyneura resembles *A. sapindina*, but the lateral veins are closer together and the leaves are brown when dry, whereas they are bluish-green or pale green in *A. sapindina*.

69. Aglaia sapindina (F. von Muell.) Harms in Engl. & Prantl, Nat. Pflanzenfam. 3(4): 298 (1896). Henty in Bot. Bull. Lae 12: 100, t. 60 (1980).

[*Celastrus micranthus* Roxb., Hort. Beng.: 86 (1814) nom. nud. *Celastrus micrantha* Roxb., Fl. Ind. 1: 625 (1832), non *Aglaia micrantha* Merrill (1905) (= *A. elliptica*). Type: Maluku, *Anon.* s.n. (L!).
Hearnia sapindina F. von Muell., Fragm. Phyt. Austr. 5: 55 (1865). Syntypes: Australia, Rockingham's Bay, *Dallachy* (BM!, BO!, G!, K!, L!, W!); C. de Candolle in A. & C. de Candolle, Monog. Phan. 1: 630 (1878).
Aglaiopsis glaucescens Miq., Ann. Mus. Bot. Lugd. Bat. 4: 58 (1868). Lectotype (designated here): Nov. Guinea, *Zippelius* (L!; isolectotypes: BO!, U!).
Hearnia glaucescens (Miq.) C. de Candolle in A. & C. de Candolle, Monog. Phan. 1: 631 (1878).
Hearnia glaucescens (Miq.) C. de Candolle var. *novaguineensis* C. de Candolle in A. & C. de Candolle, Monog. Phan. 1: 632 (1878). Lectotype (designated here): Nov. Guinea, *Zippelius* (U!).
Hearnia macrophylla C. de Candolle in A. & C. de Candolle, Monog. Phan. 1: 631 (1878). Lectotype (designated here): Moluccas [Maluku], *Lambert* (G- DC!).
Aglaia ermischii Warb. in Bot. Jahrb. 13: 345 (1891). Type: New Guinea. Deutsch Neu-Guinea, Sattelberg im Gipfelwald, ca. 3000 ft [c. 900 m], Unterholz, *O. Warburg* 20105 (BM!).
Aglaia hartmannii C. de Candolle in Bull. Herb. Boiss. Sér. II 3: 173 (1903). Lectotype (designated here): British New Guinea, fl., *Hartmann* 41 (G!).
Aglaia novaguineensis (C. de Candolle) C. de Candolle in Bull. Herb. Boiss. Sér. II 3: 173 (1903).

Aglaia gibbsiae C. de Candolle in L.S. Gibbs, Phytogeogr. & Fl. Arfak. Mts: 212 (1917). Lectotype (designated here): Dutch N.W. New Guinea, Manokoeari, track to Armbarni, 300 ft [c. 90 m], fl., Jan. 1914, *Gibbs* 6204 (G!; isolectotypes: BM!, K!, L!).

Aglaia roemeri C. de Candolle in Lorentz, Nova Guinea 8: 1015 (1914). Lectotype (designated here): Nova Guinea neerlandica meridionalis, ['Noordrivier' pr. 'Alkmaar'], fl., 29 Oct. 1909, *van Römer* 703 (L!).

Aglaia nudibacca C. de Candolle in Denkschr. Akad. Wiss. Wien Math.-Nat., 89: 566 (1913). Lectotype (designated here): Salomons-Insel [Solomon Islands], Insel Buka, fr., Sep. 1905, *Rechinger* 4414 (W!).

Aglaia rechingerae C. de Candolle in Denkschr. Akad. Wiss. Wien Math.-Nat. 89: 565 (1913). Lectotype (designated here): Salomons-Inseln, Insel Bougainville, Strandbaum in der Bucht von Kieta, fr., Sep. 1905, *Rechinger* 4067 (W!).

Aglaia brevipeduncula C. de Candolle in Lorentz, Nova Guinea 8: 1014 (1914). Lectotype (designated here): Nova Guinea, Septentrion. pr. lac. Sentani, 50 m, fl., 12 June 1911, *Gjellerup* 471 (U!; isolectotypes: BO!, K!, L!).

Aglaia gjellerupii C. de Candolle in Lorentz, Nova Guinea 8: 1014 (1914). Lectotype (designated here): Nova Guinea, Septentrion. pr. lac. Sentani, 50 m, fl., 12 June 1911, *Gjellerup* 472 (U!; isolectotypes: BO!, L!).

Aglaia porulifera C. de Candolle in Lorentz, Nova Guinea 8: 1016 (1914). Lectotype (designated here): New Guinea, Septentr. Bivak Hollandia, 5 m, fr., 5 May 1910, *Gjellerup* 120 (L!; isolectotypes: BO!, K!, U!).

Aglaia miquelii Merrill in Philipp. Jour. Sci., Bot., 11: 280 (1916). Lectotype (designated here): [Maluku], Amboina, [Batoe merah, c. 200 m], fl., July – Nov. 1913 [24 Aug. 1913], *Robinson* 1992 (L!; isolectotypes: K!, P!, PNH†).

Aglaia schlechteri Merrill & Perry, in Jour. Arn. Arb. 31: 318 (1940). Lectotype (designated here): Papua [Papua New Guinea N.E.], Kaiser-Wilhelmsland, Djamu, 300 m, fl., 4 Oct. 1907, *Schlechter* 16634 (A!; isolectotype: K!).

Aglaia rudolfi Harms in Engl. Bot. Jahrb. 72: 173 (1942). Lectotype (designated here): Dutch New Guinea, [Lorentz-Fluss] Kloofbiv. O., 30 m, fl., 18 Oct. 1912, *Pulle* 190 (L!; isolectotype: B†, BO!).

Aglaia hapalantha Harms in Engl. Bot. Jahrb. 72: 172 (1942). Lectotype (designated here): New Guinea, [N.E.], Morobe Province, Kalasa, 1700 ft [c. 500 m], fl., 3 Jan. 1938, *Clemens* 7937 (B!).

Aglaia clemensiae Merrill & Perry in Jour. Arn. Arb. 29: 158 (1948). Holotype: New Guinea [N.E.], Morobe District, vicinity of Kajabit Mission, 800 – 2000 ft [c. 250 – 600 m], fl., Aug. – Dec. 1939, *Clemens* 10691 (A! ; isotype: MICH!).

Small tree 4 – 12(– 30) m. Bole up to c. 22 cm in diameter. Bark greenish-brown, greyish-brown or reddish-brown, flaking in small scales, with scattered small black lenticels or longitudinal lines of reddish-brown lenticels; middle bark red or white, fibrous; inner bark pink or white, pale brown or white, fibrous; sapwood yellowish-brown, pink or white; either with no exudate or with some white latex.

Twigs slender, grey or greyish-green, with numerous or densely covered with pale brown or reddish-brown peltate scales which have a fimbriate margin, usually interspersed with very dark purplish-brown or blackish-brown peltate scales.

Leaves 21 – 65(– 95) cm long, 16 – 50 cm wide, obovate in outline; petiole 6 – 17.5(– 29) cm, the petiole, rhachis and petiolules with scales like those on the twigs. Leaflets (3 –)5 – 9, the laterals subopposite, all 7 – 26(– 45) cm long, 5 – 13.5(– 21) cm wide, elliptical, oblong, or obovate, pale bluish-green or yellowish-green when dry, apex acuminate, the obtuse acumen often parallel-sided and 5 – 10 mm, rounded or cuneate at the slightly asymmetrical base; veins 8 – 20 on each side of the midrib, ascending and curved upwards near the margin, nearly anastomosing; midrib prominent, lateral veins prominent or subprominent, reticulation just visible, with few to numerous scales like those on the twigs on the midrib and veins and sometimes few on the lower surface, occasionally interspersed with stellate hairs, with few to numerous dark or pale reddish-brown pits on the midrib and lower surface; petiolules (5 –)10 – 15(– 30) mm, up to 20(– 100) mm on the terminal leaflet.

Inflorescence 7 – 28 cm long, 2 – 24 cm wide, densely covered with reddish-brown stellate scales; sessile or with a short peduncle up to 5 mm, the peduncle, rhachis and branches with few to numerous peltate scales like those on the twigs, the distal branches and petiolules densely covered with stellate scales or hairs. Flowers 1 – 3 mm long, 1 – 3 mm wide, subglobose, fragrant; pedicels 0.5 – 1 mm. Calyx $1/4 - 1/2$ the length of the corolla with numerous reddish-brown stellate scales on the outer surface, divided almost to the base into 5 (or 6) subrotund lobes, obtuse at the apex, patent at anthesis. Petals 5, c. 0.6 mm long, white or yellow, elliptical or obovate-rotund, glabrous, aestivation quincuncial. Staminal tube less than $1/2$ the length of the corolla, shallowly cup-shaped with the apical margin incurved and shallowly 5- lobed; anthers c. 0.2 mm long and 0.2 mm wide, ovoid, with a pale margin, inserted just below and protruding through the aperture of the tube, pointing towards the centre of the flower. Ovary subglobose densely covered with stellate hairs; stigma ovoid with two small apical lobes; ovary and stigma together $1/2$ to as long as the staminal tube.

Infructescence 7 – 15 cm long, 3.5 – 10 cm wide; sessile or with a short peduncle up to 15 mm. Fruit 2 – 2.5 cm long, 1.5 – 2 cm in diameter, indehiscent, ellipsoid or subglobose, red, yellow or orange, the pericarp thin, soft, with numerous or densely covered with stellate scales, glabrescent. Loculi 2, each containing 0 or 1 seed; seed completely surrounded by an orange or yellow, translucent aril; testa brown; cotyledons green. Fig. 74.

DISTRIBUTION. Maluku, New Guinea, Bougainville, Australia. Fig. 75.

ECOLOGY. Found in riverine forest, secondary forest, swamps, alluvial flats, along the beach and paths, in complex mesophyll vine-forest (Queensland); on limestone, sandy clay with granite. Alt.: sea level to 3800 m. Common. In Papua New Guines, the twigs are sometimes inhabited by ants (e.g. *Hartley* 110312, NGF 28612, LAE 50327).

USES. Sometimes reaches timber size, but the sawdust may cause dermatitis (Henty, 1980).

VERNACULAR NAMES. Moluccas: Loeka Loeka Ma. New Guinea: A'usekumo, Awasekuma, Owasekuma (Orokaiva, Mumuni dial.); Bokrar

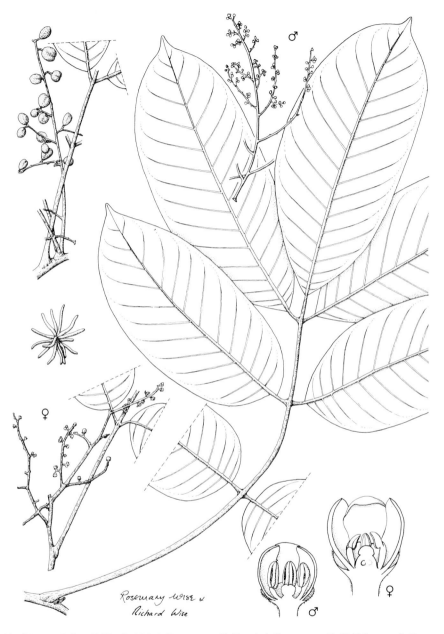

Fig. 74. *A. sapindina*. Habit x¹/₂. Male inflorescence x¹/₂. Female inflorescence x¹/₂. Half flowers x7. Young infructescence x¹/₂. Stellate hair x150.

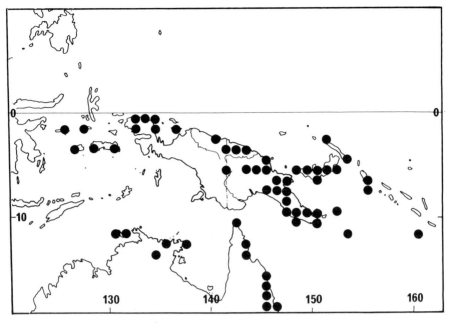

Fig. 75. Distribution of *A. sapindina*.

(Andjai); Buweh (Hattam); Lakora (W. Nanakai); Maineh (Orne, Kaiye dial.); Ornitoi (Warea); Rila Keia (Kulumo); Sa (Maibrat); Damrabuta, Kola, Naunehenehe. Solomon Is.: Ulukwalo (Kwara'ae); Pargoloi.

Representative specimens. MALUKU. Ceram, Batu Asa, sea level, fl., 18 Jan. 1918, *Kornassi* (Exp. Rutten) 842 (U!). Ambon, fr., 1859 – 1869, *de Vriese & Teijsmann* s.n. (L!).

PAPUA NEW GUINEA. Milne Bay District, Harada River, below Waigani Plantation, [10°20'N 150°17'E], c. 30 m, fr., 8 June 1964, *Henty* NGF 20516 (K!). Western Highlands District, Hagen Sub-district, Baiyer R. Bird Sanctuary [5°35'S 144°10'E], c. 1300 m, 17 July 1970, ♀ fl., *Stevens* LAE 50237 (K!). Morobe District: Lae, c. 10 m, fl., July 1944, *White, Dadswell & Smith* NGF 1573 (K!); Lae Sub-district, Lae Botanic Garden, [6°45'S 147°00'E], sea level, fl., 3 June 1970, *Millar* NGF 48562 (K!). Bougainville: c. 10 miles N. of Buin Patrol Post, nr Kugugai village, fr., 23 July 1964, *Schodde (& Craven)* 3648 (K!); c. 12 miles S. W. of Buin, nr Takuaka village, fr., 15 Sep. 1964, *Craven (& Schodde)* 476 (K!).

AUSTRALIA. Northern Territory: Melville Island, Garden Point, ♀ fl. & fr., 14 Sep. 1982, *Wightman* 191 (DNA!); Melville Island, Garden Point, fl., 15 Sep. 1977, *Must* 1638 (BRI!, CANB!, DNA!); Arnhem Land, 3 km N.E. of Port Bradshaw, fr., 10 Feb. 1988, *Wightman* 4118 (BRI!, DNA!). Queensland: Headwaters of Massey Creek, McIlwraith Range, fl., Oct. 1969, *Webb & Tracey* 9135 (BRI!); National Park Reserve 202, Lake Eacham, fr., 14 Feb. 1964, *Hyland & Volk* 3010 (BRI!, K!); Wongabel, S.F.R. 191, Cpt 2a, Exp. LT22, 780 m, 11 Jan. 1977, *Unwin* 168 (BRI!).

SOLOMON ISLANDS. N.W. Treasury Island, Kughala R., c. 90 m, yg fr., 24 April 1969, *Mauriasi & collectors* BSIP 14059 (K!, L!).

The leaves of *A. sapindina* have a characteristic mixture of peltate scales (some very dark purplish-brown) and stellate or fimbriate-peltate scales on the midrib below; with dark reddish-brown pits on lower leaflets surface and midrib; sometimes interspersed with these are pale brown stellate scales (e.g. *White* NGF 9622 (L!)). The scales on the distal branches of the inflorescence are always reddish-brown and stellate. The separation of *A. sapindina* from non-flowering specimens of *A. euryanthera* may be difficult, but the latter species usually has a greater proportion of very dark purple peltate scales on midrib and peltate scales rather than stellate scales on the inflorescence, infructescence and fruits.

70. Aglaia ceramica (Miq.) C.M. Pannell stat. nov.

Aglaia elliptica Blume var. *ceramica* Miq., Ann. Mus. Bot. Lugd. Bat. 4: 51 (1868). Lectotype (designated here): Moluccas, [Maluku], Ceram, yg fr., 1859 – 1860, *Teijsmann & de Vriese* (L!).

Small tree up to 3 m, with few branches; stem and twigs pale brown, densely covered with orange brown stellate hairs near the apex. Leaves up to 80 cm long and 70 cm wide; petiole 6 – 16 cm, the petiole, rhachis and branches densely covered with hairs like those on the twigs. Leaflets 7 – 9, 10.5 – 35 cm long, 5 – 14.5 cm wide, often bluish-green below when dry, ovate or elliptical, acuminate at apex, the obtuse acumen up to 15 mm long, rounded or cuneate at the base; with orange-brown stellate hairs numerous on or densely covering the midrib below, numerous on the lower leaflet surface and with paler stellate hairs which have fewer arms interspersed, glabrescent; veins 8 – 19 on each side of the midrib, ascending, curved upwards, not or just anastomosing, midrib prominent, lateral veins subprominent; petiolules up to 10 mm on lateral leaflets, up to 30 mm on the terminal leaflet. Male inflorescence in the axils of existing or fallen leaves, 16 – 24 cm long, 4 – 32 cm wide, with few branches and small ovate bracts c. 0.4 mm long. Flowers 1.5 mm long, 1.5 mm wide, pale yellow; pedicels 1.5 mm, the pedicel and calyx densely covered with stellate hairs or scales. Calyx 0.5 – 1 mm long, cup-shaped and divided into 5 lobes. Petals 5, aestivation quincuncial. Staminal tube 0.7 mm long, cup-shaped, aperture 0.8 mm wide, shallowly lobed; anthers 5, 0.6 mm long and 0.4 mm wide, inserted at the base of the tube. Ovary 0.2 mm high, 0.3 mm wide, subglobose, densely covered with stellate scales; stigma 0.2 mm high, 0.3 mm wide, ovoid with two small apical lobes. Female inflorescence in the axils of fallen leaves, up to 1 cm long, with branches from the base and up to 13 flowers. Female flowers 3 mm long, 2.8 mm wide; pedicels 0.5 mm, the pedicel and calyx densely covered with stellate hairs or scales. Calyx 1 mm long, cup-shaped and divided into 5 lobes. Petals 5, aestivation quincuncial, pale yellow. Staminal tube 1.5 mm long and 1.5 mm wide, cup-shaped with the margin incurved and shallowly lobed, with a few simple white hairs inside; anthers 5, 0.5 mm long, 0.6 mm wide, dark brown with pale margins when dry. Ovary 1 mm long and 1 mm wide, subglobose, densely covered with orange brown stellate hairs; loculi 2, each containing 1 ovule;

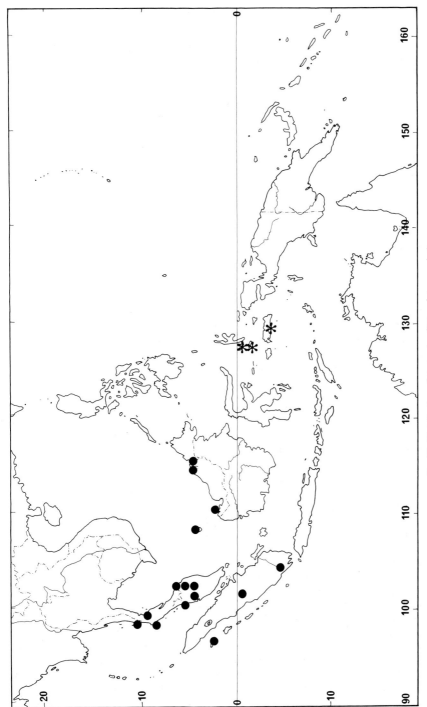

Fig. 76. Distribution of A. *tenuicaulis* ● and A. *ceramica* *.

stigma c. 0.5 mm long and 0.8 mm wide, ovoid, longitudinally ridged and with two small apical lobes.

Infructescence c. 2 cm long and 3 cm wide, (borne on the stem in the axils of about 6 fallen leaves); the first maturing at the same time as flowers on the more distal inflorescences. Fruit c. 1.5 cm long and 0.8 cm wide, yellow, obovoid, densely covered with orange brown peltate scales which have a fimbriate margin and stellate hairs on the outside; inner pericarp white. Seeds 3; aril white, edible and rather sweet; testa blackish-green.

DISTRIBUTION. Maluku only. Fig. 76.

ECOLOGY. Found in primary forest, sometimes on limestone. Alt.: 200 to 650 m.

VERNACULAR NAMES. Maluku: Langsat-langsat Hutan, Lansaba.

Representative specimens. MALUKU. Buru, W., Bara, Wae Dunn, base camp, 650 m: ♂ fl., 29 Nov. 1984, *Mogea* 5406 (L!) & ♀ fl. & fr., 29 Nov. 1984, *Mogea* 5381 (L!); N.W. Buru, Waeduna River, 350 – 400 m, yg fl., 26 Nov. 1984, *van Balgooy* 4922 (L!). Batjan, Gg Damar, Masoeroeng, 200 m, fl., 11 Aug. 1937, *Nedi* (exp. De Haan) 3 (L!). Obi, Laiwoei, Kg Baoe, fr., 3 Oct. 1937, *Nedi* (exp. De Haan) 519 (L!).

Aglaia ceramica resembles *Aglaia sapindina* in the colour of its leaves but differs in indumentum. It resembles *A. tenuicaulis* in size and its indumentum but the hairs on the lower leaflet surface are glabrescent, the inflorescences are much smaller and the infructescences are ramiflorous. The anthers in the female flowers of *Mogea* 5381, which has fruits on the same shoot as the flowers, dehisce when mature and contain a few pollen grains, some of which take up cotton blue stain.

71. Aglaia parksii A.C. Smith in Bulletin of the Torrey Botanical Club 70: 541 (1943). Holotype: Fiji, Viti Levu, Naitasiri, Tholo-i-suva, 200 m, yg fr., May – July 1927, *Parks* 20076 (BISH; isotypes: A!: photo at NY!, UC!, US!); A.C. Smith, Flora Vitiensis Nova 3: 554 (1985).

Tree up to 12 m. Twigs pale brown or greyish-brown, densely covered with orange-brown or reddish-brown stellate scales and hairs, some of which have long arms.

Leaf 38 – 60 cm long, 30 – 42 cm wide; petiole 10 – 18 m, the petiole, rhachis and petiolules with few to numerous hairs and scales like those on the twigs. Leaflets 9, the laterals subopposite, all 7 – 23 cm long, 3.5 – 7 cm wide, obovate or elliptical; rounded at the slightly asymmetrical base, the apex rounded or with a short broad rounded acumen up to 7 mm long; the midrib below and sometimes the veins with numerous to densely covered with hairs and scales like those on the twigs; veins 11 – 23 on each side of the midrib, ascending at an angle of 60 – 65° to the midrib, curving upwards near the margin and nearly or quite anastomosing, with some shorter lateral veins in between; petiolules 10 – 15 mm (– 20 mm on the terminal leaflet).

Flowers c. 2.5 mm long and 3 mm wide; pedicels 3 – 4.5 mm long. Calyx c. 1 mm long, cup-shaped, deeply divided into 5 lobes, densely covered with reddish-brown stellate scales on the outside. Petals 5, the exposed parts of the petals densely covered with scales like those on the

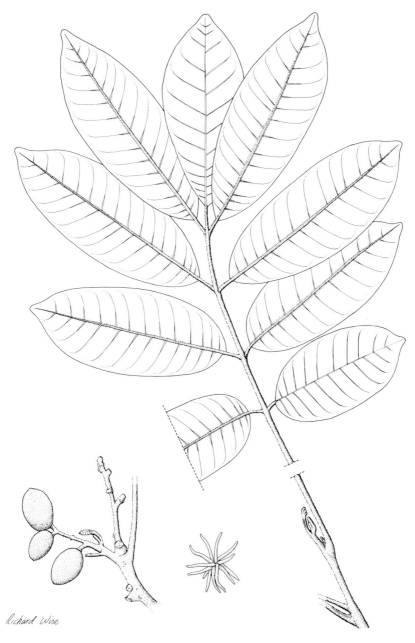

Fig. 77. *A. parksii*. Habit x½. Shoot subtending an infructescence x½. C. Stellate hair x100.

calyx, aestivation quincuncial. Staminal tube c. 1.5 mm long and 1.5 mm wide, up to 0.6 mm thick below the anthers, the aperture c. 1 mm across; anthers 5, c. 0.5 mm long and 0.4 mm wide, inserted about half way up the tube and just protruding. Ovary c. 0.6 mm long and 0.4 mm wide, columnar; locule ? 1; stigma c. 0.8 mm long and 0.3 mm wide, columnar, the apex flattened and protruding just beyond the anthers.

Infructescence 4 – 17 cm long, 4 – 7 cm wide; peduncle 0.5 – 5 cm, the peduncle, rhachis and branches with few to numerous hairs and scales like those on the twigs. Fruits 4 – 12 per infructescence, c. 2.7 cm long and 2.2 cm wide, dull orange or brown, ellipsoid, with or without a small beak, indehiscent; the pericarp thin, densely covered with orange brown or reddish-brown stellate scales and hairs; seeds 2. Fig. 77.

DISTRIBUTION. Papua New Guinea (Bougainville), Solomon Islands, Fiji.

ECOLOGY. Found in primary forest, well-drained or swampy. Alt.: 17 to 250 m.

VERNACULAR NAMES. Bougainville: Horomurini, Hurungnipira. Solomon Islands: Ulukwalo (Kwara'ae).

Representative specimens. PAPUA NEW GUINEA. Bougainville, 1 m, fl., 28 Sep. 1931, *Waterhouse* 553 (K!, L!).

SOLOMON ISLANDS. W. Choiseul, E. Mbirambira, fr., 17 Jan. 1970, *Gafui and collectors* BSIP 18860 (K!, L!). Shortland, S.W. Kaong Kopi River, fr., 25 Feb. 1969, *Runikera and collectors* BSIP 13093 (K!, L!).

FIJI. Viti Levu: Serua, hills north of Ngaloa, in drainage of Waininggere Creek, 30 – 150 m, fr., 19 Nov. – 3 Dec. 1953, *A.C. Smith* 9427 (L!, UC!); Namosi, hills bordering Wainavindrau Creek, in vicinity of Wainimakutu, 150 – 250 m, yg fr., 17 Sep. – 8 Oct. 1953, *A.C. Smith* 8856 (L!).

Aglaia parksii is recognised by the lateral veins joining the midrib at a wide angle of 60 – 65° and by the dense covering of stellate hairs or scales on the midrib of the leaflets below, sometimes with long-armed hairs interspersed and sometimes also along the veins. This species resembles *A. basiphylla*, but is more robust; the leaves and leaflets are larger and none are basal; the indumentum on the midrib is similar. *A. parksii* is also similar to *A. elliptica* and *A. conferta* except that it has a blunt leaflet apex and spreading veins and the fruit is narrowly ellipsoid in shape. *A. parksii* resembles *Aglaia meridionalis* which belongs to section *Amoora*, but differs in the structure of the indumentum. Similar specimens from the Solomon Islands belong to sect. *Aglaia*, they are included in this species, but further investigation in the field is required to establish whether the Bougainville and Solomon Islands plants belong to the same species as the Fiji one.

72. Aglaia subminutiflora C. de Candolle in Bull. Herb. Boiss. Sér. II 3: 175 (1903). Holotype: S.E., New Guinea, [meridionali-orientali], fl., *Forbes* 51 (G!).

Aglaia chalmersi C. de Candolle in Bull. Herb. Boiss. Sér. II 3: 173 (1903). Lectotype (designated here): New Guinea, fl., 1885, *Rev. J. Chalmers* in h. Cand. (G!).

Aglaia edelfeldti C. de Candolle in Bull. Herb. Boiss. Sér. II 3: 174 (1903). Holotype: [New Guinea], near Port Moresby, fr., *Edelfeldt* 33 (G!).

Aglaia stellipila C. de Candolle in Lorentz, Nova Guinea 8: 425 (1910). Lectotype (designated here): Nov. Guinea, Noordkust, [Irian Jaya, N. coast], *Atasrip* (Exp. *Wichmann*) 25 (BO!).

Aglaia parvifoliola C. de Candolle in Lorentz, Nova Guinea 8: 1017 (1914). Lectotype (designated here): Nova Guinea neerlandica meridionalis [Irian Jaya], ['Noordrivier' pr. 'Alkmaar'], fr., 7 Oct. 1909, *van Römer* 464 (L!).

Aglaia exigua Merrill & Perry in Jour. Arn. Arb. 21: 321 (1940). Lectotype (designated here): Papua [Papua New Guinea, N.E.], Kaiser-Wilhemsland, Malia, 150 m, fl., 15 Oct. 1908, *Schlechter* 18508 (A!).

Aglaia carrii Harms in Engl. Bot. Jahrb. 72: 163 (1942). Lectotype (designated here): Papua, [New Guinea. S.E.], Veiya, sea level, fl., 13 March 1935, *Carr* 11711 (L!; isolectotype: B†, BM!, K!, NY!).

Small tree to 6(– 15 m). Bole to 1.5 m, branches ascending, bark pale brown. Twigs pale pinkish-brown, densely covered with dark orange or reddish-brown stellate hairs and scales.

Leaves 12 – 35 cm long, 11 – 34 cm wide, obovate in outline; petiole 1 – 9 cm long, petiole, rhachis and petiolules with few to densely covered with hairs and scales like those on the twigs. Leaflets 3 – 11, the laterals usually subopposite, sometimes alternate, all 2.5 – 18 cm long 0.5 – 5 cm wide, obovate or elliptical, occasionally ovate, acuminate or caudate at the apex, the obtuse acumen, often parallel sided and to 15 cm long, cuneate or rounded at the asymmetrical base, with few to densely covered with peltate and stellate scales or hairs on the midrib and few to numerous on lateral veins and few or occasionally numerous on the rest of the surface, sometimes with numerous faint pits on the lower surface; veins 7 – 12 on each side of the midrib, ascending, curved upwards near the margin, not anastomosing, the midrib prominent below, lateral veins subprominent or not prominent; petiolules 1 – 10(– 16) mm.

Inflorescence 5.5 – 17 cm long, to 14 cm wide, peduncle 1 – 25 cm long, the peduncle, rhachis and branches densely covered with stellate hairs and scales. Flowers 1 – 2 mm long, 1 – 2 mm wide, subglobose, whitish-yellow to ochre yellow; subsessile or with a pedicel to 2 mm, the pedicel and calyx densely covered with pale orange-brown stellate hairs or scales. Calyx 0.5 – 1 mm long, deeply divided into 5 broad, ovate, obtuse lobes. Corolla divided almost to the base into 5 obovate lobes, aestivation quincuncial. Staminal tube 1(– 1.2) mm long, c. 1 mm wide, cup-shaped or obovoid, the aperture (0.4 –)1 mm across, entire or shallowly lobed, without hairs or with numerous pale yellow simple or stellate hairs inside; anthers 0.3 – 0.6 mm long, 0.3 – 0.5 mm wide, ovoid, brown with a pale yellow margin when dry, without hairs or with a few simple hairs, sometimes with tufts of simple white hairs at the apex, either inserted just inside the margin of the tube, protruding and pointing towards the centre of the flower or inserted c. $^1/_2$ way up the tube and just protruding, the staminal tube thickened below the anthers. Ovary (0.2 –)0.4 – 0.5 mm long, (0.2 –)0.5 mm wide, depressed globose, densely covered with pale brown or golden stellate scales or hairs; loculi 1 or 2, each containing 1 ovule; stigma (0.2 –)0.6 mm long, (0.2 –)0.4 – 0.5 mm wide, depressed-ovoid with a central apical depression.

Infructescence to 11 cm, peduncle to 6.5 few to numerous hairs or

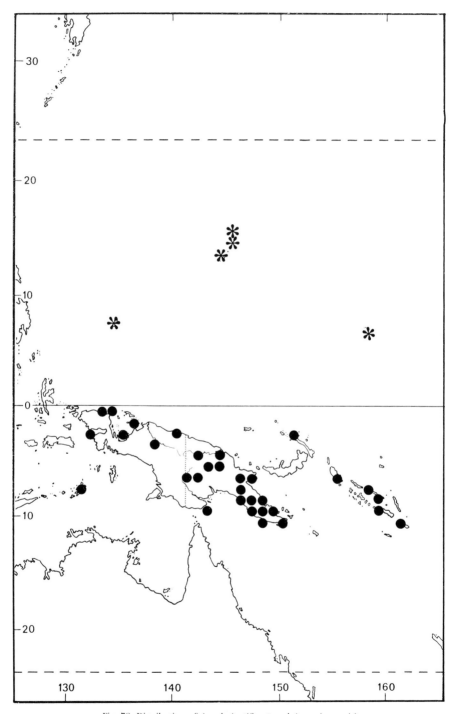

Fig. 78. Distribution of *A. subminutiflora*● and *A. mariannensis**.

scales. Fruits 1.5 – 2.5 cm long, 1.2 – 2.7 cm wide, obovoid, brown, orange or yellow, with few to densely covered with scales on the outside; loculi 2, each containing one seed.

DISTRIBUTION. New Guinea, New Ireland, ? Solomon Islands. Fig. 78.

ECOLOGY. Understorey tree found in sometimes inundated primary forest, secondary forest, riverine forest, *Castanopsis* forest; on sandy clay or basalt. Alt.: sea level to 1920 m. Frequent.

VERNACULAR NAMES: Maluku: Langsa Oetan. New Guinea: Bengraam (Hattam); Bina (Minufia, Kabubu dial.); Bufara (Onjob, Naukwate dial.); Djomo (Orokaiva, Mumuni dial.); Masambre (Biak); Miya (Wagu); Palya (Enga); Piek (Waskuk). Solomon Islands: Ulukwalu (Kwara'ae).

Representative specimens. PAPUA NEW GUINEA. Central Division, Brown River, 1 chain S.W. of 10 acre block, fl., 29 April 1955, *Gray* NGF 7158 (K!). Morobe District: about 12 miles N. of Lae, S. of Busu River, 90 m, fl., 24 Dec. 1962, *Hartley* TGH 11043 (K!); c. 18 miles W. of Lae, Oomsis Creek, 90 m, fl., 4 May 1964, *Hartley* TGH 13076 (K!, L!). Central District, Port Moresby Subdistrict, ridge S.W. of Efogi Village [9°10'S 147°39'E], 1370 m, fr., 25 Sep. 1973, *Foreman et al* LAE 52498 (K!). New Ireland District, inland from Lavongai [2°42'S 151°03'E], 15 – 30 m, fr., 30 Jan. 1967, *Coode & Cropley* NGF 29669 (K!, L!).

The following specimens may belong to *A. subminutiflora*, but these and other specimens from the Solomon Islands seem to be intermediate between this species and *A. samoensis*. SOLOMON ISLANDS. San Cristobal, Wairaha River, 5 miles from N. coast, c. 240 m, fl., 11 May 1964, *Whitmore* BSIP 4263 (K!, L!); San Cristobal, Wairaha River, 5 miles from N. coast, fr., 14 May 1964, *Whitmore* BSIP 4356 (K!, L!).

Aglaia subminutiflora and *A. cuspidata* are similar to *A. basiphylla* from Fiji. The latter species has a blunt leaflet apex and spreading veins. In all three species, the indumentum may consist of either hairs or scales or a mixture of scales and long-armed hairs. Specimens from the Solomon Islands are intermediate between *A. subminutiflora* and *A. samoensis*.

73. Aglaia basiphylla A. Gray, United States Exploring Expedition 1838 – 42, Botany Phanerogamia 1: 237. Lectotype (designated here): Fiji, Ovalau, (Herbarium of the) U.S. Exploring Expedition (under the command of Captain Wilkes) s.n. (US!; isolectotypes: A!, GH!, P!); C. de Candolle in A. & C. de Candolle, Monog. Phan. 1: 613 (1878); A.C. Smith, Flora Vitiensis Nova 3: 550 (1985).

Aglaia greenwoodii A.C.Smith in Bulletin of the Bernice Bishop Museum, Honolulu 141: 79, t. 40 (1936). Holotype: Fiji, Vanua Levu, Wainikoro, fl., 25 Feb. 1935, *Greenwood* 500A (K!; isotypes: A, K!, NY!); A.C. Smith, Flora Vitiensis Nova 3: 551 (1985).

Aglaia elegans Gillespie in Bulletin of the Bernice Bishop Museum, Honolulu 83: 11 (1931). Holotype: Fiji, Viti Levu, Naitasiri Province, woods near road beyond Tamavua village, 7 miles from Suva, 150 m, yg fr., 6 Aug. 1927, *Gillespie* 2005 (BISH; isotypes: A!, K!, NY! UC!, US!); A.C. Smith, Flora Vitiensis Nova 3: 549 (1985).

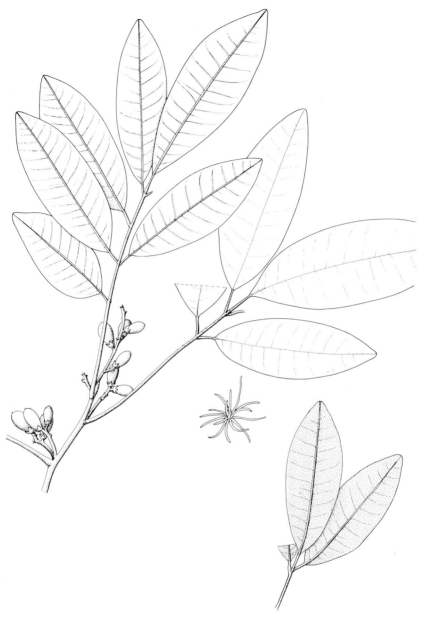

Fig. 79. *A. basiphylla*. Habit with infructescences x½. Undersurface of leaflets x½. Stellate hair x40.

Aglaia venusta A. C. Smith in Contrib. U.S. Nat. Herb. 30: 492 (1952). Holotype: Fiji, Vanua Levu, Thakaundrove, southwestern slope of Mount Mbatini, 300 – 700 m, fl., 28, 29 Nov. 1933, *A.C. Smith* 616 (U.S. 1676177) (US!; isotypes: A!, NY!, GH!, K!, P!); A.C. Smith, Flora Vitiensis Nova 3: 550 (1985).

Tree 3 – 12 m high. Twig apices densely covered with reddish-brown peltate scales which have a fimbriate margin, these sometimes concealed by a dense covering of reddish-brown stellate hairs which have long arms and are often deciduous.

Leaves up to 20 cm long; usually with a petiole up to 5.5 cm, occasionally sessile; the petiole rhachis and petiolules densely covered with scales like those on the twigs. Leaflets 5 – 9, the laterals subopposite, all 2.5 – 16 cm long, 1.5 – 5 cm wide, elliptical or obovate, sometimes ovate; rounded or cuneate at the asymmetrical base, the apex rounded or with a short broad rounded acumen; veins 8 – 14 on each side of the midrib, ascending at an angle of 65 – 80° to the midrib, curving upwards near the margin and anastomosing; sometimes with short intermediate lateral veins; the midrib below with numerous to densely covered with scales like those on the twigs and some stellate hairs which are often deciduous, the upper and lower surface with few stellate hairs and numerous tiny pits; the young leaves densely covered with stellate hairs and scales on both surfaces; petiolules 3 – 10 mm on lateral leaflets, up to 15 mm on terminal leaflet.

Flower buds 2 mm long, 2 mm wide; pedicel 1.5 mm; calyx deeply divided into 5 rounded lobes; the pedicel and calyx densely covered with stellate hairs; staminal tube shallowly cup-shaped; anthers 5. Stigma obovoid, nearly reaching the apices of the anthers, longitudinally ridged from the adpressed anthers, truncate at apex, the apex flat with a small central apical depression.

Infructescence 7.5 cm long, 4 cm wide; peduncle 1 cm, the peduncle, rhachis and branches densely covered with scales like those on the twigs; young fruits 2 cm long, 0.5 cm wide, obovoid, reddish-orange, longitudinally wrinkled with an enlarged calyx 4 mm long and often a small beak at the apex; the pericarp densely covered with dark reddish-brown stellate scales; fruit-stalks 1 – 4 mm. Fig. 79.

DISTRIBUTION. Fiji only.
ECOLOGY: Found in primary and creek forest. Alt.: 50 to 1075 m.
USES. Young trunks are used for spears, wood also used for house building.
VERNACULAR NAMES. Fiji: Kula, Tombuthe (Ra), Waithavuthavu.
Representative specimens. FIJI. Viti Levu: Lautoka, Mt Evans, c. 1050 m, yg fr., 18 Aug. 194-, *Greenwood* 1142 (UC!); Namosi, northern slopes of Korombasambasanga Range, in drainage of Wainavindrau Creek, 450 – 600 m, 28 Sep. 1953, fr., *A.C. Smith* 8713 (L!).

The apparent variability in indumentum is probably due to stellate hairs and peltate scales occurring in different proportions on different plants, but intermediates occur between the two extremes, making it difficult to recognise separate species. The sessile leaves found in the type of *A. basiphylla* do not seem to justify recognition of a separate species.

Fig. 80. *A. evansensis*. Habit x½. Habit with fruit x½. Seed x1. Stellate hair x100.

74. Aglaia evansensis A.C. Smith in Contributions from the U.S. National Herbarium 30: 497 (1952). Holotype: Fiji, Viti Levu, Mba, Eastern slopes of Mt Koroyanitu, Mt Evans Range, 950 – 1050 m, yg fr., 1 or 2 May, 1947, *Smith* 4152 (A!; isotypes: K!, L!, NY!, P!, US!); A.C. Smith, Flora Vitiensis Nova 3: 546 (1985).

Slender shrub or tree up to 8 m. Twigs pale brown, longitudinally wrinkled, with few to numerous orange brown stellate scales or peltate scales with a fimbriate margin and occasionally a few stellate hairs. Leaves simple or with 3(– 7) leaflets; when trifoliolate or imparipinnate, the laterals subopposite and the basal pair are at base of petiole and sub-sessile; with scales like those on the twigs few to numerous on the rhachis and petiolules terminal leaflet 4.5 – 13 cm long, 2 – 5 cm wide, elliptical or sometimes ovate, rounded at the apex or with a short broad round acumen, cuneate at the base, veins 9 – 13 on each side of the midrib, petiolules 10 – 20 mm; basal pair 0.5 – 3.5 cm long, 0.5 – 2 cm wide, ovate or subrotund, rounded at the apex or with a short broad round acumen, cuneate or subcordate at the base, veins 3 – 7 on each side of the midrib, petiolule 1 – 3 mm; remaining leaflets 1.5 – 6.5 cm long, 10 – 2.5 cm wide, elliptical or ovate, rounded at the apex or with a short broad round acumen, cuneate at the base, veins 6 – 9 on each side of the midrib, petiolule 1 – 3 mm; all leaflets pale green when dry, with a few scales like those on the twigs on the midrib below; veins ascending at 65 – 70° to the midrib, curved upwards near the margin and nearly or quite anastomosing and with shorter lateral veins in between, the midrib prominent and the lateral veins subprominent on the lower surface. Flowers not seen.

Infructescence up to 3.5 cm long. Fruit c. 2 cm long and 1.5 cm wide, ellipsoid, indehiscent; pericarp thin, soft, densely covered with compact reddish-brown stellate and peltate scales and a few paler stellate scales; one seed. Fig. 80.

DISTRIBUTION. Fiji Islands, known only from the isolated Mt Evans range in N.W. Viti Levu (Smith, 1985).

ECOLOGY. Forest. Alt.: 900 to 1180 m.

Representative specimens. FIJI. Viti Levu, Mba, slopes of Mt Nairosa, eastern flank of Mt Evans Range, 700 – 1050 m, fr., 26 April – 14 May 1947, *A.C. Smith* 4080 (A!); Viti Levu, Lautoka, Mt Evans, c. 1180 m, fr., Oct. 1942, *Greenwood* 1072 (A!).

A. evansensis resembles *A.basiphylla* in having a basal pair of leaflets, but it differs in its indumentum.

75. Aglaia subsessilis C.M. Pannell spec. nova inter species *Aglaiae singularis* ob folia paene subsessilia, quorum bina ima foliola parva sunt et paene rotunda. (Solum duae aliae species *Aglaiae*, *Aglaia basiphylla* A. Gray et *Aglaia evansensis* A.C. Smith, ambae ad insulam Fiji limitatae, possunt habere folia omnino vel paene subsessilia). *Aglaia subsessilis* imitatur *Aglaiam leucophyllam* King colore pallide flavovirente foliolorum siccorum, *Aglaiam ellipticam* Blume indumento. Fructus anguste ellipsoideus, longitudinaliter porcatus. Holotype. Malaysia, Sabah, Tongod District, Pinangah, Kg Saguan, 120 m, fl., 9 July 1981, *Sundaling* SAN 93651 (FHO!).

Tree up to 15 m. Outer bark thin, dark reddish-brown or white with black patches, scaly, with brown lenticels; inner bark reddish-brown or pale yellow, laminated; cambium pale yellow; sapwood pale purple, white or red and white; latex white. Twigs greyish-brown with numerous to densely covered with orange-brown stellate hairs.

Leaves trifoliolate or imparipinnate, up to 21 cm long and 32 cm wide; petiole 3 – 10(– 30) mm, the petiole, rhachis and petiolules with numerous to densely covered with hairs like those on the twigs. Leaflets 3 – 5, the laterals opposite, differing markedly in size, all pale yellowish-green when dry, with numerous orange-brown pits on the lower surface of the leaflets and the midrib on the lower surface densely covered with orange-brown stellate hairs; the lateral veins ascending, curved upwards near the margin and anastomosing; the midrib and lateral veins subprominent on lower surface. The basal pair of leaflets 2.5 – 8 cm long and 1.5 – 5 cm wide, ovate, acuminate at the apex, with the obtuse acumen parallel-sided and up to 10 mm long, rounded at the base, veins 5 – 8 on each side of the midrib, petiolules c. 2 mm; the remaining lateral leaflets, if present, 11 – 18 cm long, 4.5 – 6.5 cm wide; the terminal leaflet up to 29 cm long and 13 cm wide; these lateral and the terminal leaflets obovate, caudate at the apex, the obtuse acumen parallel-sided and up to 20(– 25) mm long, cuneate or rounded at the asymmetrical base; veins 10 – 14(– 17) on each side of the midrib; petiolules up to 3 mm on the lateral leaflets, up to 14 mm on the terminal leaflet.

Inflorescence 21 – 26 cm long, 15 – 28 cm wide; peduncle 6 – 8.5 cm, the peduncle, rhachis and branches with numerous to densely packed orange-brown stellate hairs. Flowers (not known whether male or female) 0.5 – 1.2 mm long, 0.5 – 1.2 mm wide, yellow, fragrant; pedicels c. 0.2 – 0.8(– 10) mm long, the pedicels and calyx with a few orange-brown stellate hairs or scales. Petals 5; aestivation quincuncial. Staminal tube up to 0.5 mm high, cup-shaped, the margin shallowly lobed; anthers 5, up to 0.3 mm long and 0.3 mm wide, inserted just inside the margin and pointing towards the centre of the flower. Ovary subglobose, densely covered with pale yellow stellate hairs; locule 1, containing 1 ovule; stigma ovoid, pale brown with two minute dark brown apical lobes.

Infructescence up to 14 cm long; peduncle up to 8 cm, the peduncle, rhachis and branches with numerous to densely packed hairs like those on the twigs. Fruits 1 or 2, c. 6.5 cm long and 2.5 cm wide, red when young, yellow when ripe, narrowly ellipsoid, with a short stipe up to 5 mm long and a beak up to 5 mm long; the pericarp thin and leathery when dry, with 10 longitudinal ridges, along two of which the pericarp splits when dry, densely covered with compact reddish-brown stellate hairs on the outer surface. Locule 1, containing one seed; the seed 4 – 4.2 cm long, 1.2 – 1.5 cm wide, and 0.6 – 0.7 mm thick, ellipsoid, flattened on the raphe side, the testa with a conspicuous network of veins.

DISTRIBUTION. Only known from Borneo (Sarawak and Sabah). Fig. 81.
ECOLOGY. Found in primary forest. Alt.: 430 to 830 m.
VERNACULAR NAMES. Lantupak (Dusun Kinabatangan).
Representative specimens. SARAWAK. 1st Division, Kuching, mile 38 Padawan road, Bukit Manok [1°N 110°E], fl., 7 Feb. 1984, *Mamit* S 41077

76. elliptica

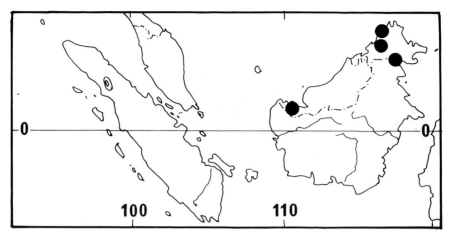

Fig. 81. Distribution of *A. subsessilis*.

(K!). SABAH. Tawau District: Ulu Imbak, Luasang, fr., 11 Aug. 1978, *Philip & Sumbing* SAN 89002 (L!); Path to Hot spring, Maria Road, fl., 22 May 1962, *Singh & Aban* SAN 30091 (FHO!, L!, SAN!). Jesselton, Sg. Arawan camp, fl., 6 Dec. 1962, *Madani* SAN 33158 (L!). Mt Kinabalu, Eastern Shoulder [6°05′N 116°36 – 40′E], c. 750 m, fl., 13 April 1961, *Chew, Corner & Stainton* RSNB 60 (FHO!, L!, SAN!). Lamag District, near Lake on Gunang Lotung, S.E. of Inarat, 400 m, fl., 10 May 1976, *Cockburn* SAN 83157 (L!).

Aglaia subsessilis is unusual in having almost sessile leaves, in which the basal pair of leaflets are small and almost round in shape. *A. basiphylla* and *A. evansensis*, both of which are confined to Fiji, are the only other species of *Aglaia* in which subsessile leaves are found. *A. subsessilis* resembles *A. leucophylla* in the pale yellowish-green colour of the leaflets when dry and *A. elliptica* in its indumentum. The fruit is narrowly ellipsoid and longitudinally ridged.

The long, narrow fruit of *A. subsessilis* resembles that of the uncommon Bornean variant of *A. simplicifolia* which was treated as a separate species, *A. sterculioides* by Kostermans. In both of these the pericarp splits open in herbarium specimens; although whether or not the ripe fruit dehisces needs to be confirmed in the field.

76. Aglaia elliptica Blume, Bijdr.: 171 (1825). Type: Java [ad montem Parang in Provincia Tjanjor], fl., Aug., *Anon.* 1367 (L!); Miq., Ann. Mus. Bot. Lugd. Bat. 4: 50 (1868); Koorders & Valeton, Atlas der Baumarten von Java, 1: t. 15 (1913); Backer and Bakhuizen, Fl. Java 2: 126 (1965); Pannell in Ng, Tree Flora of Malaya 4: 214 (1989).

Aglaia ovata Teijsm. & Binn. in Nat. Tijdschr. Ned. Ind. 27: 43 (1846).
    Type: [Java, Mount Salak, Saliem] *Anon.* s.n. (U!).
Aglaia inaequalis [inaequale] Teijsm. & Binn. in Nat. Tijdschr. Ned. Ind . 2:

Section Aglaia

305 (1851). Types: [Duizend-gebergte, Cult. Hort. Bog., III-B-42] (L!, U!); Nederl. Kruidk. Arch. 3: 409 (1885) (II(i) Add.).
Milnea lancifolia Hooker fil., in Trans. Linn. Soc. 23: 165 (1860). Lectotype (designated here): Borneo, [North coast], fl., *Lowe* s.n. (K!).
Milnea dulcis Teijsm. & Binn. in Cat. Hort. Bog.: 211 (1866) nomen, in Nat. Tijdschr. Ned. Ind. 29: 253 (1867). Lectotype (designated here): Celebes [Sulawesi, Manado Province], *Teysmann* (U!).
Aglaia reinwardtii Miq., Ann. Mus. Bot. Lugd. Bat. 4: 51 (1868). Lectotype (designated here): Celebes [Sulawesi], prope Pogowotto, Septembri 1821, *Reinwardt* s.n.; Saparoea, Julio 1821, *Anon.* s.n. (L! (both labels on same specimen), U!); C. de Candolle in A. & C. de Candolle, Monog. Phan. 1: 614 (1878).
Aglaiopsis lancifolia (Hooker fil.) Miq., Ann. Mus. Bot. Lugd. Bat. 4: 59 (1868).
Aglaia rufa Miq. var. *celebica* Miq., Ann. Mus. Bot. Lugd. Bat. 4: 49 (1868). Lectotype (designated here): Celebes [Sulawesi], [near Rottahoe, Oct. 1840], *Forsten* 460 (L!, isolectotype: U!).
Hearnia elliptica (Blume) C. de Candolle in A. & C. de Candolle, Monog. Phan. 1: 628 (1878). *Hearnia lancifolia* (Hooker fil) C. de Candolle in A. & C. de Candolle, Monog. Phan. 1: 630 (1878).
Hearnia villosa C. de Candolle in A. & C. de Candolle, Monog. Phan. 1: 632 (1878). Lectotype (designated here): Borneo, Sarawak, *Beccari* 3944 (K!; isolectotype: P!).
Aglaia lancifolia (Hooker fil.) Harms in Engl. & Prantl Pflanzenf. 3 (4) : 298, t. 163 M & N (1896); van Steenis, Rheophytes of the World: 289 (1981).
Aglaia menadonensis Koord. in Meded. 'S Lands Plantent. 19: 381, 635 (1898). Lectotype (designated here): Celebes, N., Minahassa, *Koorders* 17902β (L!).
Aglaia stapfii Koord. in Meded. 'S Lands Plantent. 19: 383, 635 (1898). Lectotype (designated here): Celebes [Sulawesi], Minahassa, Menado, fr., 23 April 1895, *Koorders* 17888β (BO!; isolectotype: L!).
Aglaia oxypetala Valet. in Icon. Bogor.: t. 86 (1901). Lectotype (designated here): Cult. in Hort. Bogor, III-B-444, [III-B-44, origin: Sulawesi] (BO!; isolectotypes: BM!, K!, L!, P!).
? *Aglaia harmsiana* Perkins in Notizbl. Königl. Bot. Gart. & Mus. Berlin, 32: 78 (1903). Type: Philippines, Mindanao, Davao District, Libulan, July, *Warburg* 14271 (B†).
Aglaia micrantha Merrill in Philipp. Gov. Lab. Bur. Bull. 29: 22 (1905). Lectotype (designated here): Philippines, Luzon, Province of Bataan, Lamao River, Mt Mariveles, fl., July 1904, *H.N. Whitford* 477 (US!; isolectotypes: K!, PNH†).
Aglaia pauciflora Merrill in Philipp. Gov. Lab. Bur. Bull. 35: 31 (1906). Lectotype (designated here): Philippines, Island of Luzon, Province Bataan, Mt Mariveles, fr., Nov. 1904, *Elmer* 6699 (K!; isolectotypes: E!, G!, PNH†).
Aglaia apoana Merrill in Philipp. Gov. Lab. Bur. Bull. 35: 30 (1906). Lectotype (designated here). Philippines, Mindanao, Davao District, Mt Apo, Oct 1904, *Copeland* s.n. (G!; isolectotype: PNH†).
Aglaia langlassei C. de Candolle in Ann. Conserv. & Jard. Bot. Genève 10: 151 (1907). Holotype: Philippines, fl., 1894, *Langlassé* (G-BOIS!).
Aglaia palawanensis Merrill in Philipp. Jour. Sci. 3: 235 (1908). Lectotype

(designated here): Philippines, Island of Palawan, fl., March – April 1906, *Foxworthy* Bur. Sci. 689 (US!; isolectotypes: K!, PNH†).
Aglaia trunciflora Merrill, in Philipp. Jour. Sci., Bot. 1914, 9: 303 (1915). Lectotype (designated here): Philippines, Leyte, Dagami, fr., Aug. 1912, *Ramos* Bur. Sci. 15232 (US!; isolectotypes: BM!, K!, L!, PNH†). ?
Aglaia lagunensis Merrill in Philipp. Journ. Sci., Bot. 1914, 9: 537 (1915). Syntypes: Philippines, Luzon, Province of Laguna, Mount Maquiling: Sep. 1913, *Villamil* For. Bur. 20497 (PNH†) & Sep. 1913, *Villamil* For. Bur. 20586 (PNH†) & Aug. 1912, *Florentino* 10324 (PNH†).
Aglaia clementis Merrill in Philipp. Jour. Sci., Bot. 13: 76 (1918). Lectotype (designated here): Borneo, Mount Kinabalu, Minitindok Gorge, fl., Nov. 1915, *M.S. Clemens* 10484 (PNH†: photo A!; isolectotypes: A!, BM!, UC!).
? *Aglaia moultonii* Merrill in Philipp. Jour. Sci., Bot. 13: 78 (1918). Lectotype (designated here): Borneo, Sarawak, Amproh river, yg fl., Feb. – June 1914, *Native collector* Bur. Sci. 2138 (PNH†: photo A!).
? *Aglaia robinsonii* Merrill in Philipp. Jour. Sci., Bot. 13: 291 (1918). Lectotype (designated here): Philippines, Luzon, Province of Tayabas, Mt Dingalan, fr., Aug. – Sep. 1916, *Ramos & Edaño* Bur. Sci. 26562 (A!; isolectotypes: K!, PNH†).
Aglaia tayabensis Merrill in Philipp. Jour. Sci., Bot. 13: 292 (1918). Type: Philippines, Luzon, Province of Tayabas, Mt Tulaog, fl., May 1917, *Ramos & Edaño* Bur. Sci. 29133 (US!; isolectotypes: K!, PNH†).
Aglaia villosa (C. de Candolle) Merrill in Jour. As. Soc. Straits, Special no.: 323 (1921).
Aglaia baramensis Merrill in Jour. As. Soc. Straits 86: 317 (1922). Lectotype (designated here): Borneo, Rio Mater, Baram, fl., 30 Oct. 1914, *Moulton* 19 (PNH†: photo A!; isolectotype: SING!); van Steenis, Rheophytes of the World: 287, t. 30 (1981).
Aglaia urdanetensis Elmer ex Merrill, Enum. Philipp. Fl. Pl. 2: 373 (1923), in obs., pro syn.: *A. diffusa* Merrill; Elmer in Leafl. Philipp. Bot. 3319 (1937), sine diagn. lat. Type no.: *Elmer* 13668, Philippine Islands, Island of Mindanao, Province of Agusan, Cabadbaran, Mt Urdaneta, fl., Sep. 1912 (A!, BM!, BO!, GH!, K!, NY!, U!, UC!, US!).
Aglaia banahaensis Elmer ex Merrill, Enum. Philipp. Fl. Pl. 2: 373 (1923), in obs. pro syn.: *A. diffusa* Merrill; Elmer in Leafl. Philipp. Bot. 9: 3281 (1937), sine diagn. lat. Type no.: *Elmer* 7522, Philippine Islands, Luzon, Tayabas Province, Lucban, fr., May 1906 (BM!, BO!, GH!, K!, NY!, US!).
Aglaia marginata Craib in Kew Bull. 1926: 343 (1926). Lectotype (designated here): Siam, Pattani, Kao Kalakiri, 700 m, fl., 11 Oct. 1923, *Kerr* 7809 (K!; isolectotypes: BM!, C!, NY!, P!).
Aglaia caulobotrys Quisumb. & Merrill in Philipp. Jour. Sci. 37: 156 (1928). Lectotype (designated here): Philippines, Mindanao, Davao Province, Mt Mayo, fl., April – May 1927, *Ramos & Edaño* Bur. Sci. 49374 (NY!; isolectotypes: A!, BO!, L!, PNH† SING!).
Aglaia havilandii Ridley in Kew Bull. 1930: 367 (1930). Holotype: Sarawak, nr Kuching, fl., 12 Jan. 1893, *Haviland* 2740/2261 (K!; isolectotype: *Haviland* 2261 SING!).
Aglaia tembelingensis M.R. Henderson in Gard. Bull. Straits Settlements 7: 94 (1933). Holotype: Malay Peninsula, State of Pahang, Tembeling,

♀ fl., 29 May 1932, *Henderson* 24805 (SING!; isotypes: FHO!, FRI!, K!).

Aglaia antonii Elmer in Leafl. Philipp. Bot. 9: 3278 (1937), sine diagn. lat. Type no.: *Elmer* 12835, Philippine Islands, Island of Palawan, Province of Palawan, Brooks Point, Addison Peak, March 1911 (A!, BM!, BO!, GH!, K!, L!, NY!, U!, UC!, US!, W!).

Aglaia querciflorescens Elmer in Leafl. Philipp. Bot. 9: 3303 (1937), sine diagn. lat. Type no.: *Elmer* 13387, Philippines, Island of Mindanao, Province of Agusan, Cabadbaran, Mt Urdaneta, fl., Aug. 1912 (A!, BM!, BO!, GH!, K!, L!, NY!, P!, U!, UC!, US!, W!).

[*Aglaia mindanaensis* Merrill ex Elmer in Leafl. Philipp. Bot. 9: 3306 (1937), in adnot.]

[*Aglaia negrosensis* Merrill ex Elmer in Leafl. Philipp. Bot. 9: 3306 (1937), in adnot.]

Aglaia sorsogonensis Elmer in Leafl. Philipp. Bot. 9: 3310 (1937) sine diagn. lat.. Type no.: *Elmer* 15158, Philippine Islands, Luzon, Irosin, Mt Bulusan, ♀ fl., Nov. 1915 (A!, BM!, BO!, G!, GH!, K!, L!, NY!, P!, SING!, U!, UC!, US!, W!).

Aglaia davaoensis Elmer in Leafl. Philipp. Bot. 9: 3289 (1937), sine diagn. lat.. Type no.: *Elmer* 10925, Philippine Islands, Island of Mindanao, District of Davao, Todaya, Mt Apo, fl., June 1909 (A!, BM!, G!, GH!, K!, L!, NY!, P!, U!, US!).

Aglaia longipetiolata Elmer in Leafl. Philipp. Bot. 9: 3295 (1937), sine diagn. lat.. Type no.: *Elmer* 11829, Philippine Islands, Island of Mindanao, District of Davao, Todaya, Mt Apo, fl., July 1909, (A!, BM!, BO!, G!, GH!, K!, L!, NY!, P!, U!, US!, W!).

Tree 2 – 20(– 40) m, with an irregularly rounded crown. Bole up to 15 m, up to 50 cm in diameter, sometimes fluted throughout, with L-shaped buttresses upwards up to 150 cm, outwards up to 100 cm and up to 45 cm thick. Bark dark reddish-brown or greenish-brown with shallow pits, inner bark magenta; sapwood pale yellow pinkish-red or dark reddish-brown; latex white. Branches patent or ascending. Twigs slender, grey, densely covered with usually reddish-brown, pale orange brown or yellowish-brown stellate hairs or scales, sometimes with pale brown or reddish-brown peltate scales which have a fimbriate margin.

Leaves imparipinnate, 15 – 65 cm long, 12 – 60 cm wide, obovate in outline; petiole 3 – 10 cm, the petiole, rhachis and petiolules densely covered with stellate hairs or scales like those on the twigs. Leaflets (5 –)7 – 11 (– 16 in rheophytic form in Borneo), the laterals subopposite or almost alternate, all 5 – 34.5 cm long, 1 – 11 cm wide, young leaves yellowish-green turning darker green when mature, usually elliptical (narrowly elliptical in the rheophytic form) or oblanceolate-oblong, rarely oblong, the apex acuminate or acuminate-caudate with the obtuse acumen 2 – 20 mm long, cuneate or rounded at the sometimes asymmetrical base, the young leaves densely covered with hairs and scales like those on the twigs on both surfaces, when mature the upper and lower surfaces sometimes pitted, with hairs or scales like those on the twigs numerous on to densely covering the midrib and sometimes the lateral veins below, few on the rest of that surface; veins 6 – 19 on each side of the midrib, ascending and curved upwards near the margin and nearly or quite anastomosing, midrib prominent and lateral veins subprominent on lower surface; petiolules 4 – 20(– 24) mm.

Male inflorescence 23 – 50 cm long, 14 – 60 cm wide, with minute linear bracteoles up to 5 mm long, triangular and 1 mm long on the distal branches; peduncle 1 – 10 cm, the peduncle, rhachis and branches with indumentum like that on the twigs. Flowers up to 6,000, 1.2 – 1.5 mm long, 1 – 1.6 mm wide, depressed-globose, smelling of citronella; pedicels 0.5 – 2 mm, clothed like the twigs. Calyx up to $\frac{1}{2}$ the length of the corolla, deeply divided into 5 broadly ovate or elliptical, obtuse lobes, densely covered with brown stellate scales on the outside. Petals usually 5, yellow obovate or sub-rotund, aestivation quincuncial. Staminal tube 0.5 – 0.75 mm long, 1 mm wide, shallowly cup-shaped, yellow, thickened inside below the insertion of the anthers, the apical margin shallowly or deeply 5-lobed (in the rheophytic form, the staminal tube is divided almost to the base into 5 lobes); anthers c. 0.4 mm long and 0.3 mm wide, yellow when immature, brown at anthesis, turning black later, ovoid, inserted just below the aperture and pointing towards the centre of the flower. Ovary subglobose densely covered with stellate hairs, loculi 2; stigma c. 0.2 mm long, 0.1 – 0.3 mm wide, ovoid or depressed-globose, with two small apical lobes or a central depression; the ovary and stigma together slightly shorter than the staminal tube.

Female inflorescence 13 – 37 cm long and 5 – 14 cm wide, with fewer branches and fewer flowers than in the male; peduncle 2 – 7 cm. Flowers 1.8 – 2.2 mm long, 2 – 2.5 mm wide, larger and less strongly scented than the male, otherwise similar.

Infructescence 5.5 – 30 cm long and 5 – 15 cm wide with few to 100 or more fruits, when numerous the fruits packed tightly together; peduncle up to 10 cm, the peduncle, rhachis, branches and fruit-stalks with numerous stellate hairs and scales. Fruits 1.5 – 3.5(– 5) cm long, 1.5 – 3(– 5) cm wide, bright pale green when young, orange when mature, obovoid or ellipsoid, indehiscent, with few to densely covered with reddish-brown stellate scales; pericarp 3 – 10 mm thick, inner surface shiny, orange, white latex present until fruit ripens, opening under pressure loculicidally along a longitudinal ridge encircling the fruit. Loculi 2; septum persistent. Seeds 1 or 2, 2.2 – 2.8 cm long, 1 – 1.4 cm across, ovoid, the inner surface flattened; aril 2 – 3 mm thick, sometimes not quite complete on the antiraphe side, pinkish-orange, translucent, sweet or acidic tasting; with two layers beneath the aril, the outer hard, chestnut brown, the inner thin and membranous, with the main vascular bundle running through the raphe and antiraphe, divaricately branching from the raphe over the sides of the seed; cotyledons equal, transverse; the radicle enclosed, shoot axis covered with pale stellate hairs. Fig. 82.

DISTRIBUTION. Burma, Thailand, Peninsular Malaysia, Sumatra, Borneo, Philippine Islands, Java, Bali, Nusa Tenggara (Flores), Sulawesi. Fig. 83.

ECOLOGY. Found in swamp forest, secondary forest, primary forest, evergreen forest, river banks, along road, edges of marshes and on periodically inundated land. Found on granite, clay, limestone, sandstone, sand. Alt.: sea level to 2000 m. Scattered to locally common.

FIELD OBSERVATIONS. Kuala Lompat, Peninsular Malaysia: each

tree bears inflorescences in the order of hundreds and ten to twenty of these may be mature at any one time. The male inflorescences are up to 50 cm long with about 6000 bright yellow flowers, crowded together in sweet smelling, hanging panicles. They are usually on the outside of the crown and conspicuous when seen from other trees at the same level. The female trees bear smaller, erect inflorescences, up to 35 cm long with about 600, less crowded flowers. The fragrance is much weaker and the flowers are larger than in the male. Both male and female flowers are globular with a small aperture at the apex when mature. Removal of the petals reveals a ring of anthers on the inner edge of a short staminal tube. At anthesis in the male and maturity of the stigma in the female, the anthers undergo characteristic colour changes, similar to those in the male, from pale yellow when immature, to grey, although they produce no pollen. The male has a tiny stigma and sterile ovary. In the female the stigma is larger and bright yellow when receptive. All the flowers in an inflorescence mature at about the same time. After anthesis, the male flowers turn brown and fall off in clumps. In the female, the corolla and staminal tube fall, together with a number of presumably unfertilized flowers. Up to 70% of the flowers in a female inflorescence produce fruits, widely separated at first and bright pale green in colour, but as they grow they become tightly packed and the mature infructescence resembles a bunch of large orange-brown grapes. The fruits ripen less simultaneously within the same infructescence, individual infructescences on the same tree mature at different times through a period of several weeks or months.

USES. Bark boiled and bathed in is used against tumours; leaves are applied to wounds (Philippine Islands, Mindanao).

VERNACULAR NAMES. Peninsular Malaysia: Peler Tupai (Malay). Sumatra: Beukeu (Malay), Badjing Talang, Ganggoe, Kajoe Rambe Tombak, Laban Abang. Borneo: Bunyau, Buyau (Punan); Kalantupak, Madam Bungau, Tapau (Dusun); Kopeng, Kuping Menghiasan (Bassap Dyak); Lambunau (Orang sungei); Langsat-gajah, Langsat-langsat, Langsat Munyit, Mata Kuching Munyit, Rambutan Munjit (Malay); Lantupak (Dusun Kinabatangan); Langsep Burung (Banjar-Malay); Lempunjan (Dyak); Mambo (Suluk); Segera (Iban); Mulah, Kabunyau, Lambungao. Philippine Islands: Sorowan (Manobo); Malasaguing. Sulawesi: Pisek.

Representative specimens. BURMA. Tavoy District [14°N, 98°30′E], fr., 18 Jan. 1927, *Parker* 2439 (K!, UC!).

THAILAND. Peninsula District, Nakhon Si Thammarat Province: Krabi, ♂ fl., 18 Aug. 1964, *Sangkhachand* 1043 (C!); Khao Luang, c. 950 – 1000 m, ♀ fl ., 25 May 1968, *van Beusekom & Phengkhlai* 1004 (AAU! C!). Klongnaka, Banong, Ka Per, 50 m, fr., 14 July 1979, *Niyomdham et al.* 304 (AAU!, C!).

PENINSULAR MALAYSIA. Perak, Btg Hijau F.R., Compt. 44, Grik District, c. 260 m, fl., Nov. 1961, *Yong* KEP 94672 (KEP!, L!). Central Kedah, Sungkup F.R., Compt. 12, c. 75 m, fl., 26 Sep. 1963, *T.D. Pennington* 7844 (FHO! , KEP!, L!). Kedah, Gunong Raya, Langkawi, Kedah, c. 30 m, fl. 18 May 1957, *Chew* CWL 164 (K!, L!). C. Pahang, Kuala Lompat, Krau Game Reserve, fr., 8 Sep. 1978, *Pannell* 1334 (FHO!). S.W. Pahang, Lesong F.R., Ulu Sg. Pukin, fl., 16 Feb. 1971,

Fig. 82. *A. elliptica* (top left). Habit with male inflorescence x¹/₂. Half flower, male x25. Infructescence x¹/₂. Fruit x¹/₂. Stellate hairs x40. *A. lawii* (bottom right). Habit x¹/₂. Infructescence x¹/₂. Flower x7. Peltate scale x50. (Reproduced from Pannell (1989) with permission from the Forest Research Institute Malaysia).

Section Aglaia

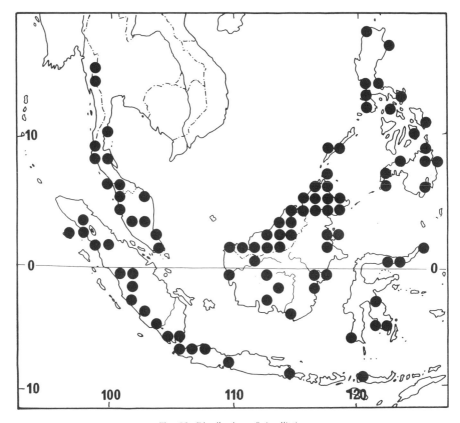

Fig. 83. Distribution of *A. elliptica*.

Suppiah FRI 11894 (K!, L!). Selangor, 15 mile Ginting Simpah, fl. 23 June 1927, *Jaamat* FRI 12857 (KEP!).

SINGAPORE. Bukit Timah Reserve, fl., 15 Aug. 1938, *Henderson* SFN 35507 (FHO!).

SUMATRA. Atjeh [Aceh], Gunung Leuser Nature Reserves, Alas River Valley, Ketambe Research Station and vicinity, c. 35 km N.N.W. of Kutacane [c. 3°40′N 97°40′E], 300 – 350 m, fr., 28 July 1979, *de Wilde & de Wilde-Duyfjes* 19227 (L!). Gajo Loeas (Gg Singamarta), fr., 2 March 1904, *Prin go Atmodjo* 161 (L!).

SARAWAK. 1st Division, Serian District, Tebakang Road, Lobang Mawang, next to Bukit Selebor [1°10′N 110°30′E], c. 250 m, fr. 28 Sept. 1968, *Paie* S 28095 (FHO!). 4th Division, Bintulu, Ulu Segan [2°90′N 113°15′E], fl., 23 Aug. 1968, *Wright* S 27169 (FHO!); Marudi, Tinjar, Ulu Sg. Dapoi [3°05′N 114°50′E], low, fl., 2 April 1965, *Ilias Paie* S 22913 (FHO!).

KALIMANTAN. Bukit Raya [112°45′E 0°45′S], c. 130 m. fr., 4 Dec. 1982, *Nooteboom* 4142 (FHO!).

PHILIPPINE ISLANDS. Leyte, Leyte Province, Palo, fl., Jan. 1906, *Elmer* 7137 (NY!). Luzon, Camarines Province, Paracale, fr., Nov. – Dec. 1918, *Ramos & Edaño* PNH 33751 (NY!). Mindoro, fr., May 1909, *Rosenbluth* 12909 (L!).

JAVA. Batavia, Tjampea, c. 200 – 300 m, st., 7 July 1898, *Koorders* 30493β (L!).

SULAWESI. Menado: Subdivision Poso, fl., 1 Sep. 1938. *Eyma* 3511 (L! U!); Subdivision Poso, between bivouac I and Borone, fl., 8 Sep. 1938, *Eyma* 3718 (U!); N., Gorontalo, fl., & fr., *Riedel* s.n. (BO!, W!).

A. elliptica is most frequently found in forest close to rivers and subject to periodic flooding. The leaves of this species sometimes resemble those of *A. odoratissima*, but they usually have more leaflets and the indumentum is of stellate hairs or stellate scales only; the peltate scales of *A. odoratissima* are absent. The indumentum is usually dense on the midrib of the lower surface of the leaflet. *A. elliptica* is usually a larger tree than *A. odoratissima*. The flowers are similar in structure in the two species, but the inflorescences of *A. elliptica* are larger and the flowers more numerous.

The fruit of *A. elliptica* has two loculi, one or both of which contains a single seed; a longitudinal ridge nearly always encircles the ripe fruit and it usually has a dense reddish-brown indumentum; the fruit is, however, sometimes small or glabrescent or lacks the longitudinal ridge in the Philippine Islands and Sulawesi. In these cases, the tree is usually smaller and the indumentum is of pale stellate scales rather than reddish- brown stellate hairs. Similar variants of *A. elliptica* are occasionally found in Peninsular Malaysia and Borneo. Some of the variation in size and conspicuousness of the longitudinal ridge is attributable to different stages in development of the fruit. Glabrescent fruits are also found in the Philippine Islands in several other species, which have a dense indumentum in other parts of their range (e.g. *A. elaeagnoidea*). Some more robust specimens from Sulawesi and Borneo have stellate scales on the midrib and the fruit has a thick woody pericarp when dry. These are included here because of the characteristic longitudinal ridge on the fruit and because there are no reliable features on which they can be recognised as a separate species or subspecies.

A rheophytic form (*A. lancifolia*) occurs only in Borneo. It usually has more numerous, narrower and often smaller leaflets than in the rest of the species and is found along riverbanks, often overhanging the water. A series of intermediates connects *A. elliptica* and *A. lancifolia*, so the latter is treated as synonym of *A. elliptica*.

77. Aglaia conferta Merrill & Perry in Jour. Arn. Arb. 21: 322 (1940). Holotype: Northeastern New Guinea, Morobe District, Sattelberg, Grotto Falls, 3000 ft [c. 900 m], fl., 5 Nov. 1935, *Clemens* 770 (A!; isotypes: G!, L!).

Aglaia cinnamomea Baker fil. in Jour. Bot., Lond. 61 Suppl.: 9 (1923).
Holotype: New Guinea [Papua New Guinea], Sogere (9°28'45"S 147°31'37" E), fl., 22 Oct. 1885, *Forbes* 197 (BM!).

Tree 5 – 23 m high. Bole 9 – 30 cm in diameter, without buttresses. Outer bark grey brown, smooth, slightly peeling; middle bark pink to

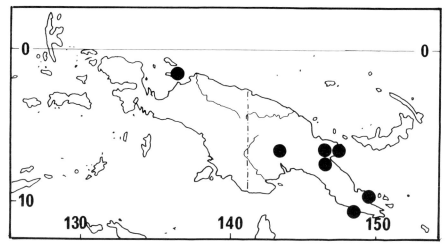

Fig. 84. Distribution of *A. conferta*.

dark red; inner bark white to red with white streaks, layered and fibrous, with a pleasant odour; some white latex; sapwood cream to reddish; heartwood reddish-brown to dark red. Twigs with pale stellate scales and reddish-brown stellate hairs, which sometimes have long arms.

Leaves 22 – 46 cm long, 14 – 42 cm wide, petiole 6 – 14 cm. Leaflets (9 –)11 – 13(– 14), the laterals usually subopposite, rarely alternate, all 7 – 21 cm long, 2.5 – 6 cm wide, usually oblanceolate, sometimes obovate or elliptical, acuminate at apex, the acute (?or obtuse) apex up to 15 mm long, rounded at the base; veins 9 – 20 ascending and curved upwards, sometimes anastomosing, the midrib prominent and veins subprominent below; petiolules 2 – 7 mm.

Inflorescence 33 cm long, 24 cm wide, sparsely branched, the peduncle, branches and pedicels densely covered with stellate scales and hairs which have long arms. Flowers 2 mm long, 2.5 mm wide, ± sessile. Calyx cup-shaped and divided into 5 obtuse lobes, densely covered with stellate hairs on the outside. Petals 5, white or yellow, aestivation quincuncial. Staminal tube c. 0.8 mm long and 0.8 mm wide, aperture c. 0.4 mm across, anthers $\frac{1}{2} - \frac{1}{3}$ the length of the tube, with pale orange stellate scales and hairs. Ovary with 2 loculi, each containing 1 ovule; stigma ovoid, with two apical lobes.

Infructescence 15 – 25 cm long, 9 – 18 cm wide; peduncle 4 cm, the peduncle, branches and pedicels with numerous stellate scales and hairs which have long arms. Fruits 5.5 cm long, 4 cm wide, yellow or brown. Pericarp fibrous, 0.5 cm thick, no dehiscence line, pericarp densely covered with peltate scales which have a fimbriate margin on the outside, inner surface shiny; loculus 1, seed 1; seed 2.7 – 3 cm long, 2 – 1.8 cm wide, 1.3 – 2 cm thick; aril thin, red or orange, with numerous veins originating from a wide raphe, almost certainly translucent when fresh;

cotyledons unequal, peltate; shoot axis with long dark reddish-brown hairs.

DISTRIBUTION. New Guinea only. Fig. 84.

ECOLOGY. Found in lowland primary forest and in secondary forest on limestone. Alt.: 15 to 1800 m.

Representative specimens. PAPUA NEW GUINEA. Morobe District, Wau Sub- district: G. Pines Logging, Leklu no 1 [7°06′S 146°35′E], c. 900 m, fr., 13 Sep. 1968, *Streiman & Katik* NGF 39109 (K!); Bulolo, Watut fall of range, c. 1100 m, fl., 11 April 1968, *Streimann & Kairo* NGF 39368 (K!).

A. conferta resembles *A. subminutiflora* in its indumentum of peltate fimbriate and long-armed stellate hairs, but it is more robust; the distinction between these two species is not clear-cut. However, the fruits and infructescence are usually stouter in *A. conferta* (more ripe fruits are needed) and there are usually fewer leaflets in *A. conferta* which tend to be oblong or elliptical in shape. *A. conferta* is like *A. elliptica*, but the aperture of the staminal tube is smaller, the fruits are smaller and the pericarp thinner except in the large woody ellipsoid fruits of NGF 39109 (which are densely covered with pale compact stellate hairs on the outside).

A. conferta also resembles *A. parksii* in vegetative characters, but differs in its flower structure, and *Aglaia meridionalis*, but that species has a trimerous flower.

78. Aglaia aherniana Perkins in Frag. Fl. Philipp.: 32 (1904). Lectotype. Philippine Islands, Luzon, Principe Province, Baler, fl., Aug. – Oct. 1903, *Merrill* 1025 (NY!; isolectotype: PNH†).

Aglaia myriantha Merrill in Philipp. Journ. Sci., Bot. 13: 295 (1918). Lectotype (designated here): Philippines, Luzon, Province of Tayabas, Mt Dingalan, fl., Aug. – Sep. 1910, *Ramos & Edaño* Bur. Sci. 26593 (US!; isolectotype: PNH†).

Aglaia irosinensis Elmer ex Merrill, Enum. Philipp. Fl. Pl. 2: 371 (1923), in obs. pro syn. *A. aherniana*. Type no.: *Elmer* 17292, Luzon, Sorsogon Province, Irosin, Mt Bulusan, fl., Sep. 1916 (A!, BM!, G!, GH!, K!, L!, U!, W!).

Tree. Twigs greyish-brown or reddish-brown, with numerous to densely covered with pale brown stellate hairs. Leaves c. 41 cm long and 22 cm wide; petiole 13 cm, the petiole, rhachis and petiolules with numerous to densely covered with hairs like those on the twigs. Leaflets 13 – 23, the laterals subopposite, all 3.5 – 12 cm long, 2 – 3.4 cm wide, usually oblong, sometimes ovate, coriaceous, acuminate at apex, the obtuse acumen 5 – 10 mm long, cuneate or rounded at the slightly asymmetrical base, with pale brown stellate hairs numerous to densely covered with on the midrib below, numerous on the upper and lower leaflet surfaces when young, glabrescent, with numerous pits on the upper and lower leaflet surfaces; veins 7 – 12 on each side of the midrib, ascending, curved upwards near the margin and just anastomosing; petiolules to 10 mm on lateral leaflets, up to 20 mm on terminal leaflet. Inflorescence 24 – 45 cm long, 14 – 26 cm wide; peduncle 5 – 8 cm, the peduncle, rhachis and branches with numerous to densely covered with scales like those on the twigs. Flower 1 – 2 mm long, 1 – 1.5 mm wide, subsessile. Calyx 0.5 mm long, deeply divided into 5 rounded lobes densely covered with pale brown stellate hairs on the outside. Petals 5, aestivation quincuncial. Staminal tube 0.7 – 1.3 mm long, 0.6 – 1.3 mm wide, obovoid, narrowed at the base, aperture 0.5 – 0.7 mm, shallowly lobed; anthers 5, 0.2 – 0.4 mm long, 0.2 – 0.3 mm wide, inserted half way up the staminal tube, the tube thickened below the insertion of the anthers. Ovary 0.3 – 0.4 mm long, 0.2 – 0.4 mm wide, ovoid densely covered with stellate hairs; locule 2 each containing 1 ovule; stigma 0.2 – 0.4 mm long, 0.3 – 0.4 mm wide, depressed-globose. Infructescence 22 – 25 cm long, 12 – 20 cm wide; peduncle 6 – 12 cm long. Fruits 0.5 cm in diameter, subglobose, dark brown when dry, with few stellate scales on the outside.

DISTRIBUTION. Philippine Islands only.

Representative specimens. PHILIPPINE ISLANDS. Sibuyan Island, Capiz Province, Magallanes (Mt Giting-Giting). fl., May 1910, *Elmer* 12343 (U!). Dinagat Island, 1901, *Ahern's coll.* 481 (NY!, US!: Paralectotypes of *A. aherniana*).

A. aherniana resembles *A. elliptica*, but has coriaceous leaflets which have numerous pits on both surfaces; the fruits are much smaller than in *A. elliptica* and they have no longitudinal ridge.

79. Aglaia barbanthera C. de Candolle in Lorentz, Nova Guinea 8: 1016 (1914). Lectotype (designated here): Nova Guinea neerlandica meridionalis [Irian Jaya], ['Noordrivier', 800 m], fl., Nov. 1909, *van Römer* 899 (L!).

Tree 2 – 15(– 20) m. Bole c. 10 cm in diameter. Bark brown; sapwood pale yellow. Twigs densely covered with dark reddish-brown peltate scales which have a fimbriate margin and some orange-brown stellate scales interspersed. Leaves 9 – 42 cm long, 7 – 24 cm wide; petiole 1.5 – 11 cm, petiole, rhachis and petiolules densely covered with scales like those on the twigs. Leaflets (7 –)9 – 11, the laterals subopposite, all 3.5 – 18 cm long, 1.5 – 6.5 cm wide, ovate or elliptical, acuminate at apex, the obtuse acumen parallel-sided and up to 1.2 mm long, usually cuneate, sometimes rounded at the base, veins 8 – 15 on each side of the midrib, ascending and curving upwards near the margin, sometimes with shorter veins in between, the midrib prominent and lateral veins subprominent below, densely covered with scales like those on the twigs on the midrib and occasionally on the veins below, with numerous faint pits on the rest of that surface; petiolules 1 – 13 mm.

Male inflorescence 18 – 22 cm long, 12 – 16 cm wide; peduncle 3 – 5 cm; female inflorescence c. 9.5 cm long and 4 cm wide; peduncle 0.5 – 2 cm, the peduncle, rhachis and branches in both sexes densely covered with scales like those on the twigs or sometimes with stellate hairs.

Male and female flowers 1 – 2 mm long and 1 – 2 mm wide, subglobose; pedicels 0.5 – 1.2 mm. Calyx 0.5 – 1 mm, cup-shaped, divided into 5 lobes, the pedicel and calyx densely covered with dark reddish-brown stellate scales or hairs. Petals 5, white or pale yellow, aestivation quincuncial. Staminal tube 0.6 – 1 mm long, 0.6 – 1 mm wide, cup-shaped or obovoid with a lobed margin; anthers 5, 0.2 – 0.6 mm long and 0.2 – 0.5

mm wide, ovoid, with pale yellow margins and either without hairs or scales or densely covered with white or pale brown simple or stellate hairs, scattered on the inside of the tube and sometimes on the outside of the tube or along the margins of the lobes, inserted in the upper half of the tube either protruding and filling the aperture or barely protruding. Ovary 0.2 – 0.4 mm long, 0.3 – 0.4 mm wide, depressed globose, densely covered with reddish-brown stellate hairs or scales, loculi 2 each containing one ovule; stigma 0.3 – 0.6 mm long, 0.2 – 0.5 mm wide, ovoid, with two apical lobes in the female, pale yellow.

Infructescence 3.5 – 12 cm long, 4 – 12 cm wide; peduncle up to 2 cm, the peduncle, rhachis and branches densely covered with scales like those on the twigs. Fruits obovoid, 1.2 cm long and 0.8 cm wide, bright orange or brown, with dense scales.

DISTRIBUTION. New Guinea only.
ECOLOGY. Found in primary, fagaceous and hill forest. Alt.: 60 to 2000 m.
Representative specimens. IRIAN JAYA. N., near Prauwen bivouac, 180 m, fl., 6 Sep. 1920, *Lam* 1083 (K!).
PAPUA NEW GUINEA. West Sepik District, Telefomin Subdistrict, Oksapmin [c. 5°20′S 142°15′E], c. 2000 m, fl., & fr., 16 Oct. 1968, *Henty, Isgar & Galore* NGF 41574 (K!). Rossel Island, Mt Rossel, 700 m, fl., 14 Oct. 1956, *Brass* 28402 (K!). Normanby Island, Mt Pabinama, 700, fl., 8 May 1956, *Brass* 25769 (K!).

A. *barbanthera* is characterised by its indumentum of dark reddish-brown peltate scales which are often interspersed with paler stellate scales. The dark colour of the dense indumentum on the midrib contrasts with the pale green of the lower surface of the leaflets.

80. Aglaia saltatorum A.C. Smith in Contrib. U.S. Nat. Herb. 30: 483 (1952). Holotype: Fiji, Vanua Mbalavu, Malatta, southern limestone section, 0 – 100 m, fl., 29 March 1934, *Smith* 1439 (U.S. National Herb. 1674954) (US!; isotypes: GH!, NY! UC!); Sykes, Contributions to the Flora of Niue: 118 (1970). A.C. Smith, Flora Vitiensis Nova 3: 548 (1985).

Aglaia edulis sensu Seeman, Flora Vitiensis: 37 (1865), non (Roxb.) Wall., non A. Gray, quoad spec. Fiji Islands, Taviuni, fr., 1860, *Seeman* 59 (GH!, K!) & fl., 1860, *Seeman* 60 (A!, BM!, CGE!, G-BOIS!, G-DC!, GH!, K!, W!).

Tree up to 5 m. Twigs densely covered with pale yellowish-brown or orange brown stellate scales or peltate scales with a fimbriate margin, the apical bud occasionally with stellate hairs interspersed.

Leaves 12 – 50 cm long; petiole 4 – 14 cm; the petiole, rhachis and petiolules densely covered with scales like those on the twigs. Leaflets 5 – 7, the laterals subopposite, all 3.5 – 23 cm long, 1.5 – 9 cm wide, pale green above and paler yellowish-green below when dry, usually ovate, sometimes elliptical, the basal pair more ovate than the rest and sometimes markedly smaller in size; acuminate at apex with the broad, obtuse acumen up to 25 mm long, rounded at the slightly asymmetrical base; the midrib on the lower surface with few to densely covered with scales like those on the twigs, few to numerous on the rest of that surface; veins 6 – 18 on each side of the midrib, ascending at an angle or 60 – 70° to the midrib, curving upwards near the margin, not or barely anastomosing, reticulation visible; petiolules 4 – 17 mm.

Male inflorescence up to 33 cm long and 30 cm wide; female inflorescence 15 cm long and 4 cm wide; the peduncle, rhachis and branches densely covered with scales like those on the twigs. Flowers 1.5 – 2 mm long, 1.5 – 2.5 mm wide, depressed globose or subglobose; pedicel 1 – 2.5 mm, becoming broader towards the base of the receptacle, densely covered with pale stellate scales. Calyx c. $^1/_2$ the length of the flower, with numerous stellate scales. Petals 5, pale yellow, thick and fleshy, densely covered with stellate scales on the outside, except near the apex, aestivation quincuncial. Staminal tube shallowly cup-shaped (Solomon Islands and female flowers from Fiji) or absent and replaced by 5 short free lobes (male flowers from Fiji) with the anthers inserted on the margin and curved over the stigma. Ovary either depressed-globose and semi-inferior or ovoid, densely covered with stellate scales; loculi 2, each containing 1 ovule; stigma small, ovoid, with 2 small lobes at the apex.

Fruits 2 – 4 cm long, 1.5 – 4.5 cm wide, subglobose or obovoid, brown or orange, indehiscent; loculi 2, each containing 1 seed. Fig. 85.

DISTRIBUTION. Solomon Islands (Santa Cruz only), Vanuatu, Fiji, Wallis Island, Tonga and Niue (probably introduced).

ECOLOGY. Found in well-drained ridge forest, primary forest and secondary forest; on limestone, clay. Alt.: 8 to 520 m.

USES. Inflorescenses used in making floral necklaces, fragrant.

VERNACULAR NAMES. Solomon Islands: Ulukwalo, Ulukwalobala (Kwara'ae). Fiji: Langkali, Langkali Thavuthavu. Tonga: Lagakali.

Representative specimens. SOLOMON ISLANDS. Santa Cruz: N.W., Mbania Area, fl., 9 Oct. 1969, *Mauriasi & collectors* BSIP 17621 (K!); S.W., Baenga Area, 4 m, fl., 16 Oct. 1969, *Mauriasi & collectors* BSIP 16612 (K!); Vanikoro Island, E. side of Saboe Bay, 365 m, fr., 18 April 1963, *Whitmore* BSIP 1747 (K!).

VANUATU. Santo, Nokovula-Kerepua, 1000 m, fl., 18 Aug. 1985, *Cabalion* 2867 (K!). Banks Group, Vanua Lava Island, 500 m, fr., 16 June 1928, *Kajewski* 482 (NY!).

FIJI. Viti Levu, Ra, vicinity of Rewasa, near Vaileka, Mataimeravula, 50 – 200 m, fr., 3 June 1941, *Degener* 15422 (L!, NY!, UC!). Kambara, limestone formation, 0 – 100 m, fl., 2 – 7 March 1934, *A.C. Smith* 1240 (UC!); Koro , sea level, fr., 1 – 8 Feb. 1934, *A.C. Smith* 1079 (UC!).

ILES DE HORN. Futuna, Vaisei, 2 m, fl., 25 Oct. 1968, *McKee* 19776 (K!).

TONGA. Tongatabu Island, Kologa, near chair of Tui Tonga, fl., & fr., June – Aug. 1926, *Setchell & Parks* 15369 (UC!).

NIUE ISLAND. Lakepa, probably cultivated, [c. 19°02'S 169°55'W] fl., 8 Sep. 1965, *Sykes* 169813 (L!).

The leaflets of *A. saltatorum* are often subcoriaceous and when dry, are pale yellow in colour with the venation often paler than the rest of the leaflet. The pale golden or orange stellate or peltate scales densely cover the midrib and are either numerous or few on the rest of the lower surface of the leaflets.

A. saltatorum resembles *A. parviflora*, but *A. parviflora* has hairy anthers

Fig. 85. *A. saltatorum*. Habit with inflorescence x¹/₂. Half flower, with cross section of ovary x12. Habit with infructescences x¹/₂. Mature fruit x¹/₂. Stellate scale x120.

and a characteristic flattened stigma which has a lobed margin. *A. basiphylla* is distinguished from *A. saltatorum* by its reddish-brown indumentum which is confined mainly to the midrib on the lower surface of the leaflet.

In Fiji some specimens are more delicate than in the rest of the range of the species and they bear a closer resemblance to *A. basiphylla*. Other specimens from Fiji, along with those from Tonga and Niue are more robust and have larger, more coriaceous leaves and numerous pale stellate scales on the lower surface of the leaflets.

81. Aglaia mariannensis Merrill in Philipp. Jour. Sci., Bot. 9: 99 (1914). Lectotype (designated here): Marianne Islands, Guam, fl., [July 1912], *Thompson*, GES 465 (US!; isolectotypes: G!, K!, L!, PNH†); Kanehira in Bot. Mag., Tokyo 45: 288 (1931); B.C.Stone in Micronesica 6: 335 (1970).

Aglaia palauensis Kanehira, Fl. Micronesica: 162, t. 61 (1933), japonica ; Kanehira in Bot. Mag., Tokyo 47: 671 (1933) latine. Syntypes: Caroline Islands, Pelew [Palau] Island, New Gaspan, fr., Aug. 1932, *Kanehira* 2126 (FU?, NY!); Kanehira in Tokyo Bot. Mag. 45: 288 (1931) pro parte, as *Aglaia mariannensis*.

? *Aglaia ponapensis* Kanehira, Fl. Micronesica: 163 (1933), japonice; Kanehira in Bot. Mag., Tokyo 47: 672 (1933), latine. Syntypes: Caroline Islands, Ponape, Niinoani-zan, fl., Aug. 1929, *Kanehira* 825 (FU); Ponape, Kolonia, fr., Aug. 1929, *Kanehira* 649 (FU); Ponape, Kity, *Kanehira* 1528 (FU); Ponape, Kolonia, Aug. 1931, *Kanehira* 1488 (FU); Kanehira in Tokyo Bot. Mag. 45: 288 (1931) pro parte, as *Aglaia mariannensis*.

Small tree up to 8 m. Twigs with numerous to densely covered with orange-brown peltate scales which have a fimbriate margin, sometimes with orange- brown stellate hairs interspersed, especially at the apex. Leaves up to 41 cm or more long and 44 cm wide; petiole 2.5 – 16 cm, the petiole, rhachis and petiolules with numerous or densely covered with hairs or scales like those on the twigs. Leaflets 3 – 7(– 10), the laterals subopposite, all 6 – 22.5 cm long, 2.5 – 9 cm wide, orange-green, pale yellowish-green or brownish-green when dry; ovate or obovate, acuminate at apex, the obtuse acumen up to 10 mm on lateral leaflets, up to 20 mm on terminal leaflet, rounded or subcordate at the slightly asymmetrical base, with numerous orange-brown or pale brown stellate scales and sometimes with pale brown stellate hairs interspersed on the midrib below, often with numerous faint orange pits on the lower surface; veins 8 – 18 on each side of the midrib, ascending at an angle of 50 – 60° to the midrib and curving upwards near the margin; petiolules 5 – 10 mm. Inflorescence 12 – 16 cm long, 6 – 12 cm wide; peduncle 1 – 4.5, the peduncle, rhachis and branches with numerous to densely covered with hairs or scales like those on the twigs. Flowers orange, 1 – 1.5 mm long, 1 – 2 mm wide, pedicel 0.5 – 2 mm, densely covered with pale brown stellate scales or orange-brown or reddish-brown stellate scales or hairs. Calyx with 5 obtuse lobes. Petals 5, aestivation quincuncial. Staminal tube c. 1 mm long and 1.6 mm wide, cup-shaped with the margin incurved and shallowly lobed; anthers 5, c. 0.6 mm long and 0.5 mm wide, inserted inside the margin and pointing towards the centre of the flower. Ovary c. 0.5 mm high and 0.4 mm wide, subglobose; stigma c. 0.5 mm long and 0.3 mm or more wide, either obovoid and with two apical lobes or depressed-globose and with a central depression. Infructescence 3.5 – 4 cm long with about 7 fruits. Fruits 2.5 cm in diameter, pericarp thin and brittle with no longitudinal ridge; loculi 2, each containing one seed; seeds c. 15 mm long, 10 mm wide and 6 mm through.

DISTRIBUTION. Marianne Islands (Saipan, Rota and Guam) and Caroline Islands (Palau and Ponape). Fig. 78.
ECOLOGY. Thickets and secondary forests, limestone cliffs.
VERNACULAR NAMES. Ponape: Marashau. Saipan: Maffnyo.
Representative specimens. MARIANNE ISLANDS. Rota, Sabana, c. 450 m, fr., 24 June 1952, *Kondo* s.n. (K!) & fl., 24 June 1952, *Kondo* s.n. (K!, L!). Saipan, fl., July 1930, *Kanehira* 920 (K!). Guam: Punte de los das Amantes, fl., 4 May 1962, *Schmull* 44 (L!); fr., October 1911, *McGregor* GES 546 (paralectotypes of *A. mariannensis*: BM!, G!, K!); Ritidian Point, slope of Mt Machanao, 180 m, yg fr., 16 April 1936, *Bryan Jr.* 1172 (K!).
CAROLINE ISLANDS. Ponape, U District, Mount Seletenreh, trail ascending northwest face, up to 600 m, fr., 6 Feb. 1965, *Stone* 5395 (L!). Pelew [Palau] Island, Aimiriik, fr., Aug. 1932, *Kanehira* 1942 (NY! paralectotype).

The indumentum in *A. mariannensis* is usually of peltate scales with stellate hairs interspersed, but the hairs are sometimes absent; in this respect *A. mariannensis* resembles *A. basiphylla*. *A. basiphylla* differs in its rounded leaf apex and the widely spreading veins. When stellate hairs are present *A. mariannensis* resembles *A. subminutiflora* and when they are absent it is sometimes difficult to separate from *A. saltatorum*. In those specimens in which the midrib is densely covered with pale brown stellate scales, *A. mariannensis* resembles the less robust form of *A. elliptica* which occurs in the Philippine Islands and Sulawesi.

This is a variable species with a scattered distribution on the larger Micronesian islands. Further field study is required to confirm this treatment or to establish whether it should be either subdivided or included in one of the species it resembles.

82. Aglaia [Aglaja] cumingiana Turcz. in Bull. Soc. Nat. Mosc. 31: 409 (1858). Lectotype (designated here). Ins. Philippinae, Luzon, Prov. Albay, fl., 1841, *Cuming* 1008 (K!; isolectotypes: BO!, OXF!, W!).

Hearnia cumingiana (Turcz) C. de Candolle in A. & C. de Candolle, Monog. Phan. 1: 629 (1878).
Aglaia tarangisi Elmer in Leafl. Philipp. Bot. 9: 3314 (1937), sine diagn. lat.. Type no.: *Elmer* 10956, Philippines, Mindanao (L!).

Tree up to 15 m, sometimes flowering at 3.5 m. Bole up to 6 m, up to 15 cm in diameter. Branches ascending, forming a narrow open crown. Outer bark greenish-grey or greyish-brown, with hoop marks, with pale lenticels in vertical rows; inner bark pale yellow to pinkish-brown; sapwood white or pale yellow; heartwood reddish-brown; latex white, copious. Twigs greyish-green, finely longitudinally wrinkled, with few scales except at the apex which is densely covered with small delicate pale brown stellate scales. Leaves 18 – 43 cm long, 16 – 38 cm wide; petiole

3 – 12 cm, the petiole, rhachis and petiolules green or greyish-green, with occasional scales like those on the twigs. Leaflets 3 – 7, the laterals subopposite, all 4 – 23 cm long, 1.5 – 7.5 cm wide, usually elliptical, sometimes ovate or obovate, shiny dark green above and pale green below when fresh, orange-brown or brownish-green below and both surfaces matt when dry, usually cuneate, sometimes rounded at the slightly asymmetrical base, acuminate at apex, the obtuse acumen up to 1(– 1.5) cm long, with very few scales like those on the twigs on the midrib below, often with numerous tiny shiny orange spots below; veins (4–)7–12 on each side of the midrib, orange-green when dry, curved upwards, anastomosing, midrib prominent lateral veins subprominent and reticulation subprominent below, all visible above; petiolules 0.5 – 1 cm long (- 3 cm on terminal leaflet), with a swelling at the base, which is black when dry.

Male inflorescence 11 – 31 cm long, 11 – 28 cm wide; peduncle 0.5 – 7 cm, the peduncle, rhachis and branches flattened, greenish-brown or pale brown when dry and with a few pale brown or nearly white scales like those on the twigs. Female inflorescence about 19 cm long and 10 cm wide; peduncle c. 7 cm long. Male and female flowers similar, 0.7 – 1.5 mm long, 1 – 1.7 mm wide, subglobose, fragrant; pedicels 0.5 – 0.8(– 1.3) mm, the calyx and pedicel with scattered stellate scales or none at all. Calyx divided into 5 subrotund lobes which have a fimbriate margin. Petals 5, yellow; aestivation quincuncial. Staminal tube dark yellow, 0.5 – 1 mm long, 0.7 – 1.5 mm wide, cup-shaped or obovoid, thickened inside below the anthers, the aperture c. 0.7 mm, sometimes with the margin incurved, anthers (4 or) 5, 0.3 – 0.5 mm long, up to 0.4 mm wide, protruding and pointing towards the centre of the flower and filling the aperture. Ovary 0.1 mm, without hairs or scales; loculi 2; stigma 0.2 – 0.3 mm long, 0.3 mm wide, ovoid or depressed globose with a central depression and lobed margin, blackish- brown when dry.

Infructescence 20 – 25 cm long, 11 – 20 cm wide; peduncle 4 – 13 cm wide, the peduncle, rhachis and branches pale orange-brown when dry, with few scales like those on the inflorescence; fruit-stalks c. 5 mm. Fruits 0.9 – 2 cm long, 0.5 – 0.8 cm wide, orange, obovoid when young, ellipsoid when ripe, (often asymmetrical when only one seed develops, and the stigma is displaced to one side during development of the fruit), with few scales like those on the twigs or none at all, pericarp orange-brown, papery thin and brittle when dry, orange or red when ripe and fresh; loculi 2, each containing 0 – 1 seed.

DISTRIBUTION. N. Borneo, Philippine Islands. Fig. 86.

ECOLOGY. Found in primary, secondary and gallery forest, and along the seashore. Soil: sandy. Alt.: sea level to 1330 m.

VERNACULAR NAMES. Borneo: Lantopak. Philippine Islands: Daonbunsikag (Pal-Tagb.).

Representative specimens. SARAWAK. 3rd Division, Bt Raya, Kapit [2°20′N 113°15′E], c. 230 m, fr., 27 Oct. 1965, *Wright* S 24706 (FHO!). SABAH. Ranau District, Mokodu River, Kg Takutan, fr., 23 May 1981, *Kalantas* SAN 93597 (FHO!). Keningau District, Mt Trusmadi Forest Reserve, c. 900 – 1200 m, fl., 23 Aug. 1977, *Madani* SAN 87189 (SAN!). Kota Belud District, Sandakan, above Ladang Nangkop Ranau,

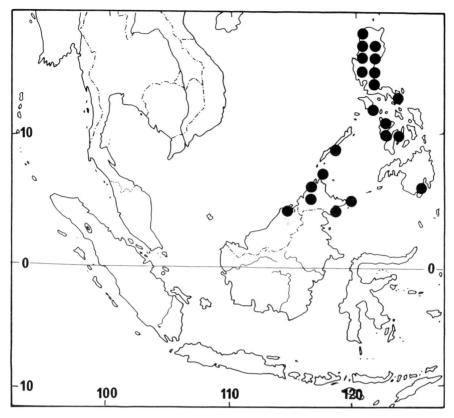

Fig. 86. Distribution of *A. cumingiana*.

c. 1000 m, fl., 10 Aug. 1960, *Meijer & Corner* SAN 23472 (SAN!). Kudat District, Pulau Balembangan: southern peninsula end, northern face, 0 – 15 m, fl., 10 April 1977, *Stone & Anderson* SAN 85525 (SAN!, UC!); southern peninsula, east of Kampung Sina, c. 6 m, fl., 9 April 1977, *Stone & Anderson* SAN 86718 (SAN!, UC!). Banguey Is., seashore, yg fr., July – Sep. 1923, *Castro & Melegrito* 1625 (UC!).

PHILIPPINE ISLANDS. Cagayan Province, Peñablanca, Luzon, fl., April 1926, *Ramos & Edaño* 46602 (W!). Bohol, fr., Aug.- Oct. 1923, *Ramos* 4327 2 (W!).

A. cumingiana resembles *A. lawii* and *A. oligophylla* in its foliage, but it is distinguished by the indumentum, inflorescence and fruit. In some respects, especially flower structure, *A. cumingiana* is like *A. parviflora*, but the latter is more robust, with larger leaves and twigs; the reticulation on the leaflets is not prominent; peltate scales with a fimbriate margin are numerous on the midrib below, while stellate scales are more abundant on the inflorescence; the fruits have dense scales on the outside.

83. Aglaia laxiflora Miq., Ann. Mus. Bot. Lugd. Bat. 4: 52 (1868). Lectotype (designated here): Borneo, Doesson, *Korthals* (L!; isolectotype: BO!); C. de Candolle in A. & C. de Candolle, Monog. Phan. 1: 610 (1878).

Tree up to 10 m (? or more). Bole up to 4 m, up to 35 cm in diameter. Bark light grey with numerous orange-brown depressions and scalloped pattern; inner bark reddish-pink, laminated; sapwood whitish-pink, with tiny rays. Twigs greenish-brown often with numerous lenticels and small pale brown stellate scales densely covering the apex.

Leaves up to 60 cm long and 48 cm wide, obovate in outline; petiole up to 11 cm, the petiole rhachis and petiolules with numerous scales like those on the twigs. Leaflets 11 – 14, the laterals alternate, all 5 – 24 cm long, 2 – 8 cm wide, narrowly elliptical or oblong, pale green above and pale brownish-green below when dry, usually cuneate, sometimes rounded at the base, acuminate at apex, the acute acumen up to 1.5 cm, with few to numerous scales like those on the twigs on the midrib below and few on the rest of that surface; veins 8 – 15 on each side of the midrib, curving upwards, not anastomosing, the reticulation subprominent above, the midrib prominent, lateral veins and reticulation subprominent below; petiolules 0.5 – 0.7 cm on lateral leaflets, up to 2 cm on terminal leaflet. Inflorescence (male only seen) up to 56 cm long and 40 cm wide; peduncle up to 14 cm, the branches slender and widely spaced giving the inflorescence a lax appearance, the peduncle, rhachis and branches with numerous scales like those on the twigs, more densely covering the distal branches. Flowers 1.1 – 1.2 mm long, 1.2 – 1.3 mm wide, subglobose, fragrant; pedicels 0.7 – 1 mm to an articulation where it easily breaks off, borne on a branch c. 0.5 mm long, pedicel with a small ovate bracteole which has a ciliate margin and occasional scales like those on the twigs. Calyx with 5 spreading, rounded lobes which have a ciliate margin and no hairs or scales. Petals 5, yellow, aestivation imbricate. Staminal tube 0.5 – 0.9 mm high, 0.9 – 1 mm wide, cup-shaped, the apical margin incurved, aperture 0.3 – 0.6 mm; anthers 5, 0.3 – 0.4 mm long, 0.3 mm wide, ovoid, dehiscent only in the lower half, darker at the apex, inserted inside the rim of the tube to $^2/_3$ up the tube, protruding and filling the aperture. Ovary 0.2 – 0.4 mm long, 0.3 mm wide, with two loculi, each containing 1 ovule with a dense ring of pale stellate scales at the junction with the stigma; stigma c. 0.2 mm long and 0.2 mm wide, ovoid with two tiny apical lobes.

Infructescence up to 18 cm; peduncle 2 – 3 cm, the peduncle, rhachis and branches with few scales like those on the twigs. Fruits 5 – 6 cm long and c. 3.5 cm wide, ellipsoid or obovoid, orange or orange-yellow when ripe, indehiscent; pericarp 3 – 5 mm thick. Loculi 2, each containing 0 – 1 seed; the fruit curved and asymmetrical when a seed fails to develop in one of the loculi. Seed c. 3.1 cm long, 1.8 cm wide, 1 cm through; the seed surrounded by an entire translucent aril 2 – 3 mm thick.

DISTRIBUTION. Borneo only: Brunei, Sabah, Kalimantan. Fig. 87.

ECOLOGY. Found in primary forest, along ridges and along river banks, alluvial or occasionally flooded Dipterocarp forest with a low canopy; sometimes on limestone, sand. Alt.: sea level to 600 m. Common.

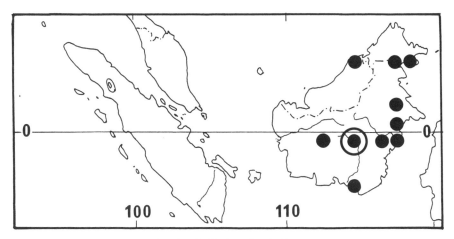

Fig. 87. Distribution of *A. laxiflora*.

VERNACULAR NAMES. Borneo: Embunjau, Perjau (Kenya); Lantupak (Dyak Kinabatangan); Bako, Katitiwar, Pendjeha.

Representative specimens. SARAWAK. Punan Busan, fl., 12 June 1971, *Geh & Samsuri* GSY 116 (SING!).

BRUNEI. Belait District, bank of Sg. Belait between Sg. Buau & Sukang [4°17'S 114°39'E], fl., 1 Nov. 1989, *Forman* (with *Blewett*) 1137 (K!).

SABAH. Lahad Datu District, Ulu Segama, Danum Valley Research Centre, 150 m, fr., 26 Feb. 1985, *Argent, Ratter, Dongop & Kumin* 108385 (K!). Tawau, Ulu Sg. Apas, Mile 15, Quoin Hill Road, fl., 11 Sep. 1962, *Aban Gibot* SA N 31245 (K!). Kalabakan, Geynawood logged area, km 24, Kalabakan road, fr., 16 Feb. 1984, *Sumbing Jimpin* SAN 101497 (K!).

KALIMANTAN. E. Kutai, Kutai Nature Reserve: alluvial, occasionally flooded forest, understorey tree, fr., (?) 12 Fcb. 1979, *Leighton* 530 (L!); 50 m, fl., 2 Aug. 1979, *Leighton* 898 (L!). Bukit Raya, 130 m, [112°47'E 0°45'S], fr., 5 Jan. 1983, *Nooteboom* 4427 (FHO!, L!). S., Pleihari, Sg. Asem-Asem, 100 m, fl., Sept. 1965 *Sauvern* 985 (K!). Tewingan, Riau Kanan, fl., 17 Sep. 1919, *Ramli* 2003 (L!).

The leaflets of *Aglaia laxiflora* are pale when dry, usually with subprominent reticulation above and below; the pale stellate scales are sparse on the leaflets, inflorescence and infructescence; the inflorescence is 'lax' and the pedicels are articulated; scales are sparse on the pedicels and absent from the calyx.

84. Aglaia pyriformis Merrill in Philipp. Jour. Sci., Bot. 13: 290 (1918). Lectotype (designated here): Philippine Islands, Luzon, Province of Tayabas, Mt Dingalan, ♀ fl. & fr., Aug. – Sep. 1916, *Ramos & Edaño* Bur. Sci. 26604 (A!; isolectotypes: BM!, K!, NY!, P!, PNH†).

Aglaia puncticulata Merrill in Philipp. Jour. Sci., Bot. 13: 290 (1919). Lectotype (designated here). Philippine Islands, Luzon, Province of Nueva Ecija, Mt Umingan, ♂ fl., Aug. – Sep. 1916, *Ramos & Edaño* Bur. Sci. 2630 6 (US!; isolectotype: PNH†).

Tree. Twigs usually densely covered with very dark reddish-brown peltate scales which have a reddish-brown fimbriate margin, with stellate scales of similar colour interspersed at the apex, sometimes with a few paler reddish-brown scales.

Leaves 18 – 40 cm long, 11 – 29 cm wide; petiole 4 – 10 cm, the petiole, rhachis and petiolules longitudinally wrinkled, densely covered with scales like those on the twigs. Leaflets 7 – 9, the laterals subopposite, all 5.5 – 17.5 cm long, 2 – 5 cm wide, pale yellowish-green above and pale brownish-yellow below when dry, oblong or obovate, acuminate at apex, with a broad obtuse acumen up to 3 mm, cuneate or rounded at the base; veins 6 – 11 on each side of the midrib, ascending and curved upwards near the margin, not anastomosing, sometimes with shorter lateral veins in between, midrib prominent and lateral veins subprominent; with scales like those on the twigs numerous on to densely covering the midrib above, few on to densely covering the veins above and either with similar scales or with scales which have a dark reddish-brown centre and a long white fimbriate margin few to numerous on the rest of the upper leaflet surface and on the lower leaflet surface, with numerous pits on both surfaces; petiolules 1 – 3 cm.

Male inflorescence 20 cm long, 11 cm wide; peduncle 5 cm. Male flower c. 1 mm long and 1 mm wide; pedicels c. 1 mm, the pedicel and calyx densely covered with dark reddish-brown peltate scales which have a long paler fimbriate margin. Calyx 0.75 mm long, cup-shaped and deeply divided into 5 sub-rotund lobes. Petals 5, aestivation quincuncial. Staminal tube 0.2 – 0.4 mm high, and 0.7 – 0.8 mm wide, shallowly cup-shaped; anthers 5, c. 0.5 mm long and 0.3 mm wide, ellipsoid, inserted on the margin of the tube and curved over the stigma. Ovary c. 0.1 mm high and 0.3 mm wide, depressed-globose, densely covered with pale brown peltate scales which have a long fimbriate margin; loculi 2; stigma c. 0.2 mm high, 0.2 mm wide, subglobose.

Female inflorescence 15 cm long and 6 cm wide; peduncle 4 cm. Female flowers c. 3 mm long and 3 mm wide; pedicels 0.2 – 1 mm, the calyx and pedicel densely covered with scales like those on the male flower. Calyx 2 – 3 mm long, cup-shaped and divided into 5 obtuse ovate lobes. Petals 5, aestivation quincuncial. Staminal tube c. 1 mm high, 1.5 mm wide with an aperture 1.5 mm across; anthers 6, 0.6 mm long and 0.4 mm wide, inserted just inside the margin and pointing towards the centre of the flower. Ovary c. 0.2 mm high and 0.5 mm wide, depressed-globose, densely covered with scales like those in the male flower; loculi 2; stigma c. 0.4 mm high and 0.3 mm wide, subglobose.

Infructescence 5 – 10 cm long. Fruit 3 cm long and 1.7 cm wide, obovoid, narrowed to a stipe 7 mm long, with scales like those on the twigs densely covering the outside; peduncle 2 mm. Loculi 2, each containing 0 or 1 seed.

DISTRIBUTION. Philippine Islands only.

One specimen only seen in addition to the types. PHILIPPINE ISLANDS. Luzon, Rizal Province, ♂ fl., March 1909, *Loher* Bur. Sci. 12815 (UC!).

A. pyriformis has coriaceous leaflets which have numerous pits, often reddish-brown in colour, on both surfaces. In two of the three collections seen (the type collections of the two names), the upper surface of the leaflets has distinctive peltate scales in which the centre is very dark reddish-brown and the long fimbriate margin is white. Merrill's two species are male (*A. puncticulata*) and female (*A. pyriformis*) plants of this species.

85. Aglaia coriacea Korth. ex Miq., Ann. Mus. Bot. Lugd. Bat. 4: 57 (1868). Lectotype (designated here): Borneo, [S., G. Bahay], *Korthals* s.n. (U!; isolectotype: L!); C. de Candolle in A. & C. de Candolle, Monog. Phan. 1: 622 (1878); Pannell in Ng, Tree Flora of Malaya 4: 214 (1989).

Small tree up to 5 m, usually unbranched with up to 10 leaves in a spiral at the apex, but occasionally with 1 – 2 branches in the upper part of the tree. Bole up to 10 cm in circumference. Bark brown with green and grey patches, with longitudinal and transverse cracks; inner bark dark pinkish-red; sapwood slightly paler than inner bark; heartwood pale pinkish-red or yellowish-brown; densely covered with reddish-brown stellate hairs at the apex. Leaves imparipinnate, up to 120 cm long and 90 cm wide, elliptical in outline; petiole up to 35 cm, patent, the rhachis descending, the petiole, rhachis and petiolules densely covered with often deciduous reddish-brown stellate hairs which have a dense cluster of short arms and a few up to 0.5 mm long.

Leaflets 7 – 15, the laterals subopposite, all 13 – 43 cm long, 4 – 9 cm wide, dark glossy green above, paler below, coriaceous, usually oblong, sometimes obovate-oblong, recurved at margin, acuminate at apex with the obtuse or acute acumen up to 25 mm long, narrowed to a shortly cuneate or rounded sometimes asymmetrical base, with reddish-brown stellate hairs which are numerous on upper surface when young but deciduous before maturity usually densely covering and conspicuous on the midrib on lower surface (but sometimes sparse) and numerous on the lower surface when young but usually deciduous before maturity; veins 11 – 33 on each side of the midrib, ascending and curved upwards near the margin, midrib prominent and lateral veins subprominent on lower surface; petiolules 1 – 35 mm.

Inflorescence up to 6 cm long and 6 cm wide, usually in the axils of the leaves, sometimes on the upper part of the stem below the lowest leaves; peduncle up to 1 cm, the peduncle, rhachis and branches densely covered with reddish-brown stellate hairs. Flowers c. 2.5 cm long and 2 cm wide, obovoid; pedicels up to 2.5 mm long densely covered with reddish-brown stellate hairs. Calyx c. $1/4$ the length of the corolla, with indumentum like the pedicels, deeply divided into 5 obtuse lobes, patent at anthesis. Petals 5, elliptical, aestivation quincuncial. Staminal tube c. $3/4$ the length of the corolla, obovoid, the aperture c. 0.6 mm in diameter and shallowly 5-lobed; anthers 5, c. $1/2$ the length of the staminal tube, ovoid, in the upper half of the tube and just protruding beyond the aperture. Ovary small, depressed-globose; stigma c. $1/3$ the length of the staminal tube, narrowly cylindrical, with two small apical lobes. Infructescence with c. 6 fruits at different stages of ripening, tightly clustered at the end of a peduncle c. 9 cm long. Fruit 2.3 – 4 cm long, 1.8 – 3.5 cm wide, ellipsoid, brown, densely covered with dark brown stellate hairs like those on the twigs; pericarp 0.5 – 1 mm thick, leathery, inner surface white. Loculi 1 – 3, septa disintegrating in ripe fruit. Seeds 2 – 3.5 cm long, 1.5 – 2

Section Aglaia

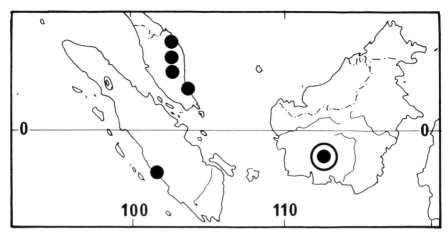

Fig. 88. Distribution of *A. coriacea*.

cm wide with inner surfaces flattened; aril 0.5 – 1 mm thick, the flesh translucent, white, sweet and juicy. Shoot axis 2 – 3 mm long, covered with dark brown stellate hairs.

DISTRIBUTION. E. & C. Peninsular Malaysia, S. Borneo (known only from the type collection). Fig. 88.

ECOLOGY. Found in primary forest. Alt.: up to 270 m.

VERNACULAR NAMES. Peninsular Malaysia: Lek Nyor (Orang Asli).

Representative specimens. PENINSULAR MALAYSIA. Krau Game Reserve, K. Lompat, fr., 22 Aug. 1979, *Pannell* 1407 (FHO!); Johore: N.E., Labis, Sg. Juasseh, fr. 28 June 1970, *Samsuri* S 282 (K!); Labis Sg. Juasseh, fl., 19 Feb. 1971, *Whitmore* FRI 15911 (FRI!).

A small unbranched tree with a terminal cluster of large pinnate leaves. The leaflets are coriaceous, when young the dark glossy green upper surface contrasts with the paler lower surface which has dense bright reddish-brown stellate hairs along the midrib and veins.

86. Aglaia odorata Loureiro, Fl. Cochinch. 1: 173 (1790). No type specimen found. Type: Rumph., Herb. Amb. 5: 28, t. 18, fig. 1 (1747), as *Camunium sinense*. Miq., Ann. Mus. Bot. Lugd. Bat. 4: 48 (1868); C. de Candolle in A. & C. de Candolle, Monog. Phan. 1: 602 (1878); Hiern in Hooker fil., Fl. Brit. India 1: 554 (1875); King in Journ. As. Soc. Bengal 64: 62 (1895); Koorders & Valeton, Atlas der Baumarten von Java, 1: t. 159 (1913); I.H. Burkill, Dictionary of Economic Products of the Malay Peninsula 1: 74 (1935); Harms in Engler & Prantl, Planzenfam., ed. 2, 19b1: 143, fig. 31 A – D, (1940); Corner, Wayside Trees Malaya 1: 456, fig. 150, 2: t. 174 (1940), ed. 3, 2: 495, t. 145 (1988); Dimitri, De la Revista de Invest. Agric. 3: 43 – 45 (1949); Backer and Bakhuizen, Fl. Java 2: 128 (1965); Huang, Pollen Flora of Taiwan: 166, t. 106: 45 – 47 (1972); Dai in Novit. Syst. Plant. Vasc. 14: 182 – 184 (1977); C.Y. Wu, Flora Yunnanica 1: 239 (1977).

Opilia odorata Sprengel, Syst. Veg. ed 6, 1: 766 (1825).
Camunium sinense [Rumph., Herb. Amb. 5: 28, t. 18, fig. 1 (1747) Roxb., Hort. Beng. 18 (1814, 'chinensis')], Roxb., Fl. Ind. 2: 425 (1824, 'chinense'); Pierre, Fl. Forest. Cochinch. Fasc. 21, sub. t. 334 (1 July 1895).
[*Aglaia pentaphylla* Kurz ex Miq., Ann. Mus. Bot. Lugd. Bat. 4: 48 (1868) nom. in syn.]
Aglaia odorata Lour. var. *microphyllina* C. de Candolle, Monog. Phan. 1: 602 (1878). Lectotype (designated here): China, *Parker* s.n. (G-DC!).
Aglaia sinensis Pierre, Fl. Forest. Cochinch. Fasc. 21, sub. t. 334 (1 July 1895).
Aglaia chaudocensis Pierre, Fl. Forest. Cochinch. Fasc. 22, ante t. 339B (1 July 1896). Lectotype (designated here): Cochinchina, ad montem Chiung Diang in proefect. Cambodiae ubi spontanea, Dec. 1867, *Pierre* 4265 (P!; isolectotypes: BM!, G!, K!, L!).
? *Aglaia chaudocensis* Pierre var. *angustifolia* Pierre, Fl. Forest. Cochinch. Fasc. 22, ante t. 339B (1 July 1896). Type: Habite la Province de Tpong au Cambodge *Pierre* 871B (P!) (? = *Aglaia oligophylla*).
? *Aglaia chaudocensis* Pierre var. *robusta* Pierre, Fl. Forest. Coch inch. Fasc. 22, ante t. 339B (1 July 1896). Holotype: Cochinchina, ad Tintinh in proef. Thudeau [habite les bords de la rivière Be, dans la province de Bien hoa], fl., May 1866, *Pierre* 4266 (P!).
Aglaia repouensis Pierre, Fl. Forest. Cochinch. Fasc. 22, ante t. 340 B (1 July 1896). Lectotype (designated here): Cochinchina, in insula Phu Quc, fl., Feb. 1874, *Pierre* 1426 (P!).
Aglaia duperreana Pierre, Fl. Forest. Cochinch. Fasc. 22, ante t. 341B (1 July 1896). Lectotype (designated here): Cochinchina [Vietnam], in proefentara Saigon culta, fl., Sep. 1869, *Pierre* 467 (P!; isolectotypes: K!, L!, NY!, P!).
Aglaia odorata var *chaudocensis* (Pierre) Pellegrin in Lecomte, Fl . Gén. L'Indo-Chine 1: 757 (1911).
Aglaia oblanceolata Craib in Kew Bull. 1926: 344 (1926). Lectotype (designated here): Siam, Nakawn Sawan, Me Wong, 200 m, fr., 28 May 1922, *Kerr* 6034 (K!; isolectotypes: BM!, SING!).

A bush or sometimes tree up to 10 m. Twigs smooth or with shallow longitudinal wrinkles, grey or green with yellowish-brown stellate scales or occasionally peltate scales which have a fimbriate margin, densely covering the apex only and scattered on the young twigs.

Leaves imparipinnate up to 16 cm long and 14 cm wide, obovate in outline; petiole up to 3.5 cm, often deeply furrowed on adaxial side, the petiole, rhachis and petiolule with occasional scales like those on the twigs, the rhachis sometimes narrowly winged. Leaflets 3 – 5(– 7), the laterals opposite, all 1.5 – 11 cm long, 1 – 7 cm wide usually obovate sometimes elliptical; subcoriaceous, shortly acuminate at apex with the obtuse acumen up to 5 mm long, cuneate or attenuate at the slightly asymmetrical base, often extending down the petiolule of the terminal leaflet and giving it the appearance of being winged, with very occasional scales on lower surface like those on the twigs; veins 5 – 9 on each side of the midrib, curved upwards, lateral and secondary veins subprominent on upper surface,

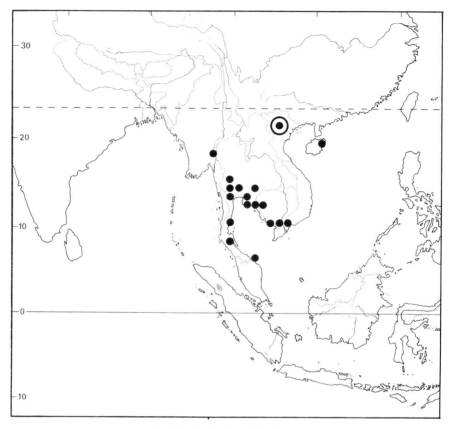

Fig. 89. Distribution of *A. odorata*.

midrib prominent, lateral and secondary veins subprominent on lower surface; sessile or with petiolules up to 0.7 mm long.

Male inflorescence up to 15 cm long and 10 cm wide with flowers widely spaced on pedicels up to 4 mm; peduncle up to 5 cm, the peduncle rhachis, branches and pedicels glabrous or with occasional scales like those on twigs. Female inflorescence smaller than the male, up to 4 cm long. Male and female flowers alike, up to 2.5 mm long, subglobose or obovoid, fragrant; pedicels up to 4 mm. Calyx $1/4 - 1/3$ the length of the corolla, ± glabrous, deeply divided into 5 subrotund lobes. Petals 5, obovate, yellow, aestivation quincuncial. Staminal tube white, nearly as long as the corolla, ovoid or narrowly cylindrical in the lower half, wider and subglobose in the upper half, incurved and shallowly 5-lobed at the apical margin; anthers 5, $1/3 - 2/3$ the length of the tube, broadly ovoid, inserted in the upper half of the tube and not quite reaching or just protruding beyond the aperture. Ovary and stigma together ovoid or narrowly ovoid, $1/3 - 1/2$ the length of the tube; ovary densely covered with yellowish-brown stellate scales; stigma longitudinally ridged, with 2 small apical lobes.

Infructescence with few obovoid fruits. Fruits brown or orange, up to 1 cm long and 0.7 cm wide, red with few to densely covered with scales like those on the twigs.

DISTRIBUTION. China (Hainan), Vietnam, Cambodia, Thailand, ? Maluku. Cultivated in India, Sri Lanka, China, Malaya, Sumatra, Java. Fig. 89.

ECOLOGY. Found in evergreen forest, secondary forest and along the coast; on clay. Alt.: 10 to 700 m. Scattered to fairly common.

USES. Commonly cultivated because of ornamental value.

VERNACULAR NAMES. Cambodia: Trayang. Thailand: Kai Tien, Kasam Nok. Peninsular Malaysia: Me Shui Lan (Chinese: fragment of rice flower). Sumatra: Patjar-tjina, Samentara-harum, Telor Blankas. Borneo: Bunga Maniran (Javanese); Mai Tsai Lan (Cantonese); Mi Sui Fa, Tjoelan. Maluku: Tjoelam.

Representative specimens. VIETNAM. fr., *Poilane* 9302 (AAU!). Annam, Province de Nghê An (Vinh), Délégation de Nga Hung, huyên de Nghia dan (Tram Lui), ♂ fl., 19 May 1914, *Fleury* in herb. *Chevalier* 32574 (AAU!, L!).

CHINA. Hainan, Ching Mai District, Pak Shik Ling and vicinity, Ku tung village: fr., 10 Nov. 1932, *Lei* 219 (L!, W!) & fr., 19 May 1933, *Lei* 677! (L!, W!).

THAILAND. Petchburi, ♂ fl., 1 Oct. 1972, *Maxwell* 72-401 (AAU!). S.E. Thailand, Rayong Province, Koh Samet Island, [12°40′N 101°25′E], low, fr., 18 March 1970, *van Beusekom & Santisuk* 3239 (AAU!).

A. odorata was described from China and Cochinchina ('Type: China. Habitat in Cochinchina & China tam agrestris, quam culta in hortis magantum, habitum speciosissima, & odore grata' (non vidi)) and the male tree is cultivated in some parts of Malesia. It is usually a small bush but sometimes a tree up to 10 m high. The 3 – 5 leaflets are small and glabrous with the secondary venation subprominent on the lower surface. *A. odorata* is planted in the grounds of Universiti Malaya, Kuala Lumpur and in Bogor Botanic Gardens; in the latter, solitary trees are clipped into a neat dome and there is a hedge about 3 m high. The female tree is rare in cultivation; the male is propagated by cuttings and bears flowers for much of the year. The flowers are strongly scented and used by the Chinese for scenting tea and by the Javanese for perfuming clothes.

87. Aglaia pleuropteris Pierre, Fl. Forest. Cochinch. Fasc. 22, ante t. 341A (1 July 1896). Lectotype (designated here): ad Chuachiang in prov. Bienhoa, Austro Cochinchina, fl. & fr. Sep. 1865, *Pierre* 1551 (P!; isolectotypes: L!, NY!).

Small tree 3 – 6 m. Twigs densely covered with brown stellate hairs or scales at the apex only.

Leaves 11 – 13 cm long, 2.5 – 3 cm wide; petiole 1 – 2.5 cm with a few pale brown peltate scales which have a fimbriate margin; the petiole and rhachis flattened and channelled on the adaxial side; the rhachis with a wing c. 1 mm wide on each side. Leaflets 11 – 15, the laterals opposite or subopposite, all 1.2 – 3 cm long, 0.6 – 1.2 cm wide, sessile or subsessile, acuminate at apex, with the short broad acumen up to 2 mm, cuneate or rounded at base, without scales or with occasional scales like those on the petiole; veins 4 – 7 on each side of the midrib, midrib prominent below, the lateral veins and reticulation subprominent on the upper and lower leaflet surfaces.

Inflorescence 4.5 – 11 cm long, 2 – 4 cm wide, with small bracts, with a few pale brown peltate scales like those on the petiole; peduncle c. 1 cm. Flower c. 2 mm long and 1.5 mm wide. Calyx cup-shaped with 5 rounded lobes. Petals 5, yellow, aestivation quincuncial. Staminal tube c. 1 mm long, 1 mm wide, cup-shaped, narrowed at the base, the aperture 0.6 mm across and shallowly lobed; anthers 5, c. 0.5 mm long and 0.4 mm wide, ovoid, inserted in the upper half of the tube, protruding and filling the aperture. Ovary 0.4 mm long, 0.3 mm wide, ovoid; locule ? 1; stigma c. 0.2 mm high and 0.15 mm wide, ovoid.

Infructescence c. 3 cm long; peduncle c. 1 cm. Fruits c. 1 cm long and 0.8 cm wide, ellipsoid, with a few scales like those on the petiole but with a longer fimbriate margin.

DISTRIBUTION. Known only from S. Vietnam and ? Cambodia.
USES: Roots used in a decoction against fever.
One collection only seen in addition to the type. ? VIETNAM or CAMBODIA. Bao Chang, fl., July 1877, *Pierre* 1802 (K!, L!).

Aglaia pleuropteris is similar to *Aglaia odorata*, but the leaves have many more leaflets.

88. Aglaia oligophylla Miq., Fl. Ind. Bat. Suppl. 1: 507 (1861). Lectotype (designated here): [Sumatra, West], Priaman, *Diepenhorst* (U!, also determined '? *Milnea montana* Jack'); Miq., Ann. Mus. Bot. Lugd. Bat. 4: 41 (1868); C. de Candolle in A. & C. de Candolle, Monog. Phan. 1: 607 (1878); Pannell in Ng, Tree Flora of Malaya 4: 222 (1989).

[*Meliacea singapureana* Wallich, Cat. 1278 (1829) nom. nud.]
Aglaia oligantha C. de Candolle in A. & C. de Candolle, Monog Phan. 1: 603 (1878). Lectotype (designated here): Ins. Philippinae, Luzon, Prov. Albay, *Cuming* 1278 (K!; isolectotypes: BM!, CGE!, FI!, G-BOIS!).
Aglaia pedicellaris C. de Candolle in A. & C. de Candolle, Monog. Phan. 1: 607 (1878). Holotype: Tenasserim [Burma] and Andamans, fl., 12 April 1899, *Helfer* 1046 (K!; isotype: W!); C. de Candolle in A. & C. de Candolle, Monog. Phan. 1: 607 (1878).
Aglaia glaucescens King in Jour. As. Soc. Beng. 64: 64 (1895). Syntypes: [S.] Andaman [Island], fl., 1884, *King's Coll.* s.n. (BM!, K!, W!).
Aglaia fusca King in Jour. As. Soc. Beng. 64: 62 (1895). Type: S. Andamans, Holidaypur, fl., 25 Apr. 1891, *King's Coll.* s.n. (K!); Pannell in Ng, Tree Flora of Malaya 4: 215 (1989).
Aglaia euphorioides Pierre, Fl. Forest. Cochinch. Fasc. 22, ante t. 338B (1 July 1896). Lectotype (designated here): in montibus Day in prov. Chaudoc, austro Cochinchina, mountain, fr., June 1876, *Pierre* 2090/1428 (P!; isolectotypes: BM!, BO!, K!).
Aglaia quocensis Pierre, Fl. Forest. Cochinch. Fasc. 22, ante t. 337B (1 July 1986). Lectotype (designated here): [Cochinchina], in montibus ad Giang Dong insulae Phu Quc, Sinus Siamensis, Nov. 1874, *Pierre* 1428 (P!; isolectotypes: BM!, K!).
Aglaia bordenii Merrill in Philipp. Gov. Lab. Bur. Bull. 17: 22 (1904).

Lectotype (designated here): Philippines, Luzon, Prov. of Bataan, Mt Mariveles, Lamao River, fl., May 1904, *Borden* For. Bur. 714 (NY!; isolectotypes: BM!, G!, K!, PNH†, SING!, US!).

Aglaia anonoides Elmer ex B.D. Jackson, *Ind. Kew.* Suppl. 10, 6 (1947) nom. in syn.

Aglaia polyantha Ridley in Kew Bull. 1930: 369 (1930). Holotype: Borneo, Sarawak, nr Kuching, fl., 6 Aug. 1894, *Haviland & Hose* 3200 (K!; isolectotype: BM!).

Aphanamixis reticulosa Kosterm. in Reinwardtia 7: 30, t. 10 (1965). Holotype: North Borneo [Sabah], Ranau District, above hot spring, 2500 ft [c. 760 m], fr., 16 Feb. 1961, *Singh* SAN 24030 (BO; isotypes: L!, SING!).

Aglaia ridleyi P.T. Li & X.M. Chen in Acta Phytotaxonomica Sinica, 22(6): 495 (1984) nom superfl. nom. illegit. pro *Aglaia polyantha* Ridley, non Pannell (1982).

Tree up to 20(– 25) m, with a narrow rounded crown. Bole up to 10 m, up to 60(– 90) cm in circumference, with small buttresses. Branches ascending or patent. Bark smooth, pinkish-brown, greyish-green or very pale grey, with orange longitudinal and round lenticels and green longitudinal streaks and transverse ridges, inner bark green or yellowish-brown; sapwood pale orange brown or pale yellow; very little white latex. Twigs slender, very pale greyish-brown, or greyish-green, densely covered with or with numerous, pale yellowish-brown stellate hairs or scales. Leaves imparipinnate, up to 40 cm long and 30 cm wide, obovate in outline; petiole up to 9 cm, the petiole, rhachis and petiolules with surface and indumentum like the twigs. Leaflets 3 – 11, the laterals opposite or subopposite, all 4.5 – 22 cm long, 2 – 9 cm wide, subcoriaceous, both surfaces rather shiny when dry, obovate or elliptical, acuminate-caudate at apex with the obtuse acumen often parallel sided and up to 10 mm long, cuneate or rounded at the asymmetrical base, with few to densely covered with pale brown stellate hairs on midrib on lower surface and occasional on the rest of the leaf; veins 5 – 10 on each side of the midrib, curved upwards, midrib slightly depressed, lateral veins subprominent and the network of secondary veins visible on upper surface, midrib prominent, lateral veins and reticulation subprominent on lower surface; petiolules up to 12 mm on lateral leaflets, up to 20 mm on terminal leaflet, enlarged at the base.

Inflorescence 10 – 20 cm long and 9 – 15 cm wide; peduncle up to 4 cm, the peduncle, rhachis and branches with surface and with numerous to densely covered with hairs like those on the twigs, the pedicels and calyx dark brown or black when dry and the calyx without or with very few pale brown stellate scales. Flowers up to 2 mm long and 2.5 mm wide, depressed-globose; pedicels up to 1.5 mm. Calyx c. $^1/_4$ the length of the corolla, black when dry, with a few scales like those on the pedicels, divided to c. $^1/_2$ way into 5 blunt lobes, which have ciliate margins. Petals 5, yellow, obovate, aestivation quincuncial. Staminal tube c. $^1/_2$ the length of the corolla, c. 1.3 mm across, depressed-globose with, the aperture c. 0.8 mm across; anthers 5, obovoid, c. $^1/_2$ the length of the tube inserted c. $^1/_2$ way up the tube and just protruding beyond the aperture. Ovary depressed-globose, densely covered with yellowish-brown stellate hairs;

Fig. 90. *A. oligophylla*. Habit with male inflorescence x½. Half flower, male x12. Habit with very young infructescence x½.

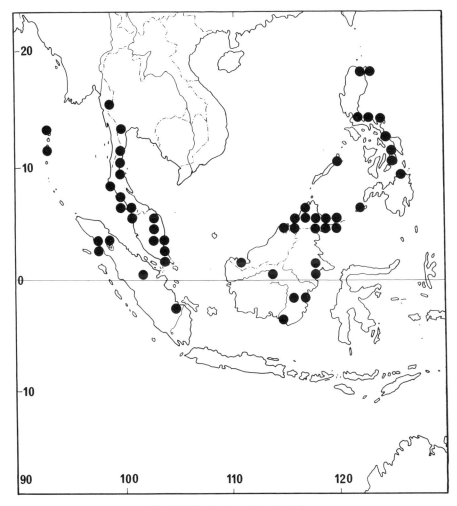

Fig. 91. Distribution of *A. oligophylla*.

stigma narrowly ovoid, densely covered with stellate hairs on the lower $2/3$, with two small shiny black glabrous apical lobes.

Infructescence with few fruits; peduncle up to 2 cm. Fruits 1 – 3 cm in diameter, subglobose; pericarp brown or yellow, either thin, hard and brittle or thick, woody and longitudinally ridged, densely covered with pale yellowish-brown stellate hairs on the outside. Loculi 1 or 2, each containing 1 seed, with a translucent gelatinous, white or brown, sweet edible aril. Fig. 90.

DISTRIBUTION. Andaman Islands, Thailand, Peninsular Malaysia, Sumatra, Borneo, Philippine Islands. Fig. 91.

ECOLOGY. Found in deciduous forest, evergreen forest, primary forest, secondary forest, swamp forest, riverine forest and in kerangas; on limestone, granite, basalt, sandy loam, clay. Alt.: sea level to 830 m. Few to rather common.

VERNACULAR NAMES. Peninsular Malaysia: Manis Sikading. Sumatra: Dukuh Hutan, Ketepan Moesang, Ketepan Moesangpoetih. Borneo: Antagiras, Mumutah (Murut); Langsat-langsat, Langsat Munyit, Sadapan Tuau (Malay); Lantupak (Dusun Kinabatangan); Segera (Iban).

Representative specimens. ANDAMAN ISLANDS. S., Namuna Ghat, fl., 22 Oct. 1990, *King's Coll.* s.n. (BO!). Holidaypur, Tusonabad, fr., 12 July 1890, *King's Coll.* s.n. (K!).

THAILAND. Satun Province, Waterfall, Koh Talutao [6°30'N, 99°45'E], 150 m, fr., 20 June 1974, *Geesink, Hattink, & Chaerenphol* 7340 (AAU!). Surat , Kaw Samui, ♂ fl., 17 Nov. 1927, *Put* 1297A (C!). Langsuan, Pang Wan, c. 100 m, ♂ fl., 15 Feb. 1927, *Kerr* 11950 (C!).

PENINSULAR MALAYSIA. Trengganu, Gunong Padang Expedition, Ulu Brang, Camp 1 nr K. Lallang, c. 180 m, fr., 15 Sep. 1969, *Whitmore* FRI 12542 (L!). Johore, Sg. Kayu, low, *Salleh* SFN 32184 (K!); Gg Panti Forest Reserve, beginning of ridge after ascent from Kg Lukut, 300 m, fr., *Pannell* 1644 (FHO!).

SUMATRA. Priaman [0°36'S 100°09'E], fl., *Diepenhorst* HB 1361 '44' (L!) . Padang Province, Ayer mancior [1°50'S 100°30'E], c. 360 m, fr. Aug. 1878. *Beccari* 810 (L!).

SABAH. Tawau, Gg Lara Forest Reserve, c. 400 m, fl., 7 July 1969, *Talip* SAN 65875 (FHO!). Kabili-Sepilok F.R., fl., 4 May 1937, *Keith* (S.H. no. 7110) FMS 41245 (FRI!). N. Borneo, Sandakan, Ranau District, c. 750 m, fr., 16 Feb. 1961, *Singh* SAN 24030 (L!).

PHILIPPINE ISLANDS. Mindanao, Agusan, Cabadbaran, Mt Urdaneta, Elmer 14116 (U!); Luzon, Sorsogon, Irosin, Mt Bulusan: *Elmer* 15273 (U!) & 16517 (U!) & 16285 (U!) & 16923 (U!) & 17177 (U!).

The leaflets of *A. oligophylla* are either small and have dense pale brown stellate hairs only on the midrib or are larger and are almost without hairs or scales on the leaflets. Both surfaces are rather shiny when dry and the secondary venation is subprominent, resembling that of *A. odorata*. It consistently differs from *A. odorata* in having terete petiolules throughout; the leaflets are usually much larger than those of *A. odorata*.

89. Aglaia simplicifolia (Bedd.) Harms in Engl. & Prantl, Pflanzenf. 3(4): 300 (1896).

Beddomea simplicifolia Bedd., Fl. Sylv. 1: t. 135 (1871). Lectotype (designated here): India, Tinnevelly hills, fr., *Beddome* '1165' (BM!).

Beddomea simplicifolia var. *parviflora* Bedd., Fl. Sylv. 1: t. 135 (1871). Type: Anamallays, [Annamallay Hills & Pulney Hills, 3 – 4000 ft [c. 900 – 1200 m]] *Beddome* 267 (BO!).

Beddomea simplicifolia var. *racemosa* Bedd., Fl. Sylv. 1: t. 135 (1871). Types: Wynad, Coorg & S. Canara (non vidi); C. de Candolle in A. & C. de Candolle, Monog. Phan. 1: 599 (1878).

? *Beddomea racemosa* Ridley in Jour. Fed. Malay States 4: 10 (1909). Type: Malaya, [Perak, Ulu Batang Padang], *Ridley* (SING?).

Aglaia meliosmoides Craib in Kew Bull. 1913: 68 (1913). Lectotype (designated here): Thailand, nr Rawng Kwang, Mê K'Mi, 210 m, fl., 14 Feb.

1912, *Kerr* 2369 (K!; isolectotypes: BM!, E!, K!); Corner in Gardens' Bull. Singapore Suppl. 1: 31 (1978); Pannell in Ng, Tree Flora of Malaya 4: 219 (1989).

Aglaia matthewsii Merrill in Philipp. Jour. Sci., Bot. 13: 79 (1918). Lectotype (designated here). British North Borneo [Sabah], fl., May – Sep. 1917, *Villamil* 368 (A!; isolectotypes: K!, PNH†, US!).

Aglaia unifoliolata Ridley in Kew Bull. 1930: 369 (1930), non Koorders. Lectotype (designated here): Borneo, Sarawak, Kuching, fl., 17 April 1893, *Haviland* 2849 (K!; isolectotypes: BM!, BO!).

Aglaia triandra Ridley in Kew Bull. 1938: 215 (1938), nom. nov. pro *A. unifoliolata* Ridley.

? *Aglaia mirabilis* Harms in Engl. & Prantl Nat. Pflanzenfam., ed. 2, 19b 1: 145, 176 (1940), nom. nov. pro *Beddomea racemosa* Ridley.

Aglaia haplophylla Harms in Notizbl. Bot. Gart. Berlin 15: 474 (1941). Lectotype (designated here): Singapore, Bukit Mandai, fl., 3rd July 1901, *Ridley* 11317 (K!).

Aglaia gagnepainiana Pellegrin in Bull. Soc. Bot. France 93: 320 (1946). Lectotype (designated here): Laos, Borikhare, Wengchan, 200 m, fl., 29 April 1932, *Kerr* 21301 (P!).

Aglaia odoardoi Merrill in Webbia 7: 312 (1950). Lectotype (designated here): Borneo, Sarawak, fl., *Beccari* 1511 (A!; isolectotypes: FI! fr., K!).

Aglaia shawiana Merrill in Webbia 7: 314 (1950). Lectotype (designated here): British North Borneo, fl., Sep. – Dec. 1917, *Agama* 461 (K!).

Aglaia heterobotrys Merrill in Jour. Arn. Arb. 35: 138 (1954). Lectotype (designated here): Sumatra, East Coast, District Kota Pinang, Subdivision Laboehan Batoe, Soengi Kanan, Si Mandi Angin, *Rahmat si Toroes* 4197 (A!; isolectotype: US!).

Aglaia sterculioides Kosterm. in Reinwardtia 7: 434 (1969). Holotype: Borneo, Amai Ambit, fr., 1893 – 94, *Hallier* 3114 (L!; isotypes: BO!, K!).

Aglaia neotenica Kosterm. in Reinwardtia 7: 433 (1969). Holotype: Borneo, Liang gagang, fr., 1893 – 94, *Hallier* 2810 (BO!; isotypes: K!, L!).

Small tree up to 8(– 20) m. Bole up to 20 cm in diameter. Outer bark greyish-brown; inner bark reddish-brown; sapwood yellow or red; latex white. Twigs greyish-brown, usually with reddish-brown stellate hairs, sometimes with peltate scales, densely covering the apex only, sparse elsewhere.

Leaves simple, 15 – 32 cm long, 4.5 – 10 m wide, acuminate or caudate at apex with the obtuse or acute acumen up to 25 mm long, cuneate at the slightly asymmetrical base, upper surface often shiny and minutely pitted, the lower surface usually with occasional stellate hairs or scales, that surface sometimes densely covered with hairs or scales, veins 11 – 18 on each side of the midrib, ascending and markedly curved upwards near the margin, midrib, lateral veins and sometimes the reticulation subprominent on upper surface, midrib prominent, lateral veins longitudinally wrinkled and barely prominent or subprominent and secondary veins usually visible on lower surface; petiole up to 4 cm, with a swelling 0.5 cm long adjacent to the lamina and with occasional hairs or scales like those on the twigs. Inflorescence up to 15 cm long and 10 cm wide, peduncle up to 1 cm, the peduncle, rhachis, branches and pedicels densely covered with stellate

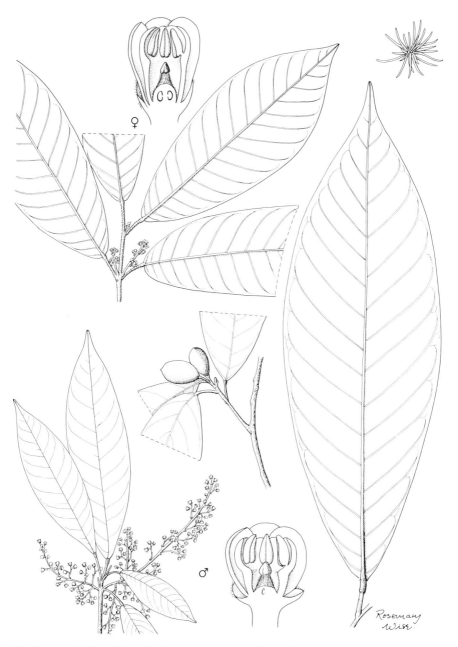

Fig. 92. *A. simplicifolia*. Habit with female inflorescences x¹/₂. Half flower, female x12. Habit with male inflorescences x¹/₂. Half flower, male x12. Large leaf (right) x¹/₂. Habit with infructescence x¹/₂. Stellate hair x70.

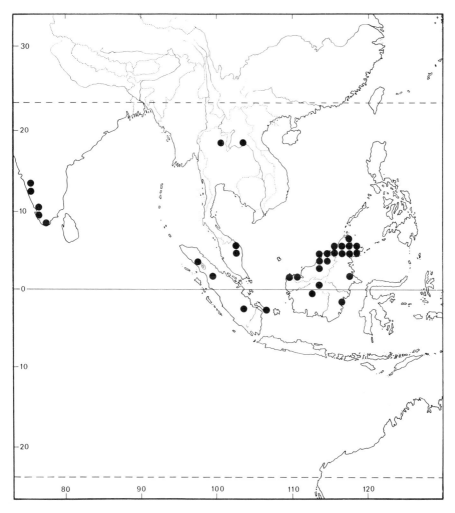

Fig. 93. Distribution of *A. simplicifolia*.

hairs like those on the twigs. Flowers up to 2 cm long, subglobose, pedicel up to 2 mm. Calyx c. $^{1}/_{3}$ the length of the corolla, deeply divided into 5 subrotund lobes which are densely covered with stellate scales. Petals 5, yellow, obovate, aestivation quincuncial. Staminal tube nearly as long as the corolla, obovoid, with a small aperture, apical margin entire; anthers c. $^{1}/_{2}$ the length of the tube, broadly ovoid, in the upper half of the tube, not or just protruding through the aperture. Ovary subglobose, densely covered with stellate scales; stigma ovoid, with two small apical lobes; ovary and stigma together c. $^{1}/_{3}$ the length of the staminal tube.

Fruits up to 4 cm long and 4 cm wide, obovoid or subglobose, brown, red, orange or pale yellow, indehiscent, with a thick woody pericarp up

Section Aglaia

to 5 mm thick and densely covered with stellate hairs on the outside; pericarp often longitudinally ridged. Loculi 1 or 2 (or sometimes 3 in India), each containing 1 seed; aril transparent, gelatinous. Fig. 92.

DISTRIBUTION. India, Laos, Thailand, Peninsular Malaysia, Sumatra, Borneo. Fig. 93.

ECOLOGY. Found in primary forest, secondary forest, evergreen forest, riverine forest and in ridge forest; understorey tree; on granite, sand, sandy loam, limestone, clay, sandstone. Alt.: 7 to 1330 m. Rare and scattered to common.

VERNACULAR NAMES. Thailand: Ham Hawk. Peninsular Malaysia: Memberas. Sumatra: Kajoe Piran. Borneo: langsat-langsat, Rengas (Malay); Segera (Iban).

Representative specimens. INDIA. Kerala State, Cannanore District, Chandanthode, 1280 m, 29 June 1965, *Ellis* 25236 (K!). Teppakulam, Karingalodathodu, fl., 7 Dec. 1961, *Anon.* 15 (K!).

THAILAND. N., Phrae, Huaytong, fl., 24 March 1961, *Phengkhlai* 72 (C!). Udawn, Pu [hill], 200 m, ?♀ fl., March 1924, *Kerr* 8604 (AAU!, C!, E!).

PENINSULAR MALAYSIA. Trengganu, Trengganu mountains, Sg. Kerbat nr K. Kerbat, Jeram Garok, N. of river, 335 m, fl. 24 June 1971, *Whitmore* FRI 20232 (FRI!). Selangor: Kajang, Sg. Lalang F.R., fl., 28 Mar. 1930, *Symington* FMS 24084 (FRI!).

SUMATRA. Atjeh [Aceh], Gunung Leuser Nature Reserves, Southern part of the reserves, Alas River valley near the mouth of the Bengkong River, c. 50 km S. of Kutacane, [c. 3°N 97°50′E], 50 – 125 m, fl., 18 July 1979, *de Wilde & de Wilde-Duyfjes* 18856 (L!).

SARAWAK. 1st Division: 25th mile, Bau – Lundu Road, Sampadi F.R., Selampit, path to Kampong, [1°60′N 110°E], c. 200 m fl., 3 July 1968, *Paie* S 26982 (FHO!); Semengoh Arboretum, mile 12, fl., 20 April 1974, *Mabberley* 1593 (FHO!). 4th Division: Miri District, Niah river, Ulu Sungei Sekaloh, [3°75′N 113°80′E], fr., 7 Dec. 1966, *Sibat ak Luang* S 27851 (FHO!); Marudi, Bok-Tisam, Bukit Mentagai [3°60′N 113°95′E], low altitude, fr., 3 May 1965, *Sibat ak Luang* S 22844 (FHO!)

KALIMANTAN. Bukit Raya [112°45′E 0°45′S], c. 130 m, fl., 28 Dec. 1982, *Nooteboom* 4420 (FHO!).

A. simplicifolia has simple leaves; the shoot apex, petiole, inflorescence and fruit are densely covered with stellate hairs. The leaves are glabrous and the secondary venation is subprominent on the upper surface of the leaves. This species resembles *A. oligophylla*, but it has simple leaves. It is variable in the texture of the leaflets which may be shiny or dull and the prominence of the midrib, lateral veins and reticulation, which are more prominent when the leaflet is shiny.

A. sterculioides is treated as a synonym of *A. simplicifolia* and appears to be indistinguishable from it except in the fruit. The infructescence of *A. sterculioides* has 1 – 3 fruits and a peduncle 4 – 13.5 cm long. The fruit is narrowly ellipsoid, 7 – 7.5 cm long, 1.7 – 2 cm wide, with a stipe 1.5 cm long and a beak 1 cm long, longitudinally ridged, splits along the most prominent ridge on the concave side, the pericarp is thin, c. 1 mm, and the seed is 3.5 cm long and 1.5 cm wide.

90. Aglaia monozyga Harms in Notizbl. Bot. Gart. Berlin. 15: 473 (1941). Lectotype (designated here): British North Borneo [Sabah], Mount Kinabalu, Tenompok, 5000 ft [c. 1500 m], fl., 24 Feb. 1932, *Clemens* 28192 (K!; isolectotypes: B†, BO!, G!, K!, L!).

Small tree 3 – 10 m, bole up to 16 cm diameter. Bark smooth, white, inner bark reddish-brown or yellowish-brown, cambium white. Twigs grey or greenish-grey, longitudinally wrinkled, densely covered with reddish-brown stellate hairs which are soon deciduous.

Leaves imparipinnate (or simple), 8 – 67 cm long, 2 – 52 cm wide, obovate in outline; petiole 2.5 – 10(– 14.5) cm long, the petiole, rhachis and petiolules densely covered with hairs like those on the twigs, deciduous. Leaflets (1 –)3 – 5, the laterals opposite, all 8 – 25(– 46) cm long and 3 – 7 (– 13) cm wide, pale green or yellowish-green when dry, coriaceous, usually lanceolate, sometimes obovate, recurved at margin when dry, densely covered with hairs like those on the twigs on both surfaces when young, deciduous, a few sometimes persisting on the midrib, the upper and lower surfaces rugulose and minutely pitted, acuminate or caudate at apex, the acute or obtuse acumen often parallel-sided and up to 2 cm long, cuneate or attenuate at the base, sometimes asymmetrical and rounded on one side; veins 6 – 15(– 23) on each side of the midrib, curved upwards, not anastomosing, the midrib prominent or subprominent on the upper and lower surfaces, the lateral veins subprominent or barely prominent on the upper and lower surfaces; petiolules 1 – 3.5(– 4) cm long, flattened or channelled on the upper side, rounded below, swollen at the base.

Inflorescences usually several per flowering shoot, the most distal often in the axil of an unopened leaf near the apex of the shoot, 11 – 3.5 cm long, 7 – 12 cm wide, peduncle up to 0.5 cm, the peduncle, rhachis and branches densely covered with reddish-brown or pale reddish-brown stellate hairs.

Inflorescence branches often with one or more bracts up to 1.3 mm long. Flowers cream. Male flower 1.2 – 2 mm long, 1.5 – 2.5 mm wide, subglobose, pedicels 0.5 – 3.5 mm, the pedicel and calyx densely covered with pale reddish-brown stellate hairs. Calyx cup-shaped, divided into 5 subrotund lobes. Petals 5, yellow or pink, aestivation quincuncial. Staminal tube cup-shaped or broadly cone-shaped 0.5 – 1.5 mm long, aperture c. 1 mm, with longitudinal thickenings forming ribs on the inside below the anthers; anthers 5, inserted on the inner margin of the tube and pointing towards the centre of the flower, c. 0.5 mm long, dehiscing by short longitudinal slits in the lower $\frac{1}{2}$ of the anthers, the apices of the anthers extending beyond these. Ovary 0.2 – 0.3 mm long depressed globose densely covered with pale brown stellate hairs which have long arms; style 0.2 mm long, stigma 0.3 – 0.4 mm long, ovoid or subglobose, or depressed globose with a shallowly lobed margin and central depression; locule 1. Female flower c. 3 mm long and 2.5 mm wide, subglobose. Calyx cup-shaped with 5 acute lobes. Petals 5, aestivation quincuncial. Staminal tube 1.5 mm long, 1.5 mm wide, cup-shaped; the aperture 1 mm; anthers as in male flowers, 0.6 – 0.7 mm long. Stigma depressed globose, with the margin lobed and with a central depression; loculi 2, each containing one ovule; otherwise like the male.

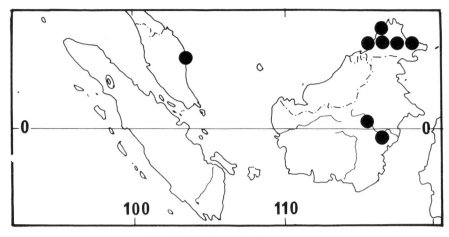

Fig. 94. Distribution of *A. monozyga*.

Infructescence up to 6 cm with 1 – 2 fruits; peduncle up to 3 cm. Fruits 1.7 – 3.5 cm long 1.3 – 3 cm in diameter, subglobose or ellipsoid and sometimes with a small beak, indehiscent, pericarp orange or orange-red densely covered with or with numerous reddish-brown hairs or scales, deciduous on the ripe fruit; aril pink.

DISTRIBUTION. E. Peninsular Malaysia (one gathering), Borneo. Fig. 94.

ECOLOGY. Found in primary forest, secondary forest, montane forest, riverine forest, and fresh water swamp forest; on sand, loam, laterite. Alt.: 17 to 2000 m.

VERNACULAR NAMES. Borneo: Beluno-beluno, Langsat-langsat (Malay); Lukut (Bisaya); Tintin (Dusun); Karang Karang.

Representative specimens. PENINSULAR MALAYSIA. Kemaman, Bt Kajang, *Corner* SFN 30370 (L!).

SABAH. Mt Kinabalu, Tenompok, c. 1500 m, fl. & fr., 8 March 1932, *Clemens* 28718 (BO!, G!, K!, L!: paralectotypes of *A. monozyga*) & fl., 29 April 1932, *J. & M.S. Clemens* 30181 (L!). Sandakan, Ulu Dusun, mile 32, fr., 16 May 1974, *Mabberley* 1687 (FHO!). Betotan, c. 25 m, fl., 26 April 1933, *Orolfo*, B.N.B. Forestry Dept. 3243 (FHO!).

KALIMANTAN. E., Central Kutei, Belajan River, Gg Kelopok, near Tabang, 50 m, fr., 13 April 1955, *Kostermans* 10441 (K!).

A. monozyga is characterised by its long petiolules, attenuate leaf base, the leaflet margin slightly recurved towards lower surface, the matt texture of lower and upper leaflet surfaces and characteristic pale green or pale yellowish-green colour of the leaves when dry.

91. – 104. Aglaia tomentosa *group*

A. tenuicaulis, A. membranifolia, A. rufinervis, A. exstipulata, A. palembanica, A. fragilis, A. brownii, A. tomentosa, A. integrifolia, A. angustifolia, A. hiernii, A. cuspidata, A. rufibarbis and *A. archboldiana* form a group of closely related species. They have reddish-brown stellate hairs which, in some species in the group, are numerous on the lower surface of the leaflets with the arms of adjacent hairs overlapping to form a continuous indumentum on that surface of the leaflets; in other species the hairs are mainly on the midrib or they are numerous on the lower surface but the arms do not overlap. The hairs are interspersed with smaller, paler, stellate hairs or, sometimes, scales. The species are separated on leaflet number, shape, hair structure and density of indumentum, correlated in some species with either flower or fruit characters.

A. tomentosa has the greatest variation in leaflet number, size and the density of indumentum. The staminal tube is usually cup-shaped and the anthers protrude beyond the aperture; it is sometimes subglobose with a small apical pore. *A. angustifolia* has numerous (13 – 21) very long narrow leaflets with indumentum like that of *A. tomentosa*. In *A. rufibarbis* the leaflets are large and obovate, the hairs are present on both leaflet surfaces and have arms up to 4 mm or even 6 mm long. The staminal tube in *A. rufibarbis* is obovoid with a minute apical pore and the anthers are included. The arms of the hairs in *A. cuspidata* are also up to 4 mm long and in *A. archboldiana* up to 1.5 mm, whereas they rarely exceed 1 mm in the other species. The leaflets of *A. hiernii* have a more dense, darker indumentum than the other species and the calyx is usually glabrous. In *A. palembanica*, the indumentum on leaves and fruits is sparse, while the number of leaflets in *A. exstipulata* is greater than in most other members of this group. *A. exstipulata* is sometimes difficult to distinguish from *A. rufinervis*; this species and *A. tenuicaulis* are also related to *A. tomentosa*.

A. tomentosa is widespread and occurs from India to New Guinea and Australia. Two species are endemic to Fiji (*A. archboldiana* and *A. fragilis*).

91. Aglaia tenuicaulis Hiern in Hooker fil., Fl. Brit. India 1: 556 (1875). Lectotype (designated here): Malaya, [? Penang], 'on the very top of the hill,' *Maingay* (Kew Dist. 335/3) (K!); C. de Candolle in A. & C. de Candolle, Monog. Phan. 1: 615 (1878); King in Jour. As. Soc. Bengal 64: 76 (1895); Ridley, Fl. Malay Penins. 1: 408 (1922); I.H. Burkill, Dictionary of Economic Products of the Malay Peninsula 1: 76 (1935); Pannell in Ng, Tree Flora of Malaya 4: 226 (1989).

Aglaia acuminatissima Teijsm. & Binn. var. *kambangana* Miq., Ann. Mus. Bot. Lugd. Bat. 4: 48 (1868). Lectotype. [Java], Noesa Kambangan, 1867, *Teysmann* (BO!; isolectotype: L!); as *A. acuminatissima* C. de Candolle in A. & C. de Candolle, Monog. Phan. 1: 625 (1878).

Small tree up to 10(– 15) m. Bole up to 30 cm in circumference; unbranched or with a few ascending branches. Bark smooth, pale brown or grey with minute longitudinal cracks; inner bark pale yellowish-brown; sapwood pale brown or pale pinkish-brown; latex white. Twigs stout, pale brown, densely covered with reddish-brown stellate hairs which have arms up to 1 mm long, with white latex.

Leaves imparipinnate up to 130 cm long and 75 cm wide, oblong in outline; petiole up to 25 cm long, the petiole, rhachis and petiolules with indumentum like the twigs. Leaflets 7 – 11, the laterals alternate or subopposite, all 15 – 45 cm long and 4 – 14 cm wide, dull dark green above, yellowish-green below, usually pale green when dry, coriaceous when old, elliptical, ovate or obovate, shortly acuminate at apex with the obtuse or acute acumen up to 15 mm long, cuneate or rounded at base, with numerous stellate hairs on lower surface which sometimes have a few long arms which overlap with adjacent hairs but usually with the arms all short and not overlapping; veins 16 – 28 on each side of the midrib, ascending and markedly curved upwards near the margin, midrib prominent and lateral veins subprominent on lower surface; petiolules up to 25 mm.

Male inflorescence up to 40 cm long and 40 cm wide. Female inflorescence up to 10 cm long and 10 cm wide; peduncle, rhachis and branches angular and longitudinally ridged. Flowers subglobose, c. 1.5 mm in diameter, fragrant; pedicels up to 2 mm. Calyx c. $1/3$ the length of the corolla, cup-shaped, densely covered with brown stellate scales on the outer surface, deeply divided into 5 lobes which are acute at the apex. Petals 5, pale yellow, obovate or elliptical, glabrous, aestivation quincuncial. Staminal tube c. $2/3$ the length of the corolla, ellipsoid, 5-lobed, with a few hairs at the base on the inner surface; anthers 5, pale yellow, ovoid c. $1/2$ the length of the tube and just protruding through the aperture. Ovary small, depressed-globose, densely covered with stellate hairs; stigma $1/2 - 2/3$ the length of the staminal tube, ovoid, obtuse at apex, black, shiny, glabrous.

Infructescence up to 15 cm long and 15 cm wide with up to 25 fruits. Fruits 1.5 – 3 cm long, up to 2.5 cm wide, ellipsoid or subglobose; pericarp yellow or orange brown on the outside, densely covered with reddish-brown stellate hairs which have many short arms and yellow on the inside. Loculi 1 or 2, each containing 1 seed; seed reddish-brown, completely surrounded by a translucent, edible, sweet aril; testa reddish-brown. Fig. 95.

DISTRIBUTION. Thailand, Malay Peninsula, Singapore, Sumatra, Bunguran Island, Borneo, ? Philippine Islands (Samar). Fig. 76.

ECOLOGY. Found in evergreen forest, riverine forest and primary forest; on sandy clay. Alt.: sea level to 1000 m. Rare to common.

VERNACULAR NAMES. Peninsular Malaysia: Bekak, Kedondong Bekak, Memberas, Sengkerlit, Tengkurok Kelang. Borneo: Battang Merah.

Representative specimens. THAILAND. Lam, Ranawng, c. 100 m, fr., 2 Feb. 1927, *Kerr* 11764 (C!).

PENINSULAR MALAYSIA. Kedah, nr Kulim, Gg Bongsu F.R., Compt 11., 100 m, ♀ fl., 24 Sep. 1963, *Pennington* 7834 (FHO!); Penang, West Hill, 835 m, ♂ fl., *Curtis* 747 (K!); Selangor/Pahang boundary, Ginting Simpah, ♀ fl. & fr., Mar. 1917, *Ridley, Robinson & Kloss* s.n. (K!); Johore, 5 miles W. of Jemaluang, fr., 16 Dec. 1963, *Pennington* 8025 (FHO!).

SINGAPORE. ♀ fl., *Lobb* s.n. (K!).

SARAWAK. Kuching, Semengoh Arboretum: Mile 12, Penrissen Road, Kuching, [1°50'N 133°33'E], 100 m, fr., 7 Mar. 1978, *Othman et al.* S 36644 (FHO!); 12th Mile, Serian Road, (Block No. 22), 50 m, fl., 5 Oct. 1983, *Paie* S 40586 (FHO!); fl. & fr., 8 Nov. 1963, *Pennington* 7952 (FHO !).

PHILIPPINE ISLANDS. Samar, Cataman, Mt Cansayao, yg fr., March – April 1951, *Sulit* PNH 14393 (L!).

A. tenuicaulis is a small, often unbranched, tree with large leaves. It is common in Peninsular Malaysia and there is considerable variation in the

Fig. 95. *A. tenuicaulis*. Leaflet x¹/₂. Male inflorescence (left & bottom) x¹/₂. Half flower, male x25. Female inflorescence x¹/₂. Half flower, female x25. Infructescence x¹/₂. Trichome x70.

size of leaflets and the density of hairs on the lower surface. The leaflets are usually large (up to 45 cm long and 14 cm wide) and the arms of adjacent hairs rarely overlap. The Borneo specimens are taller (up to 15 m) and branched; the arms of the stellate hairs are shorter than in those on specimens from the Malay Peninsula and Sumatra.

92. Aglaia membranifolia King in Jour. As. Soc. Bengal 64: 75 (1895). Syntypes: Malay Peninsula, Perak, Kinta, 300 – 500 ft [100 – 165 m], fl., Jan. 1885, *King's Coll.* 7104 (K!, P!, W!); excl. *King's Coll.* 5901 (K!); excl. *Forbes* 1679 (BM!, L!); Ridley, Fl. Malay Penins. 1: 406 (1922); Pannell in Ng, Tree Flora of Malaya 4: 221 (1989).

Tree up to 10 m. Bole up to 15 cm in circumference. Twigs stout grey with reddish-brown stellate hairs which have arms up to 1 mm long.

Leaves imparipinnate 75 – 175 cm long, up to 115 cm wide; petiole up to 25 cm, the petiole, rhachis and petiolules clothed like the twigs. Leaflets 7 – 11, the laterals subopposite, all 35 – 57 cm long and 10 – 15.5 cm wide, rather membraneous, obovate or oblong, acuminate at apex, with the obtuse or acute acumen up to 10 mm long, tapering to a sub-cordate base, with scattered hairs like those on the twigs and smaller stellate hairs with fewer arms and stellate scales interspersed on lower surface; veins 23 – 47 on each side of the midrib, ascending and markedly curved upwards near the margin, midrib prominent, lateral veins subprominent and secondary venation visible on lower surface; petiolules up to 10 mm on lateral leaflets, up to 20 mm on terminal leaflet.

Inflorescence up to 40 cm long and 12 cm wide; peduncle up to 7 cm, the peduncle rhachis, branches and pedicels with indumentum like the twigs. Flowers up to 1.5 mm in diameter, subglobose. Calyx c. $\frac{1}{3}$ the length of the corolla, densely covered with stellate hairs, deeply divided into 5 subrotund lobes. Petals 5, yellow, subrotund, aestivation quincuncial. Staminal tube c. $\frac{1}{2}$ the length of the corolla, cup-shaped slightly incurved and shallowly lobed at the apical margin; anthers c. $\frac{1}{3}$ the length of the tube, inserted just below the aperture, protruding beyond it and pointing towards the centre of the flower. Ovary subglobose densely covered with stellate hairs; stigma subglobose; ovary and stigma together c. $\frac{1}{3}$ the length of the staminal tube.

Fruits up to 2 cm long and 1.5 cm wide, pyriform, densely covered with orange-brown stellate scales. Loculi 2, each containing 1 seed.

DISTRIBUTION. Malay Peninsula, Sumatra.
ECOLOGY. Found in primary forest. Alt.: 330 to 500 m.
Representative specimens. PENINSULAR MALAYSIA. Perak: Larut, c. 60 – 90 m, fr., Jan. 1883, *King's Coll.* 3750 (K!); Larut, c. 150 – 240 m, *King's Coll.* 5159 (SING!); Larut, c. 300 – 550 m, fl., Dec. 1884, *King's Coll.* 6962 (K!, L!). Pahang, C., Bt Tapah, S. side of Krau Game Reserve, banks of Sg. Rangit, 35 m, yg fl., 10 Nov. 1969, *Whitmore* FRI 12841 (FRI!). Kuala Lumpur, Ampang F.R., st., 23 March 1978, *Pannell* 1151 (FHO!). SUMATRA. West Sumatra, nr Padang, Gunong Gaduk and Ulu Gaduk [0°55'S 100°28'E], 500 m, yg fr., 27 June 1983, *Pannell* 1939 (FHO!).

A. membranifolia resembles *A. tenuicaulis*, but the leaflets of *A. membranifolia* taper gradually to a subcordate base whereas it is rounded or cuneate in *A.*

tenuicaulis. The lateral veins are usually more numerous than in *A. tenuicaulis* and the secondary venation is conspicuous. It is known from only a few specimens from the Malay Peninsula and Sumatra.

93. Aglaia rufinervis (Blume) Bentvelzen in Acta Bot. Neerl. 11: 19 (1962); Backer and Bakhuizen, Fl. Java 2: 127 (1965).

Trichilia rufinervis Blume, Bijdr.: 164 (1825). Lectotype (designated here): Java, Mount Gede and Pangoerangoe, fr., Dec. (L!).
Heynia quinquejuga Sprengel, Syst. Veg.: 252 (1827); G. Don, Gen Syst.: 685 (1831), nom. superf. pro *Trichilia rufinervis*.
Aglaia trichostemon C. de Candolle in A. & C. de Candolle, Monog. Phan. 1: 608 (1878). Lectotype (designated here): Malaysia, Sarawak, fl., 1865 – 1868, *Beccari* 3981 (K!); King in Jour. As. Soc. Bengal 64: 77 (1895); Ridley, Fl. Malay Penins. 1: 407 (1922); I.H. Burkill, Dictionary of Economic Products of the Malay Peninsula 1: 76 (1935); Pannell in Ng, Tree Flora of Malaya 4: 227 (1989).
[*Aglaia cupanioidea* King in Jour. As. Soc. Bengal 64: 77 (1895) nom. in syn.]
Aglaia montana C. de Candolle in Ann. Conserv. & Jard. Bot. Genève 15/16: 246 (1912). Holotype: Java, Salak. Au-dessus de Tegalan-kap; chemin du sommet, c. 1250 m, fl., 10 April 1904, *Anon.* in herb. *Hochreutiner* 791 (G!); Backer and Bakhuizen, Fl. Java 2: 129 (1965).
Aglaia borneensis Merrill in Jour. As. Soc. Straits 76: 87 (1917). Lectotype (designated here): British North Borneo [Sabah], [Silimpogan], fl., Sep. – Oct. 1916, *Villamil* 247 (SING!; isolectotypes: BO!, K!, L!, P!, US!).
Aglaia winckelii Adelb. in Blumea 6: 321 (1948). Lectotype (designated here): Java, Preanger, G. Beser, Tjidadap, Tjibeber, 1000 m, 10 Nov. 1918, *Winckel* 322b (L!).

Tree up to 15 m with a small sometimes open crown. Bole up to 10 m, up to 50 cm in circumference. Bark smooth, brown or grey, with minute vertical cracks; inner bark pale orange-brown; sapwood pale orange-brown or reddish-brown; latex white. Twigs stout, dark brown, densely covered with dark brown stellate hairs.

Leaves imparipinnate, up to 100 cm long and 40 cm wide, oblanceolate in outline; petiole up to 12 cm, the petiole, rhachis and petiolules with indumentum like the twigs. Leaflets 15 – 19, the laterals sub-opposite, all 6 – 28.5 cm long, 3 – 8 cm wide, oblong or ovate or elliptical, shortly caudate at apex with the acute acumen up to 10 mm long, rounded or cuneate at the asymmetrical base, upper surface rugulose and pitted, lower surface with numerous pits and densely covered with reddish-brown stellate scales on the midrib and brown or pale brown stellate scales and hairs evenly scattered on lower surface; veins 9 – 18 on each side of the midrib, ascending and markedly curved upwards near the margin, midrib prominent, lateral veins subprominent and secondary veins visible on lower surface; petiolules up to 15(– 30) mm.

Male inflorescence up to 80 cm long and 75 cm wide, the peduncle, rhachis and branches rather flattened, shallowly longitudinally channelled, densely covered with reddish-brown stellate scales, the ultimate branches and pedicels with numerous to densely covered with scales. Flowers c.

Fig. 96. *A. rufinervis*. Habit with male inflorescence x½. Half flower, male x30. Shoot with female inflorescence x½. Infructescence x½. Stellate hair x50.

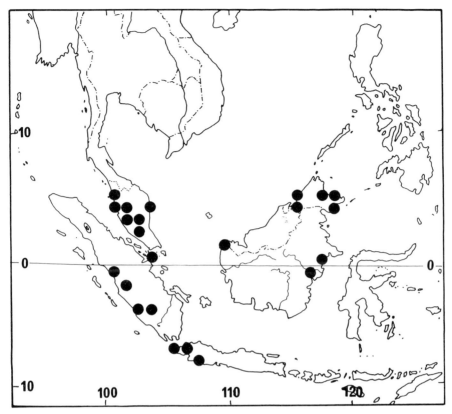

Fig. 97. Distribution of A. *rufinervis*.

1.2 mm long and 1.1 – 1.2 mm wide subglobose or slightly longer than broad, fragrant; pedicel 0.3 – 1 mm. Calyx c. 0.5 mm long with few to densely covered with reddish-brown stellate scales on the outer surface divided almost to the base into 5 rounded lobes. Petals 5, unequal, subrotund or obovoid, yellow or white; aestivation quincuncial. Staminal tube c. 0.9 mm long and 0.8 mm wide, shorter than the petals, obovoid, aperture c. 0.3 mm, shallowly 5-lobed, with pale yellow stellate hairs on the inner surface; anthers 5, c. 0.5 mm long and 0.3 mm wide, obovoid, with pale yellow stellate hairs, inserted in the upper half of the tube and just protruding through the aperture. Ovary c. 0.2 mm high and 0.3 mm wide, depressed-globose; loculus 1 containing 1 ovule; stigma c. 0.4 mm long and 0.3 mm wide, narrowly ovoid.

Infructescence up to 16 cm long and 16 cm wide. Fruits up to 20 cm in diameter, subglobose, brown or orange, densely covered with dark brown or orange-brown stellate hairs and scales on the outside; fruit-stalks up to 2 mm. Loculus 1, containing 1 seed; aril translucent, sour; testa dark brown. Fig. 96.

DISTRIBUTION. Peninsular Malaysia, Singapore, Sumatra, Borneo, Java. Fig. 97.

ECOLOGY. Found in primary and secondary forest; on coral limestone, sandy loam and clay. Alt.: sea level to 1330 m. Common.

VERNACULAR NAMES. Peninsular Malaysia: Bekak. Java: Kawauk.

Representative specimens. PENINSULAR MALAYSIA. c. 150 – 240 m, fr., Apr. 1884, *King's Coll.* 5901 (CGE!). C. Kedah, Bukit Enggang F.R., 60 m, fl. , 25 Sep. 1963, *Pennington* 7839 (FHO!).

SINGAPORE. MacRitchie Reservoir, c. 25 m, fr., 9 Dec. 1976, *Maxwell* 76-7 72 (AAU!). Bukit Timah F.R.: 15 m, fl., 13 Dec. 1963, *Pennington* 8018 (FHO!); fl., 14 Nov. 1938, *Ngadiman* SFN 35940 (FHO!, K!).

SARAWAK. 1st Division, Lundu, Gunong Gading F.R., c. 120 m, fl., 14 Nov. 1963, *Pennington* 7969 (FHO!).

SABAH. Sepilok F.R., nr Sandakan, sea level, fl., 18 Oct. 1963, *Pennington* 7898 (FHO!).

JAVA. Preanger, Takoka, fl., 3 Nov. 1896, *Koorders* 25684 (L!).

The leaves of *A. rufinervis* have 15 – 19 leaflets which distinguishes this species from *A. tenuicaulis* and *A. membranifolia*, both of which have fewer leaflets. The pale brown stellate hairs and scales scattered on the lower surface of the leaflets are often inconspicuous.

94. Aglaia exstipulata (Griffith) Theobald in F. Mason, Burmah, ed 3, 2: 583 (1883); Balakrishnan in Jour. Bombay Nat. Hist. Soc. 67: 57 (1970); Pannell in Ng, Tree Flora of Malaya 4: 215 (1989).

Euphoria [*Euphora*] *exstipulata* [*exstipulatis*] Griffith, Not. Pl. As. 4: 547 (1854). Lectotype (designated here): Burma, Mergui, fr., *Griffith* 985, (Kew Dist. 1040) (K!); C. de Candolle in A. & C. de Candolle, Monog. Phan. 1: 616 (1878) as *Euphora exstipularis*.

Aglaia longifolia Teijsm. & Binn. in Tijdschr. Ned. Ind. 27: 2 (1864). Lectotype (designated here): Java, *Teijsmann* s.n. (BO!; isolectotype: U!); C. de Candolle in A. & C. de Candolle, Monog. Phan. 1: 627 (1878).

Aglaia minutiflora Beddome var. *griffithii* Hiern in Hooker fil., Fl. Brit. Ind. 1: 557 (Feb. 1875). Lectotype (designated here): Burma, Mergui, fr., *Griffith* 985 (Kew Dist. 1040) (K!).

Aglaia griffithii (Hiern) Kurz in Jour. As. Soc. Bengal 44: 146 (1875) nom. superfl. illegit. pro *Euphoria exstipulata* Griffith; King in Jour. As. Soc. Bengal 64: 75 (1895); Ridley, Fl. Malay Penins. 1: 409 (1922); ? I.H. Burkill, Dictionary of Economic Products of the Malay Peninsula 1: 74 (1935) (states that wood used for house-building in Celebes, where this species does not occur, ? refers to *A. tomentosa*).

Tree up to 25 m, with a rounded or conical finely branched crown. Bole up to 15 m, up to 50 cm in circumference. Bark smooth, brown or greyish- brown, with lenticels in longitudinal rows; inner bark pale brown; sapwood pale yellowish-brown; latex white. Twigs slender, brown, densely covered with reddish-brown stellate hairs. Leaves imparipinnate, up to 60 cm long and 35 cm wide, oblong in outline, petiole up to 11 cm, the petiole, rhachis and petiolules with hairs like those on the twigs. Leaflets 11 – 23, the laterals subopposite, all 6 – 16 cm long, 1.5 – 4 cm wide, usually oblong, sometimes narrowly elliptical, the margin rather wavy,

Fig. 98. *A. exstipulata*. Leaf x¹/₂. Infructescence x¹/₂. Female inflorescence x¹/₂. Half flower, female x12. Male inflorescence x¹/₂. Half flower, male x12. Stellate hair x50.

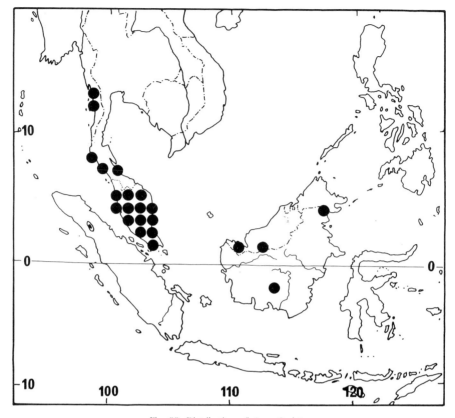

Fig. 99. Distribution of *A. exstipulata*.

acuminate or caudate at apex, with the usually acute or sometimes obtuse acumen up to 15 mm long, rounded or cuneate at the asymmetrical base, with hairs like those on the twigs on the depressed midrib on upper surface, densely covering the midrib and numerous on rest of the lower surface and with smaller, paler, fewer-rayed stellate hairs, interspersed, with numerous pits on both surfaces in Borneo; veins 7 – 16 on each side of the midrib, ascending and markedly curved upwards near the margin, midrib prominent, lateral veins subprominent on the lower surface, reticulation prominent in Sarawak; petiolules up to 10 mm on lateral leaflets, up to 20 mm on terminal leaflet.

Male inflorescence up to 40 cm long and 40 cm wide, peduncle up to 5 cm, the peduncle, rhachis, branches and pedicels densely covered with hairs like those on the twigs. Flowers minute, up to 0.8 mm in diameter, depressed-globose, fragrant of citronella; pedicels up to 0.8 mm, with minute, triangular, often deciduous bracts. Calyx c. $\frac{1}{2}$ the length of the corolla, cup-shaped, with numerous to densely covered with reddish- brown stellate hairs on the outer surface, divided almost to the base into 5 rotund, obtuse lobes. Petals 5, yellow, elliptical or obovate-rotund, glabrous, aestivation quincuncial. Staminal tube deeply divided into 5 lobes; anthers small, ovoid, pointing obliquely upwards towards the centre of the flower.

Female inflorescence smaller and less branched than in the male. Flowers larger, 1.5 mm long, obovoid. Staminal tube cup-shaped with the apical margin incurved leaving a small aperture, obscurely 5-lobed; anthers 5, minute and inserted below the rim, with a few pale yellow hairs.

Infructescence up to 25 cm long. Fruits up to 3 cm long and 1.5 cm wide, subglobose or pyriform, red, orange, brown or grey, densely covered with stellate hairs like those on the twigs. Loculi 2, each containing 1 seed; the seed surrounded by a white edible aril. Fig. 98.

DISTRIBUTION. Burma, Thailand, Vietnam, Malay Peninsula, Singapore, Borneo. Fig. 99.

ECOLOGY. Found in evergreen forest, primary forest, secondary forest, along ridges and along the road on sand, granite, shale, clay-loam. Alt.: 50 to 1400 m.

USES. Aril edible.

Representative specimens. BURMA. Mergui, *Griffith* 985 (Kew Dist. 1040) (K!); Tenasserim & Andamans fr., *Helfer* 1039 (K, W!).

VIETNAM. Annam, Tourane and vicinity, 100 km south of Hue, fl., May – July 1927, *J. & M.S. Clemens* 4209 (U!).

THAILAND, Panom Benche, Krabi, 300 m, fr., 29 Mar. 1930, *Kerr* 18760 (C!) . Lanta, Krabi, 300 m, fr., 15 April 1930, *Kerr* 18983 (C!). Klawng Ton, Satul, c. 400 m, fr., 11 March 1928, *Kerr* 14463 (C!).

PENINSULAR MALAYSIA. Penang, Pulau Bootong, 150 m, fl., *Curtis* 894 (K!, SING!). Perak, c. 450 – 600 m, fl., Sep. 1886, *King's Coll.* 10957 (K!, U!). Pahang, Kemasul F.R. near Temerloh, comp. 43. c. 60 m, fl., 8 Oct. 1963, *Pennington* 7866 (FHO!).

SINGAPORE. Bt Timah F.R., 120 m, fr., 13 Dec. 1963, *Pennington* 8019 (FHO!).

SARAWAK. 7th Division, near Sg. Apa, near foot of Bukit Bakak on eastern slope, [1°75′N 112°65′E], 400 m, fr., 4 March 1975, *Chai* S 36221 (FHO!).

A. exstipulata is intermediate in characters between *A. tomentosa* Teijsm. et Binn. and *A. palembanica* Miq. *A. tomentosa* has fewer broader leaflets which usually have a dense covering of stellate hairs on the lower surface, with the arms of adjacent hairs overlapping. In *A. exstipulata* the arms of adjacent hairs do not overlap. The leaves of *A. palembanica* usually have fewer smaller leaflets with sparser indumentum, which often consists partly of peltate scales, and the ripe fruit are smaller. *A. exstipulata* shows some variation in leaflet size and fruit shape. Specimens from high altitudes often have fewer, larger leaflets and pyriform fruits whereas those from lower elevations have subglobose fruits. The lectotype, which comes from Burma, has fewer larger leaflets, but subglobose fruits.

95. Aglaia palembanica Miq., Fl. Ind. Bat. Suppl. 1: 507 (1861). Lectotype (designated here): [Sumatra, E.], Palembang, Batu Radja, fl., 3527 H.B. (U!; isolectotypes: K!, L!); Miq., Ann. Mus. Bot. Lugd. Bat. 4: 52 (1868);

Hiern in Hooker fil. Fl. Brit. India 1: 557 (1875); C. de Candolle in A. & C. de Candolle, Monog. Phan. 1: 619 (1878); King in Jour. As. Soc. Bengal 64: 72 (1895) pro parte; Ridley, Fl. Malay. Penins. 1: 409 (1922) pro parte; Corner in Gard. Bull., Suppl. 1: 131 (1978); Pannell in Tree Flora of Malaya 4: 223 (1989).

Aglaia sipannas Miq., Fl. Ind. Bat. Suppl. 1: 197, 506 (1861). Lectotype (designated here): [Sumatra, E.], Palembang, Derma Enim, *Anon* 4792 H.B. (U!; isolectotypes: K!, L!); Miq., Ann. Mus. Bot. Lugd. Bat. 4: 53 (1868).

Aglaia pamattonis Miq., in Ann. Mus. Bot. Lugd. Bat. 4: 53 (1868). Lectotype (designated here): Borneo, [S., Mount Pamatton et Martapoera], yg fr., *Korthals* s.n. (U!; isolectotype: L!); C. de Candolle in A. & C. de Candolle, Monog. Phan. 1: 626 (1878).

Tree up to 5 m, with an irregularly rounded crown. Twigs slender, greyish-brown, densely covered with brown stellate hairs which have arms up to 0.7 mm long.

Leaves up to 36 cm long and 25 cm wide, oblong in outline; petiole up to 10 cm, the petiole, rhachis and petiolules densely covered with hairs like those on the twigs. Leaflets 9 – 13, the laterals subopposite, all 6 – 22.5 cm long, 1.3 – 3.5 cm wide, usually narrowly elliptical, sometimes lanceolate, oblong or oblanceolate, the margin slightly wavy and recurved when dry, caudate or acuminate at apex with the obtuse or acute acumen up to 15 mm long, cuneate and sometimes rounded on one side at the asymmetrical base, with stellate hairs like those on the twigs evenly scattered on lower surface and some paler stellate scales or peltate scales with a long fimbriate margin interspersed; veins 9 – 13 on each side of the midrib, ascending and markedly curved upwards near the margin, nearly or quite anastomosing, midrib prominent, lateral veins subprominent and some secondary veins visible on lower surface; petiolules up to 8 mm.

Inflorescence up to 30 cm long and 30 cm wide, the final branches up to 10 mm long and tightly packed with sessile flowers, branches clothed like the twigs. Flowers c. 1.2 mm long, subglobose or slightly longer than broad. Calyx c. 1 mm in diameter, with few or many pale brown stellate hairs on the outer surface, with 5 rotund or elliptical obtuse lobes. Petals 5, up to 1 mm long and 0.8 mm wide, yellow, elliptical or obovate, obtuse, glabrous, aestivation quincuncial. Staminal tube c. 0.8 mm long, cup-shaped with the apical margin incurved; anthers ovoid c. $1/2$ the length of the tube and just protruding beyond the aperture. Ovary depressed-globose.

Infructescence up to 10 cm long. Fruits subglobose, up to 4 mm long, brown or red, the pericarp thin and brittle and with few hairs like those on the twigs. Fig. 100.

DISTRIBUTION. S. Peninsular Malaysia, S. Sumatra, Borneo, Philippine Islands. Fig. 101.

ECOLOGY. Found on flood plains, along river banks and in primary and secondary forest; on clay, laterite, sandstone, sand, limestone. Alt.: sea-level to 450 m.

VERNACULAR NAMES. Peninsular Malaysia: Memberas. Sumatra:

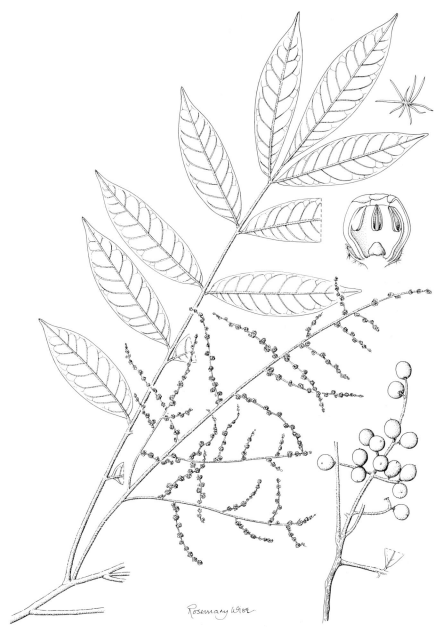

Fig. 100. *A. palembanica*. Habit with male inflorescence x¹/₂. Half flower x20, male. Infructescence x¹/₂. Stellate scale x70.

Section Aglaia

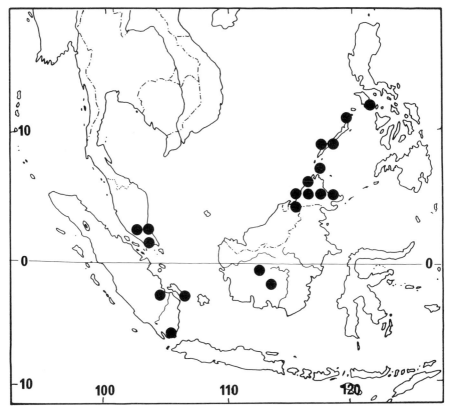

Fig. 101. Distribution of *A. palembanica*.

Pangkal pinag, Sepannas. Borneo: Koping Koping (Bajau); Lantupak (Dusun Kinabatangan); Lampasek.

Representative specimens. PENINSULAR MALAYSIA. Pahang, Jenkai F.R. near Temerloh, c. 60 m, fl., 9 Oct. 1963, *Pennington* 7869 (FHO!). Selangor, Sungei Buloh Reserve, fl., 14 April 1923, *Kaia* 8283 (FRI!).

SINGAPORE, Stagmount, up to 635 m, fl., 1909, *Ridley* 14140 (K!).

SUMATRA. Bangka: fr., 1896, *Teysmann* s.n. (L!); Toboali, fr., *Teysmann* 372. SABAH. Sandakan, Laughon (Tabilong), Kudat forest district, fl., 8 March 1951, *Cuadra* A 3177 (FRI!, SING!). Tenom District, Crocker Range, F.R., Melalap, c. 450 m, fr., 16 Nov. 1968, *Binideh* SAN 63271 (SAN!);

KALIMANTAN. Bangarmassing, fl., 1857 – 8, *Motley* (CGE!).

PHILIPPINE ISLANDS. Palawan, Taytay, fr., May 1913, *Merrill* 9317 (L!); Mindanao, Butuan Subprovince, fr., Sep. 1913, *Miranda* 20574 (L!).

A. palembanica is similar to *A. exstipulata*, but has fewer, smaller leaflets with the indumentum usually sparse. When the hairs on the lower surface of the leaflets are more numerous (as they sometimes are in the Philippine

Islands), *A. palembanica* resembles *A. tomentosa*, but it usually has more leaflets than *A. tomentosa*. The fruits of *A. palembanica* are characteristic, being small, subglobose and glabrescent. The fruits of *A. tomentosa* are sometimes glabrescent in the Philippine Islands, but they are larger than those of *A. palembanica*. Nevertheless, the distinction of *A. tomentosa* from *A. palembanica* in the Philippine Islands is some times difficult if fruits are not present.

96. Aglaia fragilis A.C.Smith in Sargentia 1: 45 (1942). Holotype: Fiji Viti Levu, Tholo North, vicinity of Nandarivatu, c. 2500 ft [c. 750 m], fl., *Degener* 14680 (A!; isotypes: K!, NY!, UC!, US!); A.C. Smith, Flora Vitiensis Nova 3: 552 (1985).

Tree up to 5 m. Young twigs densely covered with reddish-brown stellate hairs. Leaves 3 – 14 cm long, 2 – 12 cm wide; petiole 0.5 – 2.5 cm, the petiole, rhachis and petiolules densely covered with hairs like those on the twigs. Leaflets (1 –)3 – 5, the laterals subopposite, all 1.8 – 10 cm long, 1 - 3 cm wide, elliptical or ovate, occasionally obovate, apex rounded or with a short broad acumen, cuneate at the base; veins 6 – 13 on each side of the midrib, ascending at an angle of 60 – 70° to the midrib, curving upwards near the margin and anastomosing, with some shorter lateral veins in between; the lower surface with numerous hairs like those on the twigs, the arms of adjacent hairs overlapping; petiolules 1 – 3 mm on lateral leaflets, up to 20 mm on terminal leaflet.

Inflorescence up to 7 cm with widely spaced flowers; pedicels up to 5 mm; the petiole, rhachis, branches, pedicels and calyces densely covered with hairs like those on the twigs. Flowers 2.5 mm long, 3 mm across, depressed globose, pedicels 0.5 mm. Calyx divided into 5 obtuse, ovate lobes. Petals 5, thick and fleshy densely covered with stellate scales on the outside, aestivation quincuncial. Staminal tube cup-shaped, 0.8 mm long, divided almost to the base into 5 lobes; anthers 5. Ovary densely covered with stellate scales; stigma ovoid with a flattened 2-lobed apex, reaching the apices of the anthers. Fruits 1 cm long, obovoid, indehiscent; the pericarp thin, densely covered with dark reddish-brown stellate scales; 2 loculi, each containing one seed. Fig. 102.

DISTRIBUTION. Fiji only.
ECOLOGY. Forest. Alt.: 600 to 1130 m.
VERNACULAR NAME. Nauwanga.
Representative specimen. FIJI. Viti Levu, Mba (formerly Tholo North), hills east of Nandala Creek, about 3 miles south of Nandarivatu, 850 – 970 m, fr., 9 – 25 Sep. 1947, *A.C Smith* 5937 (L!).

A. fragilis resembles *A. tomentosa* in its indumentum, but its leaflets are usually smaller, the lateral veins are at a wider angle to the midrib and the apex rounded. It also resembles *A. brownii* but differs in its indumentum and venation.

97. Aglaia brownii C.M. Pannell spec. nova *Aglaiae tomentosae* Teijsm. et Binn. similis, sed differt multis squamis stellatis pallide brunneis paginam inferiorem foliolorum obtegentibus nonnullis pilis stellis interspersis; pili stellati in *Aglaia tomentosa* badii, brachiis eiusdem pili inaequa longitudine,

Fig. 102. *A. fragilis*. Habit with inflorescences x$\frac{1}{2}$. Half flower x8. Habit with fruit x$\frac{1}{2}$. Stellate hairs x40.

usque 1 mm longis, in *Aglaia brownii* pili pallidius brunnei brachiis brevioribus, cunctus plerumque simili longitudine, instructi. *Aglaia brownii* aliquantum refert *Aglaiam elaeagnoideam* (A. Juss.) Benth., a qua indumento ex pilis stellatis squamisque stellatis constante (non, ut in illa, ex squamis peltatis) differt. Holotype: Australia, Carpentaria, Cavern Island, 14 Jan. 1803, / Groote Eylandt, 15 Jan. 1803, fl., *R. Brown* 5232 (BM!)).

Small tree 2 – 12 m. Bark pale brown or grey; sapwood white. Twigs slender, densely covered with pale brown stellate scales and hairs. Leaves imparipinnate or occasionally trifoliolate, 15 – 45 cm long, 12 – 40 cm wide, obovate in outline; petiole 4 – 12 cm, the petiole, rhachis and petiolules with indumentum like the twigs. Leaflets (3 –)5 – 9, the laterals subopposite, all 4.5 – 20 cm long, 1.5 – 8 cm wide, obovate, ovate or elliptical, the leaflet margin sometimes strongly recurved (apparently when growing in dry conditions), the apex acuminate with the obtuse or acute acumen 5 – 10 mm long, cuneate or rounded at the slightly asymmetrical base, with pale brown stellate scales on the lower surface, densely covering the midrib, numerous elsewhere, sometimes interspersed with compact stellate hairs in which all the arms are usually of similar length; veins 9 – 15 on each side of the midrib, ascending and curved upwards near the margin, midrib prominent, lateral veins subprominent and reticulation often white or pale brown on the lower surface when dry; petiolules 3 – 10 mm.

Inflorescence 7 – 20 cm long, 3 – 27 cm wide, in the axils of 1 to 3 leaves near the twig apex; peduncle 1 – 4 cm, the peduncle, rhachis and branches with indumentum like the twigs. Flowers (not known whether male or female) c. 1 mm long, 1 – 1.5 mm wide, strongly perfumed; pedicels 0.5 – 2 mm densely covered with pale brown stellate hairs. Calyx c. 0.5 mm long, cup-shaped, deeply divided into 5, broadly ovate, obtuse lobes, densely covered with pale brown stellate hairs on the outer surface. Petals 5, obovate or subrotund, yellow; aestivation quincuncial. Staminal tube c. 0.5 mm long, cup-shaped, the apical margin shallowly 5-lobed; anthers (3 –)5(or 6), c. 0.4 mm long and 0.4 mm wide, ovoid, brown with a pale yellow margin when dry, inserted inside the margin of the tube and pointing towards the centre of the flower, sometimes in New Guinea with numerous pale simple hairs on the margin of the staminal tube and on the anthers. Ovary c. 0.2 mm high and 0.4 mm wide, depressed-globose, densely covered with pale brown, stellate hairs; loculi 1 or 2, each containing 0 or 1 ovule; stigma c. 0.3 mm long and 0.4 mm wide, ovoid or depressed-globose with numerous shallow longitudinal ridges.

Infructescence 9 – 19.5 cm long, 3 – 4 cm wide; peduncle 2 – 6.5 cm, the peduncle, rhachis and branches densely covered with hairs and scales like those on the twigs. Fruits c. 2 cm long and 1.5 cm wide, orange when ripe, obovoid or subglobose, indehiscent, densely covered with hairs and scales like those on the twigs; pericarp thin, brittle when dry; loculi 2, each containing 1 seed.

DISTRIBUTION. New Guinea, Australia (Northern Territory and Queensland). Fig. 103.

ECOLOGY. Found on rocks and dunes near the sea, monsoon forest on coastal dunes, fringing coastal woods, semi-deciduous notophyll vine

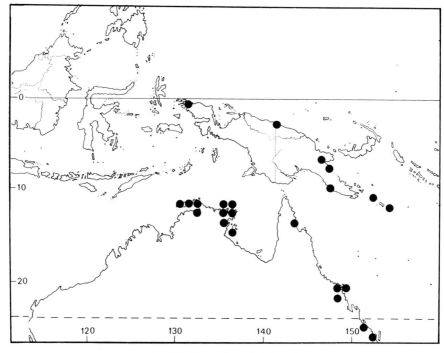

Fig. 103. Distribution of *A. brownii*.

forest at edge of beach close to mangroves, eucalyptus forest. Grows on dark brown organic soil, aeolian sands, stabilised dune or laterite.

Representative specimens. PAPUA NEW GUINEA. Morobe District, Wakaia, mountain valley, c. 1000 m, [7°50'S 147°10'E], fl., 15 Jan. 1966, *Frodin* NGF 26426 (L!, NY!). Central District, c. 43 miles S.E. of Port Moresby, Tavai Creek, [147°20'E 9°30'S], c. 100 m, fl., 4 May 1967, *Pullen* 6904 (K!, UC!).

AUSTRALIA. Northern Terrritory: Arnhem land, Yirrkala, fl., 14 Aug. 1948, *Specht* 886 (BRI!, K!, L!); Gulf of Carpentaria, Groote Eylandt, Hemple Bay, fr., 5 May 1948, *Specht* 368 (BRI!, K!, L!, PERTH!); West Alligator River, st., May 1978, *Webb & Tracey* 12434 (BRI!, QRS!). Queensland: Chester River [13°40'S 143°25'], 60 m, yg fl., 26 July 1977, *Hyland* 94 81 (FHO! ex QRS); Carlisle Island, north of MacKay, 30 m, fr., 6 Sep. 1986, *Sharpe & Batianoff* 4483 (BRI!, CANB!); Agnes Waters, Round Hill, fr., July 1988, *Randall & Sproule* 576 (BRI!).

Aglaia brownii resembles *A. tomentosa* but differs from it in having numerous pale brown stellate scales on the lower surface of the leaflet with some stellate hairs interspersed. The stellate hairs of *A. tomentosa* are reddish-brown, with the arms of the same hair differing in length and up to 1 mm long, whereas in *Aglaia brownii* the hairs are paler brown arms and they have shorter arms which are usually all of similar length. *Aglaia brownii* bears some resemblance to *A. elaeagnoidea* but differs from it in having an indumentum of stellate hairs and scales rather than the peltate scales which are typical of *A. elaeagnoidea*.

Aglaia brownii was included in *A. elaeagnoidea* by Bentham, but it differs markedly from that species in having a stellate indumentum.

98. Aglaia tomentosa Teijsm. & Binn. in Nat. Tijdschr. Ned. Ind. 27: 43 (1864) Lectotype (designated here): [Sumatra], Bangka [Island], Plangas Djiboes, *Teysmann* (BO!); Pannell in Ng, Tree Flora of Malaya 4: 226 (1989).

Argophilum pinnatum Blanco, Fl. Filip., ed 1: 186 (1837); ed. 2: 131 (1845); ed. 3: 235 (1877), non *Aglaia pinnata* (L.) Druce (1914). Neotype: Philippines, Alabat Island, 23 Dec. 1916, *Merrill*, Species Blancoanae 10 55 (A!; isoneotypes: BO!, GH!, L!, W!).

Aglaia rufa sensu Miq., Ann. Mus. Bot. Lugd. Bat. 4: 49 (1868) quoad descript. excl. *Milnea dulcis* Teijsm. et Binn.

Aglaia rufa Miq., Ann. Mus. Bot. Lugd. Bat. 4: 49 (1868) superfl. nom. illegit. pro *Milnea dulcis* Teijsm. et Binn. Lectotype (designated here): Borneo [? Sumatra], *Korthals* s.n. (K!, L!); C. de Candolle in A. C. de Candolle, Monog. Phan. 1: 613 (1878).

Aglaia zippelii Miq., Ann. Mus. Bot. Lugd. Bat. 4: 54 (1868). Lectotype (designated here): Nova Guinea, fl., *Zippelius* (U!; isolectotype: L!); C. de Candolle in A. & C. de Candolle, Monog. Phan. 1: 624 (1878).

Aglaia minutiflora Bedd., Icon. Pl. Ind. Or. 1: 44, t. 193 (1874). Lectotype (designated here): India, Tinnevelly ghats, fl., *Beddome* '1137'(BM!); Hiern in Hooker fil., Fl. Brit. India 1: 557 (1875) pro parte; C. de Candolle in A. & C. de Candolle, Monog. Phan. 1: 616 (1878).

[*Aglaia polyantha* Bedd., Icon. Pl. Ind. Or.: 44 (1874) nom. in syn.]

Aglaia minutiflora Bedd. var. *travancorica* Hiern in Hooker fil., Fl. Brit. Ind. 1: 557 (1875). Lectotype (designated here): India, Travancore, fl., March 1872, *Beddome* 246 (K!).

Aglaia cordata Hiern in Hooker fil., Fl. Brit. India 1: 557 (1875). Lectotype (designated here): [Peninsular Malaysia], Malacca, 1868, *Maingay* 2969 (Kew Dist. 335/2) (K!); C. de Candolle in A. & C. de Candolle, Monog. Phan. 1: 618 (1878) pro parte; King in Jour. As. Soc. Bengal 64: 73 (1895) pro parte; Ridley Fl. Malay Penins. 1: 409 (1922) pro parte; I.H. Burkill, Dictionary of Economic Products of the Malay Peninsula 1: 73 (1935); Pannell in Ng, Tree Flora of Malaya 4: 214 (1989).

Milnea harmandiana Pierre, Fl. Forest. Cochinch. Fasc. 21, ante t. 333 (1 July 1895). Lectotype (designated here): [Laos, plateau d'] Attopeu [sur la rive du Mékong], fr., March 1872, *Harmand* 1213 in herb. *Pierre* 5740 (P!).

Aglaia harmandiana Pierre, Fl. For. Cochinch. Fasc. 21, sub t. 333 (1895).

[*Aglaia palembanica* Miq. var. *borneensis* Miq. ex Koorders in Meded . 'S Lands Plantent. 19: 382 (1898) nom. nud.].

Aglaia dyeri Koord. in Meded. 'S Lands Plantent. 19: 634 (1898). Lectotype (designated here): Celebes [Sulawesi], Minahassa, Menado, fl., 8 March 1895, *Koorders* 17919 (BO!; isolectotypes: K!, L!).

? *Aglaia ramuensis* Harms in K. Schum. & Lauterb., Fl. Deutsch. Südsee:

3 86 (1900). Type: Kaiser Wilhelmsland [New Guinea], Bismarck-Gebirge, Ramu Expedition, 7 July 1899, *Rodatz u. Klink* 232 (B†).

Aglaia glomerata Merrill in Philipp. Gov. Lab. Bur. Bull. 35: 30 (1906). Lectotype (designated here): Philippines, Island of Masbate, fl., Oct. 1904, *Clark* 2524 (NY!; isolectotypes: BO!, PNH†, SING!, US!).

Aglaia pinnata (Blanco) Merrill, Sp. Blancoanae: 212 (1918), non *Aglaia pinnata* (L.) Druce (1914) (= *Vitex* sp.).

Aglaia ferruginea C.T. White & Francis in Proc. Roy. Soc. Queensl. 35: 66 (1923) 66. Holotype: Australia, Queensland, Atherton, Tableland, fl., 8 Jan. 1918, *C.T. White* s.n. (BRI!, photo FHO; isotype: K!).

Aglaia kabaensis Baker fil. in Jour. Bot., Lond. 62, Suppl.: 19 (1924). Lectotype (designated here): Sumatra. S., foot of Kaba [volcano, Palembang], fr., 1881 – 82, *Forbes* s.n. (BM!; isolectotypes: K!, L!).

Aglaia palembanica Miq. var. *longifolia* Craib, Florae Siamensis Enumeratio 1: 258 (1926). Type: Siam, Pattani, Banang Sta [6°16′N 101°15′E], c. 100 m, fl., 28 July 1923, *Kerr* 7400 (K!).

Aglaia elaphina Merrill & Perry in Jour. Arn. Arb. 1940 21: 316 (1940). Lectotype (designated here): Nordöstliches Neu-Guinea, Morobe Distrikt, Kulungbutu, 5300 ft [c. 1600 m], fr., 7 June 1937, *Clemens* 6551 (A!).

Usually a small tree, sometimes up to 15(– 23) m, with an irregularly rounded crown. Bole up to 9 m, up to 20 cm in diameter; branches ascending or patent. Outer bark pale reddish-brown or grey with green patches, with longitudinal cracks and lenticels in longitudinal rows; inner bark yellow, fibrous or granular; sapwood pale brown or pinkish-brown; latex white. Twigs slender, pale brown, with longitudinal and horizontal cracks and densely covered with reddish-brown or sometimes orange-brown stellate hairs which have arms up to 1 mm long.

Leaves imparipinnate, 13 – 60 cm long, 13 – 50 cm wide, obovate in outline; petiole up to 13 cm long, the petiole, rhachis and petiolules with indumentum like the twigs. Leaflets 5 – 11(– 13), the laterals opposite or subopposite, all 2.5 – 32 cm long, 1.5 – 11 cm wide, narrowly elliptical, elliptical or obovate, often recurved at the margin when dry, acuminate or caudate at apex with the obtuse or acute acumen up to 35 mm long, tapering to a cuneate, rounded or cordate asymmetrical base, with hairs like those on the twigs usually absent but sometimes densely covering the midrib on upper surface, numerous on to densely covering the midrib and veins and numerous on the rest of the lower surface, the arms of adjacent hairs usually overlapping, with smaller paler hairs which have fewer and shorter arms interspersed on the surface in between; veins 5 – 25 on each side of the midrib, ascending and markedly curved upwards near the margin, nearly or quite anastomosing, midrib prominent, lateral veins subprominent and secondary veins visible on lower surface; sessile or with petiolules up to 10 mm on lateral leaflets, up to 20 mm on the terminal leaflet.

Inflorescence up to 9 – 18 cm long, 3 – 22 cm wide, with minute linear bracteoles; peduncle 1 – 3 cm, the peduncle, rhachis and branches with indumentum like the twigs. Flowers on the final branches, fragrant, 1 – 4 mm long, 1 – 4 mm in diameter, subglobose, sessile. Calyx up to $\frac{1}{2}$ the length of the corolla densely covered with stellate hairs on the outside, deeply divided into 5 acute or obtuse and subrotund lobes which have ciliate margins. Petals 5, white or yellow, subrotund or obovate, yellow, glabrous, aestivation quincuncial. Staminal tube c. $\frac{1}{2}$ the length of the corolla, either cup-shaped, slightly incurved and shallowly 5-lobed at the apical margin or subglobose, c. 1 mm in diameter with the aperture c. 0.4 mm across; anthers 5, $\frac{1}{2}$ to as long as the length of the tube, broadly ovoid, inserted near the base or just below the margin of the tube, usually protruding, curved and pointing towards the centre of the flower. Ovary depressed-globose, densely covered with stellate hairs; stigma subglobose, longitudinally ridged, black and shiny; the ovary and stigma together $\frac{1}{3} - \frac{2}{3}$ the length of the staminal tube. The female inflorescence is smaller and with fewer branches than the male.

Infructescence 5 – 19 cm long and 15 cm wide, with up to 15 fruits; peduncle c. 1 cm, with indumentum like the twigs. Fruits 1.6 – 2.5 cm long, 1.2 – 1.7 cm in diameter, yellow, subglobose or pyriform, with indumentum like the twigs; fruit-stalks up to 5 mm. Loculi 1 – 2, each containing 0 – 1 seed. The seed with a complete orange, red or brown, gelatinous, translucent, acidic-tasting aril; testa brown.

DISTRIBUTION. S. India, Vietnam, Laos, Thailand, Peninsular Malaysia, Singapore, Sumatra, Borneo, Philippine Islands, Nusa Tenggara (Flores), Sulawesi, New Guinea, Australia. Fig. 104.

ECOLOGY. Found in evergreen forest, primary forest, secondary forest, riverine forest, montane forest, ridge forest; sometimes periodically inundated; on sandstone, alluvial, granite, limestone, sand, loam, laterite, clay. Alt.: sea level to 2000 m. Scattered to common. Fruit eaten by monkeys.

VERNACULAR NAMES. Peninsular Malaysia: Buah Patung (Temuan); Kalban (Batek); Memberas, Pasak. Sumatra: Awa Saeloe Saeloe Datan, Boenjouw, Kajoe Si Rah-rah Batoe, Lado Lado Seding, Langsat Langsat, Telar. Borneo: Kalambiao (Dusun Kalabakan); Kalambunau, Ombuakat (Tengara); Lambunau Burong, Lantupak (Dusun Kinabatangan); Langsat Munyit (Malay); Lombunou (Sungei); Menabawen (Murut); Punyau (Punan); Segera (Iban); Alin, Belajang, Bilajang, Bilajang Katjang, Buku Rantau, Bunjauw, Bunyo Koik, Djarampang Hoetan, Gimpangu, Lagah. Philippine Islands: Arangnang (Dum); Balubud, Denden, Maliamad, Malubakay (Sub.); Maybosug (Yakan). Sulawesi: Malasot. New Guinea: Borar (Kebar); Gun (Daga).

Representative specimens. INDIA. Udumanparai, Anamalais, fr., 2 May 1903, *Barber* (K!); Izerpadi, Coimbatore, fl., 20 Oct. 1901, *Barber* 3830 (K!).

VIETNAM. Annam, trail to Asclep. falls, Hoi-Mit., 40 km N. of Tourane, fl., 10 – 22 Aug, 1927, *J. & M.S. Clemens* 4209 (U!, UC!).

THAILAND. Ranong, Klongnake, Ka Per [9°58'N 98°38'E], 50 m, fr., 14 July 1979, *Niyomdham et al.* 297 (AAU!, C!). Ranong [9°20'N 98°38'E], 80 m, fl., 17 Nov. 1973, *Santisuk* 595 (AAU!). Peninsular District, Phangnga Province; Hill along new road, c. 30 km E. of Takua Pa, fl., 11 May 1968, *van Beusekom & Phengkhlai* 703 (AAU!, C!); Kao lak, S. of Takuapa, [9°00'N 98°10'E], sea level, fr., 5 May 1973, *Geesink & Santisuk* 5193 (AAU! C! , L!).

PENINSULAR MALAYSIA. C. Kedah, Sungkup F.R. Compt. 12, 75 m, fr., 26 Sep. 1963, *Pennington* 7841 (FHO!); Perak, Larut, 600 – 750 m, fr., July 1884, *King's Coll.* 6360 (K!, SING!). Pahang: Jenkai F.R. near Temerloh, 60 m, ♀ fl., 9 Oct. 1963, *Pennington* 7871 (FHO!); Temerloh Kemasul F.R., fr., 22 Feb. 1966, *Kochummen* KEP 98583 (FRI!); Rompin, Gg Lesong, fl., 10 April 1956, *Sabui* KEP 80989 (FRI!). Selangor, Ulu Gombak F.R., 600 m, fl., 12 Sep. 1963, *Pennington* 7813 (FHO!); Ulu Langat, Gunong Nacha, Kg Panso n, fr., 29 Sep. 1958, Phytochemical Survey of Malaya, *Gadeh anak Umbai for Hillard* KL 946 (FRI!, K) Johore: Labis, Ulu Sg. Segamat, base camp, 300 m, fl., *Samsuri & Ahmad* SA 726 (FRI!).

SINGAPORE. *Cantley* s.n. (SING!).

SUMATRA. Aceh: Simalur Island, fl., 21 Feb. 1918, *Achmad* 259 (U!); Gunung Leuser Nature Reserve, Ketambe, valley of Lau Alas, near tributary of Sg Ketambe, 35 km N.W. of Kutatjane, 200 – 400 m, fr., 15 Sep. 1973, *Rijksen* 15973 (L!). W. Sumatra, Barisan mountain range, Muro Kalumpi, Sg Kwantan, nr Sijunjung, [0°36'S 100°59'E], 350 m, fr., 1 March 1974 *de Vogel* 2767 (L!). Jambi Province, Kerinci District, Sungeipenuh to Bangko road, between Temiai and Muaraimat, N.W. slope of Bt Langayang [2°15'S 101°42'E], 550 m: ♀ fl. & fr., 8 Sep. 1983, *Pannell* 2034 (FHO!) & ♂ fl. , 8 Sep. 1983, *Pannell* 2036 (FHO!) & ♀ fl., 8 Sep. 1983, *Pannell* 2039 (FHO!). Bangka Island: *Ackeringa* (U!: paralectotype of *A. tomentosa*); fr. 31 Dec. 1904, (herb. *Hochreutiner* 139, *Budding* s.n. (BO!, L!); Lobok besar, 20 m fl., 24 Aug. 1949, *Kostermans & Anta* 52 (BO!, L!).

ANAMBAS & NATUNA ISLANDS. Bunguran Island, Gg Ranai, fr., 9 April 1928, *van Steenis* 1101 (U!).

SARAWAK. 1st Division, Datu P.F. [2°05'N 109°20'E], 90 m, ♂ fl. 23 May 1980, *Bernard Lee* S 41940 (FHO!). 2nd Division, Gunong Silantek, Ulu Sg . Silantek Kiri, 85th mile, S'ggang Road [1°05'N 111°08'E], 900 m, ♀ fl., 29 Aug. 1980, *Ilias Paie* S 42630 (FHO!). Hose mountains, Ulu Temiai, Mujong [2°25'N 113°63'E], 950 m, fr., 23 March 1964, *Ashton* S 16532 (FHO!). Miri, S. Ukong [4°30'N, 114°15'E], 10 m, fr., 25 May 1964, *Othman Haron* S 21401 (FHO!, L!). Baram District, Entogut River, fl., De c. 1894, *Hose* 371 (CGE!, L!). Gat, Upper Rejang River, Mt Majau, ♀ fl., 22 July 1929, *M.S. & J. Clemens* 21702 (field no. 6086) (NY!).

SABAH. Sandakan, Sepilok F.R.: Jalan Hujong Tanjong, 3 m, fl., 9 May 1974, *Mabberley* 1673 (FHO!, L!); Research plot 18, mile 15, fl., 14 May 1974, *D.J. Mabberley* 1683 (FHO!, SAN!). Beluran District, Labuk, Bukit Mangkanana [5°90'N 117°14'E], 300 m, fl., 20 Aug. 1982, *Dewol Sundaling* SAN 90525 (FHO!, L!). Tawau District, East of Brantian camp [4°28'N 118°00'E], fr., & ♀ fl., 29 April 1970, *Lentoh* SAN 69298 (L!, SAN!). mile 32 – 33 Ranau Road, Tenompok F.R. fr., 29 Oct. 1963, *Pennington* 7947 (FHO!). Lahad Datu District, Ulu Sg. Danum near Kuala Sg. Segama [5°05'N 117°55'E], ♀ fl. & fr., 31 Aug. 1976, *Stone et al.* SAN 85169 (SAN!).

KALIMANTAN. Gg Damoes, 1893 – 1894, ♂ fl., *Hallier* 426 (L!, U!); Kalimantan Selatan, Djaro Dam, c. 10 km NE of Muara Vja [1°50'S 115°40'E], 280 m, fr., 15 Nov. 1971, *de Vogel* 804 (L!); Kalimantan Timur, Sekatak, west of Tarakan, sea level, fr., 4 Feb. 1979, *Murata et al* B-4641 (BO!, L!); Sampit region, Sg. Penjahuan, N. of Sampit, 10 m, fr., 17 March 1948, *Kostermans* 4701 (L!, NY!).

PHILIPPINE ISLANDS. Tarawakan [10°N 120°E], fl., 13 Nov. 1961,

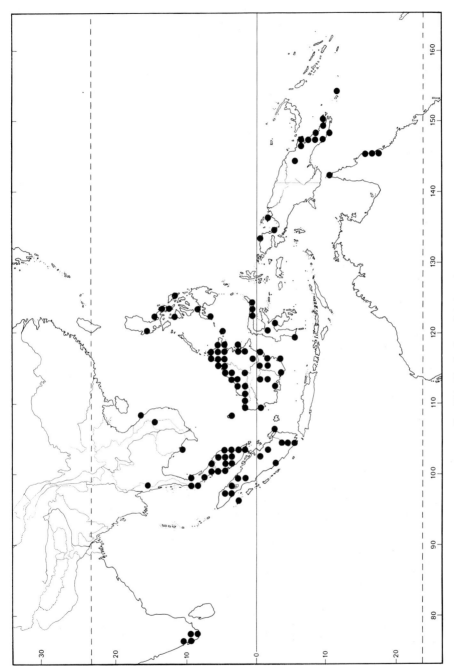

Fig. 104. Distribution of A. *tomentosa*.

Olsen 749 (FHO!). Luzon, Province of Rizal, fl., Aug. 1911, *Ramos* Bur. Sci. 1055 (U!). Basilan Island, Zamboanga Province, fl., 5 – 18 Jan. 1941, *Ebalo* 875 (L!, NY!, UC!).

SULAWESI. N., Bolaan Mongondo, between Pinogaluman and Pindol [0°46'N 123°58'E], 40 m: fl., 19 Oct. 1973, *de Vogel* 2553 (FHO!, L!) & fl., 21 Oct. 1973, *de Vogel* 2579 (FHO!, L!).

IRIAN JAYA. Lorentz, Alkmaar, 45 m, *Branderhorst* 366 (U!).

PAPUA NEW GUINEA. Morobe District, Wau Subdistrict, Bulolo-Watut Divide [7°15'S 146°37'E], 1700 m, fr., 7 April 1970, *Kairo* NGF 47661 (K!). Eastern Highlands District, Kainantu Subdistrict, Head of Arona valley, above Kassam Pass [6°10'S 146°00'E], 1350 m, fl., 9 Dec. 1968, *Womersley & Vandenberg* NGF 37415 (K!). Central District, Abau Subdistrict, Cape Rodney [10°07'S 148°18'E], 60 m, fr., 21 June 1968, *Henty* NGF 38583 (K!).

AUSTRALIA. Queensland, Barron, 6 miles south of Atherton, Reserve no. 191, 800 m: fl., 15 Sep. 1966, *McWhirter* 17 (FHO!) & fr. 15 Sep. 1966, *McWhirter* 18 (FHO!).

Aglaia tomentosa is one of the most widespread and variable species of *Aglaia* and it has about 13 closely related species, each more restricted in its range than *A. tomentosa*.

99. Aglaia integrifolia C.M. Pannell spec. nova *Aglaiae tomentosae* Teijsm. et Binn. similis sed arbor minor, foliis simplicibus; paginam inferiorem foliolorum obtegunt multi pili stellati longibrachiati pallide aurantiaco-brunnei vel badii pilis minoribus pallidioribus interspersis. Holotype: Papua New Guinea, Gulf District, Saw Mountains, near junction of Tauri & Kapau rivers, 500 m, fl., 16 March 1966, *Craven & Schodde* 1057 (K!; isotype: L!).

Small tree up to 3 m, with few branches. Twigs dark greyish-brown, densely covered with reddish-brown stellate hairs. Leaves simple, 12 – 23 cm long, 3 – 9 cm wide, dark dull or glossy green on upper leaflet surface, paler on lower surface, elliptical, acuminate at apex, the acute acumen up to 3 cm long, rounded to a subcordate base, with hairs like those on the twigs numerous on the lower surface of the leaf, the arms of adjacent hairs overlapping and with smaller, paler hairs interspersed; veins 13 – 32 on each side of the midrib, ascending, curved upwards near the margin but not anastomosing, with shorter lateral veins about a quarter of the length of the main lateral veins in between, the midrib prominent and the lateral veins subprominent on the lower surface; petiole 2 – 4 cm densely covered with hairs like those on the twigs.

Inflorescence 1.5 – 2 cm long, with linear bracteoles up to 3 mm long; the peduncle, rhachis and branches densely covered with hairs like those on the twigs. Flowers (not known whether male or female) 2 mm long and 4 mm wide, sessile or with pedicels up to 0.5 mm; pedicels and calyx with a few stellate hairs which have arms up to 1 mm long. Staminal tube c. 1 mm long and 1 mm wide, obovoid, aperture c. 1 mm wide; anthers 5, pale, half the length of the tube, inserted just inside the margin and pointing towards the centre of the flower; ovary depressed-globose, densely covered with pale stellate scales; stigma ¾ the length of the tube, broadly ovoid.

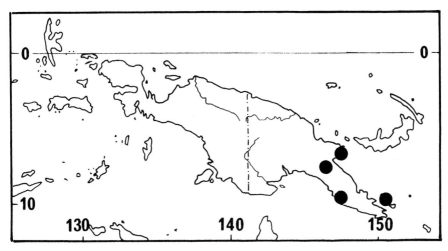

Fig. 105. Distribution of *A. integrifolia*.

Infructescences in the axils of leaves near the apex of the shoot. Fruits solitary or in pairs, subglobose, densely covered with hairs like those on the twigs.

DISTRIBUTION. Papua New Guinea only. Fig. 105.
ECOLOGY. Found in lowland and deciduous hill forest. Alt.: up to 585 m alt.
Representative specimens. PAPUA NEW GUINEA. Central District: Port Moresby Sub-district, near Veimauri River, Kuriva Forestry area [9°05′S 147°05′E], fl. & fr., 5 May 1971, *Streimann & Kairo* LAE 51520 (L!); Port Moresby, 6 km W. of Veimauri River, near Veimauri plantation, fr., 4 July 1974, *Mabberley* 1777 (FHO!, L!).

Aglaia integrifolia resembles *A. tomentosa* except that it is a smaller tree and has simple leaves. The lower surface of the leaflets have numerous pale orange-brown or reddish-brown stellate hairs, which have long arms, and smaller, paler hairs interspersed. *A. integrifolia* is the only species in the *A. tomentosa* group which has simple leaves.

100. Aglaia angustifolia (Miq.) Miq., Ann. Mus. Bot. Lugd. Bat. 4: 55 (1868). C. de Candolle in A. & C. de Candolle, Monog. Phan. 1: 617 (1878); Corner in Gardens' Bull. Suppl. 1: 131 (1978); Pannell in Ng, Tree Flora of Malaya 4: 211 (1989).

Hartighsea angustifolia Miq., Fl. Ind. Bat. Suppl. 1, 196, 504 (1861). Lectotype (designated here): Indonesia, [Sumatra, E.], Loeboe-Aloeng, [*Teijsmann*] 689 HB (U!).
? *Aglaia angustifolia* (Miq.) Miq. var. *horsefieldiana* C. de Candolle in A. & C. de Candolle, Monog. Phan. 1: 617 (1878). Type: Java, *Horsefield* (BM?).

Hearnia beccariana C. de Candolle in A. & C. de Candolle, Monog. Phan. 1: 629 (1878). Holotype: Borneo [Sarawak], fl., 1865 – 1868 [1872], *Beccari* 3604 (K!; isotype: G!).

Aglaia beccariana (C. de Candolle) Harms in Engl. & Prantl. Pflanzenf. 3(4): 298 (1896).

Aglaia stenophylla Merrill in Philipp. Jour. Sci., Bot. 11: 185 (1916). Lectotype (designated here): Philippines, Samar, Catubig River, fr., Feb. – March 1916, *Ramos* Bur. Sci. 24182 (A!; isolectotype: BM!, BO!, K!, L!, NY!, PNH†, US!).

Small tree up to 3.5(– 8) m, unbranched or rarely forked. Stem or bole up to 10 cm in diameter. Outer bark smooth, greyish-green or pale brown, with large lenticels; inner bark pale orange or brown; sapwood pale yellowish-brown, becoming orange or brown towards the centre; latex white. Twigs stout densely covered with reddish-brown stellate hairs which have arms up to 1 mm long. Leaves imparipinnate, up to 100 cm long and 80 cm wide; petiole up to 15 cm, the petiole, rhachis and petiolules clothed like the twigs.

Leaflets 13 – 21(– 25), the laterals subopposite, all 15 – 40 cm long, 1 – 4 cm wide, linear-lanceolate, the apex caudate, acute, the base asymmetrical, rounded or subcordate, with numerous to densely covered with hairs like those on the twigs on the midrib on lower surface, few to numerous on the veins, sparse to numerous on the lamina with the arms of adjacent hairs usually overlapping and with smaller paler hairs with fewer arms on the surface in between; veins 18 – 25 on each side of the midrib, curved upwards, midrib prominent on lower surface, lateral veins subprominent, secondary veins usually visible; petiolules 0 – 5 mm. Inflorescence up to 20 cm long and 12 cm wide; peduncle up to 3 cm, the peduncle, rhachis, branches and pedicels with numerous to densely covered with brown hairs like those on the twigs. Flowers c. 1.5 mm long, subglobose, sweetly fragrant; pedicels up to 1.5 mm. Calyx c. $1/2$ the length of the corolla, deeply divided into 5 elliptical lobes with numerous stellate scales on the outside. Petals 5, bright yellow or white, subrotund or obovate, aestivation quincuncial. Staminal tube c. $2/3$ the length of the corolla, pale yellow, subglobose, irregularly lobed at the apical margin, anthers 5, c. $1/2$ the length of the tube, ovoid, just protruding through the aperture. Ovary subglobose, densely covered with stellate scales; stigma c. $2/3$ the length of the tube, narrowly ovoid with two small apical lobes.

Infructescence up to 10 cm long; peduncle c. 5 mm, the peduncle, rhachis, branches and fruit-stalks with indumentum like the inflorescence.

Fruits up to 2 cm in diameter, subglobose, yellowish-green or yellowish-brown; pericarp soft and mealy, densely covered with reddish-brown stellate hairs on the outside. Loculi 2, each containing 1 seed; the seed pale yellowish-green with a complete translucent aril. Fig. 106.

DISTRIBUTION. S. Peninsular Malaysia (Johore), N. & C. Sumatra, S. & W. Borneo, Bunguran Island, Philippine Islands (known only from the type collection of *Aglaia stenophylla*). Fig. 107.

ECOLOGY. Found in primary forest, lowland and hill forest, swamp forest, sometimes along rivers. roadside, kerangas forest on sandy and yellow-red loamy soil. Often common. Alt.: 35 to 1450 m.

VERNACULAR NAMES. Pasak Bumis, Segera.

Fig. 106. *A. augustifolia*. Habit with male inflorescence x½. Half flower, male x15. Infructescence x½. Stellate hair x20.

Section Aglaia

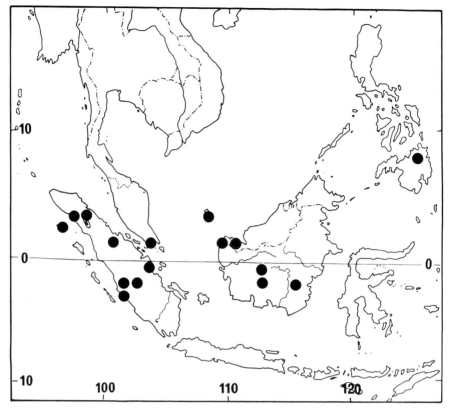

Fig. 107. Distribution of *A. angustifolia*.

Representative specimens. PENINSULAR MALAYSIA. Johore: 17 miles from Jemaluang, Mersing F.R. logging track, c. 30 m, fl., 16 Dec. 1963, *Pennington* 8026 (FHO!); Panti F.R., Kota Tinggi, c. 130 m, fr., 10 July 1961, *Yong* KEP 99212 (FRI!).

SUMATRA. Sumatera Barat [West Sumatra], eastern foothills of Barisan range, logging road S.W. from Sungaidareh, Sungei Mimpi, Ulu Batang Hari, P.T. Pasar Besar logging concession, base camp [1°01′S 101°27′], 150 m, ♂ fl., 14 June 1983, *Pannell* 1892 (FHO!).

ANAMBAS & NATUNA ISLAND. Bunguran Island, Gg Ranai, 150 m, *van Steenis* 1112 (SING!).

SARAWAK. 1st Division: Gunong Gading F.R. Lundu, c. 200 m, ♀ fl. & fr., 14 Nov. 1963, *Pennington* 7973 (FHO!); Semengoh F.R., fl., 21 Apr. 1974, *Mabberley* 1605 (FHO!); Bako National Park, nr Kuching, Telok Baku path, c. 15 m, fr., 9 Nov. 1963, *Pennington* 7962 (FHO!).

BRUNEI. Andulau Forest Reserve, Compartment 6, c. 45 m, fl., 2 May 1957, *Wood et al* SAN 17565 (K!, L!).

KALIMANTAN. Bukit Raya, [0°41′S 112°45′E], c. 250 m, fl., 30 Jan. 1983, *Nooteboom* 4776 (FHO!).

A. angustifolia is a small, unbranched tree. The large leaves have 13-21 very long and narrow leaflets. The indumentum on the lower surfaces of the leaflets is usually like that of *A. tomentosa* but may be less dense on old leaves.

101. Aglaia hiernii King in Jour. As. Soc. Bengal 64: 74 (1895). Syntypes: Malacca [Malaya], *Maingay* 2493 (Kew Dist. 335) (K!); Perak, Gopeng, 300 – 500 ft [c. 90 – 150 m], fl., April 1884, *King's Coll.* 5976 (K!); Perak, Larut, <100 ft, [c. 30 m], fl., Oct. 1884, *King's Coll.* 6706 (CGE!, SING!, U!); 10877 (SING!); Ridley, Fl. Malay Penins. 1: 408 (1922); I.H. Burkill, Dictionary of Economic Products of the Malay Peninsula 1: 74 (1935); Corner in Gard. Bull. Suppl. 1: 131 (1978); Pannell in Ng, Tree Flora of Malaya 4: 218 (1989).

[*Aglaia cordata* form 1 of Hiern in Hooker fil., Fl. Brit. India 1: 557 (1875) quoad *Maingay* 2493].
Aglaia curtisii King in Journ. As. Soc. Beng. 64: 71 (1895). Syntypes: [Peninsular Malaysia], Pangkore, Tulloh Tua, fr., July 1888, *Curtis* 1627 (K!); Malay Peninsula, Perak, 500 – 800 ft [c. 150 – 240 m], fl., July 1885, *King's Coll.* 7786 (BM!, G!, K!, P!).
Aglaia caudatifoliolata Merrill in Univ. Calif. Publ. Bot. 15: 126 (1929). Lectotype (designated here): British North Borneo, Elphinstone Province, Tawao, fl., Oct. 1922 – March 1923, *Elmer* 21572 (UC!; isolectotypes: A!, BM!, BO!, G!, GH!, K!, L!, NY!, P!, U!).
Aglaia ochneocarpa Merrill in Contrib. Arn. Arb. No. 8: 83 (1934). Lectotype (designated here): Sumatra, East Coast, Asahan Province, Masihi Forest Reserve, fl., 22 Oct., 1932, *Krukoff* 4140 (NY!).

Tree up to 30 m, with broad rounded crown. Bole up to 100 cm in circumference. Bark greenish-brown or grey with fine longitudinal lines of lenticels; inner bark green; sapwood green, pink, pale yellow or white; latex white. Branches ascending or patent. Twigs fairly stout, yellowish-brown, densely covered with dark reddish-brown stellate hairs with arms up to 1 mm long. Leaves imparipinnate, up to 70 cm long and 60 cm wide, obovate in outline; petiole up to 18 cm, the rhachis, petiole and petiolules densely covered with stellate hairs like those on the twigs. Leaflets (7 –)9(– 13), the laterals opposite, all 7 – 30 cm long, 4 – 11 cm wide, yellowish-green when young, usually markedly obovate, sometimes elliptical or oblong, recurved at the margin for up to 1 cm when dry, shortly caudate at apex with the acute acumen up to 10(– 15) mm, rounded at the base, when young the upper surface pale brown stellate and darker reddish-brown stellate hairs densely covering the midrib, numerous on the lateral veins and a few interspersed with stellate scales on the upper surface, both deciduous before maturity, lower surface with reddish-brown stellate hairs which have arms up to 1 mm long densely covering the midrib and numerous on the surface with the arms of adjacent hairs usually overlapping, with numerous pale brown stellate scales or hairs which have few ascending arms interspersed; veins 12 – 25 on each side of the midrib, ascending and markedly curved upwards near the margin, midrib prominent, lateral veins subprominent; petiolules up to 10 mm on lateral leaflets, up to 25 mm on terminal leaflet.

Inflorescence up to 35 cm long and 35 cm wide; peduncle up to

5 cm, the peduncle, rhachis and branches clothed like the twigs, the final branches up to 7 mm long and tightly packed with sessile flowers. Flowers c. 1 mm in diameter, subglobose, fragrant. Calyx c. $\frac{1}{4}$ the length of the corolla, cup-shaped, glabrous, with occasional stellate scales (in Borneo indumentum dense), divided nearly to the base into 4 or 5 sub-rotund lobes. Petals 5, 0.5 – 0.7 mm long, sub-rotund or obovate, dark yellow, aestivation quincuncial. Staminal tube shorter than the corolla, cup- shaped with the apical margin incurved leaving a small aperture; anthers 5, ovoid, inserted just below the margin of the tube, apices with pale stellate hairs which cover the aperture of the tube. Ovary 0.3 mm across densely covered with reddish-brown stellate hairs; stigma glabrous, black when dry.

Infructescence c. 35 cm long, with few fruits. Peduncle, rhachis and branches stout. Fruits 5 cm long and 3.5 cm wide, obovoid or ellipsoid, brown or yellow; pericarp 2 – 4 mm thick, woody when dry and densely covered with stellate hairs or scales on the outside. Locule 1, containing 1 seed; seed with aril c. 2.9 cm long, 1.9 cm wide and 1.7 cm through; aril c. 1.5 mm thick, translucent, pale orange, sweet and edible.

DISTRIBUTION. Peninsular Malaysia, Sumatra, Borneo. Fig. 108.

ECOLOGY. Found in primary forest, secondary forest and in old belukar; on granite, sand, sandstone, clay, clay-loam. Alt.: 20 to 1700 m.

VERNACULAR NAMES: Sumatra: Balikangin, Madang Palapah. Borneo: Jalungang Sasak (Malay); Labonoh (Kelabit); Lantupak (Kadazan); Pulu (Kayan); Segera (Iban).

Representative specimens. PENINSULAR MALAYSIA. Selangor, Rawang, Bt Kutu, fl., 9 Sep. 1936, *Kassim* KEP 119802 (FRI!). Johore, E., Sg. Kayu, low, fl., 16 Oct. 1936, *Kiah* SFN 32089 (FRI!).

SUMATRA. Aceh: Simalur Is., fl., 29 June 1918, *Achmad* 512 (U!); Takigeum , 1000 m, fr., 8 Jan. 1932. *W.N. & C.M. Bangham* 746 (A!). Palembang, Lematang Ilir, c. 75 m, fl., 17 Nov. 1924, *Boschproefst* T 965 (U!)

SARAWAK. 1st Division, Semengoh, experimental plots, ♂ fl., 27 April 1974, *Mabberley* 1630 (FHO!). 3rd Division, Bukit Raya, 2½ hours upstream from Kapit, c. 200 m, fr., 27 Nov. 1963, *Pennington* 8009 (FHO!). 7th Division, Bt Bara, Sg. Bera-an, Belaga [2°75′N 113°80′E], 415 m, ♀ fl. & yg fr., 26 Aug. 1978, *Bernard Lee* S 39837 (FHO!).

SABAH. Sandakan District, mile 57 Telupid road, c. 30 m, 12 Oct. 1977, yg fl. & fr., *Tarmiji & Paul* SAN 87647 (SAN!). Tambunan District, Trusmadi F.R., above Ulu Koingaran, c. 1520 m, fl., 1 Nov. 1964, *Mikil* SAN 41780 (SAN!).

A. hiernii is based on Maingay 2493 (Kew Dist. 353), and King's Coll. 5976, 6706, 10877. The Maingay specimen is selected as the lectotype (vide *A. cordata*). The leaflets of *A. hiernii* have a dense indumentum of dark reddish-brown stellate hairs on the lower surface. The flowers are sessile and the calyx glabrous. It appears that at high altitudes the leaves are larger and the indumentum less dense. A young tree from Maxwell's Hill (*Pannell* 1353), had very large leaves. The fruits are large and have a thick woody pericarp which distinguishes this species from others in the *tomentosa* group. *A. hiernii* is less distinct in Borneo because *A. tomentosa* may have glabrous calyx and *A. hiernii* a hairy one.

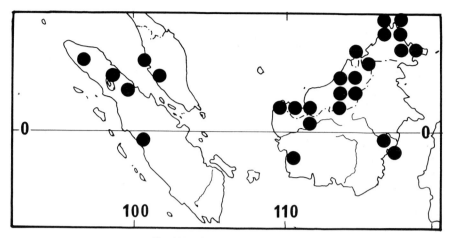

Fig. 108. Distribution of *A. hiernii*.

Some specimens from Sumatra and Borneo are intermediate between *A. hiernii* and *A. exstipulata* and are provisionally placed in *A. hiernii*. If these specimens are correctly excluded from *A. exstipulata*, that species has not been recorded from Sumatra.

102. Aglaia cuspidata C. de Candolle in Lorentz, Nova Guinea 8: 426 (1910). Lectotype (designated here): Nova Guinea neerlandica meridionalis, fluv. Lorentz, prope Bivak Alkmaar, 130 m, fr., 9 April 1908, *Branderhorst* 366 (L!; isolectotypes: BO!, K!).

Tree up to 12 m high; bole up to 8 m, up to 20 cm in diameter. Outer bark grey brown; middle bark red; inner bark pink. Sapwood cream; heartwood pinkish-brown. Twigs with numerous reddish-brown stellate hairs which have arms up to 6 mm long, dense on young twigs but soon glabrescent, dark reddish-brown stellate scales interspersed.

Leaves 44 cm long, 52 cm wide, petiole 8 – 17 cm long, the petiole rhachis and petiolules with dense hairs like those on the twigs. Leaflets 5 – 7, the laterals subopposite, all 11 – 26 cm long, 4 – 8 cm wide, usually elliptical, sometimes obovate, acuminate at apex, the acumen acute and up to 20 mm long, tapering to a subcordate base, with dense hairs like those on the twigs on the midrib below; veins 12 – 19 on each side of the midrib, ascending and curved upwards, anastomosing, the midrib prominent and lateral veins subprominent, reticulation visible. Petiolules c. 5 mm. Infructescence 37 cm long, sparsely branched, peduncle up to 24 cm long, slender and flexible. Fruits c. 2.5 – 3 cm long and 1.6 – 3 cm wide, orange, obovoid or subglobose, pericarp thin and brittle when dry, with numerous hairs like those on the twigs; locule 1, seed 1, c. 1.9 cm long, 1.7 cm wide and 1.4 cm thick, no aril seen, testa vascularised.

DISTRIBUTION. Papua New Guinea only. Fig. 109.
ECOLOGY. Found in rainforest. Alt.: c. 366 m.

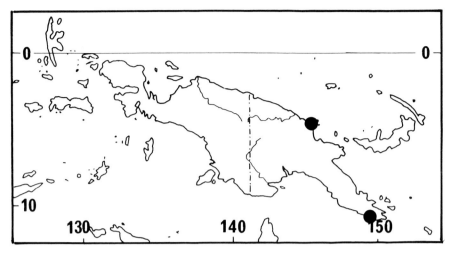

Fig. 109. Distribution of *A. cuspidata*.

Representative specimens. PAPUA NEW GUINEA. Milne Bay District, M.I. road to Mt Suckling, Rabaraba, [9°37'S 149°10'E], 366 m, fr., 9 June 1972, *Katik* NGF 46938 (K!, L!). Madang District, Josephstaal [4°45'S 145°00'E], c. 75 m, fr., 2 Sep. 1958, *White* NGF 10248 (K!, L!)

The long-armed stellate hairs of *A. cuspidata* resemble those of *A. rufibarbis*, but they are sparse on the leaflets and glabrescent on the fruits. The fruits are like those of *A. tomentosa*, with a thin brittle pericarp.

103. Aglaia rufibarbis Ridley in Jour. As. Soc. Straits Branch 75: 17 (1917). Holotype: Malaya, State of Malacca, [Mt Ophir], fl., [*Cantley*] 25 (K!); Ridley, Fl. Malay Penin. 1: 409 (1922); I.H. Burkill, Dictionary of Economic Products of the Malay Peninsula 1: 75 (1935); Pannell in Ng, Tree Flora of Malaya 4: 225 (1989).

[*A. rufa* sensu Ridley in Jour. As. Soc. Straits Branch 54: 32 (1910) quoad descript.]

Tree up to 10 m, with an ovoid crown. Bole up to 4 m, up to 21 cm in circumference. Bark usually grey and pale brown, sometimes dark brown with dark grey patches, with longitudinal cracks; inner bark green, pale yellowish-brown or orange-brown, with longitudinal striations; sapwood pale yellowish-brown or orange-brown; latex white. Branches suberect, ascending or patent with the apical shoots ascending. Twigs slender, pale brown, densely covered with reddish-brown stellate hairs which have arms up to 4 mm long.

Leaves imparipinnate, up to 85 cm long and 65 cm wide, obovate in outline; petiole 10 – 22 cm long, the petiole, rhachis and petiolules clothed like the twigs. Leaflets (5 –)7 – 9, the laterals opposite, the terminal and uppermost pair of lateral leaflets usually longer than the remainder, all (4 –)9 – 41 cm long, (3 –)4.5 – 11.5 cm wide, yellowish-green when young, turning dark green above when mature, obovate or elliptical, acuminate or caudate at apex, with the obtuse or acute acumen 5 – 25 mm long, tapering to a cuneate or subcordate, asymmetrical base, with stellate hairs like those on the twigs on both surfaces but more frequent on the lower surface, sometimes white on young leaves, with smaller paler hairs which have fewer arms interspersed; veins (7 or) 8 – 23 on each side of the midrib, ascending, markedly curved upwards near the margin and usually anastomosing, midrib and lateral veins prominent on lower surface; petiolules 2 – 7 mm on lateral leaflets, up to 40 mm on terminal leaflet. Male inflorescence up to 40 cm long and 40 cm wide, with bracts up to 3 cm long on the rhachis and up to 1.5 cm on the branches, with numerous stellate hairs; peduncle up to 6 cm, the peduncle, rhachis and branches densely covered with hairs like those on the twigs. Flowers minute, up to 1.2 mm in diameter, subglobose, fragrant; pedicels up to 2 mm, with numerous stellate hairs. Calyx $\frac{1}{3}$ – $\frac{1}{2}$ the length of the corolla, with numerous stellate hairs, deeply divided into 5 narrow, acute lobes. Petals 5 (or 6), c. 1 mm long, pale yellow, elliptical, glabrous, aestivation quincuncial. Staminal tube nearly the length of the corolla, obovoid with a minute apical pore c. 0.2 mm across; anthers less than $\frac{1}{4}$ the length of the tube, yellow when immature, brown at anthesis, turning black later, broadly ovoid, inserted in the uppermost $\frac{1}{3}$ and included within the tube. Ovary depressed-globose; stigma subglobose with two small apical lobes; ovary and stigma together c. $\frac{1}{3}$ the length of the staminal tube.

Female inflorescence up to 6 cm long and 6 cm wide, with fewer branches; peduncle up to 2 cm; the bracteoles more dense; the flowers fewer but slightly larger and arranged amongst the bracts; otherwise like the male.

Infructescence with persistent bracts and up to 10 fruits. Young fruits c. 2 cm long and 2 cm wide, subglobose, green, densely covered with often deciduous stellate hairs like those on the twigs, the longer arms often breaking off leaving a dense cluster of short arms; pericarp brittle, readily torn open, c. 1 mm thick, outer layer green, inner surface smooth, white and shiny. Loculus 1, containing 1 seed. Fig. 110.

DISTRIBUTION. Peninsular Malaysia, Borneo. Fig. 111.
ECOLOGY. Found in primary forest. Alt.: 100 to 250 m.
FIELD OBSERVATIONS. Kuala Lompat, Krau Game Reserve, Peninsular Malaysia: The branches are ascending and give the tree a conical shape. The leaves are borne in spirals on the apical shoots; they are large with 7 – 9 soft, bright green leaflets which increase in size towards the terminal leaflet and characteristically hang down from the rhachis when young. When in flower, there are usually fewer than 10 inflorescences on a tree. The male inflorescences are large, with up to 10,000 minute flowers. The staminal tube has a pin-prick aperture and at maturity the petals separate slightly, forming an opening at the apex of the flower. Their citronella scent is strongest during the day; it lessens at dusk and remains low through the night until after dawn. The female inflorescences are little- branched, with rarely more than 100 flowers. They are partly concealed among long bracts and bracteoles on which there are numerous reddish-brown hairs. The flowers have very little perfume. The inflorescences on every flowering tree in the population mature and die

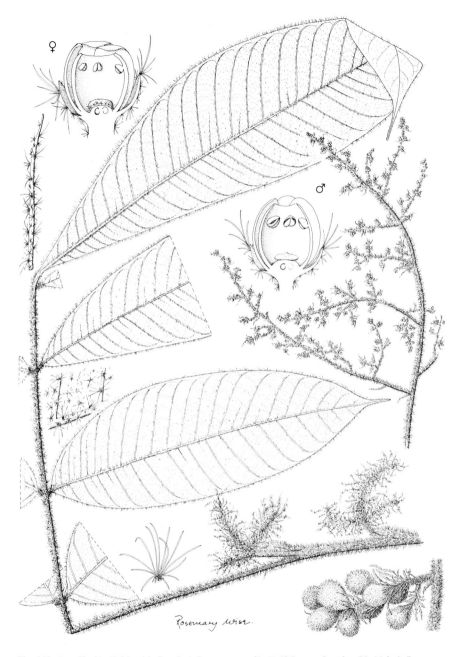

Fig. 110. *A. rufibarbis*. Habit with female inflorescences x½. Half flower, female x20. Male inflorescence x½. Half flower, male x20. Infructescence (bottom right) x½. Bract from infructescence (upper left) x3. Detail of lower leaflet surface showing indumentum of stellate hairs x½. Stellate hair x4.

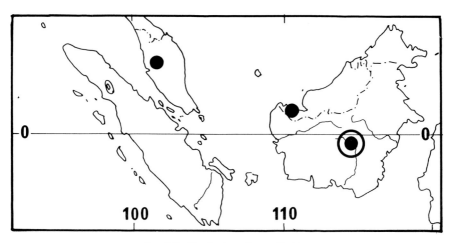

Fig. 111. Distribution of *A. rufibarbis*.

within about two weeks. The infructescences have up to ten 1-seeded, indehiscent fruits which develop more or less synchronously. The seed in the single mature fruit of this species seen was unusual in having no fleshy outer layer: the aril and outer membranous seed coat (? testa) were confined to a small area around the hilum, an inner membranous seed coat (? tegmen) was thin and papery with the main vascular bundle running through the raphe and antiraphe and branching divaricately from the raphe over the sides of the seed. The cotyledons were equal, transverse, green outside, yellow within; the shoot axis was covered with pale yellow stellate hairs.

VERNACULAR NAMES. Peninsular Malaysia: Jambu Gulong, Gempar Miam.

Representative specimens. PENINSULAR MALAYSIA. Pahang, Krau Game Reserve, Kuala Lompat [3°40′N 102°20′E]: 75 m, ♀ fl., 1 July 1981, *Pannell* 1573 (FHO!) & ♂ fl., 27 June 1981, *Pannell* 1551 (FHO!).

SARAWAK, 1st Division: Semengoh Forest Reserve, fl., 19 April 1974, *Mabberley* 1582 (FHO!); Gunong Matang, 220 m, fl., 28 April 1974, *Mabberley* 1631 (FHO!).

KALIMANTAN. Bukit Mehipit, 500 m, st., 7 Dec. 1924, *Winkler* 637 (BO!).

The indumentum of stellate hairs with long arms, up to 4 mm (which at first sight appear to be simple), is unmistakable. The twigs, rhachis and both surfaces of the leaflets have an indumentum of reddish-brown or white stellate hairs which have very long, rather sharp arms up to 4 mm, hence the specific epithet *rufibarbis*. The holotype bears a note stating that the fruit is edible and sweet, which suggests the presence of an aril surrounding the seed.

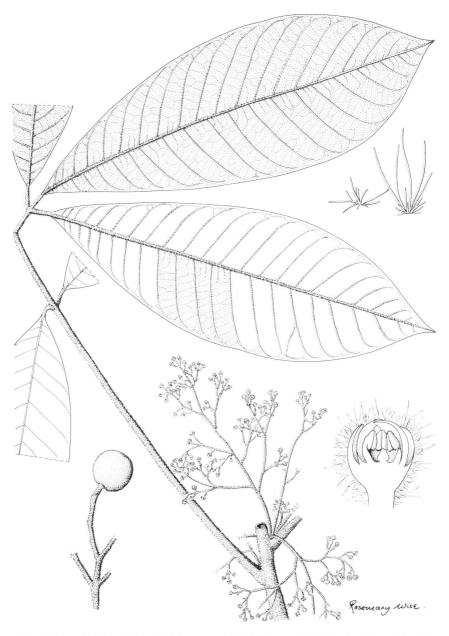

Fig. 112. *A. archboldiana*. Habit with inflorescence x½. Half flower x20. Fruit x½. Stellate hairs x25.

104. Aglaia archboldiana A.C. Smith in Sargentia 1: 44 (1942). Holotype: Fiji, Viti Levu, Serua, Ngaloa, fl. & fr., 20 Nov. – 31 Dec. 1940, *Degener & Ordonez* 13705 (A!; isotypes: K!, L!, MICH!, NY!, UC!, US!); A.C. Smith, Flora Vitiensis Nova 3: 552 (1985).

Tree up to 15 m. Twigs densely covered with orange-brown stellate hairs which have arms up to 1.5 mm long.
Leaf up to 57 cm, obovate in outline; petiole 8 – 20 cm. Leaflets (5 –)7(– 9), the laterals opposite or subopposite, all 14 – 30 cm long, 4 – 11.5 cm wide, obovate; apex acuminate, the short broad obtuse acumen up to 5 mm long; rounded or cuneate at the asymmetrical base, lower surface with (few to) numerous hairs like those on the twigs, the arms of adjacent hairs overlapping, and with numerous paler hairs interspersed; veins 12 – 20 on each side of the midrib, ascending at an angle of 60 – 70° to the midrib and curving upwards towards the margin, with some shorter intermediate lateral veins and reticulation visible; petiolules 5 – 15 mm, up to 20 mm on the terminal leaflet.
Inflorescence sessile, the ? male 11 – 25 cm long and 21 – 25 cm wide, the female about 3 cm long and 3 cm wide; with hairs like those on the twigs numerous on the branches. Flower 2.5 – 3 mm long, 2.5 – 3 mm wide, subglobose; pedicels 0 – 3 mm. Calyx 1.5 – 2.5 mm long, deeply divided into 5 lobes, the pedicel and calyx densely covered with hairs which have arms up to 1 mm. Petals 5, aestivation quincuncial. Staminal tube cup- shaped, 2 mm across, deeply divided into 5 lobes; anthers 5, 1.5 mm long, ovoid, attached near the apex of the staminal tube and pointing towards the centre of the flower. Ovary subglobose, with densely covered with stellate scales or hairs; loculi 2, each containing 1 ovule; stigma ovoid with longitudinal ridges where adpressed to the anthers and tiny bilobed apex.
Infructescence 9 – 13 cm long and c. 9 cm wide. Fruits c. 4 cm in diameter, subglobose, indehiscent; the pericarp thin; loculi 2, each containing 1 seed. Fig. 112.

DISTRIBUTION. Fiji Islands only.
ECOLOGY. Forest. Alt.: up to 970 m.
Representative specimens. FIJI. Viti Levu, Serua, hills west of Waivunu Creek, between Ngaloa and Korovou, 50 – 150 m, fl., 23 Dec. – 7 Dec. 1953, *A.C. Smith* 9231 (L!); Viti Levu, Tholo North, Mount Matomba, Nandala, vicinity of Nandarivatu, 750 – 900 m, fr., 15 – 18 Feb. 1941, *Degener* 14506 (A!).

A. archboldiana is endemic to Fiji and is distinguished from the non-Fijian species in the *A. tomentosa* group by its leaflets which have a short broad obtuse acumen and the lateral veins ascending at an angle of 60 – 70° to the midrib and by its sessile inflorescence. It has much larger leaves and is more robust than *A. fragilis*, which is the other species in this found in Fiji.

NAMES NOT PLACED, FOR WHICH NO TYPE SPECIMEN SEEN

Aglaia allocotantha Harms in Engl. Bot. Jahrb. 72: 175 (1942). Type: New Guinea, N.E., Leonhard-Schultze-Fluss, 400 m, 27 June 1912, *Ledermann* 77 25 (B†).

Aglaia araeantha Harms in Engl. Bot. Jahrb. 72: 162 (1942). Type: New Guinea, N.E., Hunsteingebirge, 1050 m, 20 Aug 1912, *Ledermann* 8467 (B†).

Aglaia bamleri Harms in K. Schum. & Lauterb., Fl. Deutsch. Südsee: 385 (1900). Type: New Guinea, Kaiser-Wilhelmsland, Sattelberg, 20 Jan. 1899, *Bamler* 50 (B†).

Aglaia bergmannii O. Warburg in Bot. Jahrb. 8: 346 (1891). Type: New Guinea, bei Stephansort an der Astrolabebay (B†).

Aglaia caroli Harms in Engl. Bot. Jahrb. 72: 176 (1942). Type: New Guinea. N.E., Lager am Aprilfluss, 21 Nov. 1912, *Ledermann* 9799 (B†).

Aglaia dasyclada How & T.C. Chen in Acta Phytotax. Sinica 4: 21, t.2 (1955) (? = *Dysoxylum hongkongense*). Syntypes: China, Hainan: *Fenzel* 238 (IBSC), *Fenzel* 707 (IBSC), *Fenzel* 7156 (IBSC); Ngai Hsien, July 1933, *C. Wang* 32950 (IBSC); Tin-On Hsien, 1300 m, Dec. 1933, *C. Wang* 36563 (IBSC); Tin-On Hsien, *S.P. Ko* 52233; Ngai Hsien, June 1933, *H.Y. Liang* 62229 (IBSC); Chong-Kam Hsien, Sep. – Oct. 1933, *H.Y. Liang* 63089 (IBSC), Chong-Kam Hsien, Sep. – Oct. 1933, *H.Y. Liang* 63794 (IBSC); Po-Ting Hsien, Hing-Lung, 400 m, June 1935, *F.C. How* 73302 (IBSC). China, Yunnan: *C.W. Wang* s.n. (IBSC), *C.W. Wang* 80662 (IBSC).

Aglaia everettii Merrill in Philipp Journ. Sci. 4: 271 (1909). Syntypes: Philippine Islands, Negros, Province of Negros Occidental, March 1907, *Everett* For. Bur. 7319 (PNH†); Negros, Cadiz, March 1908, *Danao* For. Bur. 15035 (PNH†); Cebu, Mount Licos, Feb. 1907, *Everett* For. Bur. 6452 (PNH†).

Aglaia gracillima Harms in Engl. Bot. Jahrb. 72: 171 (1942). Syntypes: New Guinea. N.E., 14 – 1500 m, 10 Aug. 1913, *Ledermann* 12790 (B†) & 31 July 1913, *Ledermann* 12422 (B†).

[*Aglaia hiernii* Koord. in Meded. 'S Lands Plantent. 19: 381, non King, nom. nud.]

Aglaia huberti Harms in Notizbl. Bot. Gart. Berlin, 15: 472 (1941). Type : Borneo, Hayoep, 15 Juni 1908, *Hubert Winkler* 2479 (B†).

[*Aglaia koordersii* Jain et Bennet in Indian Journal of Forestry, 1986, 9(3): 271 (1987), nom. nov. pro *Aglaia hiernii* Koord.]

Aglaia ledermannii Harms in Engl. Bot. Jahrb. 72: 165 (1942). Type: New Guinea, N., Sepik Lager 3 (Frieda), 3 – 400 m, 3 June 1912, *Ledermann* 7484 (B†).

Aglaia leptoclada Harms in Engl. Bot. Jahrb. 72: 166 (1942). Type: New Guinea, N.E., Waube-Bach, 150 m, 10 March 1908, *Schlechter* 17409 (B†).

Aglaia maboroana Harms in Engl. Bot. Jahrb. 72: 173 (1942). Type: New Guinea, N.E., Maboro, c. 1000 m, 23 May 1909, *Schlechter* 19553 (B†).

Aglaia peekelii Harms in Engl. Bot. Jahrb. 72: 167 (1942). Type: Bismarck Islands, Neu-Mecklenburg, Namatanai, Galipapaul, 17 Aug. 1910, *Peekel* 55 6 (B†).

Aglaia phaeogyna Harms in Engl. Bot. Jahrb. 72: 167 (1942). Syntypes: New Guinea, N.E., Etappenberg, 850 m, 4 Oct 1912, *Ledermann* 9022 (B†); 14 – 1500 m, 30 July 1913, *Ledermann* 12401 (B†); Lager 18, Aprilfluss, 2 – 40 0 m, frucht, 12 Nov. 1912, *Ledermann* 9640 (B†).

Aglaia pycnoneura Harms in Engl. Bot. Jahrb. 72: 179 (1942). Type: New Guinea. N.E., Aprilfluss, Standlager, 100 m, 10 Nov. 1912, *Ledermann* 8622 (B†).

Aglaia ramuensis Harms in K. Schum. & Lauterb., Fl. Deutsch. Südsee: 386 (1900). Type: New Guinea, Kaiser-Wilhelmsland, Bismarck-Gebirge, Ramu-Expedition, *Rodatz & Klink* n. 232, am 7 Juli 1899 (B†).

Aglaia rodatzii Harms in K. Schum. & Lauterb., Fl. Deutsch. Südsee: 386 (1900). Syntypes: New Guinea, Kaiser-Wilhelmsland, Bismarck-Gebirge, Ramu- Expedition, June – July 1879, fr., *Rodatz & Klink* 211 (B†) & fl. buds, *Rodatz & Klink* 123 (B†).

Aglaia schraderiana Harms in Engl. Bot. Jahrb. 72: 168 (1942). Type: New Guinea, N.E., Shraderberg, 2070 m, 7 June 1913, *Ledermann* 12115 (B†).

Aglaia schultzei Harms in Engl. Bot. Jahrb. 72: 177 (1942). Type: New Guinea, N.E., Sepik-Bivak, 1 Nov. 1910, *Leonhard Schultze* 244 (B†).

Aglaia schumanniana Harms in Engl. & Prantl, Nat. Pflanzenfam., ed. 2, 19b1: 176 (1940), nom. nov. pro *A. simplicifolia* Harms.

Names not placed, no type specimen seen

Aglaia simplicifolia Harms in K. Schum. & Lauterb., Fl. Deutsch. Südsee: 386 (1900), non Beddome (1895). Syntypes: New Guinea, Kaiser-Wilhelmsland, Gogol-Oberlauf, Nov. 1890, *Lauterbach* 1130 (B†) & *Lauterbach* 1132 (B†).

Aglaia spaniantha Harms in Engl. Bot. Jahrb. 72: 168 1942). Syntypes: New Guinea, N.E., Hauptlager, Malu, am Sepik, 12 April 1912, *Ledermann* 6992 (B†); Lager 3 (Frieda), 3 June 1912, *Ledermann* 7502 (B†); Hauptlager, Malu, 23 Mar. 1912, *Ledermann* 6749 (B†) & 16 July 1912, *Ledermann* 7931 (B†) & 21 Jan. 1913, *Ledermann* 10668 (B†).

Aglaia steinii Harms in Engl. Bot. Jahrb. 62: 178 (1942). Type: New Guinea, N.W., Waigeu, 11 June 1931, *Stein* 258 (B†).

Aglaia trichostoma Harms in Engl. Bot. Jahrb. 72: 169 (1942). Type: New Guinea, N.E., Aprilfluss, Standlager, 20 June 1912, *Ledermann* 8807 (B†).

Aglaia urophylla Harms in Engl. Bot. Jahrb. 72: 174 (1942). Syntypes: New Guinea, N.E., 14 – 1500 m, 24 Aug. 1913, *Ledermann* 13108 (B†) & 3 Aug. 1913, *Ledermann* 12546 (B†).

Aglaia vilamilii Merrill in Philipp. Journ. Sci. Bot., 1914, 15: 536 (1915). Type: Philippine Islands, Mindanao, *Villamil* For. Bur. 21866 (PNH†).

Aglaia vulpina Harms in Engl. Bot. Jahrb. 72: 175 (1942). Type: New Guinea, N.E., Lager 18, Aprilfluss, 22 Nov. 1912, *Ledermann* 9820 (B†).

Amoora lactescens Kurz in Jour. As. Soc. Beng. 44: 20 (1875). Type: Burma, Martaban, east of Toungoo, in herb. *Kurz* 1381 (CAL).

Amoora macrocarpa Merrill in Philipp. Gov. Lab. Bur. Bull. 17: 24 (1904) . Type: Philippine Islands, *Merrill* 3731 (PNH†).

Amoora moulmeiniana C. de Candolle in A. & C. de Candolle, Monog. Phan. 1: 584 (1878). Type: Burma, Moulmein, *Philippi* in herb. Berol. (B†).

Milnea montana Jack in Trans. Linn Soc. 14: 118 (1825) nom. provis. Type: Sumatra, near Bencoolen; (see Mabberley, *Blumea* 31: 143 (1985)).

Selbya montana M.J. Roemer, Syn. Hesper: 89, 126 (1846) (= *Milnea montana* Jack).

REJECTED NAMES

Aglaia anamallayana (Bedd.) Kosterm. in Reinwardtia 7: 257, t. 10 (1966) = **Reinwardtiodendron anamalaiense** (Bedd.) Mabb. (see Mabberley in *Blumea* 31: 145 (1985)).

Aglaia aphanamixis Pellegrin in Lecomte, Fl. Gén. Indo-Chine 1: 767 (1911), nom. superfl. = **Aphanamixis polystachya** (Wall.) R.N. Parker (see Mabberley in *Blumea* 31: 137 (1985)).

Aglaia aphanamixis var. *frutescens* (C. de Candolle) Pellegrin in Lecomte, Fl. Gén. Indo-Chine 1: 768 (1911) = **Aphanamixis polystachya** (Wall.) R.N. Parker (see Mabberley in *Blumea* 31: 137 (1985)).

Aglaia aquea (Jack) Kosterm. in Reinwardtia 7: 234, t.4 (1966) = **Lansium domesticum** Correa *agg.* (see Mabberley in *Blumea* 31: 141 (1985)).

Aglaia beddomei (Kosterm.) Jain & Gaur in J. Econ. Tax. Bot. (1985) 7: 4 65 (1986) = **Aphanamixis polystachya** (Wall.) R.N. Parker

Aglaia breviracemosa (Kosterm.) Kosterm. in Reinwardtia 1: 233 (1966) = **Lansium breviracemosum** Kosterm. (see Mabberley in *Blumea* 31: 143 (1985)).

Aglaia chartacea Kosterm. in Reinwardtia 7: 261 (1966) = **Lepisanthes tetraphylla** (Vahl) Radlk. (Sapindaceae) (see Mabberley in *Blumea* 31: 143 (1985)).

Aglaia cochinchinensis (Pierre) Pellegrin in Lecomte, Fl. Gén. Indo-Chine 1: 769 (1911); non Pierre (1895) = **Aphanamixis polystachya** (Wall.) R.N. Parker (see Mabberley in *Blumea* 31: 137 (1985)).

Aglaia decandra Wall. in Roxb., Fl. Ind., ed. Carey, 2: 427 (1824); Wall. Cat. n. 1276 (1829) = **Sphaerosacme decandra** (Wall.) Pennington (see Pennington & Styles in Blumea 22: 489 (1975).

Aglaia domestica (Correa) Pellegrin in Lecomte, Fl. Gén. Indo-Chine 1: 766 (1911) nom. illegit. = **Lansium domesticum Correa** *agg.* (see Mabberley in *Blumea* 31: 141 (1985)).

Aglaia dookoo Griff., Notulae 4: 505 (1854) = **Lansium domesticum** Correa *agg.* (see Mabberley, *Blumea* 31: 141 (1985)).

Aglaia dubia (Merrill) Kosterm. in Reinwardtia 1: 254 (1966) = **Reinwardtiodendron humile** (Hassk.) Mabb. (see Mabberley in *Blumea* 31: 145 (1985)).

Aglaia halmaheirae Miq., Ann. Mus. Bot. Lugd. Bat. 4: 58 (1868) = **Dysoxylum arborescens** (Blume) Miq.

Aglaia iloilo (Blanco) Merrill in Philipp. Jour. Sci., Bot. 9: 533 (1914). Species Blancoanae 867 (BM!, K!, L!, P!). Blanco's description of *Melia ioilo* Blanco, Fl. Filip., ed. 2: 241 (1845) does not match *Aglaia* ('corolla with linear petals, 3 times as long as broad; staminal tube cylindrical', transl. B.T. Styles), therefore the basionym is excluded and Merrill's illustrative specimens which belong to *Aglaia argentea* are inappropriate, = ?

Aglaia intricatoreticulata Kosterm. in Reinwardtia 7: 259, t. 12 (1966) = ? *Lansium domesticum* Correa *agg.* (see Mabberley, *Blumea* 31: 142 (1985)) or ? = *Lepisanthes* (Sapindaceae) (see Mabberley in *Blumea* 31: 140 (1985))

Aglaia janowskyi Harms in Engl. Bot. Jahrb. 72: 176 (1942) = **Aphanamixis polystachya** (Wall.) R.N. Parker (see Mabberley in *Blumea* 31: 138 (1985)).

Aglaia kinabaluensis Kosterm. in Reinwardtia 1: 253, t. 7 (1966) = **Reinwardtiodendron kinabaluense** (Kosterm.) Mabb. (see Mabberley in *Blumea* 31: 145 (1985)).

Aglaia kostermansii (Prijanto) Kosterm. in Reinwardtia 7: 256, t. 9 (1966) = **Reinwardtiodendron kostermansii** (Prijanto) Mabb. (see Mabberley in *Blumea* 31: 145 (1985)).

Aglaia macrophylla Teijsm. & Binn. in Tijdschr. Ned. Ind. 27: 42 (1864) = **Dysoxylum arborescens** (Blume) Miq.

Aglaia membranacea Kosterm. in Reinwardtia 7: 260 (1966) = **Lansium membranaceum** (Kosterm.) Mabb. (see Mabberley in *Blumea* 31: 141 (1985)).

Aglaia merrillii Elmer in Leafl. Philipp. Bot. 9: 3298 (1937), sine diagn. lat. = **Lansium domesticum** Correa *agg.* (see Mabberley in *Blumea* 31: 142 (1985)).

Aglaia multijuga Seem. Fl. Vit. 37 (1865) = **Dysoxylum seemanii** Gillespie.

Aglaia nivea Elmer ex Merrill, Enum. Philipp. Fl. Pl. 2: 375 (1923), in obs. pro syn. *Aglaia iloilo* (Blanco) Merrill, = ?

Aglaia oligosperma (Pierre) Pellegrin in Lecomte, Fl. Gén. Indo-Chine 1: 775 (1911) = **Dysoxylum alliaceum** (Blume) Blume.

Aglaia papuana (Merrill et Perry) Harms in Engl. Bot. Jahrb. 72: 161 (1942) = **Dysoxylum papuanum** (Merrill et Perry) Mabberley (see Mabberley in *Blumea* 31: 130 (1985)).

Aglaia pinnata (L.) Druce in Rep. Exch. Club Brit. Isles, 1913, 3: 413 (1914) = ?

Aglaia polystachya Wall. in Roxb., Fl. Ind. 2: 429 (1824) = **Aphanamixis polystachya** (Wall.) R.N. Parker (see Mabberley in *Blumea* 31: 136 (1985)).

Aglaia pseudolansium Kosterm. in Reinwardtia 7: 252 (1966) = **Reinwardtiodendron cinereum** (Hiern) Mabb. (see Mabberley in *Blumea* 31: 144 (1985)).

Aglaia reinwardtiana Kosterm. in Reinwardtia 7: 230, t. 1 (1966) = **Reinwardtiodendron celebicum** Koord. (see Mabberley in *Blumea* 31: 145 (1985)).

Aglaia sepalina (Kosterm.) Kosterm. in Reinwardtia 7: 258, t. 11 (1966) = **Lansium domesticum** Correa *agg.* (see Mabberley in *Blumea* 31: 142 (1985)).

Aglaia somalensis Chiovenda Resultati Scientifici della Missione Stefanini-Paoli nella Somalia Italiana 1: 204 (1916) = **Sorindeia madagascariensis** Du Pet-Thou.

Aglaia steensii Kosterm. in Reinwardtia 7: 232, t. 2 (1966) (excl. spec. Meijer = *Lansium membranaceum*) = **Lansium domesticum** Correa *agg.* (see Mabberley in *Blumea* 31: 142 (1985)).

Aglaia zollingeri C. de Candolle in Bull. Herb. Boiss. 2: (1894) 579. Type: Java, *Zollinger* 2846 (G! holo) = ?.

Amoora amboinensis Miq., Ann. Mus. Bot. Lugd. Bat. 4: 36 (1868) = **Aphanamixis polystachya** (Wall.) R.N. Parker (see Mabberley in *Blumea* 31: 137 (1985)).

Amoora aphanamixis J.A. & J.H. Schultes, Syst. 7: 1621 (1830), nom. superfl. = **Aphanamixis polystachya** (Wall.) R.N. Parker (see Mabberley in *Blumea* 31: 136 (1985)).

Amoora aphanamixis var. *pubescens* Miq., Ann. Mus. Bot. Lugd. Bat. 4: 34 (1868) = **Aphanamixis polystachya** (Wall.) R.N. Parker (see Mabberley in *Blumea* 31: 136 (1985)).

Amoora balansaeana [balanseana] C. de Candolle in A. & C. de Candolle, Monog. Phan. 1: 590 (1878) = **Anthocarpa nitidula** (Benth.) Pennington in sched. ex Mabb. (see Mabberley in *Blumea* 31: 133 (1985)).

Amoora beddomei Kostermans in Acta Botanica Neerlandica 31: 133 (1982) = **Aphanamixis polystachya** (Wall.) R.N. Parker (see Mabberley in *Blumea* 31: 138 (1985)).

Amoora borneensis Miq., Ann. Mus. Bot. 4: 36 (1868) = **Aphanamixis borneensis** (Miq.) Merrill (see Mabberley in *Blumea* 31: 138 (1985)).

Amoora caesifolia Elmer in Leafl. Philipp. Bot. 9: 3321 (1937), sine diagn. lat. = **Chisocheton ceramicus** (Miq.) C. de Candolle (see Mabberley, Bull. Br. Mus. Nat. Hist. (Bot.) 6: 361 (1979)).

Amoora championii (Thwaites) Hooker fil. ex Thwaites, Enum. Pl. Ceyl.: 409 (1864). Hiern in Hooker fil., Fl. Brit. Ind. 1: 562 (1875); C. de Candolle in A. & C. de Candolle, Monog. Phan. 1: 591 (1878) = **Pseudocarapa championii** (Thwaites) Hemsley. Type: Ceylon, Central Province, up to 4000 ft, *Anon.* in *Thwaites* 1193 (G!, G-BOIS!, G-DC!, *Pseudocarapa*); (excl. *Anon.* in *Thwaites* 405 (G-BOIS!, G-DC!) = *A. apiocarpa*)

Amoora cumingiana C. de Candolle in A. & C. de Candolle, Monog. Phan. 1: 580 (1878) = **Aphanamixis polystachya** (Wall.) R.N. Parker (see Mabberley in *Blumea* 31: 136 (1985)).

Amoora cupulifera Merrill in Philipp. Jour. Sci. Bot. 9: 365 (1914) = **Chisocheton ceramicus** (Miq.) C. de Candolle (see Mabberley in Bull. Br. Mus. Nat. Hist. (Bot.) 6: 361 (1979)).

Amoora decandra (Wall.) Hiern in Hooker fil., Fl. Brit. Ind. 1: 562 = **Sphaerosacme decandra** (Wall.) Pennington (see Pennington & Styles in Blumea 22: 489 (1975).

Amoora elmeri Merrill in Philipp. Gov. Lab. Bur. Bull. 29: 23 (1905) = **Aphanamixis polystachya** (Wall.) R.N. Parker (see Mabberley in *Blumea* **31**: 138 (1985)).

Amoora ficiformis Wight, Illustr. Ind. Bot. 1: 147 (1831) = **Dysoxylum ficiforme** (Wight) Gamble. Type: India, Tamil Nadu, Salem District, Shevaroy Hills near Courtallum, *Wight* 412 (K!, L!).

Amoora forbesii S. Moore in Jour. Bot., Lond. 64, Suppl.: 4 (1926). Type: Sumatra, S., Moera Mengkoelem, R. Rawas, Palembang, 1500 ft, *Forbes* 3043 (BM!) = ?

Amoora fulva Merrill in Philipp. Jour. Sci., Bot. 11: 187 (1916) = **Chisocheton mendozai** Hildebr. (see Mabberley in Bull. Br. Mus. Nat. Hist. (Bot.) 6: 359 (1979)).

Amoora grandifolia Walp., Repert. 1: 429 (1842) = **Aphanamixis polystachya** (Wall.) R.N. Parker (see Mabberley in *Blumea* 31: 137 (1985)).

Amoora grandifolia var. *pubescens* (Miq.) C. de Candolle in A. & C. de Candolle, Monog. Phan. 1: 581 (1878) = **Aphanamixis polystachya** (Wall.) R.N. Parker (see Mabberley in *Blumea* 31: 137 (1985)).

Amoora janowskyi (Harms) Kosterm. in Reinwardtia 7: 265 (1966) = **Aphanamixis polystachya** (Wall.) R.N. Parker (see Mabberley in *Blumea* 31: 138 (1985)).

Amoora lauterbachii (Harms) C. de Candolle, Bull. Herb. Boiss. Sér. II 3: 170 (1903) + **Aphanamixis polystachya** (Wall.) R.N. Parker (see Mabberley in *Blumea* 31: 138 (1985)).

Amoora macrocalyx (Harms) C. de Candolle, Bull. Herb. Boiss. Sér. II 3: 170 (1903) = **Aphanamixis polystachya** (Wall.) R.N. Parker (see Mabberley in *Blumea* 31: 138 (1985)).

Amoora macrophylla Nimmo in J. Graham, Cat. Bomb. Pl.: t. 31 (1839) = **Aphanamixis polystachya** (Wall.) R.N. Parker (see Mabberley in *Blumea* 31: 137 (1985)).

Amoora megalophylla C. de Candolle in Bull. Herb. Boiss. 2: 577 (1894) = **Aphanamixis polystachya** (Wall.) R.N. Parker (see Mabberley in *Blumea* 31: 136 (1985)).

Amoora megalophylla var. *frutescens* C. de Candolle in Bull. Herb. Boiss. 2: 578 (1894) = **Aphanamixis polystachya** (Wall.) R.N. Parker (see Mabberley in *Blumea* 31: 136 (1985)).

Amoora mindorensis Merrill in Philipp. Jour. Sci. 26: 459 (1925) = **Chisocheton ceramicus** (Miq.) C. de Candolle (see Mabberley in Bull. Br. Mus. Nat. Hist. (Bot.) 6: 361 (1979)).

Amoora myrmecophila O. Warb. in Bot. Jahrb. 18: 194 (1894) = **Aphanamixis polystachya** (Wall.) R.N. Parker (see Mabberley in *Blumea* 31: 137 (1985)).

Amoora naumannii C.DC. in Bot. Jahrb. 7: 461 (1886) = ? *Dysoxylum* sp. (see Mabberley in *Blumea* 31: 131 (1985)).

Amoora nitidula Benth., Fl. Austral. 1: 383 (1863) = **Anthocarapa nitidula** Benth.) Pennington in sched. ex Mabb. (see Mabberley in *Blumea* 31: 133 (1985)).

Amoora oligosperma Pierre, Fl. Forest. Cochinch. Fasc. 22: t. 345A (1 July 1897) = **Dysoxylum alliaceum** (Bl.) Bl. Type: Cochinchina. La province de Bien hoa près de Chiua Chiang, *Pierre* 4303 (E!, G!, L!, NY! P!,).

Amoora perrotetiana (A. Juss.) Steudel, Nomencl. ed. 2, 1: 89 (1840) = **Aphanamixis polystachya** (Wall.) R.N. Parker (see Mabberley in *Blumea* 31: 137 (1985)).

Amoora polillensis C.B. Robinson, Philipp. Jour. Sci., Bot. 6: 206 (1911) = **Aphanamixis polystachya** (Wall.) R.N. Parker (see Mabberley in *Blumea* 31: 138 (1985)).

Amoora polystachya (Wall.) Wight & Arn. ex Steudel, Nomencl., ed. 2, 1: 78 (1840) = **Aphanamixis polystachya** (Wall.) R.N. Parker (see Mabberley in *Blumea* 31: 136 (1985)).

Amoora racemosa Ridley in Jour. Fed. Mal. States Mus. 10: 88 (1920) = **Lansium domesticum** Correa *agg.* (see Mabberley in *Blumea* 31: 142 (1985)).

Amoora rohituka (Roxb.) Wight & Arn. in Wight, Cat.: 24 (1833) = **Aphanamixis polystachya** (Wall.) R.N. Parker (see Mabberley in *Blumea* 31: 136 (1985)).

Amoora roxburghiana Korth. ex Blume, Mus. Bot. Lugd. Bat. 1: 211 (1850) nom. in syn. = **Santiria laevigata** Blume.

Amoora salomoniensis C. de Candolle in Bot. Jahrb. 1: 461 (1886) = ? **Xylocarpus granatum** Koenig (see Mabberley in *Blumea* 31: 130 (1985)).

Amoora sogerensis Baker fil. in Jour. Bot., Lond. 61, Suppl.: 8 (1923) = **Aphanamixis polystachya** (Wall.) R.N. Parker (see Mabberley in *Blumea* 31: 138 (1985)).

Amoora sumatrana Miq., Ann. Mus. Bot. Lugd. Bat. 4: (1868) = **Aphanamixis sumatrana** (Miq.) Ridley (see Mabberley in *Blumea* 31: 139 (1985)).

Amoora timorensis (A. Juss.) Wight & Arn. ex Steudel, Nomencl. ed 2, 1: 78 (1840) = **Aphanamixis polystachya** (Wall.) R.N. Parker (see Mabberley in *Blumea* 31: 137 (1985)).

Amoora tomentosa Korth. ex Blume, Mus. Bot. Lugd. Bat. 1: 211 (1850) nom. in syn. = **Santiria tomentosa** Blume.

Amoora vieillardii C. de Candolle in A. & C. de Candolle, Monog. Phan. 1: 591 (1878) = **Anthocarapa nitidula** (Benth.) Pennington in sched. ex Mabb. (see Mabberley, *Blumea* 31: 133 (1985)).

Andersonia rohituka Roxb. Hort. Beng.: 87 (1814), nom. nud. 'rohitoka'; Fl. Ind., ed Carey, 2: 213 (1832) = **Aphanamixis polystachya** (Wall.) R.N. Parker (see Mabberley in *Blumea* 31: 136 (1985)).

Hearnia balansae C. de Candolle in Bull. Herb. Boiss. 2: 580 (1894) = ? **Canarium tonkinense** Engler.

BIBLIOGRAPHY

Bachmann, O. (1866). Untersuchungen über die systematische bedeutung der schildhaare. Flora 69: 387 – 400, 403 – 415, 428 – 448, t. V11 – X.
Backer, C.A. & Bakhuizen van den Brink, R.C. (1965). *Amoora* & *Aglaia*. In Flora of Java 2: 125 – 129.
Baker, H.G. (1978). Chemical aspects of the pollination biology of woody plants in the tropics. In P.B. Tomlinson and M.H. Zimmermann (eds), Tropical trees as living systems: 57 – 82. Cambridge.
Bawa, K.S. & Opler, P.A. (1975). Dioecism in tropical forest trees. Evolution 29: 167 – 179.
Blume, C.L. (1825). Bijdragen tot de flora van Nederlandsch Indië 4: 169–172.
Burkill, I.H. (1935). A dictionary of the economic products of the Malay Peninsula 1: 72 – 76, 137 – 138.
Candolle, C. de (1878). *Amoora* & *Aglaia*. In A. & C. de Candolle, Monographiae Phanerogamarum 1: 579 – 592, 601 – 628.
Corner, E.J.H. (1988). Wayside trees of Malaya, ed 2. 2 vols. Singapore.
German-Ramirez, M.T. & Styles, B.T. (1978). Revision taxonomica del género *Cedrela* P. Br. I. – *C. oaxacensis* C.DC. & Rose, *C. salvadorensis* Standl. y *C. tonduzii* C.DC. en México y Centro América. Turrialba 28: 261 – 274.
Hallé, F. & Mabberley, D.J. (1977). Corner's architectural model. Gardens' Bull. Singapore 28: 175 – 181.
Hallé, F. & Oldeman, R.A.A. (1970). Essai sur l'architecture et la dynamique de croissance des arbres tropicaux. Paris. Translated by Stone, B.C. (1975). An essay on the architecture and dynamics of growth of tropical trees. Kuala Lumpur.
Harms, H. (1896) *Amoora* & *Aglaia*. In A. Engler & K. Prantl, Die Natürlichen Pflanzenfamilien III, 4: 297 – 300, 1 fig.
Harms, H. (1940). *Aglaia* & *Amoora*. In A. Engler & K. Prantl, op. cit., ed 2, 19b1: 128 – 129, 140 – 147.
Henty, E.E. (1980). Harmful plants in Papua New Guinea. Bot. Bull. Lae 12: 98 – 103.
Hiern, W.P. (1875). *Aglaia* & *Amoora*. In J.D. Hooker, Flora of British India. 1: 554 – 557, 559 – 562.
Hutchinson, J. & Dalziel, M.D. (1958). Flora of West Tropical Africa, ed. 2 (revised by R.W.J. Keay), 1(2): 697 – 709.
Keay, R.J.W., Onochie, C.F.A. & Stanfield, D.P. (1964). Meliaceae. Nigerian Trees 2: 257 – 284. Ibadan.
Khosla, P.K. & Styles, B.T. (1975). Karyological studies and chromosomal evolution in Meliaceae. Silv. Genet. 24: 33 – 84.
King, G. (1895). Materials for a flora of the Malayan Peninsula. Jour. As. Soc. Bengal 64: 51 – 80.

Koorders, S.H. & Valeton, Th. (1896). *Amoora* & *Aglaia*. In Boomsarten van Java, Bidjrage no. 3. Mededelingen uit's Lands Plantentuin 16: 117 – 178.
Koorders, S.H. (1898). *Aglaia*. In Verslag eener Botanische Dienstreis door de Minahasa tevens eerste overzicht der flora van N.O. Celebes. Mededelingen uit's Lands Plantentuin 19: 380 – 384, 633 – 626.
Kostermans, A.J.G.H. (1966). A monograph of *Aglaia* sect. *Lansium* Kosterm. (Meliaceae). Reinwardtia 7: 221 – 282.
Mabberley, D.J. (1979). The species of Chisocheton (Meliaceae). Bull. Br. Mus. Nat. Hist. (Bot.) 6: 301 – 386.
Mabberley, D.J. (1985). Florae malesianae praecursores LXVII. Meliaceae (divers genera). Blumea 31: 129 – 152.
Mabberley, D.J. (1988). Meliaceae. Flore de la Nouvelle-Calédonie et Dépendances, 15: 17 – 89.
Mabberley, D.J. & Pannell, C.M. (1989). Meliaceae. In F.S.P. Ng (ed.), Tree flora of Malaya 4: 199 – 260. Longman, Malaysia.
Mabberley, D.J. & Pannell, C.M. (in prep.). Meliaceae. In A. George (ed.), Flora of Australia.
Mabberley, D.J. & Pannell, C.M. (in prep.). Meliaceae. A revised handbook to the flora of Ceylon.
Mehra, P.N. & Khosla, P.K. (1969). In IOPB chromosome number reports XX, Taxon 18: 310 – 315.
Mehra, P.N., Sareen, T.S. & Khosla, P.K. (1972). Cytological studies in Himalayan Meliaceae. Jour. Arn. Arb. 53: 558 – 568.
Metcalfe, C.R. & Chalk, L. (1950). Meliaceae. Anatomy of the Dicotyledons 1: 349 – 358. Oxford.
Miquel, F.A.W. (1861). *Aglaia*. In Flora van Nederlandsch Indië. Suppl . 1: 506 – 508. Leipzig.
Miquel, F.A.W. (1868). *Amoora* and *Aglaia*. In Monographia Meliacearum Archipelagi Indici. Annales Musei Botanici Lugdano-Batavi 4: 34 – 58. Leipzig.
Nanda, A., Iyengar, M.A., Narayan, C.S. & Kulkarni, D.R. 1987. Investigations of the rootbark of *Aglaia odoratissima*. Fitoterapia 58: 189 – 191.
Netolitzky, F. (1932). In K. Linsbauer (ed.) Handbuch der Pflanzenanatomie 4: Berlin (ed. 1; for ed. 2 see Uphof, 1962).
Ng, F.S.P. (1978). Strategies of establishment in Malayan forest trees. In P.B. Tomlinson and M.H. Zimmermann (eds). Tropical trees as living systems: 129 – 162. Cambridge.
Pannell, C.M. (1980). Taxonomic and ecological studies in *Aglaia* (Meliaceae). Unpublished D. Phil. thesis, Bodleian Library, Oxford.
Pannell, C.M. (1982). *Aglaia*. In D.J. Mabberley, Notes on Malesian Meliaceae for the 'Tree flora of Malaya'. The Malaysian Forester 45: 448 – 455.
Pannell, C.M. (1989a). Spreading the seeds – sowing the future forest. Chapter 5 in L. Silcock (editor), The rainforests – A celebration. London: Barrie & Jenkins.
Pannell, C.M. (1989b). The role of animals in natural regeneration and the management of equatorial rain forests for conservation and timber production. Commonwealth Forestry Review 68(4): 309–313.
Pannell, C.M. & Kozioł, M.J. (1987). Arillate seeds and vertebrate dispersal

in *Aglaia* (Meliaceae): a study of ecological and phytochemical diversity. Philosophical Transactions of the Royal Society. B. 316: 303 – 333.

Pannell, C.M. & White, F. (1988). Patterns of speciation in Africa, Madagascar, and the tropical Far East: regional faunas and cryptic evolution in vertebrate-dispersed plants. In Proceedings of the AETFAT Congress, 1985. Monographs in Systematic Botany from the Missouri Botanical Garden. 25: 639 – 659.

Pellegrin, F. (1911). Sur les genres *Aglaia*, *Amoora* et *Lansium*. Not Syst. (Phanerogamie) 1: 284 – 290.

Pennington, T.D. (1969). Materials for a monograph of the Meliaceae I. A revision of the genus Vavaea. Blumea 17: 351 – 366.

Pennington, T.D. (1981). Meliaceae (with accounts of Swietenioideae by B.T. Styles and Chemotaxonomy by D.A.H. Taylor). Flora Neotropica, Monograph no. 28. New York.

Pennington, T.D. & Styles, B.T. (1975). A generic monograph of the Meliaceae. Blumea 22: 419 – 540.

Pennington, T.D. (1990). Sapotaceae. Flora Neotropica, Monograph no. 52. New York.

Ragonese, A.E. & Garcia, A.L. (1971). Indicacion de trabajos de mejoramiento en Paraiso. Actas del Primer Congreso Forestal Argentino, 1969: 307 – 309.

Ridley, H.N. (1922). Flora of the Malay Peninsula 1: 398-410. London.

Roe, K.E. (1971). Terminology of hairs in the genus *Solanum*. Taxon 20 : 501 – 508.

Seyani, J.H. (1991). The genus *Dombeya* (Sterculiaceae) in continental Africa. Opera Botanica Belgica 2. Meise, Belgium.

Smith, A.C. (1985). Meliaceae. Flora Vitiensis Nova 3: 527 – . Lawai, Kauai, Hawaii.

Solereder, H. (1908). Translated by L.A. Boodle & F.E. Fritsch, revised by D.H. Scott, Systematic anatomy of the dicotyledons. Oxford.

Stace, C.A. (1965). Cuticular studies as an aid to plant taxonomy. Bull. Br. Mus. nat. Hist. (Bot.) 4: 1 – 78.

Staner, P. & Gilbert, G. (1958). Meliaceae. Flore du Congo Belge 1: 147 - 213.

Steenis, C.G.G.J. van (1971). Plant conservation in Malaysia. Bull. Jard. Bot. Nat. Belg. 41: 189 – 202.

Styles, B.T. & Khosla, P.K. (1976). Cytology and reproductive biology of Meliaceae. In J. Burley & B.T. Styles (eds), Tropical trees. Variation, breeding and conservation: 61 – 67. London.

Takhtajan, A. (1986). Floristic regions of the world. Berkeley, California.

Theobald, W.L., Krahulik, J.L. & Rollins, R.C. (1979). Trichome description and classification. In C.R. Metcalfe & L. Chalk, Anatomy of dicotyledons, ed. 2, 1: 40 – 53.

Uphof, J.C. Th. (1962). In W. Zimmermann & P.G. Ozenda (eds) Handbuch der Pflanzenanatomie (Encyclopaedia of plant anatomy), ed. 2, 4: 1 – 206. Berlin (see Netolitzky, 1932 for ed. 1).

White, F. (1962). Geographic variation and speciation in Africa with particular reference to *Diospyros*. Systematics Assoc. Publ. no. 4: 71 – 103.

White, F. (1986). The taxonomy, chorology and reproductive biology of southern African Meliaceae and Ptaeroxylaceae. Bothalia 16: 143 – 168.

White, F. & Styles, B.T. (1963). Meliaceae. Flora Zambesiaca 2: 285 – 319.

White, F. & Styles, B.T. (1986). Meliaceae. Flora of Southern Africa. 18(3): 39 – 61.
White, F. & Styles, B.T. (1991) Meliaceae. In R.M. Polhill (ed.), Flora of Tropical East Africa. Rotterdam: A.A. Balkema.
Wodehouse, R.P. (1935). Pollen Grains: 106–109. London.
Wu, C.Y. (1977). *Amoora* & *Aglaia*. In Flora Yunnanica 1: 231–241.

APPENDIX 1. REGIONAL SPECIES LISTS

Endemic species are marked with an asterisk (*)

INDIA
A. apiocarpa
*A. bourdillonii
A. edulis
A. elaeagnoidea
A. korthalsii
A. lawii
*A. malabarica
A. perviridis
A. simplicifolia
A. spectabilis
A. tomentosa
(A. odorata: cultivated)

SRI LANKA
A. apiocarpa
A. elaeagnoidea
(A. odorata: cultivated)

BHUTAN
A. edulis
A. korthalsii
A. lawii
A. perviridis

BANGLADESH
A. chittagonga
A. cucullata
A. perviridis

NICOBAR ISLANDS,
GREAT COCOS,
ANDAMAN ISLANDS
A. argentea
A. crassinervia
A. edulis
A. korthalsii
A. lawii
A. odoratissima
A. oligophylla

A. perviridis
A. silvestris
A. spectabilis

BURMA
A. crassinervia
A. cucullata
A. edulis
A. elliptica
A. eximia
A. exstipulata
A. forbesii
A. korthalsii
A. lawii
A. odoratissima
A. silvestris
A. spectabilis

INDOCHINA
A. cucullata
A. edulis
A. elaeagnoidea
A. exstipulata
A. grandis
A. korthalsii
A. lawii ?
A. leptantha
A. macrocarpa
A. odorata
A. perviridis
*A. pleuropteris
A. silvestris
A. simplicifolia
A. spectabilis
A. tomentosa

CHINA
A. edulis
A. lawii
A. perviridis

363

A. spectabilis
A. teysmanniana
(Aglaia odorata: cultivated)

TAIWAN
A. elaeagnoidea
A. lawii
A. rimosa

THAILAND
A. argentea
A. aspera
A. chittagonga
A. crassinervia
A. cucullata
A. edulis
A. elaeagnoidea
A. elliptica
A. erythrosperma
A. eximia
A. exstipulata
A. forbesii
A. grandis
A. korthalsii
A. lawii
A. leptantha
A. leucophylla
A. odorata
A. odoratissima
A. oligophylla
A. pachyphylla
A. perviridis
A. silvestris
A. simplicifolia
A. spectabilis
A. tenuicaulis
A. teysmanniana
A. tomentosa

PENINSULAR MALAYSIA
A. angustifolia
A. argentea
A. aspera
A. coriacea
A. crassinervia
A. cucullata
A. densitricha
A. edulis
A. elaeagnoidea
A. elliptica
A. erythrosperma

A. eximia
A. exstipulata
A. forbesii
A. foveolata
A. glabrata
A. grandis
A. hiernii
A. korthalsii
A. lawii
A. leptantha
A. leucophylla
A. macrocarpa
*A. macrostigma
A. malaccensis
A. membranifolia
A. monozyga
A. multinervis
A. odoratissima
A. oligophylla
A. pachyphylla
A. palembanica
A. perviridis
A. rubiginosa
A. rugulosa
A. rufibarbis
A. rufinervis
A. scortechinii
A. silvestris
A. simplicifolia
A. speciosa
A. spectabilis
A. squamulosa
A. tenuicaulis
A. teysmanniana
A. tomentosa
A. variisquama
A. yzermannii
(Aglaia odorata: cultivated)

SINGAPORE
A. cucullata
A. elliptica
A. exstipulata
A. leptantha
A. macrocarpa
A. palembanica
A. rubiginosa
A. rufinervis
A. tenuicaulis
A. tomentosa

Regional species lists

SUMATRA
A. angustifolia
A. argentea
A. aspera
A. crassinervia
A. cucullata
A. edulis
A. elaeagnoidea
A. elliptica
A. erythrosperma
A. eximia
A. forbesii
A. foveolata
A. glabrata
A. hiernii
A. korthalsii
A. lawii
A. leptantha
A. leucophylla
A. macrocarpa
A. malaccensis
A. membranifolia
A. multinervis
A. odoratissima
A. oligophylla
A. pachyphylla
A. palembanica
A. rubiginosa
A. rufinervis
A. rugulosa
A. silvestris
A. simplicifolia
A. speciosa
A. spectabilis
A. squamulosa
A. tenuicaulis
A. teysmanniana
A. tomentosa
A. yzermannii
(Aglaia odorata: cultivated)

BORNEO
A. angustifolia
A. argentea
A. aspera
*A. densisquama
A. coriacea
A. crassinervia
A. cucullata
A. cumingiana
A. edulis

A. elaeagnoidea
A. elliptica
A. erythrosperma
A. exstipulata
A. forbesii
A. foveolata
A. glabrata
A. grandis
A. hiernii
A. korthalsii
A. lancilimba
A. lawii
*A. laxiflora
A. leptantha
A. leucophylla
A. luzoniensis
A. macrocarpa
A. malaccensis
A. monozyga
A. multinervis
A. odoratissima
A. oligophylla
A. pachyphylla
A. palembanica
*A. ramotricha
*A. rivularis
A. rubiginosa
A. rufibarbis
A. rufinervis
A. rugulosa
A. scortechinii
A. silvestris
A. simplicifolia
A. speciosa
A. spectabilis
A. squamulosa
*A. subsessilis
A. tenuicaulis
A. teysmanniana
A. tomentosa
A. variisquama

PHILIPPINE ISLANDS
*A. aherniana
A. angustifolia
A. argentea
A. aspera
A. crassinervia
A. cucullata
A. cumingiana
*A. costata

A. edulis
A. elaeagnoidea
A. elliptica
A. eximia
A. glabrata
A. grandis
A. korthalsii
A. lancilimba
A. lawii
A. leptantha
A. leucophylla
A. luzoniensis
? A. macrocarpa
A. malaccensis
A. odoratissima
A. oligophylla
A. pachyphylla
A. palembanica
*A. pyriformis
A. rimosa
A. silvestris
A. smithii
A. spectabilis
A. squamulosa
A. tenuicaulis
A. teysmanniana
A. tomentosa
(A. odorata: cultivated)

JAVA & BALI
A. argentea
A. aspera
A. cucullata
A. edulis
A. elaeagnoidea
A. elliptica
A. eximia
A. lancilimba
A. lawii
A. leptantha
A. macrocarpa
A. odoratissima
A. pachyphylla
A. rufinervis
A. silvestris
A. speciosa
A. teysmanniana
(Aglaia odorata: cultivated)
(Aglaia tomentosa: cultivated)

NUSA TENGGARA
A. argentea
A. edulis
A. elliptica
A. korthalsii
A. lancilimba
A. lawii
A. leptantha
A. leucophylla
A. odoratissima
A. smithii
A. spectabilis
A. squamulosa
A. tomentosa

SULAWESI
A. argentea
A. edulis
A. elaeagnoidea
A. elliptica
A. eximia
A. grandis
A. korthalsii
A. lancilimba
A. lawii
A. luzoniensis
A. macrocarpa
A. odoratissima
A. pachyphylla
A. rimosa
A. silvestris
A. smithii
A. speciosa
A. spectabilis
A. squamulosa
A. teysmanniana
A. tomentosa

MALUKU
A. argentea
*A. ceramica
A. edulis
A. elaeagnoidea
A. eximia
A. glabrata
A. lawii
A. leptantha
A. macrocarpa
A. multinervis
A. parviflora

A. rimosa
A. sapindina
A. silvestris
A. smithii ?
A. speciosa
A. teysmanniana

NEW GUINEA
*A. agglomerata
A. argentea
A. aspera
*A. barbanthera
A. brassii
A. brownii
*A. conferta
*A. cremea
A. cucullata
*A. cuspidata
A. elaeagnoidea
A. euryanthera
*A. flavescens
A. flavida
*A. integrifolia
A. lawii
*A. lepidopetala
*A. lepiorrhachis
*A. leucoclada
A. parviflora
*A. polyneura
*A. penningtoniana
*A. puberulanthera
A. rimosa
A. samoensis
A. sapindina
A. silvestris
A. smithii
A. spectabilis
A. subcuprea
A. subminutiflora
A. teysmanniana
A. tomentosa

NEW BRITAIN
A. aspera
A. flavida
A. lawii
A. parviflora
A. rimosa
A. samoensis
A. sapindina
A. silvestris

A. spectabilis
A. teysmanniana

NEW IRELAND
A. samoensis
A. sapindina
A. subcuprea
A. subminutiflora

AUSTRALIA
A. argentea
*A. australiensis
A. brassii
A. brownii
A. elaeagnoidea
A. euryanthera
*A. meridionalis
A. sapindina
A. spectabilis
A. tomentosa

BOUGAINVILLE &
SOLOMON ISLANDS
A. argentea
A. brassii
A. elaeagnoidea
A. flavida
A. lawii
A. parksii
A. parviflora
*A. rubrivenia
A. saltatorum
A. samoensis
A. sapindina
A. silvestris
A. spectabilis
A. subcuprea
A. subminutiflora

VANUATU
A. elaeagnoidea
A. saltatorum
A. samoensis

FIJI
*A. amplexicaulis
*A. archboldiana
*A. basiphylla
*A. evansensis
*A. fragilis
*A. gracilis

A. parksii
A. saltatorum
*A. unifolia
*A. vitiensis

NEW CALEDONIA
A. elaeagnoidea

WALLIS ISLANDS &
ILES DE HORN
A. saltatorum
A. samoensis

TONGA
*A. heterotricha
A. saltatorum

SAMOA
A. elaeagnoidea
A. samoensis

MARIANNE ISLANDS &
CAROLINE ISLANDS
*A. mariannensis

APPENDIX 2. A NEW SPECIES FROM INDIA

by N. Sasidharan
Kerala Forest Research Institute, Peechi 680 653, Trichur, Kerala State, India. (KFRI).

105. Aglaia malabarica N. Sasidharan spec. nova foliis eis stirpis magnifoliae *Aglaiae apiocarpae* (Thwaites) Hiern similibus sed squamis integris non fimbriatis et magis numerosis (squamis minus numerosis eis stirpis parvifoliae *Aglaiae apiocarpae*), pericarpio fructus porcis pluribus irregularibus longitudinalibus praedito crassiore eo *Aglaiae apiocarpae* cuius pericarpium semper est laeve, structura tubi staminalis et structura squamarum eis *Aglaia silvestris* (M. Roemer) Merrill similibus sed foliolis minus numerosis et forma plus ovatis eis speciminum occidentalium *Aglaiae silvestri* qui Indiam non incolit. Holotype: India, Kerala State, Trichur, Velakettupara, Peechi, c. 500 m, ♀ fl., 12 Dec. 1989, *Sasidharan* 5528 (MH!, isotype: KFRI!).

Tree 20 – 25 m. Bark surface smooth, 4 – 5 mm thick, brown; inner bark reddish-brown. Twigs pale brown or greyish-brown with longitudinal wavy ridges, densely covered with entire peltate scales c. 0.2 mm in diameter and which have a very dark reddish-brown or orange-brown centre and paler margin.

Leaves imparipinnate, 27 – 36 cm long, 21 – 38 cm wide, obovate in outline; petiole 4.5 – 8.5 cm, the petiole, rhachis and petiolules ridged and with indumentum like the twigs. Leaflets 7 – 11, the laterals alternate, all 6 – 21 cm long, 2.2 – 8 cm wide, ovate or oblong, slightly asymmetrical and sometimes curved, acuminate at apex with the usually acute, sometimes obtuse, acumen 10 – 20 mm long, rounded at the asymmetrical base, upper surface with numerous pits and with scales like those on the twigs few to numerous on the midrib and lateral veins and sometimes on the rest of that surface, lower surface with few to numerous scales; veins 10 – 19 on each side of the midrib, ascending and curved upwards near the margin, not or just anastomosing, midrib prominent and lateral veins subprominent on lower surface; petiolules 10 – 20 mm on lateral leaflets, up to 25 mm on terminal leaflets.

Male inflorescence c. 20 cm long and 15 cm wide; peduncle c. 4.5 cm, the peduncle, rhachis and branches with surface and indumentum like the twigs. Flowers 2.5 – 3 mm long, c. 2.5 mm wide; pedicels 1 – 1.5 mm, the pedicels and calyx with dense peltate scales which have a short fimbriate margin. Calyx c. 0.5 mm, cup-shaped, divided into 5 rounded lobes. Petals 5(– 7), yellow, aestivation quincuncial. Staminal tube c. 2 mm long, 1.6 – 2 mm wide, the aperture 0.4 – 0.5 mm in diameter, entire; anthers 5, 0.8 – 1 mm long, c. 0.5 mm wide, ovoid, inserted half to two thirds of the way up the staminal tube and included. Ovary c. 0.5 mm high and 0.8 mm wide, depressed globose, locules 3, each containing 1 ovule; stigma c. 0.5 mm high and 0.6 mm wide, subglobose with an apical depression.

Female inflorescence c. 5.5 cm long and 4 cm wide; peduncle c. 2 mm, the peduncle, rhachis and branches with surface and indumentum like the twigs. Flowers c. 4.5 mm long and 3.5 mm wide; pedicels 3 – 4.5 mm, the pedicels and calyx with dense peltate scales which have a short fimbriate margin. Calyx c. 3.5 mm long, deeply cup-shaped, thick and fleshy at the base, divided into 5 rounded lobes. Corolla c. 2 mm long, 2.6 – 2.8 mm wide, aestivation quincuncial; petals 5, yellow. Staminal tube c. 2.5 mm long, 2 mm wide, the aperture 0.4 – 0.6 mm in diameter, entire; anthers 5(– 7), c. 1 mm long and 0.5 mm wide, ovoid, inserted about two thirds of the way up the staminal tube and included. Ovary c. 0.7 mm high and 1 mm wide, depressed globose, with dense reddish-brown peltate scales which have a fimbriate margin, locules 3, each containing 1 ovule; stigma c. 0.7 mm high and 0.8 mm wide, subglobose with an apical depression.

Infructescence c. 8 cm long with one fruit borne on a stout peduncle; peduncle c. 3.5 cm long, c. 0.5 cm wide near the base and 1 cm wide at the insertion of the fruit, with surface and indumentum like the twigs. Fruits 3.5 – 4.5 cm long, 2.7 – 4 cm wide, obovoid, with an apical depression, indehiscent, the pericarp with numerous irregular longitudinal ridges and dense scales like those on the twigs on the outside, c. 6 mm thick and granular; loculi 3, with 0 – 1 seed in each locule; seeds 1.7 – 2.5 cm long, c. 1.5 cm wide and 0.7 – 1 cm thick, completely surrounded with a thin, pale pink, translucent aril. Fig. 113.

DISTRIBUTION. Known only from Trichur, Kerala, India. Fig. 40.

ECOLOGY. Fairly common in the evergreen forests of Northern Kerala (Malabar), where the trees can be recognised by their golden-brown young leaves. Alt.: up to 500 m.

ETYMOLOGY. The species is named after the type locality.

VERNACULAR NAME. Churannakil.

Representative specimens. INDIA. Kerala, Trichur, Velakettupara, Peechi: c. 500 m, ♂ fl., 12 Dec. 1989, *Sasidharan* 5531 (KFRI!, MH!) & c. 500 m, fr., 6 April 1988, *Sasidharan* 5027 (KFRI!, MH!) & c. 325 m, fr., 17 July 1990, *Sasidharan* 5657 (FHO!) & *Sasidharan* 5068 (KFRI!, MH!).

The leaves of *A. malabarica* resemble the large-leaved form of *A. apiocarpa* but the scales are entire rather than fimbriate and are more numerous (although less numerous than in the smaller-leaved form of *A. apiocarpa*). The fruit has a thicker pericarp than that of *A. apiocarpa* and has numerous irregular longitudinal ridges, whereas in *A. apiocarpa* the pericarp is always smooth. The structure of the staminal tube and of the scales resemble those of *A. silvestris*, but the leaflets are fewer in number and more ovate in shape than those of *A. silvestris* in the western part of its range (which does not include India).

ACKNOWLEDGEMENTS. I am deeply indebted to Dr C.M. Pannell for confirming the identity of the species and for various other help. The encouragement and facilities provided by the Director, Kerala

Forest Research Institute, is gratefully acknowledged. Caroline Watterston prepared the plate. I am grateful to M.J.E. Coode for translating the diagnosis into Latin.

Fig. 113. *Aglaia malabarica*. Habit with male inflorescence x$\frac{1}{2}$. Half flower, male x7. Female inflorescence x$\frac{1}{2}$. Half flower, female x7. Fruit x$\frac{1}{2}$. Peltate scale x50.

NEW TAXA

Aglaia australiensis C.M. Pannell, spec. nova, p. 63.
Aglaia brownii C.M. Pannell, spec. nova, p. 327.
Aglaia ceramica (Miq.) C.M. Pannell, comb. nova, p. 262.
Aglaia densisquama C.M. Pannell, spec. nova, p. 133.
Aglaia densitricha C.M. Pannell, spec. nova, p. 90.
Aglaia erythrosperma C.M. Pannell, spec. nova, p. 76.
Aglaia foveolata C.M. Pannell, spec. nova, p. 211.
Aglaia integrifolia C.M. Pannell, spec. nova, p. 336.
Aglaia macrocarpa (Miq.) C.M. Pannell, comb. nova, p. 65.
Aglaia malabarica N. Sasidharan, spec. nova, p. 369.
Aglaia meridionalis C.M. Pannell, nom. novum, p. 88.
Aglaia multinervis C.M. Pannell, nom. novum, p. 84.
Aglaia penningtoniana C.M. Pannell, spec. nova, p. 94.
Aglaia ramotricha C.M. Pannell, spec. nova, p. 115.
Aglaia rugulosa C.M. Pannell, spec. nova, p. 73.
Aglaia subsessilis C.M. Pannell, spec. nova, p. 273.
Aglaia variisquama C.M. Pannell, spec. nova, p. 153.

INDEX

A. *abbreviata* C.Y. Wu = 29
A. *acariaeantha* Harms = 55
A. *acida* Koord. & Val. = 58
A. *acuminata* Merrill = 47
A. *acuminatissima* Teijsm. & Binn. = 54
A. *acuminatissima* Teijsm. & Binn. var. *kambangana* Miq. = 91
A. *affinis* Merrill = 60
A. agglomerata Merrill & Perry 34
A. *agusanensis* Elmer ex Merrill = 57
A. aherniana Perkins 78
A. *allocotantha* Harms, n.v.
A. *alternifoliola* Merrill = 15
A. amplexicaulis A.C. Smith 65
A. *anamallayana* (Bedd.) Kosterm., excl.
A. *ancolana* Miq. = 21
A. *andamanica* Hiern = 15
A. angustifolia (Miq.) Miq. 100
A. *angustifolia* (Miq.) Miq. var. *horsefieldiana* C. de Candolle = 100
A. *annamensis* Pellegrin = 49
A. *anonoides* Elmer ex Merrill = 88
A. *antonii* Elmer = 76
A. *aphanamixis* Pellegrin, excl.
A. apiocarpa (Thwaites) Hiern 37
A. *apoana* Merrill = 76
A. *aquatica* (Pierre) Harms = 36
A. *aquea* (Jack) Kosterm., excl.
A. *araeantha* Harms, n.v.
A. archboldiana A.C. Smith 104
A. argentea Blume 22
A. *argentea* auct. = 21
A. *argentea* Blume var. *angustata* Miq. = 22
A. *argentea* Blume var. *borneensis* Miq. = 22
A. *argentea* Blume var. *cordulata* C. de Candolle = 22
A. *argentea* Blume var. *curtisii* King = 22
A. *argentea* Blume var. *eximia* Miq. = 21
A. *argentea* Blume var. *hypoleuca* (Miq.) Miq. = 22
A. *argentea* Blume var. *latifolia* Miq. = 21
A. *argentea* Blume var. *microphylla* Miq. = 22
A. *argentea* Blume var. *multijuga* Koord. & Val. = 22
A. *argentea* Blume var. *splendens* Koord. & Val. = 22
A. *argentea* Blume var. *stellatipilosa* Adelb. = 21
A. *argentea* Blume var. *superba* Miq. = 22
A. aspera Teijsm. & Binn. 54
A. *aspera* Teijsm. & Binn. var. *sumatrana* Baker fil. = 54
A. *attenuata* H.L. Li = 15
A. australiensis C.M. Pannell 2

A. *axillaris* A.C. Smith = 44
A. *badia* Merrill = 30
A. *baillonii* (Pierre) Pellegrin = 47
A. *bamleri* Harms, n.v.
A. *banahaensis* Elmer ex Merrill = 76
A. *baramensis* Merrill = 76
A. barbanthera C. de Candolle 79
A. *barbatula* Koord. & Valet. = 19
A. *barberi* Gamble = 58
A. basiphylla A. Gray 73
A. *batjanica* Miq. = 32
A. *bauerleni* C. de Candolle = 22
A. *beccariana* (C. de Candolle) Harms = 100
A. *beccarii* C. de Candolle = 15
A. *beddomei* (Kosterm.) Jain & Gaur, excl.
A. *bergmannii* Warb., n.v.
A. *bernardoi* Merrill = 17
A. *betchei* C. de Candolle = 42
A. *bicolor* Merrill = 30
A. *boanana* Harms = 25
A. *bordenii* Merrill = 88
A. *borneensis* Merrill = 93
A. bourdillonii Gamble 20
A. *brachybotrys* Merrill = 15
A. brassii Merrill & Perry 64
A. *brevipeduncula* C. de Candolle = 69
A. *brevipetiolata* Merrill = 61
A. *breviracemosa* (Kosterm.) Kosterm., excl.
A. brownii C.M. Pannell 97
A. *bulusanensis* Elmer ex Merrill = 32
A. *cagayanensis* Merrill = 15
A. *calelanensis* Elmer = 54
A. *cambodiana* Pierre = 58
A. *canarensis* Gamble = 48
A. *canariifolia* Koord. = 29
A. *caroli* Harms, n.v.
A. *carrii* Harms = 72
A. *caudatifoliolata* Merrill = 101
A. *cauliflora* Koord. = 36
A. *caulobotrys* Quisumb. & Merrill = 76
A. *cedreloides* Harms = 47
A. *celebica* Koord. = 36
A. ceramica (Miq.) C.M. Pannell 70
A. *chalmersi* C. de Candolle = 72
A. *chartacea* Kosterm., excl.
A. *chaudocensis* Pierre = 86
A. *chaudocensis* Pierre var. *angustifolia* Pierre = 86
A. *chaudocensis* Pierre var. *robusta* Pierre = 86
A. chittagonga Miquel 28
A. *cinerea* King = 53
A. *cinnamomea* Baker fil. = 76

A. clarkii Merrill = 19
A. clemensiae Merrill & Perry = 69
A. clementis Merrill = 76
A. cochinchinensis Pierre = 47
A. cochinchinensis Pellegrin, excl.
A. conferta Merrill & Perry 77
A. confertiflora Merrill = 36
A. congylos Kostermans = 37
A. copelandii Elmer = 47
A. cordata Hiern = 98
A. coriacea Korth. ex Miq. 85
A. costata Merrill 33
A. crassinervia Kurz ex Hiern 53
A. cremea Merrill & Perry 50
A. cucullata (Roxb.) Pellegrin 1
A. cumingiana Turcz. 82
A. cupanioidea King = 85
A. cuprea Elmer = 23
A. cupreolepidota Merrill = 29
A. curranii Merrill = 58
A. curtisii King = 101
A. cuspidata C. de Candolle 102
A. cuspidella Ridley = 60
A. dasyclada How & T.C. Chen, n.v.
A. davaoensis Elmer = 76
A. decandra Wall., excl.
A. densisquama C.M. Pannell 24
A. densitricha C.M. Pannell 12
A. denticulata Turcz. = 32
A. diepenhorstii Miq. = 60
A. diffusa Merrill = 58
A. diffusiflora Merrill = 32
A. discolor Merrill = 22
A. doctersiana Harms = 34
A. domestica Pellegrin, excl.
A. dookoo Griffith, excl.
A. dubia (Merrill) Kosterm., excl.
A. duperreana Pierre = 86
A. dyeri Koord. = 98
A. dysoxylifolia Koord. = 36
A. dysoxylonoides Koord. = 36
A. edelfeldti C. de Candolle = 72
A. edulis (Roxb.) Wall. 58
A. edulis sensu A. Gray = 42
A. edulis Seeman = 80
A. elaeagnoidea (A. Juss.) Benth. 29
A. elaeagnoidea (A. Juss.) Benth. var. *beddomei* (Gamble) K.K.N. Nair = 29
A. elaeagnoidea (A. Juss.) Benth. var. *bourdillonii* (Gamble) K.K.N. Nair = 20
A. elaeagnoidea (A. Juss.) Benth. var. *courtallensis* (Gamble) K.K.N. Nair = 29
A. elaeagnoidea (A. Juss.) Benth. var. *formosana* Hayata ex Matsumura & Hayata = 29
A. elaeagnoidea (A. Juss.) Benth. var. *glabrescens* Valeton = 29
A. elaeagnoidea (A. Juss.) Benth. var. *pallens* Merrill = 29
A. elaphina Merrill & Perry = 98
A. elegans Gillespie = 73
A. elliptica Blume 76
A. elliptica Blume var. *ceramica* Miq. = 70
A. elliptifolia Merrill = 32
A. elmeri Merrill = 57
A. ermischii Warb. = 69

A. erythrosperma C.M. Pannell 7
A. euphorioides Pierre = 88
A. euryanthera Harms 67
A. euryphylla Koord. & Val. = 15
A. eusideroxylon Koord & Valet. = 15
A. evansensis A.C. Smith 74
A. everettii Merrill, n.v.
A. exigua Merrill & Perry = 72
A. eximia Miq. 21
A. exstipulata (Griffith) Theobald 94
A. ferruginea C.T. White & Francis = 98
A. flavescens C. de Candolle 40
A. flavida Merrill & Perry 3
A. forbesiana C. de Candolle = 55
A. forbesii King 51
A. formosana (Hayata ex Matsumara & Hayata) Hayata = 29
A. forstenii Miq. = 47
A. foveolata C.M. Pannell 52
A. fragilis A.C.Smith 41
A. fraseri Ridley = 60
A. fusca King = 88
A. gagnepainiana Pellegrin = 89
A. gamopetala Merrill = 49
A. ganggo Miq. = 47
A. gibbsiae C. de Candolle = 69
A. gigantea (Pierre) Pellegrin = 8
A. gjellerupii C. de Candolle = 69
A. glabrata Teijsm. & Binn. 39
A. glabriflora Hiern = 49
A. glabrifolia Merrill = 49
A. glaucescens King = 88
A. glomerata Merrill = 98
A. goebeliana Warb. = 32
A. gracilis A.C. Smith 43
A. gracillima Harms, n.v.
A. grandifoliola Merrill = 15
A. grandis Korth. in Miq. 17
A. grata Wall. ex Voigt = 29
A. greenwoodii A.C.Smith = 73
A. griffithii (Hiern) Kurz = 94
A. halmaheirae Miq., excl.
A. hapalantha Harms = 69
A. haplophylla Harms = 89
A. haplophylla A.C. Smith = 45
A. harmandiana Pierre = 98
A. harmsiana Perkins = 76
A. hartmannii C. de Candolle = 69
A. haslettiana Haines = 15
A. havilandii Ridley = 76
A. helmsleyi Koord. = 17
A. hemsleyi Koord. = 17
A. heptandra Koord & Valet. = 16
A. heterobotrys Merrill = 89
A. heteroclita King = 57
A. heterophylla Merrill = 60
A. heterotricha A.C. Smith 56
A. hexandra Turcz. = 32
A. hiernii King 101
A. hiernii Koord & Valet., n.v.
A. hiernii Viswa. & Ramach. = 8
A. hoanensis Pierre = 29
A. huberti Harms, n.v.
A. humilis King = 51
A. hypoleuca Miq. = 22
A. ignea Valeton ex K. Heyne = 13

Index

A. *iloilo* (Blanco) Merrill, excl.
A. *inaequalis* Teijsm. & Binn. = 76
A. *indica* (Hooker fil.) Harms = 58
A. *insignis* Schwartz = 57
A. integrifolia C.M. Pannell 99
A. intricatoreticulata Kosterm., excl.
A. *irosinensis* Elmer ex Merrill = 78
A. *jainii* Viswanathan & Ramachandran = 15
A. janowskyi Harms, excl.
A. *javanica* Koord. & Valet. ex Koord. = 22
A. *kabaensis* Baker fil. = 98
A. *khasiana* Hiern = 58
A. kinabaluensis Kosterm., excl.
A. *kingiana* Ridley = 48
A. *koordersii* Jain & Bennet, n.v.
A. korthalsii Miq. 36
A. *korthalsii* (Miq.) Pellegrin = 15
A. kostermansii (Prijanto) Kosterm., excl.
A. *kunstleri* King = 57
A. *laevigata* Merrill = 49
A. *lagunensis* Merrill = 76
A. *lanceolata* Merrill = 32
A. *lancifolia* (Hooker fil.) Harms = 76
A. lancilimba Merrill 26
A. *langlassei* C. de Candolle = 76
A. *lanuginosa* King = 17
A. *latifolia* Miq. = 58
A. *latifolia* Miq. var. *teysmanniana* Koord. & Val. = 58
A. *lauterbachiana* Harms = 40
A. lawii (Wight) Saldanha ex Ramamoorthy 15
A. laxiflora Miq. 83
A. *ledermannii* Harms, n.v.
A. *leeuwenii* Harms = 34
A. lepidopetala Harms 10
A. *lepidota* Miq. = 29
A. *lepidota* Miq. var. *paupercula* Miq. = 29
A. lepiorrhachis Harms 27
A. leptantha Miq. 49
A. *leptantha* Miq. var. *borneensis* C. de Candolle = 49
A. *leptoclada* Harms, n.v.
A. leucoclada C. de Candolle 46
A. leucophylla King 57
A. *littoralis* Zippelius ex Miq. = 15
A. *littoralis* Talbot = 29
A. *llanosiana* C. de Candolle = 32
A. *loheri* Merrill = 32
A. *longifolia* Teijsm. & Binn. = 94
A. *longipetiolata* Elmer = 76
A. *longipetiolulata* Baker fil. = 36
A. luzoniensis (Vidal) Merrill & Rolfe 61
A. *luzoniensis* (Vidal) Merrill & Rolfe var. *trifoliata* Merrill & Rolfe = 60
A. *maboroana* Harms, n.v.
A. *macrobotrys* Turcz. = 32
A. macrocarpa (Miq.) C.M. Pannell 4
A. *macrophylla* Teijsm. & Binn., excl.
A. macrostigma King 59
A. *magnifoliola* C. de Candolle = 58
A. *maiae* Bourdillon = 48
A. *maingayi* (Hiern) King = 15
A. malabarica N. Sasidharan, Appendix 2

A. malaccensis (Ridley) Pannell 5
A. *mannii* (King ex Brandis) Jain & Gaur = 47
A. *marginata* Craib = 76
A. mariannensis Merrill 81
A. *matthewsii* Merrill = 89
A. *megistocarpa* Merrill = 19
A. *meliosmoides* Craib = 89
A. membranacea Kosterm., excl.
A. membranifolia King 92
A. *menadonensis* Koord. = 76
A. meridionalis C.M. Pannell 11
A. *merostela* Pellegrin = 17
A. merrillii Elmer, excl.
A. *micrantha* Merrill = 76
A. *micropora* Merrill = 47
A. *midnaporensis* Carey ex Voigt = 29
A. *minahassae* Koord. = 58
A. *mindanaensis* Merrill ex Elmer = 76
A. *minutiflora* Bedd. = 98
A. *minutiflora* Bedd. var. *griffithii* Hiern = 94
A. *minutiflora* Bedd. var. *macrophylla* C. de Candolle = 54
A. *minutiflora* Bedd. var. *travancorica* Hiern = 98
A. *miquelii* Merrill = 69
A. *mirabilis* Harms = 89
A. *mirandae* Merrill = 57
A. *monophylla* Perkins = 61
A. monozyga Harms 90
A. *montana* C. de Candolle = 93
A. *montrouzieri* Pierre = 58
A. *motleyana* Stapf ex Ridley = 58
A. *moultonii* Merrill = 76
A. *mucronulata* C. de Candolle = 58
A. *multiflora* Merrill = 47
A. *multifoliola* Merrill = 22
A. multijuga Seem., excl.
A. multinervis C.M. Pannell 9
A. *myriantha* Merrill = 78
A. *myristicifolia* C. de Candolle = 54
A. *negrosensis* Merrill ex Elmer = 76
A. neotenica Kosterm. = 89
A. nivea Elmer ex Merrill, excl.
A. *novoguineensis* C. de Candolle = 69
A. *nudibacca* C. de Candolle = 69
A. *oblanceolata* Craib = 86
A. *obliqua* C.T. White & Francis = 47
A. *oblonga* Pierre = 58
A. *ochneocarpa* Merrill = 101
A. *odoardoi* Merrill = 89
A. odorata Lour. 86
A. *odorata* Lour. var. *chaudocensis* (Pierre) Pellegrin = 86
A. *odorata* Lour. var. *microphyllina* C. de Candolle = 86
A. odoratissima Blume 60
A. *odoratissima* auct. = 29
A. *odoratissima* Blume var. *forbesii* Baker fil. = 60
A. *odoratissima* Blume var. *parvifolia* Koord. & Val. = 60
A. *odoratissima* Blume var. *pauciflora* Koord. & Val. = 60
A. *oligantha* C. de Candolle = 88

A. oligocarpa Miq. = 15
A. oligophylla Miq. 88
A. ovata Tiejsm. & Binn. = 76
A. oxypetala Valet. = 76
A. pachyphylla Miq. 19
A. palauensis Kanehira = 81
A. palawanensis Merrill = 76
A. palembanica Miq. 95
A. palembanica Miq. var. *borneensis* Miq. = 98
A. palembanica Miq. var. *longifolia* Craib = 98
A. pallens Merrill = 29
A. pallida Merrill = 57
A. pamattonis Miq. = 95
A. paniculata Kurz = 60
A. papuana (Merrill & Perry) Harms, excl.
A. parksii A.C. Smith 71
A. parviflora C. de Candolle 55
A. parvifolia Merrill = 29
A. parvifoliola C. de Candolle = 72
A. pauciflora Merrill = 76
A. pedicellaris C. de Candolle = 88
A. pedicellata (Hiern) Kosterm. = 15
A. peekelii Harms, n.v.
A. penningtoniana C.M. Pannell 14
A. pentaphylla Kurz ex Miq. = 86
A. perfulva Elmer ex Merrill = 17
A. perviridis Hiern 48
A. perviridis Hiern var. *sikkimiana* C. de Candolle = 48
A. phaeogyna Harms, n.v.
A. pinnata (L.) Druce, excl.
A. pinnata (Blanco) Merrill = 98
A. pirifera Hance = 58
A. pleuropteris Pierre 87
A. poilanei Pellegrin = 29
A. polyantha Bedd. = 98
A. polyantha Ridley = 88
A. polyneura C. de Candolle 68
A. polyphylla Miq. = 54
A. polystachya Wall., excl.
A. ponapensis Kanehira = 81
A. porulifera C. de Candolle = 69
A. poulocondorensis Pellegrin = 29
A. procera C. de Candolle = 55
A. pseudolansium Kosterm., excl.
A. psilopetala A.C. Smith = 42
A. puberulanthera C. de Candolle 66
A. puncticulata Merrill = 84
A. pycnocarpa Miq. = 4
A. pycnoneura Harms, n.v.
A. pyramidata Hance = 47
A. pyricarpa Baker fil. = 53
A. pyriformis Merrill 84
A. pyrrholepis Miq. = 47
A. querciflorescens Elmer = 76
A. quocensis Pierre = 88
A. racemosa Ridley = 15
A. ramosii Quisumb. = 30
A. ramotricha C.M. Pannell 18
A. ramuensis Harms = 98
A. rechingerae C. de Candolle = 69
A. reinwardtiana Kosterm., excl.
A. reinwardtii Miq. = 76
A. repouensis Pierre = 86

A. reticulata Elmer ex Merrill = 32
A. ridleyi (King) Pannell = 8
A. ridleyi P.T. Li & X.M. Chen = 88
A. rimosa (Blanco) Merrill 32
A. rivularis Merrill 63
A. rizalensis Merrill = 61
A. robinsonii Merrill = 76
A. rodatzii Harms, n.v.
A. roemeri C. de Candolle = 69
A. roxburghiana (Wight & Arn.) Miq. = 29
A. roxburghiana (Wight & Arn.) Miq. var. *angustata* Miq. = 29
A. roxburghiana (Wight & Arn.) Miq. var. *balica* Miq. = 29
A. roxburghiana (Wight & Arn.) Miq. var. *beddomei* Gamble = 29
A. roxburghiana (Wight & Arn.) Miq. var. *brachystachya* C. de Candolle = 29
A. roxburghiana (Wight & Arn.) Miq. var. *courtallensis* Gamble = 29
A. roxburghiana (Wight & Arn.) Miq. var. *obtusa* C. de Candolle = 29
A. roxburghiana (Wight & Arn.) Miq. var. *paupercula* Miq. = 29
A. rubescens (Hiern) C.M. Pannell = 4
A. rubiginosa (Hiern) C.M. Pannell 13
A. rubra Ridley = 66
A. rubrivenia Merrill & Perry 41
A. rudolfi Harms = 69
A. rufa Miq. = 98
A. rufa Miq. var. *celebica* Miq. = 76
A. rufibarbis Ridley 103
A. rufinervis (Blume) Bentvelzen 93
A. rugosa Pierre = 58
A. rugulosa C.M. Pannell 6
A. salicifolia Ridley = 62
A. saltatorum A.C. Smith 80
A. samarensis Merrill = 58
A. samoensis A. Gray 42
A. sapindina (F. von Muell.) Harms 69
A. schlechteri Merrill & Perry = 69
A. schraderiana Harms, n.v.
A. schultzei Harms, n.v.
A. schumanniana Harms, n.v.
A. sclerocarpa C. de Candolle = 15
A. scortechinii King 38
A. sepalina (Kosterm.) Kosterm., excl.
A. sexipetala Griffith = 54
A. shawiana Merrill = 89
A. sibuyanensis Elmer ex Merrill = 15
A. silvestris (M. Roemer) Merrill 47
A. simplex Merrill = 57
A. simplicifolia (Bedd.) Harms 89
A. simplicifolia Harms, n.v.
A. sinensis Pierre = 86
A. sipannas Miq. = 95
A. smithii Koord. 30
A. somalensis Chiov., excl.
A. sorsogonensis Elmer = 76
A. spaniantha Harms, n.v.
A. spanoghei Blume ex Miq. = 29
A. speciosa Blume 35
A. speciosa Teijsm. & Binn. = 22
A. speciosa Blume var. *macrophylla* C. de Candolle = 35
A. spectabilis (Miq.) Jain & Bennet 8

Index

A. splendens (Koord. & Valet.) Koord. & Valet. = 22
A. squamulosa King 23
A. stapfii Koord. = 76
A. steensii Kosterm., excl.
A. steinii Harms, n.v.
A. stellatotomentosa Merrill = 17
A. stellipila C. de Candolle = 72
A. stenophylla Merill = 100
A. sterculioides Kosterm. = 89
A. stipitata P.T. Li & X.M. Chen = 15
A. subcuprea Merrill & Perry 25
A. subgrisea Miq. = 16
A. subminutiflora C. de Candolle 72
A. submonophylla Miq. = 15
A. subsessilis C.M. Pannell 75
A. subviridis Elmer ex Merrill = 32
A. sulingi Blume = 58
A. talbotii Sundararaghavan = 29
A. tamilnadensis Nair & Rajan = 15
A. tarangisi Elmer = 82
A. tayabensis Merrill = 76
A. tembelingensis M.R. Henderson = 76
A. tenuicaulis Hiern 91
A. tenuifolia H.L. Li = 15
A. testicularis C.Y.Wu = 58
A. tetrapetala Pierre = 15
A. teysmanniana (Miq.) Miq. 16
A. tomentosa Teijsm. & Binn. 98
A. triandra Ridley = 89
A. trichostemon C. de Candolle = 93
A. trichostoma Harms, n.v.
A. trimera Merrill = 15
A. trimera Ridley = 4
A. tripetala Merrill = 1
A. triplex Ridley = 4
A. trunciflora Merrill = 76
A. tsangii Merrill = 15
A. turczaninowii C. de Candolle = 15
A. ulawaensis Merrill & Perry = 55
A. umbrina Elmer ex Merrill = 33
A. undulata Miq. = 58
A. unifolia P.T. Li & X.M. Chen 46
A. unifoliolata Koord. = 61
A. unifoliolata Ridley = 89
A. urdanetensis Elmer ex Merrill = 76
A. urophylla Harms, n.v.
A. variisquama C.M. Pannell 31
A. venusta A. C. Smith = 73
A. verrucosa Pierre = 58
A. versteeghii Merrill & Perry = 25
A. vilamilii Merrill, n.v.
A. villosa (C. de Candolle) Merrill = 76
A. vitiensis A.C. Smith 44
A. vitiensis A.C. Smith var. *minor* A.C. Smith = 44
A. vitiensis A.C. Smith var. *vitiensis* A.C. Smith = 44
A. vulpina Harms, n.v.
A. wallichii Hiern = 29
A. wangii H.L. Li = 15
A. wangii H.L. Li var. *macrophylla* H.L.Li = 15
A. whitmeei C. de Candolle = 42
A. winckelii Adelb. = 93
A. yunnanensis H.L. Li = 15

A. yunnanensis (H.L. Li) C.Y. Wu var. *macrophylla* H.L. Li = 15
A. yzermannii Boerl. & Koord. 62
A. zippelii Miq. = 98
A. zollingeri C. de Candolle, excl.
Aglaiopsis glaucescens Miq. = 69
A. lancifolia (Hooker) fil. Miq. = 76
Amoora aherniana Merrill = 1
A. amboinensis Miq., excl.
A. aphanamixis J.A. & J.H. Schultes, excl.
A. aphanamixis J.A. & J.H. Schultes var. frutescens (C. de Candolle) Pellegrin, excl.
A. aphanamixis J.A. & J.H. Schultes var. pubescens Miq., excl.
A. auriculata Miq. = 1
A. balansaeana C. de Candolle, excl.
A. beddomei Kosterm., excl.
A. borneensis Miq., excl.
A. caesifolia Elmer, excl.
A. calcicola C.Y. Wu & H. L. Li ex C.Y. Wu = 15
A. canarana (Turcz.) Hiern = 15
A. championii Benth. & Hooker fil., excl.
A. chittagonga (Miq.) Hiern = 28
A. conduplifolia Elmer = 1
A. cucullata Roxb. = 1
A. curtispica L.S. Gibbs = 15
A. cumingiana C. de Candolle, excl.
A. cupulifera Merrill, excl.
A. dasyclada (How & T.C. Chen) C.Y. Wu, n.v.
A. decandra Hiern, excl.
A. dysoxyloides Kurz = 15
A. elmeri Merrill, excl.
A. ferruginea C.T. White = 11
A. ficiformis Wight, excl.
A. forbesii S. Moore, excl.
A. fulva Merrill, excl.
A. ganggo (Miq.) Kurz = 47
A. gigantea Pierre = 8
A. grandifolia Walp., excl.
A. grandifolia Walp. var. pubescens (Miq.) C. de Candolle, excl.
A. janowskyi (Harms) Kosterm., excl.
A. korthalsii Miq. = 15
A. lactescens Kurz, n.v.
A. lanceolata Hiern = 9
A. lauterbachiana (Harms) C. de Candolle = 40
A. lawii (Wight) Beddome = 15
A. lepidota Merrill = 15
A. macrocalyx (Harms) C. de Candolle, excl.
A. macrocarpa Merrill, n.v.
A. macrophylla Nimmo, excl.
A. maingayi Hiern = 15
A. malaccensis Ridley = 5
A. mannii King ex Brandis = 47
A. megalophylla C. de Candolle, excl.
A. megalophylla C. de Candolle var. frutescens C. de Candolle, excl.
A. mindorensis Merrill, excl.
A. moulmeiniana C. de Candolle, n.v.
A. myrmecophila O. Warb., excl.
A. naumannii C. de Candolle, excl.

378

Index

A. nitidula Benth., excl.
A. oligosperma Pierre, excl.
A. ouangliensis (Lveill) C.Y. Wu = 15
A. perrotetiana (A. Juss.) Steudel, excl.
A. polillensis C.B. Robinson, excl.
A. polystachya (Wall.) Wight & Arn., excl.
A. poulocondorensis (Pellegrin) Harms = 29
A. racemosa Ridley, excl.
A. ridleyi King = 8
A. rohituka (Roxb.) Wight & Arn., excl.
A. roxburghiana Korth. ex Blume, excl.
A. rubescens Hiern = 4
A. rubiginosa Hiern = 13
A. salomoniensis C. de Candolle, excl.
A. sogerensis Baker fil., excl.
A. spectabilis Miq. = 8
A. stellata C.Y. Wu = 16
A. stellatosquamosa C.Y. Wu & H.L. Li = 8
A. sumatrana Miq., excl.
A. tetrapetala (Pierre) Pellegrin = 15
A. tetrapetala (Pierre) Pellegrin var. macrophylla (H.L. Li) C.Y. Wu = 15
A. teysmanniana Miq. = 16
A. timorensis (A. Juss.) Wight & Arn. ex Steudel, excl.
A. tomentosa Korth. ex Blume, excl.
A. trichanthera Koord. & Val. = 4
A. tsangii (Merrill) X.M. Chen = 15
A. verrucosa C. de Candolle, excl.
A. vieillardi C. de Candolle, excl.
A. wallichii King = 8
A. yunnanensis (H.L. Li) C.Y.Wu = 15
Andersonia cucullata Roxb. = 1
A. rohituka Roxb., excl.
Aphanamixis chittagonga (Miq.) Haridasan & Rao = 15
A. reticulosa Kosterm. = 88
A. trichanthera (Koord. & Val.) Koord. =4
A. wallichii (King) Haridasan & Rao = 8
Argophilum pinnatum Blanco = 98
Beddomea indica Hooker fil. = 58
B. luzoniensis Vidal = 61
B. racemosa Ridley = 89
B. simplicifolia Bedd. = 89
B. simplicifolia Bedd. var. parviflora Bedd. = 89
B. simplicifolia Bedd. var. racemosa Bedd. = 89
Camunium bengalense Buch.-Ham. ex Wall. = 58
C. sinensis Pierre = 86
Celastrus micranthus Roxb. = 69
Chisocheton sumatranus Baker fil. = 53
Didymocheton obliquum (Gillespie) Harms = 43
Dysoxylum obliquum Gillespie = 43
Epicharis baillonii Pierre = 47
E. exarillata Nimmo = 15
E. macrocarpa Miq. = 4
Euphora exstipulatis Griffith = 94
E. exstipularis C. de Candolle = 94
Ficus ouangliensis Lveill = 15

F. vaniotii Lveill = 15
Hartighsea angustifolia Miq. =100
Hearnia aquatica Pierre = 36
H. beccariana C. de Candolle = 100
H. balansae C. de Candolle, excl.
H. cumingiana (Turcz.) C. de Candolle = 82
H. elliptica (Blume) C. de Candolle = 76
H. glaucescens (Miq.) C. de Candolle = 69
H. glaucescens (Miq.) C. de Candolle var. novaguineensis C. de Candolle = 69
H. lancifolia (Hooker fil.) C. de Candolle = 76
H. macrophylla C. de Candolle = 69
H. sapindina F. von Muell. = 69
H. sarawakana C. de Candolle = 36
H. villosa C. de Candolle = 76
Heynia quinquejuga Sprengel = 93
Lansium pedicellatum Hiern = 15
L. silvestre Rumph. = 47
L. silvestre M. Roemer = 47
Lepiaglaia baill(i)oni (Pierre) Pierre = 47
L. montrouzieri Pierre = 58
L. pyramidata (Hance) Pierre = 47
L. tetrapetala Pierre = 15
Melia argentea Reinwardt = 22
M. iloilo Blanco, excl.
Meliacea singapureana Pierre = 88
Merostela grandifolia Pierre = 17
M. grandis Pierre = 17
Milnea apiocarpa Thwaites = 37
M. blumei Teijsm. & Binn. = 60
M. cambodiana Pierre = 58
M. dulcis Teijsm. & Binn. = 76
M. edulis Roxb. = 58
M. harmandiana Pierre = 98
M. lancifolia Hooker fil. = 76
M. montana Jack, n.v.
M. montana Teijsm. & Binn. = 16
M. pirifera (Hance) Pierre = 58
M. racemosa (Kostel.) M. Roemer = 58
M. roxburghiana Wight & Arn. = 29
M. rugosa Pierre = 58
M. sulingi (Blume) Teijsm. & Binn. = 58
M. undulata Wall. ex Miq. = 58
M. verrucosa Pierre = 58
Nemedra elaeagnoidea A. Juss. = 29
N. nimmonii Dalzell = 15
Nimmonia lawii Wight = 15
Nyalelia racemosa Dennst. = 58
N. racemosa Dennst. ex Kostel. = 58
Opilia odorata Sprengel = 86
Oraoma canarana Turcz. = 15
Portesia rimosa Blanco = 32
Sapindus lepidotus Wall. = 29
Selbya montana M. Roemer, n.v.
Sphaerosacme laxa Wallich = 1
S. paniculata sensu Miq. = 1
S. rohituka Wallich = 1 & 8
Trichilia rimosa (Blanco) Blanco = 32
T. rufinervis Blume = 93
Walsura lanceolata Wall. = 29

- SIEMENTEN LEVIÄMISET: 18-19
- BEDTINSTION SEMI-HYPOGEAL. 10% TROPICAL RAINFOREST TREE SPECIES IN MALAYA HAVE ↑ (20)
- KROMOSOMILUKUMÄÄRÄ vaihtelee suuresti alueen ja kasvien sis. (21)
  • $1n^{1} - n = 16 ... 360$
- Suuri osa eristetty reeluissa ja usgulidae kaikki vain $(?)$
  • compete matrix, isoen member caryologica ja osalle alueista
  • siks, wide species concept
- [illegible]